Lectures on General Relativity, Cosmology and Quantum Black Holes (Second Edition)

Online at: https://doi.org/10.1088/978-0-7503-5824-8

Lectures on General Relativity, Cosmology and Quantum Black Holes (Second Edition)

Badis Ydri
Department of Physics, Annaba University, Annaba, Algeria

IOP Publishing, Bristol, UK

ISBN   978-0-7503-5824-8 (ebook)
ISBN   978-0-7503-5822-4 (print)
ISBN   978-0-7503-5825-5 (myPrint)
ISBN   978-0-7503-5823-1 (mobi)

DOI   10.1088/978-0-7503-5824-8

Version: 20250601

IOP ebooks

British Library Cataloguing-in-Publication Data: A catalogue record for this book is available from the British Library.

Published by IOP Publishing, wholly owned by The Institute of Physics, London

IOP Publishing, No.2 The Distillery, Glassfields, Avon Street, Bristol, BS2 0GR, UK

US Office: IOP Publishing, Inc., 190 North Independence Mall West, Suite 601, Philadelphia, PA 19106, USA

Contents

Prologue

Introduction

These notes originate from a series of formal courses delivered during the academic years 2012–2013 (General Relativity and Classical Black Holes), 2014–2015 (Cosmology), and 2016–2017 (Quantum Black Holes) to second-year Master's students in theoretical physics at Badji Mokhtar Annaba University.

The selection of topics and references reflects the personal choices and interests of the author. The primary references include: (1) Wald, for general relativity and differential geometry; (2) Carroll, for classical black holes and advanced cosmology; (3) Mukhanov, whose treatment of inflationary cosmology is arguably among the best, particularly for theoretical physicists; (4) Birrell and Davies, an outstanding reference for quantum field theory (QFT) on curved spacetimes; and (5) Hartle, for a lucid introduction to both elementary and observational cosmology. Additional references that proved helpful include Weinberg (especially on gauge fixing) and Dodelson (for observational cosmology). I also wish to acknowledge the pedagogical and insightful lecture notes of 't Hooft, Jacobson, Liddle, Harlow, and Baumann.

This is not a general relativity textbook in the traditional sense. Our treatment of this foundational topic is necessarily brief, and many important subjects are omitted. A summary of the required differential geometry is included in the appendix.

The principal objective of these notes was initially to explore both cosmology and black holes. However, as the lectures progressed, the focus shifted toward the deeper conceptual challenges posed by black holes, particularly their connection to quantum mechanics and quantum gravity. We therefore undertake a critical study of two central pillars of modern theoretical physics: (i) inflationary cosmology and (ii) quantum black holes and the information loss problem. While cosmology is treated thoroughly, greater emphasis is placed on quantum black holes as a gateway to understanding the interplay between gravitation and quantum theory.

In particular, our treatment of the information loss problem—addressed in chapters 6 and 7—is, in my view, systematic and substantially complete. These two chapters constitute the core of the book. In chapter 6, we present a first-principles derivation of Hawking radiation and a detailed analysis of black hole thermodynamics and the traditional formulation of the information loss paradox. In chapter 7, we go further by examining black hole evaporation within the framework of the anti-de Sitter/conformal field theory (AdS/ CFT) correspondence. This treatment, building on the recent breakthroughs by Almheiri et al and Penington, provides a concrete realization of the ER = EPR paradigm. It incorporates key elements such as quantum extremal surfaces, the island conjecture, and replica wormholes, ultimately yielding a consistent Page curve and resolving the paradox within a well-defined holographic setting. While topics such as the firewall proposal are only briefly mentioned, the approach developed here offers a compelling and physically transparent resolution of the information loss problem, firmly grounded in gauge/gravity duality.

With regard to inflation, we progress to the point where one can compute the location of the first acoustic peak in the cosmic microwave background (CMB) spectrum. Other significant discussions include the cosmological constant, its interpretation as dark energy, and various computations of vacuum energy using multiple methods. A foundational introduction to quantum field theory on curved spacetimes is also provided.

These notes further include a comprehensive exposition of the Arnowitt–Deser–Misner (ADM) formalism, along with a brief introduction to quantum gravity in the context of Horava–Lifshitz gravity, which we regard as a promising and conceptually coherent approach to a renormalizable quantum theory of gravity. Quantum gravity is explored further in the final chapter through two additional and arguably more promising frameworks: Loop Quantum Gravity and the Banks–Fischler–Shenker–Susskind (BFSS) matrix model, each offering a radically different perspective on the unification of quantum mechanics and gravitation.

Organization

This book is organized as follows. Chapter 1 provides a comprehensive summary of the essential concepts of general relativity, including the equivalence principle, relativistic mechanics, curvature tensors, Einstein's equations, and the Hilbert–Einstein action. Chapter 2 is devoted to classical black hole solutions in general relativity—Schwarzschild, Reissner–Nordström, Kerr, and Kerr–Newman—as well as their causal structure and thermodynamic properties. Chapter 3 presents the observational foundations of modern cosmology, including the expanding Universe, redshift, cosmological distances, and the Friedmann equations. Chapter 4 develops the theory of inflation, addressing the horizon and flatness problems, and includes a detailed treatment of cosmological perturbation theory and the origin of CMB anisotropies.

Chapter 5 introduces quantum field theory on curved backgrounds, the cosmological constant problem, and various computations of vacuum energy. It also contains an in-depth presentation of the ADM formalism and an introduction to Horava–Lifshitz quantum gravity. Chapter 6, forming one of the cores of the book, develops a systematic derivation of Hawking radiation, the Unruh effect, the information loss paradox, and black hole thermodynamics from first principles. Chapter 7, which together with chapter 6 constitutes the central focus of this work, explores the resolution of the information loss paradox through recent developments in gauge/gravity duality, including AdS^2/CFT_1, quantum extremal surfaces, the island conjecture, and replica wormholes.

Finally, chapter 8 presents two further approaches to quantum gravity: loop quantum gravity and the BFSS matrix model. This chapter provides a comparative view of background-independent quantization, spin networks, and holographic matrix formulations of M-theory. An appendix summarizes the necessary background in differential geometry for readers less familiar with the mathematical formalism.

Proposal

In addition to its broad pedagogical scope, this book offers a self-contained and conceptually unified treatment of black holes, which may be read independently as a complete study. This thematic arc unfolds in three major parts:

1. **Classical black holes (chapter 2):**
   The first part develops the geometry and physics of classical black holes within the framework of general relativity. Topics include the Schwarzschild, Reissner–Nordström, Kerr, and Kerr–Newman solutions, their causal structure, event horizons, thermodynamics, and processes such as free fall and the Penrose process. This material assumes familiarity with the essential tools of general relativity presented in chapter 1.

2. **The information loss problem (chapter 6):**
   The second part addresses the quantum aspects of black holes, focusing on the derivation of Hawking radiation, the Unruh effect, black hole thermodynamics, entanglement entropy, and the Page curve. The apparent contradiction between unitarity and black hole evaporation—the information loss paradox—is analyzed in depth. This part assumes prior familiarity with quantum field theory in curved spacetime, as introduced in chapter 5.

3. **Quantum resolution via gauge/gravity duality (chapter 7):**
   The third part presents a modern resolution of the information paradox using gauge/gravity duality. Within the AdS^2/CFT_1 correspondence, we explore quantum extremal surfaces, the island conjecture, and replica wormholes, culminating in a consistent computation of the Page curve. This framework provides a concrete realization of the ER = EPR paradigm. Just as general relativity is necessary for the classical treatment, a working knowledge of holographic duality and AdS/CFT—summarized from the author's other works on quantum field theory and matrix models—is necessary to fully engage with this part.

Together, these three parts offer a layered and coherent exploration of black holes, from classical geometry to quantum gravity.

Author biography

Badis Ydri

Badis Ydri is a professor of theoretical physics at the Institute of Physics, Annaba University, Algeria. He earned his PhD from Syracuse University, New York, USA, in 2001. He is also a research associate at the Dublin Institute for Advanced Studies, Ireland. His postdoctoral work includes a Marie Curie fellowship at Humboldt University in Berlin, Germany, and a Hamilton fellowship at the Dublin Institute for Advanced Studies, Ireland. His general areas of expertise encompasses quantum field theory, general relativity, string theory, and philosophy of physics.

His ongoing research explores: 1) matrix quantum mechanics approaches to quantum black holes and quantum gravity, 2) gauge/gravity duality and M-theory, 3) noncommutative geometry and matrix models, 4) renormalization group equation and Monte Carlo methods, 5) artificial intelligence in computational physics, 6) quantum philosophy, and 7) hard physical philosophy of consciousness and being. He is the author of seven books in theoretical physics.

IOP Publishing

Lectures on General Relativity, Cosmology and Quantum Black Holes (Second Edition)

Badis Ydri

Chapter 1

General relativity essentials

The goal in this first chapter is to go in an efficient way through the essential material of general relativity and get quickly to Einstein's equations. A classic text on general relativity is Wald [1] and a much newer text which has become a classic in its own right is Carroll [2]. We have also benefited from t'Hooft [3] and from many other texts and lecture notes on the subject which are cited in subsequent chapters.

1.1 Equivalence principle

The classical (Newtonian) theory of gravity is based on the following two equations. The gravitational potential Φ generated by a mass density ρ is given by Poisson's equations (with G being Newton's constant)

$$\nabla^2 \Phi = 4\pi G \rho. \tag{1.1}$$

The force exerted by this potential Φ on a particle of mass m is given by

$$\vec{F} = -m \vec{\nabla} \Phi. \tag{1.2}$$

These equations are obviously not compatible with the special theory of relativity. The above first equation will be replaced, in the general relativistic theory of gravity, by Einstein's equations of motion while the second equation will be replaced by the geodesic equation. From the above two equations we see that there are two measures of gravity: $\nabla^2 \Phi$ measures the source of gravity while $\vec{\nabla} \Phi$ measures the effect of gravity. Thus $\vec{\nabla} \Phi$, outside a source of gravity where $\rho = \nabla^2 \Phi = 0$, need not vanish. The analogs of these two different measures of gravity, in general relativity, are given by the so-called Ricci curvature tensor $R_{\mu\nu}$ and Riemann curvature tensor $R_{\mu\nu\alpha\beta}$, respectively.

   The basic postulate of general relativity is simply that gravity is geometry. More precisely gravity will be identified with the curvature of spacetime which is taken to

doi:10.1088/978-0-7503-5824-8ch1   1-1

be a pseudo-Riemannian (Lorentzian) manifold. This can be made more precise by employing the two guiding 'principles' which led Einstein to his equations. These are:

- *The weak equivalence principle:* This states that all particles fall the same way in a gravitational field which is equivalent to the fact that the inertial mass is identical to the gravitational mass. In other words, the dynamics of all free particles, falling in a gravitational field, is completely specified by a single worldline. This is to be contrasted with charged particles in an electric field which obviously follow different worldlines depending on their electric charges. Thus, at any point in spacetime, the effect of gravity is fully encoded in the set of all possible worldlines, corresponding to all initial velocities, passing at that point. These worldlines are precisely the so-called geodesics.

  In measuring the electromagnetic field we choose 'background observers' who are not subject to electromagnetic interactions. These are clearly inertial observers who follow geodesic motion. The worldline of a charged test body can then be measured by observing the deviation from the inertial motion of the observers.

  This procedure cannot be applied to measure the gravitational field since by the equivalence principle gravity acts the same way on all bodies, i.e. we cannot insulate the 'background observers' from the effect of gravity so that they provide inertial observers. In fact, any observer will move under the effect of gravity in exactly the same way as the test body.

  The central assumption of general relativity is that we cannot, even in principle, construct inertial observers who follow geodesic motion and measure the gravitational force. Indeed, we assume that the spacetime metric is curved and that the worldlines of freely falling bodies in a gravitational field are precisely the geodesics of the curved metric. In other words, the 'background observers' who are the geodesics of the curved metric coincide exactly with motion in a gravitational field.

  Therefore, gravity is not a force since it cannot be measured but is a property of spacetime. Gravity is in fact the curvature of spacetime. The gravitational field corresponds thus to a deviation of the spacetime geometry from the flat geometry of special relativity. But infinitesimally each manifold is flat. This leads us to Einstein's equivalence principle: in small enough regions of spacetime, the non-gravitational laws of physics reduce to special relativity since it is not possible to detect the existence of a gravitational field through local experiments.

- *Mach's principle:* This states that all matter in the Universe must contribute to the local definition of 'inertial motion' and 'non-rotating motion'. Equivalently the concepts of 'inertial motion' and 'non-rotating motion' are meaningless in an empty universe. In the theory of general relativity the distribution of matter in the Universe, indeed, influences the structure of spacetime. In contrast, the theory of special relativity asserts that 'inertial motion' and 'non-rotating motion' are not influenced by the distribution of matter in the Universe.

Therefore, in general relativity the laws of physics must:

(1) reduce to the laws of physics in special relativity in the limit where the metric $g_{\mu\nu}$ becomes flat or in a sufficiently small region around a given point in spacetime.

(2) be covariant under general coordinate transformations which generalizes the covariance under Poincaré found in special relativity. This means in particular that only the metric $g_{\mu\nu}$ and quantities derived from it can appear in the laws of physics.

In summary, general relativity is the theory of space, time and gravity in which spacetime is a curved manifold M, which is not necessarily R^4, on which a Lorentzian metric $g_{\mu\nu}$ is defined. The curvature of spacetime in this metric is related to the stress–energy–momentum tensor of the matter in the Universe, which is the source of gravity, by Einstein's equations which are schematically given by equations of the form

$$\text{curvature} \propto \text{source of gravity.} \tag{1.3}$$

This is the analog of (1.1). The worldlines of freely falling objects in this gravitational field are precisely given by the geodesics of this curved metric. In small enough regions of spacetime, curvature vanishes, i.e. spacetime becomes flat, and the geodesics become straight. Thus, the analog of (1.2) is given schematically by an equation of the form

$$\text{worldline of freely falling objects} = \text{geodesic.} \tag{1.4}$$

1.2 Relativistic mechanics

In special relativity spacetime has the manifold structure R^4 with a flat metric of Lorentzian signature defined on it. In special relativity, as in pre-relativity physics, an inertial motion is one in which the observer or the test particle is non-accelerating which obviously corresponds to no external forces acting on the observer or the test particle. An inertial observer at the origin of spacetime can construct a rigid frame where the grid points are labeled by $x^1 = x$, $x^2 = y$ and $x^3 = z$. Furthermore, she/he can equip the grid points with synchronized clocks which give the reading $x^0 = ct$. This provides a global inertial coordinate system or reference frame of spacetime where every point is labeled by (x^0, x^1, x^2, x^3). The labels have no intrinsic meaning but the interval between two events A and B defined by $-(x_A^0 - x_B^0)^2 + (x_A^i - x_B^i)^2$ is an intrinsic property of spacetime since its value is the same in all global inertial reference frames. The metric tensor of spacetime in a global inertial reference frame $\{x^\mu\}$ is a tensor of type $(0, 2)$ with components $\eta_{\mu\nu} = (-1, +1, +1, +1)$, i.e. $ds^2 = -(dx^0)^2 + (dx^i)^2$. The derivative operator associated with this metric is the ordinary derivative, and as a consequence the curvature of this metric vanishes. The geodesics are straight lines. The time-like geodesics are precisely the worldlines of inertial observables.

Let t^a be the tangent of a given curve in spacetime. The norm $\eta_{\mu\nu}t^\mu t^\nu$ is positive, negative and zero for space-like, time-like and light-like(null) curves, respectively. Since material objects cannot travel faster than light, their paths in spacetime must be time-like. The proper time along a time-like curve parameterized by t is defined by

$$c\tau = \int \sqrt{-\eta_{\mu\nu}t^\mu t^\nu}\,dt. \tag{1.5}$$

This proper time is the elapsed time on a clock carried on the time-like curve. The so-called 'twin paradox' is the statement that different time-like curves connecting two points have different proper times. The curve with maximum proper time is the geodesic connecting the two points in question. This curve corresponds to inertial motion between the two points.

The 4-vector velocity of a massive particle with a 4-vector position x^μ is $U^\mu = dx^\mu/d\tau$ where τ is the proper time. Clearly, we must have $U^\mu U_\mu = -c^2$. In general, the tangent vector U^μ of a time-like curve parameterized by the proper time τ will be called the 4-vector velocity of the curve and it will satisfy

$$U^\mu U_\mu = -c^2. \tag{1.6}$$

A free particle will be in an inertial motion. The trajectory will therefore be given by a time-like geodesic given by the equation

$$U^\mu \partial_\mu U^\nu = 0. \tag{1.7}$$

Indeed, the operator $U^\mu \partial_\mu$ is the directional derivative along the curve. The energy–momentum 4-vector p^μ of a particle with rest mass m is given by

$$p^\mu = mU^\mu. \tag{1.8}$$

This leads to (with $\gamma = 1/\sqrt{1 - \vec{u}^2/c^2}$ and $\vec{u} = d\vec{x}/dt$)

$$E = cp^0 = m\gamma c^2, \quad \vec{p} = m\gamma\vec{u}. \tag{1.9}$$

We also compute

$$p^\mu p_\mu = -m^2 c^2 \Leftrightarrow E = \sqrt{m^2 c^4 + \vec{p}^2 c^2}. \tag{1.10}$$

The energy of a particle as measured by an observed whose velocity is v^μ is then clearly given by

$$E = -p^\mu v_\mu. \tag{1.11}$$

1.3 Differential geometry primer

1.3.1 Metric manifolds and vectors

Metric manifolds: An n-dimensional manifold M is a space which is locally flat, i.e. locally looks like R^n, and furthermore can be constructed from pieces of R^n sewn together smoothly. A Lorentzian or pseudo-Riemannian manifold is a manifold with the notion of 'distance', equivalently 'metric', included. 'Lorentzian' refers to the signature of the metric which in general relativity is taken to be $(-1, +1, +1, +1)$ as opposed to the more familiar/natural 'Euclidean' signature given by $(+1, +1, +1, +1)$ valid for Riemannian manifolds. The metric is usually denoted by $g_{\mu\nu}$ while the line element (also called metric in many instances) is written as

$$ds^2 = g_{\mu\nu} dx^\mu dx^\nu. \tag{1.12}$$

For example, Minkowski spacetime is given by the flat metric

$$g_{\mu\nu} = \eta_{\mu\nu} = (-1, +1, +1, +1). \tag{1.13}$$

Another extremely important example is Schwarzschild spacetime given by the metric

$$ds^2 = -\left(1 - \frac{R_s}{r}\right) dt^2 + \left(1 - \frac{R_s}{r}\right)^{-1} dr^2 + r^2 d\Omega^2. \tag{1.14}$$

This is quite different from the flat metric $\eta_{\mu\nu}$ and as a consequence the curvature of Schwarzschild spacetime is non-zero. Another important curved space is the surface of the two-dimensional sphere on which the metric, which appears as a part of the Schwarzschild metric, is given by

$$ds^2 = r^2 d\Omega^2 = r^2(d\theta^2 + \sin^2\theta d\phi^2). \tag{1.15}$$

The inverse metric will be denoted by $g^{\mu\nu}$, i.e.

$$g_{\mu\nu} g^{\nu\lambda} = \eta_\mu^\lambda. \tag{1.16}$$

Charts: A coordinate system (a chart) on the manifold M is a subset U of M together with a one-to-one map $\phi: U \longrightarrow R^n$ such that the image $V = \phi(U)$ is an open set in R^n, i.e. a set in which every point $y \in V$ is the center of an open ball which is inside V. We say that U is an open set in M. Hence we can associate with every point $p \in U$ of the manifold M the local coordinates $(x^1,..., x^n)$ by

$$\phi(p) = (x^1,..., x^n). \tag{1.17}$$

Vectors: A curved manifold is not necessarily a vector space. For example the sphere is not a vector space because we do not know how to add two points on the sphere to get another point on the sphere. The sphere which is naturally embedded in R^3 admits at each point p a tangent plane. The notion of a 'tangent vector space' can be constructed for any manifold which is embedded in R^n. The tangent vector space at a point p of the manifold will be denoted by V_p.

There is a one-to-one correspondence between vectors and directional derivatives in R^n. Indeed, the vector $v = (v^1,..., v^n)$ in R^n defines the directional derivative $\sum_\mu v^\mu \partial_\mu$ which acts on functions on R^n. These derivatives are clearly linear and satisfy the Leibniz rule. We will therefore define tangent vectors at a given point p on a manifold M as directional derivatives which satisfy linearity and the Leibniz rule.

These directional derivatives can also be thought of as differential displacements on the spacetime manifold at the point p.

This can be made more precise as follows. First, we define s smooth curve on the manifold M as a smooth map from R into M, viz $\gamma: R \longrightarrow M$. A tangent vector at a point p can then be thought of as a directional derivative operator along a curve which goes through p. Indeed, a tangent vector T at $p = \gamma(t) \in M$, acting on smooth functions f on the manifold M, can be defined by

$$T(f) = \frac{d}{dt}(f \circ \gamma(t))|_p. \tag{1.18}$$

In a given chart ϕ the point p will be given by $p = \phi^{-1}(x)$ where $x = (x^1,..., x^n) \in R^n$. Hence $\gamma(t) = \phi^{-1}(x)$. In other words, the map γ is mapped into a curve $x(t)$ in R^n. We have immediately

$$T(f) = \frac{d}{dt}(f \circ \phi^{-1}(x))|_p = \sum_{\mu=1}^{n} X_\mu(f) \frac{dx^\mu}{dt}|_p. \tag{1.19}$$

The maps X_μ act on functions f on the Manifold M as

$$X_\mu(f) = \frac{\partial}{\partial x^\mu}(f \circ \phi^{-1}(x)). \tag{1.20}$$

These can be checked to satisfy linearity and the Leibniz rule. They are obviously directional derivatives or differential displacements since we may make the identification $X_\mu = \partial_\mu$. Hence these vectors are tangent vectors to the manifold M at p. The fact that arbitrary tangent vectors can be expressed as linear combinations of the n vectors X_μ shows that these vectors are linearly independent, span the vector space V_p and that the dimension of V_p is exactly n. Equation (1.19) can then be rewritten as

$$T = \sum_{\mu=1}^{n} X_\mu T^\mu. \tag{1.21}$$

The components T^μ of the vector T are therefore given by

$$T^\mu = \frac{dx^\mu}{dt}|_p. \tag{1.22}$$

1.3.2 Geodesics

The length l of a smooth curve C with tangent T^μ on a manifold M with Riemannian metric $g_{\mu\nu}$ is given by

$$l = \int dt \sqrt{g_{\mu\nu} T^\mu T^\nu}. \tag{1.23}$$

The length is parametrization-independent. Indeed, we can show that

$$l = \int dt \sqrt{g_{\mu\nu} T^\mu T^\nu} = \int ds \sqrt{g_{\mu\nu} S^\mu S^\nu}, \quad S^\mu = T^\mu \frac{dt}{ds}. \tag{1.24}$$

In a Lorentzian manifold, the length of a space-like curve is also given by this expression. For a time-like curve for which $g_{ab}T^aT^b < 0$ the length is replaced with the proper time τ which is given by $\tau = \int dt\sqrt{-g_{ab}T^aT^b}$. For a light-like (or null) curve for which $g_{ab}T^aT^b = 0$ the length is always 0.

We consider the length of a curve C connecting two points $p = C(t_0)$ and $q = C(t_1)$. In a coordinate basis the length is given explicitly by

$$l = \int_{t_0}^{t_1} dt\sqrt{g_{\mu\nu}\frac{dx^\mu}{dt}\frac{dx^\nu}{dt}}. \tag{1.25}$$

The variation in l under an arbitrary smooth deformation of the curve C which keeps the two points p and q fixed is given by

$$\begin{aligned}
\delta l &= \frac{1}{2}\int_{t_0}^{t_1} dt\left(g_{\mu\nu}\frac{dx^\mu}{dt}\frac{dx^\nu}{dt}\right)^{-\frac{1}{2}}\left(\frac{1}{2}\delta g_{\mu\nu}\frac{dx^\mu}{dt}\frac{dx^\nu}{dt} + g_{\mu\nu}\frac{dx^\mu}{dt}\frac{d\delta x^\nu}{dt}\right) \\
&= \frac{1}{2}\int_{t_0}^{t_1} dt\left(g_{\mu\nu}\frac{dx^\mu}{dt}\frac{dx^\nu}{dt}\right)^{-\frac{1}{2}}\left(\frac{1}{2}\frac{\partial g_{\mu\nu}}{\partial x^\sigma}\delta x^\sigma\frac{dx^\mu}{dt}\frac{dx^\nu}{dt} + g_{\mu\nu}\frac{dx^\mu}{dt}\frac{d\delta x^\nu}{dt}\right) \\
&= \frac{1}{2}\int_{t_0}^{t_1} dt\left(g_{\mu\nu}\frac{dx^\mu}{dt}\frac{dx^\nu}{dt}\right)^{-\frac{1}{2}}\left(\frac{1}{2}\frac{\partial g_{\mu\nu}}{\partial x^\sigma}\delta x^\sigma\frac{dx^\mu}{dt}\frac{dx^\nu}{dt} - \frac{d}{dt}(g_{\mu\nu}\frac{dx^\mu}{dt})\delta x^\nu + \frac{d}{dt}(g_{\mu\nu}\frac{dx^\mu}{dt}\delta x^\nu)\right).
\end{aligned} \tag{1.26}$$

We can assume without any loss of generality that the parametrization of the curve C satisfies $g_{\mu\nu}(dx^\mu/dt)(dx^\nu/dt) = 1$. In other words, we choose dt^2 to be precisely the line element (interval) and thus $T^\mu = dx^\mu/dt$ is the 4-velocity. The last term in the above equation becomes obviously a total derivative which vanishes by the fact that the considered deformation keeps the two end points p and q fixed. We get then

$$\begin{aligned}
\delta l &= \frac{1}{2}\int_{t_0}^{t_1} dt\delta x^\sigma\left(\frac{1}{2}\frac{\partial g_{\mu\nu}}{\partial x^\sigma}\frac{dx^\mu}{dt}\frac{dx^\nu}{dt} - \frac{d}{dt}(g_{\mu\sigma}\frac{dx^\mu}{dt})\right) \\
&= \frac{1}{2}\int_{t_0}^{t_1} dt\delta x^\sigma\left(\frac{1}{2}\frac{\partial g_{\mu\nu}}{\partial x^\sigma}\frac{dx^\mu}{dt}\frac{dx^\nu}{dt} - \frac{\partial g_{\mu\sigma}}{\partial x^\nu}\frac{dx^\nu}{dt}\frac{dx^\mu}{dt} - g_{\mu\sigma}\frac{d^2x^\mu}{dt^2}\right) \\
&= \frac{1}{2}\int_{t_0}^{t_1} dt\delta x^\sigma\left(\frac{1}{2}\left(\frac{\partial g_{\mu\nu}}{\partial x^\sigma} - \frac{\partial g_{\mu\sigma}}{\partial x^\nu} - \frac{\partial g_{\nu\sigma}}{\partial x^\mu}\right)\frac{dx^\mu}{dt}\frac{dx^\nu}{dt} - g_{\mu\sigma}\frac{d^2x^\mu}{dt^2}\right) \\
&= \frac{1}{2}\int_{t_0}^{t_1} dt\delta x_\rho\left(\frac{1}{2}g^{\rho\sigma}\left(\frac{\partial g_{\mu\nu}}{\partial x^\sigma} - \frac{\partial g_{\mu\sigma}}{\partial x^\nu} - \frac{\partial g_{\nu\sigma}}{\partial x^\mu}\right)\frac{dx^\mu}{dt}\frac{dx^\nu}{dt} - \frac{d^2x^\rho}{dt^2}\right).
\end{aligned} \tag{1.27}$$

By definition, geodesics are curves which extremize the length l. The curve C extremizes the length between the two points p and q if and only if $\delta l = 0$. This leads immediately to the equation

$$\Gamma^\rho_{\mu\nu}\frac{dx^\mu}{dt}\frac{dx^\nu}{dt} + \frac{d^2x^\rho}{dt^2} = 0. \tag{1.28}$$

This equation is called the geodesic equation. It is the relativistic generalization of Newton's second law of motion (1.2). The Christoffel symbols are defined by

$$\Gamma^{\rho}_{\mu\nu} = -\frac{1}{2}g^{\rho\sigma}\left(\frac{\partial g_{\mu\nu}}{\partial x^{\sigma}} - \frac{\partial g_{\mu\sigma}}{\partial x^{\nu}} - \frac{\partial g_{\nu\sigma}}{\partial x^{\mu}}\right). \tag{1.29}$$

In the absence of curvature we will have $g_{\mu\nu} = \eta_{\mu\nu}$ and hence $\Gamma = 0$. In other words, the geodesics are locally straight lines.

Since the length between any two points on a Riemannian manifold (and between any two points which can be connected by a space-like curve on a Lorentzian manifold) can be arbitrarily long we conclude that the shortest curve connecting the two points must be a geodesic as it is an extremum of length. Hence the shortest curve is the straightest possible curve. The converse is not true: a geodesic connecting two points is not necessarily the shortest path.

Similarly, the proper time between any two points which can be connected by a time-like curve on a Lorentzian manifold can be arbitrarily small and thus the curve with the greatest proper time, if it exists, must be a time-like geodesic as it is an extremum of proper time. On the other hand, a time-like geodesic connecting two points is not necessarily the path with maximum proper time.

1.3.3 Tensors

Tangent (contravariant) vectors: Tensors are a generalization of vectors. Let us start then by giving a more precise definition of the tangent vector space V_p. Let $\mathcal{F}$ be the set of all smooth functions f on the manifold M, i.e. $f: M \longrightarrow R$. We define a tangent vector v at the point $p \in M$ as a map $v: \mathcal{F} \longrightarrow R$ which is required to satisfy linearity and the Leibniz rule. In other words,

$$v(af + bg) = av(f) + bv(g), \quad v(fg)$$
$$= f(p)v(g) + g(p)v(f), \quad a, b \in R, \quad f, g \in \mathcal{F}. \tag{1.30}$$

The vector space V_p is simply the set of all tangents vectors v at p. The action of the vector v on the function f is given explicitly by

$$v(f) = \sum_{\mu=1}^{n} v^{\mu} X_{\mu}(f), \quad X_{\mu}(f) = \frac{\partial}{\partial x^{\mu}}(f \circ \phi^{-1}(x)). \tag{1.31}$$

In a different chart ϕ' we will have

$$X'_{\mu}(f) = \frac{\partial}{\partial x'^{\mu}}(f \circ \phi'^{-1})|_{x'=\phi'(p)}. \tag{1.32}$$

We compute

$$\begin{aligned} X_{\mu}(f) &= \frac{\partial}{\partial x^{\mu}}(f \circ \phi^{-1})|_{x=\phi(p)} \\ &= \frac{\partial}{\partial x^{\mu}}f \circ \phi'^{-1}(\phi' \circ \phi^{-1})|_{x=\phi(p)} \\ &= \sum_{\nu=1}^{n}\frac{\partial x'^{\nu}}{\partial x^{\mu}}\frac{\partial}{\partial x'^{\nu}}(f \circ \phi'^{-1}(x'))|_{x'=\phi'(p)} \\ &= \sum_{\nu=1}^{n}\frac{\partial x'^{\nu}}{\partial x^{\mu}}X'_{\nu}(f). \end{aligned} \tag{1.33}$$

This is why the basis elements X_μ may be thought of as the partial derivative operators $\partial/\partial x^\mu$. The tangent vector v can be rewritten as $v = \sum_{\mu=1}^{n} v^\mu X_\mu = \sum_{\mu=1}^{n} v'^\mu X'_\mu$. We conclude immediately that

$$v'^\nu = \sum_{\nu=1}^{n} \frac{\partial x'^\nu}{\partial x^\mu} v^\mu. \tag{1.34}$$

This is the transformation law of tangent vectors under the coordinate transformation $x^\mu \longrightarrow x'^\mu$.

Cotangent dual (covariant) vectors or 1-forms: Let V_p^* be the space of all linear maps ω^* from V_p into R, viz $\omega^*: V_p \longrightarrow R$. The space V_p^* is the so-called dual vector space to V_p where addition and multiplication by scalars are defined in an obvious way. The elements of V_p^* are called dual vectors. The dual vector space V_p^* is also called the cotangent dual vector space at p and the vector space of one-forms at p. The elements of V_p^* are then called cotangent dual vectors. Another form of nomenclature is to refer to the elements of V_p^* as covariant vectors, as opposed to the elements of V_p which are referred to as contravariant vectors.

The basis $\{X^{\mu*}\}$ of V_p^* is called the dual basis to the basis $\{X_\mu\}$ of V_p. The basis elements of V_p^* are given by vectors $X^{\mu*}$ defined by

$$X^{\mu*}(X_\nu) = \delta_\nu^\mu. \tag{1.35}$$

We have the transformation law

$$X^{\mu*} = \sum_{\nu=1}^{n} \frac{\partial x^\mu}{\partial x'^\nu} X^{\nu*\prime}. \tag{1.36}$$

From this result we can think of the basis elements $X^{\mu*}$ as the gradients dx^μ, viz

$$X^{\mu*} \equiv dx^\mu. \tag{1.37}$$

Let $v = \sum_\mu v^\mu X_\mu$ be an arbitrary tangent vector in V_p, then the action of the dual basis elements $X^{\mu*}$ on v is given by

$$X^{\mu*}(v) = v^\mu. \tag{1.38}$$

The action of a general element $\omega^* = \sum_\mu \omega_\mu X^{\mu*}$ of V_p^* on v is given by

$$\omega^*(v) = \sum_\mu \omega_\mu v^\mu. \tag{1.39}$$

Again we conclude the transformation law

$$\omega'_\nu = \sum_{\nu=1}^{n} \frac{\partial x^\mu}{\partial x'^\nu} \omega_\mu. \tag{1.40}$$

Generalization: A tensor T of type (k, l) over the tangent vector space V_p is a multilinear map form $(V_p^* \times V_p^* \times \cdots \times V_p^*) \times (V_p \times V_p \times \cdots \times V_p)$ into R given by

$$T: V_p^* \times V_p^* \times \cdots \times V_p^* \times V_p \times V_p \times \cdots \times V_p \longrightarrow R. \tag{1.41}$$

The domain of this map is the direct product of k cotangent dual vector space V_p^* and l tangent vector space V_p. The space $\mathcal{T}(k, l)$ of all tensors of type (k, l) is a vector space of dimension $n^k \cdot n^l$ since $\dim V_p = \dim V_p^* = n$.

The tangent vectors $v \in V_p$ are therefore tensors of type $(1, 0)$, whereas the cotangent dual vectors $v \in V_p^*$ are tensors of type $(0, 1)$. The metric g is a tensor of type $(0, 2)$, i.e. a linear map from $V_p \times V_p$ into R, which is symmetric and nondegenerate.

1.4 Curvature tensor

1.4.1 Covariant derivative

A covariant derivative is a derivative which transforms covariantly under coordinates transformations $x \longrightarrow x'$. In other words, it is an operator ∇ on the manifold M which takes a differentiable tensor of type (k, l) to a differentiable tensor of type $(k, l + 1)$. It must clearly satisfy the obvious properties of linearity and the Leibniz rule but also satisfies other important rules such as the torsion-free condition given by

$$\nabla_\mu \, \nabla_\nu \, f = \nabla_\nu \, \nabla_\mu \, f, \quad f \in \mathcal{F}. \tag{1.42}$$

Furthermore, the covariant derivative acting on scalars must be consistent with tangent vectors being directional derivatives. Indeed, for all $f \in \mathcal{F}$ and $t^\mu \in V_p$ we must have

$$t^\mu \, \nabla_\mu \, f = t(f) \equiv t^\mu \partial_\mu f. \tag{1.43}$$

In other words, if ∇ and $\tilde{\nabla}$ are two covariant derivative operators, then their action on scalar functions must coincide, viz

$$t^\mu \, \nabla_\mu \, f = t^\mu \, \tilde{\nabla}_\mu f = t(f). \tag{1.44}$$

We compute now the difference $\tilde{\nabla}_\mu(f\omega_\nu) - \nabla_\mu \, (f\omega_\nu)$ where ω is some cotangent dual vector. We have

$$\begin{aligned} \tilde{\nabla}_\mu(f\omega_\nu) - \nabla_\mu \, (f\omega_\nu) &= \tilde{\nabla}_\mu f. \, \omega_\nu + f \, \tilde{\nabla}_\mu \, \omega_\nu - \nabla_\mu \, f. \, \omega_\nu - f \, \nabla_\mu \, \omega_\nu \\ &= f(\tilde{\nabla}_\mu \omega_\nu - \nabla_\mu \, \omega_\nu). \end{aligned} \tag{1.45}$$

We use without proof the following result. Let ω_ν' be the value of the cotangent dual vector ω_ν at a nearby point p', i.e. $\omega_\nu' - \omega_\nu$ is zero at p. Since the cotangent dual vector ω_ν is a smooth function on the manifold, then for each $p' \in M$, there must exist smooth functions $f_{(\alpha)}$ which vanish at the point p and cotangent dual vectors $\mu_\nu^{(\alpha)}$ such that

$$\omega_\nu' - \omega_\nu = \sum_\alpha f_{(\alpha)} \mu_\nu^{(\alpha)}. \tag{1.46}$$

We compute immediately

$$\tilde{\nabla}_\mu(\omega'_\nu - \omega_\nu) - \nabla_\mu (\omega'_\nu - \omega_\nu) = \sum_\alpha f_{(\alpha)} (\tilde{\nabla}_\mu \mu_\nu^{(\alpha)} - \nabla_\mu \mu_\nu^{(\alpha)}). \qquad (1.47)$$

This is 0 since by assumption $f_{(\alpha)}$ vanishes at p. Hence we get the result

$$\tilde{\nabla}_\mu \omega'_\nu - \nabla_\mu \omega'_\nu = \tilde{\nabla}_\mu \omega_\nu - \nabla_\mu \omega_\nu. \qquad (1.48)$$

In other words, the difference $\tilde{\nabla}_\mu \omega_\nu - \nabla_\mu \omega_\nu$ depends only on the value of ω_ν at the point p although both $\tilde{\nabla}_\mu \omega_\nu$ and $\nabla_\mu \omega_\nu$ depend on how ω_ν changes as we go away from point p since they are derivatives. Putting this differently we say that the operator $\tilde{\nabla}_\mu - \nabla_\mu$ is a linear map which takes cotangent dual vectors at a point p into tensors, of type $(0, 2)$, at p and not into tensor fields defined in a neighborhood of p. We write

$$\nabla_\mu \omega_\nu = \tilde{\nabla}_\mu \omega_\nu - C_{\mu\nu}^\gamma \omega_\gamma. \qquad (1.49)$$

The tensor $C_{\mu\nu}^\gamma$ stands for the map $\tilde{\nabla}_\mu - \nabla_\mu$ and it is clearly a tensor of type $(1, 2)$. By setting $\omega_\mu = \nabla_\mu f = \tilde{\nabla}_\mu f$ we get $\nabla_\mu \nabla_\nu f = \tilde{\nabla}_\mu \tilde{\nabla}_\nu f - C_{\mu\nu}^\gamma \nabla_\gamma f$. By employing now the torsion-free condition (1.42) we get immediately

$$C_{\mu\nu}^\gamma = C_{\nu\mu}^\gamma. \qquad (1.50)$$

Let us consider now the difference $\tilde{\nabla}_\mu(\omega_\nu t^\nu) - \nabla_\mu (\omega_\nu t^\nu)$ where t^ν is a tangent vector. Since $\omega_\nu t^\nu$ is a function we have

$$\tilde{\nabla}_\mu(\omega_\nu t^\nu) - \nabla_\mu (\omega_\nu t^\nu) = 0. \qquad (1.51)$$

From the other hand, we compute

$$\tilde{\nabla}_\mu(\omega_\nu t^\nu) - \nabla_\mu (\omega_\nu t^\nu) = \omega_\nu(\tilde{\nabla}_\mu t^\nu - \nabla_\mu t^\nu + C_{\mu\gamma}^\nu t^\gamma). \qquad (1.52)$$

Hence, we must have

$$\nabla_\mu t^\nu = \tilde{\nabla}_\mu t^\nu + C_{\mu\gamma}^\nu t^\gamma. \qquad (1.53)$$

For a general tensor $T_{\nu_1 \dots \nu_l}^{\mu_1 \dots \mu_k}$ of type (k, l) the action of the covariant derivative operator will be given by the expression

$$\nabla_\gamma T_{\nu_1 \dots \nu_l}^{\mu_1 \dots \mu_k} = \tilde{\nabla}_\gamma T_{\nu_1 \dots \nu_l}^{\mu_1 \dots \mu_k} + \sum_i C_{\gamma d}^{\mu_i} T_{\nu_1 \dots \nu_l}^{\mu_1 \dots d \dots \mu_k} - \sum_j C_{\gamma\nu_j}^d T_{\nu_1 \dots d \dots \nu_l}^{\mu_1 \dots \mu_k}. \qquad (1.54)$$

1.4.2 Parallel transport

Let C be a curve with a tangent vector t^μ. Let v^μ be some tangent vector defined at each point on the curve. The vector v^μ is parallelly transported along the curve C if and only if

$$t^\mu \nabla_\mu v^\nu \big|_{\text{curve}} = 0. \qquad (1.55)$$

If t is the parameter along the curve C then $t^\mu = dx^\mu/dt$ are the components of the vector t^μ in the coordinate basis. The parallel transport condition reads explicitly

$$\frac{dv^\nu}{dt} + \Gamma^\nu_{\mu\lambda} t^\mu v^\lambda = 0. \tag{1.56}$$

By demanding that the inner product of two vectors v^μ and w^μ is invariant under parallel transport we obtain, for all curves and all vectors, the condition

$$t^\mu \, \nabla_\mu \, (g_{\alpha\beta} v^\alpha w^\beta) = 0 \Rightarrow \nabla_\mu \, g_{\alpha\beta} = 0. \tag{1.57}$$

Thus, given a metric $g_{\mu\nu}$ on a manifold M the most natural covariant derivative operator is the one under which the metric is covariantly constant.

There exists a unique covariant derivative operator ∇_μ which satisfies $\nabla_\mu \, g_{\alpha\beta} = 0$. The proof goes as follows. We know that $\nabla_\mu \, g_{\alpha\beta}$ is given by

$$\nabla_\mu \, g_{\alpha\beta} = \tilde{\nabla}_\mu \, g_{\alpha\beta} - C^\gamma_{\mu\alpha} g_{\gamma\beta} - C^\gamma_{\mu\beta} g_{\alpha\gamma}. \tag{1.58}$$

By imposing $\nabla_\mu \, g_{\alpha\beta} = 0$ we get

$$\tilde{\nabla}_\mu g_{\alpha\beta} = C^\gamma_{\mu\alpha} g_{\gamma\beta} + C^\gamma_{\mu\beta} g_{\alpha\gamma}. \tag{1.59}$$

Equivalently,

$$\tilde{\nabla}_\alpha g_{\mu\beta} = C^\gamma_{\alpha\mu} g_{\gamma\beta} + C^\gamma_{\alpha\beta} g_{\mu\gamma}. \tag{1.60}$$

$$\tilde{\nabla}_\beta g_{\mu\alpha} = C^\gamma_{\mu\beta} g_{\gamma\alpha} + C^\gamma_{\alpha\beta} g_{\mu\gamma}. \tag{1.61}$$

Immediately, we conclude that

$$\tilde{\nabla}_\mu g_{\alpha\beta} + \tilde{\nabla}_\alpha \, g_{\mu\beta} - \tilde{\nabla}_\beta \, g_{\mu\alpha} = 2 C^\gamma_{\mu\alpha} g_{\gamma\beta}. \tag{1.62}$$

In other words,

$$C^\gamma_{\mu\alpha} = \frac{1}{2} g^{\gamma\beta} (\tilde{\nabla}_\mu g_{\alpha\beta} + \tilde{\nabla}_\alpha \, g_{\mu\beta} - \tilde{\nabla}_\beta \, g_{\mu\alpha}). \tag{1.63}$$

This choice of $C^\gamma_{\mu\alpha}$ which solves $\nabla_\mu \, g_{\alpha\beta} = 0$ is unique. In other words, the corresponding covariant derivative operator is unique. The most important case corresponds to the choice $\tilde{\nabla}_a = \partial_a$ for which case C^c_{ab} is denoted Γ^c_{ab} and is called the Christoffel symbol.

Equation (1.56) is almost the geodesic equation. Recall that geodesics are the straightest possible lines on a curved manifold. Alternatively, a geodesic can be defined as a curve whose tangent vector t^μ is parallelly transported along itself, viz $t^\mu \, \nabla_\mu \, t^\nu = 0$. This reads in a coordinate basis as

$$\frac{d^2 x^\nu}{dt^2} + \Gamma^\nu_{\mu\lambda} \frac{dx^\mu}{dt} \frac{dx^\lambda}{dt} = 0. \tag{1.64}$$

This is precisely (1.28). This is a set of n coupled second order ordinary differential equations with n unknown $x^\mu(t)$. We know, given appropriate initial conditions $x^\mu(t_0)$ and $dx^\mu/dt\,|_{t=t_0}$, that there exists a unique solution. Conversely, given a tangent vector t^μ at a point p of a manifold M there exists a unique geodesic which goes through p and is tangent to t^μ.

1.4.3 The Riemann curvature tensor

Definition: The parallel transport of a vector from point p to point q on the manifold M is actually path-dependent. This path-dependence is directly measured by the so-called Riemann curvature tensor. The Riemann curvature tensor can be defined in terms of the failure of successive operations of differentiation to commute. Let us start with an arbitrary tangent dual vector ω_a and an arbitrary function f. We want to calculate $(\nabla_a \nabla_b - \nabla_b \nabla_a)\omega_c$. First we have

$$\nabla_a \nabla_b (f\omega_c) = \nabla_a \nabla_b f . \omega_c + \nabla_b f \nabla_a \omega_c + \nabla_a f \nabla_b \omega_c + f \nabla_a \nabla_b \omega_c. \quad (1.65)$$

Similarly,

$$\nabla_b \nabla_a (f\omega_c) = \nabla_b \nabla_a f . \omega_c + \nabla_a f \nabla_b \omega_c + \nabla_b f \nabla_a \omega_c + f \nabla_b \nabla_a \omega_c. \quad (1.66)$$

Thus,

$$(\nabla_a \nabla_b - \nabla_b \nabla_a)(f\omega_c) = f(\nabla_a \nabla_b - \nabla_b \nabla_a)\omega_c. \quad (1.67)$$

We can follow the same set of arguments which led from (A.58) to (A.62) to conclude that the tensor $(\nabla_a \nabla_b - \nabla_b \nabla_a)\omega_c$ depends only on the value of ω_c at the point p. In other words $\nabla_a \nabla_b - \nabla_b \nabla_a$ is a linear map which takes tangent dual vectors into tensors of type $(0, 3)$. Equivalently, we can say that the action of $\nabla_a \nabla_b - \nabla_b \nabla_a$ on tangent dual vectors is equivalent to the action of a tensor of type $(1, 3)$. Thus we can write

$$(\nabla_a \nabla_b - \nabla_b \nabla_a)\omega_c = R_{abc}{}^d \omega_d. \quad (1.68)$$

The tensor $R_{abc}{}^d$ is precisely the Riemann curvature tensor. We compute explicitly

$$\begin{aligned}
\nabla_a \nabla_b \omega_c &= \nabla_a(\partial_b \omega_c - \Gamma^d_{bc}\omega_d) \\
&= \partial_a(\partial_b \omega_c - \Gamma^d_{bc}\omega_d) - \Gamma^e_{ab}(\partial_e \omega_c - \Gamma^d_{ec}\omega_d) - \Gamma^e_{ac}(\partial_b \omega_e - \Gamma^d_{be}\omega_d) \\
&= \partial_a \partial_b \omega_c - \partial_a \Gamma^d_{bc} . \omega_d - \Gamma^d_{bc}\partial_a \omega_d - \Gamma^e_{ab}\partial_e \omega_c \\
&\quad + \Gamma^e_{ab}\Gamma^d_{ec}\omega_d - \Gamma^e_{ac}\partial_b \omega_e + \Gamma^e_{ac}\Gamma^d_{be}\omega_d.
\end{aligned} \quad (1.69)$$

Thus,

$$(\nabla_a \nabla_b - \nabla_b \nabla_a)\omega_c = \left(\partial_b \Gamma^d_{ac} - \partial_a \Gamma^d_{bc} + \Gamma^e_{ac}\Gamma^d_{be} - \Gamma^a_{bc}\Gamma^d_{ae}\right)\omega_d. \quad (1.70)$$

We get then the components

$$R_{abc}{}^d = \partial_b \Gamma^d_{ac} - \partial_a \Gamma^d_{bc} + \Gamma^e_{ac}\Gamma^d_{be} - \Gamma^e_{bc}\Gamma^d_{ae}. \quad (1.71)$$

The action on tangent vectors can be found as follows. Let t^a be an arbitrary tangent vector. The scalar product $t^a \omega_a$ is a function on the manifold and thus

$$(\nabla_a \nabla_b - \nabla_b \nabla_a)(t^c \omega_c) = 0. \qquad (1.72)$$

This leads immediately to

$$(\nabla_a \nabla_b - \nabla_b \nabla_a)t^d = -R_{abc}{}^d t^c \qquad (1.73)$$

Generalization of this result and the previous one to higher order tensors is given by the following equation

$$(\nabla_a \nabla_b - \nabla_b \nabla_a)T^{d_1 \ldots d_k}_{c_1 \ldots c_l} = -\sum_{i=1}^{k} R_{abe}{}^{d_i} T^{d_1 \ldots e \ldots d_k}_{c_1 \ldots c_l} + \sum_{i=1}^{l} R_{abc_i}{}^{e} T^{d_1 \ldots d_k}_{c_1 \ldots e \ldots c_l}. \qquad (1.74)$$

Properties: We state without proof the following properties of the curvature tensor:
- Anti-symmetry in the first two indices:

$$R_{abc}{}^d = -R_{bac}{}^d. \qquad (1.75)$$

- Anti-symmetrization of the first three indices yields 0:

$$R_{[abc]}{}^d = 0, \quad R_{[abc]}{}^d = \frac{1}{3}(R_{abc}{}^d + R_{cab}{}^d + R_{bca}{}^d). \qquad (1.76)$$

- Anti-symmetry in the last two indices:

$$R_{abcd} = -R_{abdc}, \quad R_{abcd} = R_{abc}{}^e g_{ed}. \qquad (1.77)$$

- Symmetry if the pair consisting of the first two indices is exchanged with the pair consisting of the last two indices:

$$R_{abcd} = R_{cdab}. \qquad (1.78)$$

- Bianchi identity:

$$\nabla_{[a}R_{bc]d}{}^e = 0, \quad \nabla_{[a}R_{bc]d}{}^e = \frac{1}{3}(\nabla_a R_{bcd}{}^e + \nabla_c R_{abd}{}^e + \nabla_b R_{cad}{}^e). \qquad (1.79)$$

- The so-called Ricci tensor R_{ac}, which is the trace part of the Riemann curvature tensor, is symmetric, viz

$$R_{ac} = R_{ca}, \quad R_{ac} = R_{abc}{}^b. \qquad (1.80)$$

- The Einstein tensor can be constructed as follows. By contracting the Bianchi identity and using $\nabla_a g_{bc} = 0$ we get

$$g_e{}^c(\nabla_a R_{bcd}{}^e + \nabla_c R_{abd}{}^e + \nabla_b R_{cad}{}^e) = 0 \Rightarrow \nabla_a R_{bd} + \nabla_e R_{abd}{}^e - \nabla_b R_{ad} = 0. \quad (1.81)$$

By contracting now the two indices b and d we get

$$g^{bd}(\nabla_a R_{bd} + \nabla_e R_{abd}{}^e - \nabla_b R_{ad}) = 0 \Rightarrow \nabla_a R - 2 \nabla_b R_a{}^b = 0. \qquad (1.82)$$

This can be put in the form

$$\nabla^a G_{ab} = 0. \qquad (1.83)$$

The tensor G_{ab} is called Einstein tensor and is given by

$$G_{ab} = R_{ab} - \frac{1}{2}g_{ab}R. \qquad (1.84)$$

The so-called scalar curvature R is defined by

$$R = R_a{}^a. \qquad (1.85)$$

1.5 The stress–energy–momentum tensor

1.5.1 The stress–energy–momentum tensor

We will mostly be interested in continuous matter distributions which are extended macroscopic systems composed of a large number of individual particles. We will think of such systems as fluids. The energy, momentum and pressure of fluids are encoded in the stress–energy–momentum tensor $T^{\mu\nu}$ which is a symmetric tensor of type (2, 0). The component $T^{\mu\nu}$ of the stress–energy–momentum tensor is defined as the flux of the component p^μ of the 4-vector energy–momentum across a surface of constant x^ν.

Let us consider an infinitesimal element of the fluid in its rest frame. The spatial diagonal component T^{ii} is the flux of the momentum p^i across a surface of constant x^i, i.e. it is the amount of momentum p^i per unit time per unit area traversing the surface of constant x^i. Thus, T^{ii} is the normal stress which we also call pressure when it is independent of direction. We write $T^{ii} = P_i$. The spatial off-diagonal component T^{ij} is the flux of the momentum p^i across a surface of constant x^j, i.e. it is the amount of momentum p^i per unit time per unit area traversing the surface of constant x^j which means that T^{ij} is the shear stress.

The component T^{00} is the flux of the energy p^0 through the surface of constant x^0, i.e. it is the amount of energy per unit volume at a fixed instant of time. Thus T^{00} is the energy density, viz $T^{00} = \rho c^2$ where ρ is the rest-mass density. Similarly, T^{i0} is the flux of the momentum p^i through the surface of constant x^0, i.e. it is the i momentum density times c. The T^{0i} is the energy flux through the surface of constant x^i divided by c. They are equal by virtue of the symmetry of the stress–energy–momentum tensor, viz $T^{0i} = T^{i0}$.

1.5.2 Perfect fluid

We begin with the case of 'dust' which is a collection of a large number of particles in spacetime at rest with respect to each other. The particles are assumed to have the same rest mass m. The pressure of the dust is obviously 0 in any direction since there is no motion of the particles, i.e. the dust is a pressureless fluid. The 4-vector velocity of the dust is the constant 4-vector velocity U^μ of the individual particles. Let n be the number density of the particles, i.e. the number of particles per unit volume as measured in the rest frame. Clearly $N^i = nU^i = n(\gamma u_i)$ is the flux of the particles, i.e. the number of particles per unit area per unit time in the x^i direction. The 4-vector number-flux of the dust is defined by

$$N^\mu = nU^\mu. \tag{1.86}$$

The rest-mass density of the dust in the rest frame is clearly given by $\rho = nm$. This rest-mass density times c^2 is the $\mu = 0$, $\nu = 0$ component of the stress–energy–momentum tensor $T^{\mu\nu}$ in the rest frame. We remark that $\rho c^2 = nmc^2$ is also the $\mu = 0$, $\nu = 0$ component of the tensor $N^\mu p^\nu$ where N^μ is the 4-vector number-flux and p^μ is the 4-vector energy–momentum of the dust. We define therefore the stress–energy–momentum tensor of the dust by

$$T^{\mu\nu} = N^\mu p^\nu = (nm)U^\mu U^\nu = \rho U^\mu U^\nu. \tag{1.87}$$

The next fluid of paramount importance is the so-called perfect fluid. This is a fluid determined completely by its energy density ρ and its isotropic pressure P in the rest frame. Hence $T^{00} = \rho c^2$ and $T^{ii} = P$. The shear stresses T^{ij} ($i \neq j$) are absent for a perfect fluid in its rest frame. It is not difficult to convince ourselves that stress–energy–momentum tensor $T^{\mu\nu}$ is given in this case in the rest frame by

$$T^{\mu\nu} = \rho U^\mu U^\nu + \frac{P}{c^2}(c^2\eta^{\mu\nu} + U^\mu U^\nu) = (\rho + \frac{P}{c^2})U^\mu U^\nu + P\eta^{\mu\nu}. \tag{1.88}$$

This is a covariant equation and thus it must also hold, by the principle of minimal coupling (see below), in any other global inertial reference frame. We give the following examples:

- Dust: $P = 0$.
- Gas of photons: $P = \rho c^2/3$.
- Vacuum energy: $P = -\rho c^2 \Leftrightarrow T^{ab} = -\rho c^2 \eta^{ab}$.

1.5.3 Conservation law

The stress–energy–momentum tensor $T^{\mu\nu}$ is symmetric, viz $T^{\mu\nu} = T^{\nu\mu}$. It must also be conserved, i.e.

$$\partial_\mu T^{\mu\nu} = 0. \tag{1.89}$$

This should be thought of as the equation of motion of the perfect fluid. Explicitly this equation reads

$$\partial_\mu T^{\mu\nu} = \partial_\mu(\rho + \frac{P}{c^2}).\ U^\mu U^\nu + (\rho + \frac{P}{c^2})(\partial_\mu U^\mu.\ U^\nu + U^\mu\partial_\mu U^\nu) + \partial^\nu P = 0. \tag{1.90}$$

We project this equation along the 4-vector velocity by contracting it with U_ν. We get (using $U_\nu \partial_\mu U^\nu = 0$)

$$\partial_\mu(\rho U^\mu) + \frac{P}{c^2}\partial_\mu U^\mu = 0. \tag{1.91}$$

We project the above equation along a direction orthogonal to the 4-vector velocity by contracting it with P_ν^μ given by

$$P_\nu^\mu = \delta_\nu^\mu + \frac{U^\mu U_\nu}{c^2}.$$ (1.92)

Indeed, we can check that $P_\nu^\mu P_\lambda^\nu = P_\lambda^\mu$ and $P_\nu^\mu U^\nu = 0$. By contracting equation (1.90) with P_ν^λ we obtain

$$\left(\rho + \frac{P}{c^2}\right) U^\mu \partial_\mu U_\lambda + \left(\eta_{\nu\lambda} + \frac{U_\nu U_\lambda}{c^2}\right) \partial^\nu P = 0.$$ (1.93)

We consider now the non-relativistic limit defined by

$$U^\mu = (c, u_i), \quad |u_i| \ll 1, \quad P \ll \rho c^2.$$ (1.94)

The parallel equation (1.91) becomes the continuity equation given by

$$\partial_t \rho + \vec{\nabla} \cdot (\rho \vec{u}) = 0.$$ (1.95)

The orthogonal equation (1.93) becomes Euler's equation of fluid mechanics given by

$$\rho(\partial_t \vec{u} + (\vec{u} \cdot \vec{\nabla}) \vec{u}) = - \vec{\nabla} P.$$ (1.96)

1.5.4 Minimal coupling

The laws of physics in general relativity can be derived from the laws of physics in special relativity by means of the so-called principle of minimal coupling. This consists in writing the laws of physics in special relativity in tensor form and then replacing the flat metric $\eta_{\mu\nu}$ with the curved metric $g_{\mu\nu}$ and the derivative operator ∂_μ with the covariant derivative operator ∇_μ. This recipe works in most cases.

For example, take the geodesic equation describing a free particle in special relativity given by $U^\mu \partial_\mu U^\nu = 0$. Geodesic motion in general relativity is given by $U^\mu \nabla_\mu U^\nu = 0$. These are the geodesics of the curved metric $g_{\mu\nu}$ and they describe freely falling bodies in the corresponding gravitational field.

The second example is the equation of motion of a perfect fluid in special relativity which is given by the conservation law $\partial^\nu T_{\nu\lambda} = 0$. In general relativity this conservation law becomes

$$\nabla^\nu T_{\nu\lambda} = 0.$$ (1.97)

Also, by applying the principle of minimal coupling, the stress–energy–momentum tensor $T_{\mu\nu}$ of a perfect fluid in general relativity is given by equation (1.88) with the replacement $\eta \longrightarrow g$, viz

$$T_{\mu\nu} = \left(\rho + \frac{P}{c^2}\right) U_\mu U_\nu + P g_{\mu\nu}.$$ (1.98)

1.6 Einstein's equation

Although local gravitational forces cannot be measured by the principle of equivalence, i.e. since the spacetime manifold is locally flat, relative gravitational forces, the so-called tidal gravitational forces, can still be measured by observing the relative acceleration of nearby geodesics. This effect is described by the geodesic deviation equation.

1.6.1 Tidal gravitational forces

Let us first start by describing tidal gravitational forces in Newtonian physics. The force of gravity exerted by an object of mass M on a particle of mass m a distance r away is $\vec{F} = -\hat{r}GMm/r^2$ where $\hat{r}$ is the unit vector pointing from M to m and r is the distance between the center of M and m. The corresponding acceleration is $\vec{a} = -\hat{r}GM/r^2 = -\vec{\nabla}\Phi$, $\Phi = -GM/r$. We assume now that the mass m is spherical of radius Δr. The distance between the center of M and the center of m is r. The force of gravity exerted by the mass M on a particle of mass dm a distance $r \pm \Delta r$ away on the line joining the centers of M and m is given by $\vec{F} = -\hat{r}GMdm/(r \pm \Delta r)^2$. The corresponding acceleration is

$$\vec{a} = -\hat{r}GM\frac{1}{(r + \Delta r)^2} = -\hat{r}GM\frac{1}{r^2} + \hat{r}GM\frac{2\Delta r}{r^3} + \cdots \tag{1.99}$$

The first term is precisely the acceleration experienced at the center of the body m due to M. This term does not affect the observed acceleration of particles on the surface of m. In other words, since m and everything on its surface is in a state of free fall with respect to M, the acceleration of dm with respect to m is precisely the so-called tidal acceleration, and is given by the second term in the above expansion, viz

$$\vec{a_t} = \hat{r}GM\frac{2\Delta r}{r^3} + \cdots$$
$$= -(\Delta\vec{r}.\vec{\nabla})(\vec{\nabla}\Phi). \tag{1.100}$$

1.6.2 Geodesic deviation equation

In a flat Euclidean geometry, two parallel lines remain always parallel. This is not true in a curved manifold. To see this more carefully we consider a one-parameter family of geodesics $\gamma_s(t)$ which are initially parallel and see what happens to them as we move along these geodesics when we increase the parameter t. The map $(t, s) \longrightarrow \gamma_s(t)$ is smooth, one-to-one, and its inverse is smooth, which means in particular that the geodesics do not cross. These geodesics will then generate a 2D surface on the manifold M. The parameters t and s can therefore be chosen to be the coordinates on this surface. This surface is given by the entirety of the points $x^\mu(s, t) \in M$. The tangent vector to the geodesics is defined by

$$T^\mu = \frac{\partial x^\mu}{\partial t}. \tag{1.101}$$

This satisfies therefore the equation $T^\mu \, \nabla_\mu \, T^\nu = 0$. The so-called deviation vector is defined by

$$S^\mu = \frac{\partial x^\mu}{\partial s}. \tag{1.102}$$

The product $S^\mu ds$ is the displacement vector between two infinitesimally nearby geodesics. The vectors T^μ and S^μ commute because they are basis vectors. Hence we must have $[T, S]^\mu = T^\nu \, \nabla_\nu \, S^\mu - S^\nu \, \nabla_\nu \, T^\mu = 0$ or equivalently

$$T^\nu \, \nabla_\nu \, S^\mu = S^\nu \, \nabla_\nu \, T^\mu. \tag{1.103}$$

This can be checked directly by using the definition of the covariant derivative and the way it acts on tangent vectors and equations (1.101) and (1.102).

The quantity $V^\mu = T^\nu \, \nabla_\nu \, S^\mu$ expresses the rate of change of the deviation vector along a geodesic. We will call V^μ the relative velocity of infinitesimally nearby geodesics. Similarly, the relative acceleration of infinitesimally nearby geodesics is defined by $A^\mu = T^\nu \, \nabla_\nu \, V^\mu$. We compute

$$\begin{aligned}
A^\mu &= T^\nu \, \nabla_\nu \, V^\mu \\
&= T^\nu \, \nabla_\nu \, (T^\lambda \, \nabla_\lambda \, S^\mu) \\
&= T^\nu \, \nabla_\nu \, (S^\lambda \, \nabla_\lambda \, T^\mu) \\
&= (T^\nu \, \nabla_\nu \, S^\lambda). \, \nabla_\lambda \, T^\mu + T^\nu S^\lambda \, \nabla_\nu \, \nabla_\lambda \, T^\mu \\
&= (S^\nu \, \nabla_\nu \, T^\lambda). \, \nabla_\lambda \, T^\mu + T^\nu S^\lambda (\nabla_\lambda \, \nabla_\nu \, T^\mu - R_{\nu\lambda\sigma}{}^\mu T^\sigma) \\
&= S^\lambda \, \nabla_\lambda \, (T^\nu \, \nabla_\nu \, T^\mu) - R_{\nu\lambda\sigma}{}^\mu T^\nu S^\lambda T^\sigma \\
&= - R_{\nu\lambda\sigma}{}^\mu T^\nu S^\lambda T^\sigma.
\end{aligned} \tag{1.104}$$

This is the geodesic deviation equation. The relative accelaration of infinitesimally nearby geodesics is 0 if and only if $R_{\nu\lambda\sigma}{}^\mu = 0$. Geodesics will accelerate towards, or away from, each other if and only if $R_{\nu\lambda\sigma}{}^\mu \neq 0$. Thus, initially parallel geodesics with $V^\mu = 0$ will fail generically to remain parallel.

1.6.3 Einstein's equation

We will assume that, in general relativity, the tidal acceleration of two nearby particles is precisely the relative acceleration of infinitesimally nearby geodesics given by equation (1.104), viz

$$\begin{aligned}
A^\mu &= - R_{\nu\lambda\sigma}{}^\mu T^\nu S^\lambda T^\sigma \\
&= - R_{\nu\lambda\sigma}{}^\mu U^\nu \Delta x^\lambda U^\sigma.
\end{aligned} \tag{1.105}$$

This suggests, by comparing with (1.100), we make the following correspondence

$$R_{\nu\lambda\sigma}{}^\mu U^\nu U^\sigma \leftrightarrow \partial_\lambda \partial^\mu \Phi. \tag{1.106}$$

Thus

$$R_{\nu\mu\lambda}{}^\mu U^\nu U^\lambda \leftrightarrow \Delta\Phi. \tag{1.107}$$

By using the Poisson's equation (1.1) we get then the correspondence

$$R_{\nu\mu\sigma}{}^{\mu}U^{\nu}U^{\sigma} \leftrightarrow 4\pi G\rho. \tag{1.108}$$

From the other hand, the stress–energy–momentum tensor $T^{\mu\nu}$ provides the correspondence

$$T_{\nu\sigma}U^{\nu}U^{\sigma} \leftrightarrow \rho c^{4}. \tag{1.109}$$

We expect therefore an equation of the form

$$\frac{R_{\nu\mu\sigma}{}^{\mu}}{4\pi G} = \frac{T_{\nu\sigma}}{c^{4}} \Leftarrow R_{\nu\sigma} = \frac{4\pi G}{c^{4}}T_{\nu\sigma}. \tag{1.110}$$

This is the original equation proposed by Einstein. However, it has the following problem. From the fact that $\nabla^{\nu}G_{\nu\sigma} = 0$, we get immediately $\nabla^{\nu}R_{\nu\sigma} = \nabla_{\sigma} R/2$, and as a consequence $\nabla^{\nu}T_{\nu\sigma} = c^{4}\nabla_{\sigma} R/8\pi G$. This result is in direct conflict with the requirement of the conservation of the stress–energy–momentum tensor given by $\nabla^{\nu}T_{\nu\sigma} = 0$. An immediate solution is to consider instead the equation

$$G_{\nu\sigma} = R_{\nu\sigma} - \frac{1}{2}g_{\nu\sigma}R = \frac{8\pi G}{c^{4}}T_{\nu\sigma}. \tag{1.111}$$

The conservation of the stress–energy–momentum tensor is now guaranteed. Furthermore, this equation is still in accord with the correspondence $R_{\nu\sigma}U^{\nu}U^{\sigma} \leftrightarrow 8\pi G\rho$. Indeed, by using the result $R = -4\pi GT/c^{4}$ we can rewrite the above equation as

$$R_{\nu\sigma} = \frac{8\pi G}{c^{4}}\left(T_{\nu\sigma} - \frac{1}{2}g_{\nu\sigma}T\right). \tag{1.112}$$

We compute $R_{\mu\nu}U^{\mu}U^{\nu} = (8\pi G/c^{4})(T_{\mu\nu}U^{\mu}U^{\nu} + c^{2}T/2)$. By keeping only the $\mu = 0$, $\nu = 0$ component of $T_{\mu\nu}$ and neglecting the other components, the right-hand side is exactly $4\pi G\rho$ as it should be.

1.6.4 Newtonian limit

The Newtonian limit of general relativity is defined by the following three requirements:
  (1) The particles are moving slowly compared with the speed of light.
  (2) The gravitational field is weak so that the curved metric can be expanded about the flat metric.
  (3) The gravitational field is static.

Geodesic equation: We begin with the geodesic equation, with the proper time τ as the parameter of the geodesic, is

$$\Gamma^{\rho}_{\mu\nu}\frac{dx^{\mu}}{d\tau}\frac{dx^{\nu}}{d\tau} + \frac{d^{2}x^{\rho}}{d\tau^{2}} = 0. \tag{1.113}$$

The assumption that particles are moving slowly compared to the speed of light means that

$$|\frac{d\vec{x}}{d\tau}| \ll c|\frac{dt}{d\tau}|. \tag{1.114}$$

The geodesic equation becomes

$$c^2\Gamma^{\rho}_{00}\left(\frac{dt}{d\tau}\right)^2 + \frac{d^2x^{\rho}}{d\tau^2} = 0. \tag{1.115}$$

We recall the Christoffel symbols

$$\Gamma^{d}_{ab} = \frac{1}{2}g^{dc}(\partial_a g_{bc} + \partial_b g_{ac} - \partial_c g_{ab}). \tag{1.116}$$

Since the gravitational field is static we have

$$\Gamma^{d}_{00} = -\frac{1}{2}g^{dc}\partial_c g_{00}. \tag{1.117}$$

The second assumption that the gravitational field is weak allows us to decompose the metric as

$$g_{ab} = \eta_{ab} + h_{ab}, \quad |h_{ab}| \ll 1. \tag{1.118}$$

Thus

$$\Gamma^{d}_{00} = -\frac{1}{2}\eta^{dc}\partial_c h_{00}. \tag{1.119}$$

The geodesic equation becomes

$$\frac{d^2x^{\rho}}{d\tau^2} = \frac{c^2}{2}\eta^{dc}\partial_c h_{00}\left(\frac{dt}{d\tau}\right)^2. \tag{1.120}$$

In terms of components this reads

$$\frac{d^2x^0}{d\tau^2} = \frac{c^2}{2}\eta^{00}\partial_0 h_{00}\left(\frac{dt}{d\tau}\right)^2 = 0. \tag{1.121}$$

$$\frac{d^2x^i}{d\tau^2} = \frac{c^2}{2}\eta^{ii}\partial_i h_{00}\left(\frac{dt}{d\tau}\right)^2 = \frac{c^2}{2}\partial_i h_{00}\left(\frac{dt}{d\tau}\right)^2. \tag{1.122}$$

The first equation says that $dt/d\tau$ is a constant. The second equation reduces to

$$\frac{d^2x^i}{dt^2} = \frac{c^2}{2}\partial_i h_{00} = -\partial_i\Phi, \quad h_{00} = -\frac{2\Phi}{c^2}. \tag{1.123}$$

Einstein's equations: Now we turn to the Newtonian limit of Einstein's equation $R_{\nu\sigma} = 8\pi G(T_{\nu\sigma} - \frac{1}{2}g_{\nu\sigma}T)/c^4$ with the stress–energy–momentum tensor $T_{\mu\nu}$ of a perfect fluid as a source. The perfect fluid is describing the Earth or the Sun. The stress–energy–momentum tensor is given by $T_{\mu\nu} = (\rho + P/c^2)U_\mu U_\nu + Pg_{\mu\nu}$. In the Newtonian limit this can be approximated by the stress–energy–momentum tensor of dust given by $T_{\mu\nu} = \rho U_\mu U_\nu$ since in this limit pressure can be neglected as it comes from motion which is assumed to be slow. In the rest frame of the perfect fluid we have $U^\mu = (U^0, 0, 0, 0)$ and since $g_{\mu\nu} U^\mu U^\nu = -c^2$ we get $U^0 = c(1 + h_{00}/2)$ and $U_0 = c(-1 + h_{00}/2)$ and as a consequence

$$T^{00} = \rho c^2(1 + h_{00}), \quad T_{00} = \rho c^2(1 - h_{00}). \tag{1.124}$$

The inverse metric is obviously given by $g^{00} = -1 - h_{00}$ since $g^{\mu\nu}g_{\nu\rho} = \delta^\mu_\rho$. Hence

$$T = -\rho c^2. \tag{1.125}$$

The $\mu = 0$, $\nu = 0$ component of Einstein's equation is therefore

$$R_{00} = \frac{4\pi G}{c^2}\rho(1 - h_{00}). \tag{1.126}$$

We recall the Riemann curvature tensor and the Ricci tensor

$$R_{\mu\nu\sigma}{}^\lambda = \partial_\nu\Gamma^\lambda_{\mu\sigma} - \partial_\mu\Gamma^\lambda_{\nu\sigma} + \Gamma^\delta_{\mu\sigma}\Gamma^\lambda_{\nu\delta} - \Gamma^\delta_{\nu\sigma}\Gamma^\lambda_{\mu\delta}. \tag{1.127}$$

$$R_{\mu\sigma} = R_{\mu\nu\sigma}{}^\nu. \tag{1.128}$$

Thus (using in particular $R_{000}{}^0 = 0$)

$$R_{00} = R_{0i0}{}^i = \partial_i\Gamma^i_{00} - \partial_0\Gamma^i_{i0} + \Gamma^e_{00}\Gamma^i_{ie} - \Gamma^e_{i0}\Gamma^i_{0e}. \tag{1.129}$$

The Christoffel symbols are linear in the metric perturbation and thus one can neglect the third and fourth terms in the above equation. We get then

$$R_{00} = \partial_i\Gamma^i_{00} = -\frac{1}{2}\Delta h_{00}. \tag{1.130}$$

Einstein's equation reduces therefore to Newton's equation, viz

$$-\frac{1}{2}\Delta h_{00} = \frac{4\pi G}{c^2}\rho \Rightarrow \Delta\Phi = 4\pi G\rho. \tag{1.131}$$

1.7 Killing vectors and maximally symmetric spaces

A spacetime which is spatially homogeneous and spatially isotropic is a spacetime in which the space is maximally symmetric. A maximally symmetric space is a space with the maximum number of isometries, i.e. the maximum number of symmetries of the metric. These isometries are generated by the so-called Killing vectors.

As an example, if $\partial_\sigma g_{\mu\nu} = 0$, for some fixed value of σ, then the translation $x^\sigma \longrightarrow x^\sigma + a^\sigma$ is a symmetry and thus it is an isometry of the curved manifold M with metric $g_{\mu\nu}$. This symmetry will be naturally associated with a conserved quantity. To see this let us first recall that the geodesic equation can be rewritten in terms of the 4-vector energy–momentum $p^\mu = mU^\mu$ as $p^\mu \nabla_\mu p_\nu = 0$. Explicitly

$$m\frac{dp_\nu}{dt} = \Gamma^\lambda_{\mu\nu} p^\mu p_\lambda$$

$$= \frac{1}{2}\partial_\nu g_{\mu\rho} \cdot p^\mu p^\rho.$$

(1.132)

Thus if the metric is invariant under the translation $x^\sigma \longrightarrow x^\sigma + a^\sigma$, then $\partial_\sigma g_{\mu\nu} = 0$ and as a consequence the momentum p_σ is conserved as expected.

For obvious reasons we must rewrite the condition which expresses the symmetry under $x^\sigma \longrightarrow x^\sigma + a^\sigma$ in a covariant fashion. Let us thus introduce the vector $K = \partial_{(\sigma)}$ via its components which are given (in the basis in which $\partial_\sigma g_{\mu\nu} = 0$) by

$$K^\mu = (\partial_{(\sigma)})^\mu = \delta^\mu_\sigma.$$

(1.133)

Clearly then $p_\sigma = p_\mu K^\mu$. Since $\partial_\sigma g_{\mu\nu} = 0$ we must have $dp_\sigma/dt = 0$ or equivalently $d(p_\mu K^\mu)/dt = 0$. This means that the directional derivative of the scalar quantity $p_\mu K^\mu$ along the geodesic is 0, viz

$$p^\nu \nabla_\nu (p_\mu K^\mu) = 0.$$

(1.134)

We compute

$$p^\nu \nabla_\nu (p_\mu K^\mu) = p^\mu p^\nu \nabla_\mu K_\nu = \frac{1}{2} p^\mu p^\nu (\nabla_\mu K_\nu + \nabla_\nu K_\mu).$$

(1.135)

We obtain therefore the so-called Killing equation

$$\nabla_\mu K_\nu + \nabla_\nu K_\mu = 0.$$

(1.136)

Thus for any vector K which satisfies the Killing equation $\nabla_\mu K_\nu + \nabla_\nu K_\mu = 0$ the momentum $p_\mu K^\mu$ is conserved along the geodesic with tangent p. The vector K is called a Killing vector. The Killing vector K generates the isometry which is associated with the conservation of $p_\mu K^\mu$. The symmetry transformation under which the metric is invariant is expressed as infinitesimal motion in the direction of K.

Let us check that the vector $K^\mu = \delta^\mu_\sigma$ satisfies the Killing equation. Immediately, we have $K_\mu = g_{\mu\sigma}$ and

$$\nabla_\mu K_\nu + \nabla_\nu K_\mu = \partial_\mu g_{\nu\sigma} + \partial_\nu g_{\mu\sigma} - 2\Gamma^\rho_{\mu\nu} g_{\rho\sigma}$$

$$= \partial_\sigma g_{\mu\nu}$$

$$= 0.$$

(1.137)

Thus if the metric is independent of x^σ then the vector $K^\mu = \delta^\mu_\sigma$ will satisfy the Killing equation. Conversely, if a vector satisfies the killing equation then one can always find a basis in which the vector satisfies $K^\mu = \delta^\mu_\sigma$. However, if we have more than one Killing vector we cannot find a single basis in which all of them satisfy $K^\mu = \delta^\mu_\sigma$.

Some of the properties of Killing vectors are:

$$\nabla_\mu \ \nabla_\nu \ K^\lambda = R_{\nu\mu\rho}{}^\lambda K^\rho. \tag{1.138}$$

$$\nabla_\mu \ \nabla_\nu \ K^\mu = R_{\nu\mu} K^\mu. \tag{1.139}$$

$$K^\mu \nabla_\mu \ R = 0. \tag{1.140}$$

The last identity in particular shows explicitly that the geometry does not change under a Killing vector.

The isometries of R^n with flat Euclidean metric are n independent translations and $n(n-1)/2$ independent rotations (which form the group of $SO(n)$ rotations). Hence R^n with flat Euclidean metric has $n + n(n-1)/2 = n(n+1)/2$ isometries. This is the number of Killing vectors on R^n with flat Euclidean metric which is the maximum possible number of isometries in n dimensions. The space R^n is therefore called maximally symmetric space. In general, a maximally symmetric space is any space with $n(n+1)/2$ Killing vectors (isometries). These spaces have the maximum degree of symmetry. The only Euclidean maximally symmetric spaces are planes R^n with 0 scalar curvature, spheres S^n with positive scalar curvature and hyperboloids H^n with negative scalar curvature[1].

The curvature of a maximally symmetric space must be the same everywhere (translations) and the same in every direction (rotations). More precisely, a maximally symmetric space must be locally fully characterized by a constant scalar curvature R and furthermore must look like the same in all directions, i.e. it must be invariant under all Lorentz transformations at the point of consideration.

In the neighborhood of a point $p \in M$ we can always choose an inertial reference frame in which $g_{\mu\nu} = \eta_{\mu\nu}$. This is invariant under Lorentz transformations at p. Since the space is maximally symmetric the Riemann curvature tensor $R_{\mu\nu\lambda\rho}$ at p must also be invariant under Lorentz transformations at p. This tensor must therefore be constructed from $\eta_{\mu\nu}$, the Kronecker delta $\delta_{\mu\nu}$ and the Levi-Civita tensor $\varepsilon_{\mu\nu\lambda\rho}$ which are the only tensors which are known to be invariant under Lorentz transformations. However, the curvature tensor satisfies $R_{\mu\nu\lambda\gamma} = -R_{\nu\mu\lambda\gamma}$, $R_{\mu\nu\lambda\gamma} = -R_{\mu\nu\gamma\lambda}$, $R_{\mu\nu\lambda\gamma} = R_{\lambda\gamma\mu\nu}$, $R_{[\mu\nu\lambda]\gamma} = 0$ and $\nabla_{[\mu}R_{\nu\lambda]\gamma\rho} = 0$. The only combination formed out of $\eta_{\mu\nu}$, $\delta_{\mu\nu}$ and $\varepsilon_{\mu\nu\lambda\rho}$ which satisfies these identities is $R_{\mu\nu\lambda\gamma} = \kappa(\eta_{\mu\lambda}\eta_{\nu\gamma} - \eta_{\mu\gamma}\eta_{\nu\lambda})$ with κ a constant. This tensorial relation must hold in any other coordinate system, viz

$$R_{\mu\nu\lambda\gamma} = \kappa(g_{\mu\lambda}g_{\nu\gamma} - g_{\mu\gamma}g_{\nu\lambda}). \tag{1.141}$$

We compute $R_{\mu\nu\lambda}{}^\gamma = \kappa(g_{\mu\lambda}\delta^\gamma_\nu - \delta^\gamma_\mu g_{\nu\lambda})$, $R_{\mu\lambda} = R_{\mu\nu\lambda}{}^\nu = \kappa(n-1)g_{\mu\lambda}$ and hence $R = \kappa n(n-1)$. In other words the scalar curvature of a maximally symmetric space is a constant over the manifold. Thus the curvature of a maximally symmetric space must be of the form

---

[1] The corresponding maximally symmetric Lorentzian spaces are Minkowski spaces M^n ($R = 0$), de Sitter spaces dS^n ($R > 0$) and anti-de Sitter spaces AdS^n ($R < 0$).

$$R_{\mu\nu\lambda\gamma} = \frac{R}{n(n-1)}(g_{\mu\lambda}g_{\nu\gamma} - g_{\mu\gamma}g_{\nu\lambda}). \tag{1.142}$$

Conversely if the curvature tensor is given by this equation with R constant over the manifold then the space is maximally symmetric.

1.8 The Hilbert–Einstein action

Einstein's equations for general relativity read

$$R_{\mu\nu} - \frac{1}{2}g_{\mu\nu}R = 8\pi G T_{\mu\nu}. \tag{1.143}$$

The dynamical variable is obviously the metric $g_{\mu\nu}$. The goal is to construct an action principle from which Einstein's equations follow as the Euler–Lagrange equations of motion for the metric. This action principle will read as

$$S = \int d^n x \; \mathcal{L}(g). \tag{1.144}$$

The first problem with this way of writing is that both $d^n x$ and $\mathcal{L}$ are tensor densities rather than tensors. We digress briefly to explain this important different.

Let us recall the familiar Levi-Civita symbol in n dimensions defined by

$$\begin{aligned}
\tilde{\varepsilon}_{\mu_1...\mu_n} &= +1 \quad \text{even permutation} \\
&= -1 \quad \text{odd permutation} \\
&= \quad 0 \quad \text{otherwise.}
\end{aligned} \tag{1.145}$$

This is a symbol and not a tensor since it does not change under coordinate transformations. The determinant of a matrix M can be given by the formula

$$\tilde{\varepsilon}_{\nu_1...\nu_n}\det M = \tilde{\varepsilon}_{\mu_1...\mu_n} M^{\mu_1}_{\nu_1} ... M^{\mu_n}_{\nu_n}. \tag{1.146}$$

By choosing $M^{\mu}_{\nu} = \partial x^{\mu}/\partial y^{\nu}$ we get the transformation law

$$\tilde{\varepsilon}_{\nu_1...\nu_n} = \det\frac{\partial y}{\partial x} \; \tilde{\varepsilon}_{\mu_1...\mu_n} \frac{\partial x^{\mu_1}}{\partial y^{\nu_1}} ... \frac{\partial x^{\mu_n}}{\partial y^{\nu_n}}. \tag{1.147}$$

In other words, $\tilde{\varepsilon}_{\mu_1...\mu_n}$ is not a tensor because of the determinant appearing in this equation. This is an example of a tensor density. Another example of a tensor density is $\det g$. Indeed from the tensor transformation law of the metric $g'_{\alpha\beta} = g_{\mu\nu}(\partial x^{\mu}/\partial y^{\alpha})(\partial x^{\nu}/\partial y^{\beta})$ we can show in a straightforward way that

$$\det g' = (\det\frac{\partial y}{\partial x})^{-2} \; \det g. \tag{1.148}$$

The actual Levi-Civita tensor can then be defined by

$$\varepsilon_{\mu_1...\mu_n} = \sqrt{\det g} \; \tilde{\varepsilon}_{\mu_1...\mu_n}. \tag{1.149}$$

Next under a coordinate transformation $x \longrightarrow y$ the volume element transforms as

$$d^n x \longrightarrow d^n y = \det\frac{\partial y}{\partial x} \ d^n x. \tag{1.150}$$

In other words, the volume element transforms as a tensor density and not as a tensor. We verify this important point in our language as follows. We write

$$
\begin{aligned}
d^n x &= dx^0 \wedge dx^1 \wedge \cdots \wedge dx^{n-1} \\
&= \frac{1}{n!}\tilde{\varepsilon}_{\mu_1 \dots \mu_n} dx^{\mu_1} \wedge \cdots \wedge dx^{\mu_n}.
\end{aligned} \tag{1.151}
$$

Recall that a differential p-form is a $(0, p)$ tensor which is completely antisymmetric. For example, scalars are 0-forms and dual cotangent vectors are 1-forms The Levi-Civita tensor $\varepsilon_{\mu_1 \dots \mu_n}$ is a 4-form. The differentials dx^μ appearing in the second line of equation (8.23) are 1-forms and hence under a coordinate transformation $x \longrightarrow y$ we have $dx^\mu \longrightarrow dy^\mu = dx^\nu \partial y^\mu / \partial x^\nu$. By using this transformation law we can immediately show that dx^n transforms to $d^n y$ exactly as in equation (8.22).

It is not difficult to see now that an invariant volume element can be given by the n-form defined by the equation

$$dV = \sqrt{\det g} \ d^n x. \tag{1.152}$$

We can show that

$$
\begin{aligned}
dV &= \frac{1}{n!}\sqrt{\det g} \ \tilde{\varepsilon}_{\mu_1 \dots \mu_n} dx^{\mu_1} \wedge \cdots \wedge dx^{\mu_n} \\
&= \frac{1}{n!}\varepsilon_{\mu_1 \dots \mu_n} dx^{\mu_1} \wedge \cdots \wedge dx^{\mu_n} \\
&= \varepsilon_{\mu_1 \dots \mu_n} dx^{\mu_1} \otimes \cdots \otimes dx^{\mu_n} \\
&= \varepsilon.
\end{aligned} \tag{1.153}
$$

In other words, the invariant volume element is precisely the Levi-Civita tensor. In the case of the Lorentzian signature we replace $\det g$ with $-\det g$.

We go back now to equation (8.16) and rewrite it as

$$
\begin{aligned}
S &= \int d^n x \ \mathcal{L}(g) \\
&= \int d^n x \sqrt{-\det g} \ \hat{\mathcal{L}}(g).
\end{aligned} \tag{1.154}
$$

Clearly $\mathcal{L} = \sqrt{-\det g} \ \hat{\mathcal{L}}$. Since the invariant volume element $d^n x \sqrt{-\det g}$ is a scalar, the function $\hat{\mathcal{L}}$ must also be a scalar and as such can be identified with the Lagrangian density.

We use the result that the only independent scalar quantity which is constructed from the metric and which is at most second order in its derivatives is the Ricci scalar R. In other words, the simplest choice for the Lagrangian density $\hat{\mathcal{L}}$ is

$$\hat{\mathcal{L}}(g) = R. \tag{1.155}$$

The corresponding action is called the Hilbert–Einstein action. We compute

$$\delta S = \int d^n x \, \delta \sqrt{-\det g} \; g^{\mu\nu} R_{\mu\nu} + \int d^n x \sqrt{-\det g} \; \delta g^{\mu\nu} R_{\mu\nu}$$
$$+ \int d^n x \sqrt{-\det g} \; g^{\mu\nu} \delta R_{\mu\nu}. \tag{1.156}$$

We have

$$\begin{aligned}
\delta R_{\mu\nu} &= \delta R_{\mu\rho\nu}{}^{\rho} \\
&= \partial_\rho \delta \Gamma^\rho_{\mu\nu} - \partial_\mu \delta \Gamma^\rho_{\rho\nu} + \delta(\Gamma^\lambda_{\mu\nu}\Gamma^\rho_{\rho\lambda} - \Gamma^\lambda_{\rho\nu}\Gamma^\rho_{\mu\lambda}) \\
&= (\nabla_\rho \delta \Gamma^\rho_{\mu\nu} - \Gamma^\rho_{\rho\lambda}\delta\Gamma^\lambda_{\mu\nu} + \Gamma^\lambda_{\rho\mu}\delta\Gamma^\rho_{\lambda\nu} + \Gamma^\lambda_{\rho\nu}\delta\Gamma^\rho_{\lambda\mu}) \\
&\quad - \Big(\nabla_\mu \delta\Gamma^\rho_{\rho\nu} - \Gamma^\rho_{\mu\lambda}\delta\Gamma^\lambda_{\rho\nu} + \Gamma^\lambda_{\mu\rho}\delta\Gamma^\rho_{\lambda\nu} \\
&\quad + \Gamma^\lambda_{\mu\nu}\delta\Gamma^\rho_{\rho\lambda}\Big) + \delta(\Gamma^\lambda_{\mu\nu}\Gamma^\rho_{\rho\lambda} - \Gamma^\lambda_{\rho\nu}\Gamma^\rho_{\mu\lambda}) \\
&= \nabla_\rho \, \delta\Gamma^\rho_{\mu\nu} - \nabla_\mu \, \delta\Gamma^\rho_{\rho\nu}.
\end{aligned} \tag{1.157}$$

In the second line of the above equation we have used the fact that $\delta\Gamma^\rho_{\mu\nu}$ is a tensor since it is the difference of two connections. Thus

$$\begin{aligned}
\int d^n x \sqrt{-\det g} \; g^{\mu\nu}\delta R_{\mu\nu} &= \int d^n x \sqrt{-\det g} \; g^{\mu\nu}\big(\nabla_\rho\delta\Gamma^\rho_{\mu\nu} - \nabla_\mu \, \delta\Gamma^\rho_{\rho\nu}\big) \\
&= \int d^n x \sqrt{-\det g} \; \nabla_\rho \, \big(g^{\mu\nu}\delta\Gamma^\rho_{\mu\nu} - g^{\rho\nu}\delta\Gamma^\mu_{\mu\nu}\big).
\end{aligned} \tag{1.158}$$

We compute also (with $\delta g_{\mu\nu} = -g_{\mu\alpha}g_{\nu\beta}\delta g^{\alpha\beta}$)

$$\begin{aligned}
\delta\Gamma^\rho_{\mu\nu} &= \frac{1}{2}g^{\rho\lambda}(\nabla_\mu\delta g_{\nu\lambda} + \nabla_\nu \, \delta g_{\mu\lambda} - \nabla_\lambda \, \delta g_{\mu\nu}) \\
&= -\frac{1}{2}(g_{\nu\lambda}\,\nabla_\mu \, \delta g^{\lambda\rho} + g_{\mu\lambda}\,\nabla_\nu \, \delta g^{\lambda\rho} - g_{\mu\alpha}g_{\nu\beta}\,\nabla^\rho \, \delta g^{\alpha\beta}).
\end{aligned} \tag{1.159}$$

Thus

$$\int d^n x \sqrt{-\det g} \; g^{\mu\nu}\delta R_{\mu\nu} = \int d^n x \sqrt{-\det g} \; \nabla_\rho \, (g_{\mu\nu}\,\nabla^\rho \, \delta g^{\mu\nu} - \nabla_\mu \, \delta g^{\mu\rho}). \tag{1.160}$$

By Stokes's theorem this integral is equal to the integral over the boundary of spacetime of the expression $g_{\mu\nu}\,\nabla^\rho \, \delta g^{\mu\nu} - \nabla_\mu \, \delta g^{\mu\rho}$ which is 0 if we assume that the metric and its first derivatives are held fixed on the boundary. The variation of the action reduces to

$$\delta S = \int d^n x \, \delta \sqrt{-\det g} \; g^{\mu\nu} R_{\mu\nu} + \int d^n x \sqrt{-\det g} \; \delta g^{\mu\nu} R_{\mu\nu}. \tag{1.161}$$

Next we use the result

$$\delta\sqrt{-\det g} = -\frac{1}{2}\sqrt{-\det g} \; g_{\mu\nu}\delta g^{\mu\nu}. \tag{1.162}$$

Hence

$$\delta S = \int d^n x \sqrt{-\det g} \; \delta g^{\mu\nu} (R_{\mu\nu} - \frac{1}{2} g_{\mu\nu} R).$$ (1.163)

This will obviously lead to Einstein's equations in vacuum which is partially our goal. We want also to include the effect of matter which requires considering the more general actions of the form

$$S = \frac{1}{16\pi G} \int d^n x \; \sqrt{-\det g} \; R + S_M.$$ (1.164)

$$S_M = \int d^n x \; \sqrt{-\det g} \; \hat{\mathcal{L}}_M.$$ (1.165)

The variation of the action becomes

$$\begin{aligned}
\delta S &= \frac{1}{16\pi G} \int d^n x \sqrt{-\det g} \; \delta g^{\mu\nu} \left(R_{\mu\nu} - \frac{1}{2} g_{\mu\nu} R \right) + \delta S_M \\
&= \int d^n x \sqrt{-\det g} \; \delta g^{\mu\nu} \left[\frac{1}{16\pi G} \left(R_{\mu\nu} - \frac{1}{2} g_{\mu\nu} R \right) + \frac{1}{\sqrt{-\det g}} \frac{\delta S_M}{\delta g^{\mu\nu}} \right].
\end{aligned}$$ (1.166)

In other words,

$$\frac{1}{\sqrt{-\det g}} \frac{\delta S}{\delta g^{\mu\nu}} = \frac{1}{16\pi G} \left(R_{\mu\nu} - \frac{1}{2} g_{\mu\nu} R \right) + \frac{1}{\sqrt{-\det g}} \frac{\delta S_M}{\delta g^{\mu\nu}}.$$ (1.167)

Einstein's equations are therefore given by

$$R_{\mu\nu} - \frac{1}{2} g_{\mu\nu} R = 8\pi G T_{\mu\nu}.$$ (1.168)

The stress–energy–momentum tensor must therefore be defined by the equation

$$T_{\mu\nu} = -\frac{2}{\sqrt{-\det g}} \frac{\delta S_M}{\delta g^{\mu\nu}}.$$ (1.169)

As a first example, we consider the action of a scalar field in curved spacetime given by

$$S_\phi = \int d^n x \sqrt{-\det g} \left[-\frac{1}{2} g^{\mu\nu} \nabla_\mu \phi \nabla_\nu \phi - V(\phi) \right].$$ (1.170)

The corresponding stress–energy–momentum tensor is calculated to be given by

$$T_{\mu\nu}^{(\phi)} = \nabla_\mu \phi \nabla_\nu \phi - \frac{1}{2} g_{\mu\nu} g^{\rho\sigma} \nabla_\rho \phi \nabla_\sigma \phi - g_{\mu\nu} V(\phi).$$ (1.171)

As a second example we consider the action of the electromagnetic field in curved spacetime given by

$$S_A = \int d^n x \sqrt{-\det g} \left[-\frac{1}{4} g^{\mu\nu} g^{\alpha\beta} F_{\mu\nu} F_{\alpha\beta} \right]. \tag{1.172}$$

In this case the stress–energy–momentum tensor is calculated to be given by

$$T_{\mu\nu}^{(A)} = F^{\mu\lambda} F_\lambda^\nu - \frac{1}{4} g^{\mu\nu} F_{\alpha\beta} F^{\alpha\beta}. \tag{1.173}$$

1.9 Exercises

Exercise 1:
- The metric on the sphere is given by

$$d\Omega^2 = d\theta^2 + \sin^2\theta \, d\phi^2. \tag{1.174}$$

Compute the non-zero components of the Christoffel symbol.
- Compute the non-zero components of the Riemann tensor and the Ricci tensor. Compute the Ricci scalar.
- Recall that the metric in polar coordinates on R^3 is given by

$$ds^2 = dr^2 + r^2 d\Omega^2. \tag{1.175}$$

The components of this metric are independent of ϕ. Determine the Killing vector associated with rotation around the z axis with angle ϕ.
- Determine the Killing vectors associated with rotations on the sphere. Hint: use ∂_x, ∂_y and ∂_z as basis elements.

Solution 1:

- $$\Gamma^\theta_{\phi\phi} = -\sin\theta\cos\theta, \quad \Gamma^\phi_{\theta\phi} = \cot\theta.$$

- $$R^\theta_{\phi\theta\phi} = \sin^2\theta, \quad R_{\theta\phi\theta\phi} = \sin^2\theta.$$
$$R_{\theta\theta} = 1, \quad R_{\phi\phi} = \sin^2\theta, \quad R_{\theta\phi} = 0.$$
$$R = 2.$$

- $$R = \partial_\phi = -y\partial_x + x\partial_y = (-y, x, 0).$$

- $$T = (\vec{r} \times \vec{\partial})_x = (0, -z, y).$$
$$S = (\vec{r} \times \vec{\partial})_y = (z, 0, -x).$$

Exercise 2: The metric on the hyperboloid H^2 (Poincaré half-plane) is given by

$$ds^2 = \frac{r^2}{y^2}(dx^2 + dy^2). \tag{1.176}$$

Compute the length of the line segment between (x_0, y_1) and (x_0, y_2).
   Solution 2:

$$l = \int_{(x_0, y_1)}^{(x_0, y_2)} \sqrt{\frac{r^2}{y^2}dy^2} = r \ln \frac{y_2}{y_1}.$$

Exercise 3:
  • We consider the metric given by

$$ds^2 = -dt^2 + a^2(t)(dx^2 + dy^2 + dz^2). \tag{1.177}$$

   Compute the non-zero components of the Christoffel symbol.
  • Write down the zero component of the geodesic equation.
  • Write down the condition for null geodesics and use it to solve the zero component of the null geodesic equation.
  • What is the energy of a photon as measured by a comoving observer in this expanding observer.
  • Write down the relation between the values of the photon energy at two different scales a_1 and a_2.
  • The matter distribution in the Universe is assumed to be a perfect fluid. Write down the non-zero components of the energy–momentum tensor in this universe as seen by a comoving observer.
  • By using the conservation law of the energy–momentum tensor $\nabla_\mu T^{\mu\nu} = 0$ and the equation of state $P = w\rho$ derive the form of the mass density ρ.

Solution 3:

  •
$$\Gamma^0_{ij} = a\dot{a}\delta_{ij}, \quad \Gamma^i_{0j} = \frac{\dot{a}}{a}\delta_{ij}.$$

  •
$$\frac{d^2t}{d\lambda^2} + a\dot{a}\left(\frac{d\vec{x}}{d\lambda}\right)^2 = 0.$$

  • Null geodesics

$$a^2\frac{d\vec{x}^2}{d\lambda^2} = \frac{dt^2}{d\lambda^2}.$$

By combing the above two equations we have

$$\frac{d^2t}{d\lambda^2} + \frac{\dot{a}}{a}\left(\frac{dt}{d\lambda}\right)^2 = 0.$$

The solution is

$$\frac{dt}{d\lambda} = \frac{\omega_0}{a}.$$

- If $U^\mu = (1, 0, 0, 0)$ is the 4-vector velocity of the comoving observer then

$$E = -p^\mu U_\mu = \frac{dt}{d\lambda} = \frac{\omega_0}{a}.$$

-

$$\frac{E_1}{E_2} = \frac{a_2}{a_1}.$$

-

$$T^{\mu\nu} = \left(\rho, \frac{P}{a^2}, \frac{P}{a^2}, \frac{P}{a^2}\right).$$

-

$$\rho = a^{-3(1+w)}.$$

Exercise 4:
- The Schwarzschild metric is given by

$$ds^2 = -\left(1 - \frac{2GM}{r}\right)dt^2 + \left(1 - \frac{2GM}{r}\right)^{-1}dr^2 + r^2d\Omega^2. \qquad (1.178)$$

Compute the time translation and rotational Killing vectors in this spacetime.
- Compute the energy and the angular momentum of a particle moving in this spacetime.
- Show that the following quantity

$$\varepsilon = g_{\mu\nu}\frac{dx^\mu}{d\lambda}\frac{dx^\nu}{d\lambda} \qquad (1.179)$$

is conserved along a geodesic.

- Write down explicitly the above conserved quantity in Schwarzschild space-time. Derive the effective potential.
- Determine the light cones of the Schwarzschild metric. What happens at $r = 2GM$.

Solution 4:

- $$K^\mu = (\partial_t)^\mu = \delta^\mu_t = (1, 0, 0, 0), \quad K_\mu = g_{\mu t} = \left(-1 + \frac{2GM}{r}, 0, 0, 0\right).$$

$$R^\mu = (\partial_\phi)^\mu = \delta^\mu_\phi = (0, 0, 0, 1), \quad R_\mu = g_{\mu\phi} = (0, 0, 0, r^2 \sin^2\theta).$$

- $$E = -K^\mu \frac{dx_\mu}{d\lambda} = \left(1 - \frac{2GM}{r}\right)\frac{dt}{d\lambda}.$$

$$L = R^\mu \frac{dx_\mu}{d\lambda} = r^2 \sin^2\theta \frac{d\phi}{d\lambda}.$$

- $$\frac{d\varepsilon}{d\lambda} = -\frac{dx^\alpha}{d\lambda}\frac{dx^\beta}{d\lambda}\frac{dx^\rho}{d\lambda} \nabla_\rho g_{\alpha\beta} = 0.$$

- $$\varepsilon = \frac{E^2}{1 - \frac{2GM}{r}} - \frac{1}{1 - \frac{2GM}{r}}\left(\frac{dr}{d\lambda}\right)^2 - \frac{L^2}{r^2}.$$

$$\frac{1}{2}\left(\frac{dr}{d\lambda}\right)^2 + V(r) = \mathcal{E}, \quad \mathcal{E} = \frac{1}{2}(E^2 - \varepsilon), \quad V(r) = -\frac{\varepsilon GM}{r} + \frac{L^2}{2r^2} - \frac{GML^2}{r^3}.$$

- $$\frac{dt}{dr} = \pm \frac{1}{1 - \frac{2GM}{r}}.$$

Exercise 5:
- Write down the electromagnetic field strength tensor $F_{\mu\nu}$ and the inhomogeneous Maxwell's equation in curved spacetime.
- Let $g = \det g_{\mu\nu}$. Show that

$$\frac{\partial g}{g} = g^{\mu\nu}\partial g_{\mu\nu}. \tag{1.180}$$

- Show that the inhomogeneous Maxwell's equation can be put in the form

$$\partial_\mu(\sqrt{-g}\,F^{\mu\nu}) = -\sqrt{-g}\,J^\nu. \tag{1.181}$$

- Write down the law of conservation of charge in curved spacetime.

References

[1] Wald R M 1984 *General Relativity* (Chicago, IL: University of Chicago Press) p 491
[2] Carroll S M 2004 *Spacetime and Geometry: An Introduction to General Relativity* (San Francisco, CA: Addison-Wesley) p 513
[3] 't Hooft G 2010 Introduction to general relativity www.phys.uu.nl/~thooft/lectures/gen-rel_2010.pdf

IOP Publishing

Lectures on General Relativity, Cosmology and Quantum Black Holes (Second Edition)

Badis Ydri

Chapter 2

Black holes

The goal in this second chapter is to review the theory of classical black holes in some detail. As in the previous chapter, the primary references here, with the order reversed, are Carroll [1], which contains a very pedagogical presentation of the subject, and Wald [2]. Again, the lectures by 't Hooft [3] on this topic were extremely useful.

2.1 Spherical star

2.1.1 The Schwarzschild metric

We consider a matter source which is both static and spherically symmetric. Clearly a static source means that the components of the metric are all independent of time. By requiring also that the physics is invariant under time reversal, i.e. under $t \longrightarrow -t$, the components g_{0i} which provide spacetime cross terms in the metric must be absent. We have already found that the most general spherically symmetric metric in 3D is of the form

$$d\vec{u}^2 = e^{2\beta(r)}dr^2 + r^2 d\Omega^2. \tag{2.1}$$

The most general static and spherically symmetric metric in 4D is therefore of the form

$$ds^2 = -e^{2\alpha(r)}c^2 dt^2 + d\vec{u}^2 = -e^{2\alpha(r)}c^2 dt^2 + e^{2\beta(r)}dr^2 + r^2 d\Omega^2. \tag{2.2}$$

We need to determine the functions $\alpha(r)$ and $\beta(r)$ from solving Einstein's equations. First we need to evaluate the Christoffel symbols. We find

doi:10.1088/978-0-7503-5824-8ch2

$$\Gamma^0{}_{0r} = \partial_r \alpha$$
$$\Gamma^r{}_{00} = \partial_r \alpha e^{2(\alpha - \beta)}, \quad \Gamma^r{}_{rr} = \partial_r \beta,$$
$$\Gamma^r{}_{\theta\theta} = -re^{-2\beta}, \quad \Gamma^r{}_{\phi\phi} = -re^{-2\beta} \sin^2 \theta$$
$$\Gamma^\theta{}_{r\theta} = \frac{1}{r}, \quad \Gamma^\theta{}_{\phi\phi} = -\sin\theta \cos\theta$$
$$\Gamma^\phi{}_{r\phi} = \frac{1}{r}, \quad \Gamma^\phi{}_{\theta\phi} = \frac{\cos\theta}{\sin\theta}.$$

$$(2.3)$$

The non-zero components of the Riemann curvature tensor are

$$R_{0rr}{}^0 = -R_{r0r}{}^0 = \partial_r^2 \alpha + (\partial_r \alpha)^2 - \partial_r \beta \partial_r \alpha$$
$$R_{0\theta\theta}{}^0 = -R_{\theta0\theta}{}^0 = re^{-2\beta} \partial_r \alpha$$
$$R_{0\phi\phi}{}^0 = -R_{\phi0\phi}{}^0 = re^{-2\beta} \partial_r \alpha \sin^2 \theta.$$

$$(2.4)$$

$$R_{0r0}{}^r = -R_{r00}{}^r = (\partial_r^2 \alpha + (\partial_r \alpha)^2 - \partial_r \beta \partial_r \alpha)e^{2(\alpha-\beta)}$$
$$R_{r\theta\theta}{}^r = -R_{\theta r\theta}{}^r = -re^{-2\beta} \partial_r \beta$$
$$R_{r\phi\phi}{}^r = -R_{\phi r\phi}{}^r = -re^{-2\beta} \partial_r \beta \sin^2 \theta.$$

$$(2.5)$$

$$R_{0\theta0}{}^\theta = -R_{\theta00}{}^\theta = \frac{1}{r}\partial_r \alpha e^{2(\alpha-\beta)}$$
$$R_{r\theta r}{}^\theta = -R_{\theta rr}{}^\theta = \frac{1}{r}\partial_r \beta$$
$$R_{\theta\phi\phi}{}^\theta = -R_{\phi\theta\phi}{}^\theta = \sin^2 \theta(e^{-2\beta} - 1).$$

$$(2.6)$$

$$R_{0\phi0}{}^\phi = -R_{\phi00}{}^\phi = \frac{1}{r}\partial_r \alpha e^{2(\alpha-\beta)}$$
$$R_{r\phi r}{}^\phi = -R_{\phi rr}{}^\phi = \frac{1}{r}\partial_r \beta$$
$$R_{\theta\phi\theta}{}^\phi = -R_{\phi\theta\theta}{}^\phi = 1 - e^{-2\beta}.$$

$$(2.7)$$

We compute immediately the non-zero components of the Ricci tensor as follows

$$R_{00} = R_{0r0}{}^r + R_{0\theta0}{}^\theta + R_{0\phi0}{}^\phi = \left(\partial_r^2 \alpha + (\partial_r \alpha)^2 - \partial_r \beta \partial_r \alpha + \frac{2}{r}\partial_r \alpha\right)e^{2(\alpha-\beta)}$$

$$R_{rr} = R_{r0r}{}^0 + R_{r\theta r}{}^\theta + R_{r\phi r}{}^\phi = -\partial_r^2 \alpha - (\partial_r \alpha)^2 + \partial_r \beta \partial_r \alpha + \frac{2}{r}\partial_r \beta$$

$$R_{\theta\theta} = R_{\theta0\theta}{}^0 + R_{\theta r\theta}{}^r + R_{\theta\phi\theta}{}^\phi = e^{-2\beta}(r\partial_r \beta - r\partial_r \alpha - 1) + 1$$

$$R_{\phi\phi} = R_{\phi0\phi}{}^0 + R_{\phi r\phi}{}^r + R_{\phi\theta\phi}{}^\theta = \sin^2 \theta[e^{-2\beta}(r\partial_r \beta - r\partial_r \alpha - 1) + 1].$$

$$(2.8)$$

We compute also the scalar curvature

$$R = -2e^{-2\beta}\left(\partial_r^2\alpha + (\partial_r\alpha)^2 - \partial_r\beta\partial_r\alpha + \frac{2}{r}(\partial_r\alpha - \partial_r\beta) + \frac{1}{r^2}(1 - e^{2\beta})\right). \quad (2.9)$$

Now we are in a position to solve Einstein's equations outside the static spherical source (the star). In the absence of any other matter sources in the region outside the star Einstein's equations read

$$R_{\mu\nu} = 0. \tag{2.10}$$

We have immediately three independent equations

$$\partial_r^2\alpha + (\partial_r\alpha)^2 - \partial_r\beta\partial_r\alpha + \frac{2}{r}\partial_r\alpha = 0$$

$$\partial_r^2\alpha + (\partial_r\alpha)^2 - \partial_r\beta\partial_r\alpha - \frac{2}{r}\partial_r\beta = 0 \tag{2.11}$$

$$e^{-2\beta}(r\partial_r\beta - r\partial_r\alpha - 1) + 1 = 0.$$

By subtracting the first two conditions we get $\partial_r(\alpha + \beta) = 0$ and thus $\alpha = -\beta + c$ where c is some constant. By an appropriate rescaling of the time coordinate we can redefine the value of α as $\alpha + c'$ where c' is an arbitrary constant. We can clearly choose this constant such that $\alpha = -\beta$. The third condition in the above equation (2.11) becomes then

$$e^{2\alpha}(2r\partial_r\alpha + 1) = 1. \tag{2.12}$$

Equivalently

$$\partial_r(re^{2\alpha}) = 1. \tag{2.13}$$

The solution is (with R_s is some constant)

$$e^{2\alpha} = 1 - \frac{R_s}{r}. \tag{2.14}$$

The first and the second conditions in equation (2.11) take now the form

$$\partial_r^2\alpha + 2(\partial_r\alpha)^2 + \frac{2}{r}\partial_r\alpha = 0 \tag{2.15}$$

We compute

$$\partial_r\alpha = \frac{R_s}{2(r^2 - R_s r)}, \quad \partial_r^2\alpha = -\frac{R_s(2r - R_s)}{2(r^2 - R_s r)^2}. \tag{2.16}$$

In other words the form (2.14) is indeed a solution.

The Schwarzschild metric is the metric corresponding to this solution. This is the most important spacetime after Minkowski spacetime. It reads explicitly

$$ds^2 = -\left(1 - \frac{R_s}{r}\right)c^2dt^2 + \left(1 - \frac{R_s}{r}\right)^{-1}dr^2 + r^2d\Omega^2. \tag{2.17}$$

In the Newtonian limit we know that (with Φ the gravitational potential and M the mass of the spherical star)

$$g_{00} = -\left(1 + 2\frac{\Phi}{c^2}\right) = -\left(1 - \frac{2GM}{c^2 r}\right).$$
(2.18)

The g_{00} component of the Schwarzschild metric should reduce to this form for very large distances which here means $r \gg R_s$. By comparison we obtain

$$R_s = \frac{2GM}{c^2}.$$
(2.19)

This is called the Schwarzschild radius. We stress that M can be thought of as the mass of the star only in the weak field limit. In general M will also include gravitational binding energy. In the limit $M \longrightarrow 0$ or $r \longrightarrow \infty$ the Schwarzschild metric reduces to the Minkowski metric. This is called asymptotic flatness.

The powerful Birkhoff's theorem states that the Schwarzschild metric is the unique vacuum solution (static or otherwise) to Einstein's equations which is spherically symmetric. See for example the discussion in [2].

We remark that the Schwarzschild metric is singular at $r = 0$ and at $r = R_s$. However, only the singularity at $r = 0$ is a true singularity of the geometry. For example we can check that the scalar quantity $R^{\mu\nu\alpha\beta}R_{\mu\nu\alpha\beta}$ is divergent at $r = 0$, whereas it is perfectly finite at $r = R_s$. Indeed, the divergence of the Ricci scalar or any other higher order scalar such as $R^{\mu\nu\alpha\beta}R_{\mu\nu\alpha\beta}$ at a point is a sufficient condition for that point to be singular. We say that $r = 0$ is an essential singularity.

The Schwarzschild radius $r = R_s$ is not a true singularity of the metric and its appearance as such only reflects the fact that the chosen coordinates are behaving badly at $r = R_s$. We say that $r = R_s$ is a coordinate singularity. Indeed, it should appear like any other point if we choose a more appropriate coordinates system. It will, on the other hand, specify the so-called event horizon when the spherical sphere becomes a black hole.

2.1.2 Particle motion in Schwarzschild spacetime

We start by rewriting the Christoffel symbols (2.3) as

$$\Gamma^0{}_{0r} = \frac{R_s}{2r(r - R_s)}$$

$$\Gamma^r{}_{00} = \frac{R_s(r - R_s)}{2r^3}, \quad \Gamma^r{}_{rr} = -\frac{R_s}{2r(r - R_s)},$$

$$\Gamma^r{}_{\theta\theta} = -r + R_s, \quad \Gamma^r{}_{\phi\phi} = (-r + R_s)\sin^2\theta$$
(2.20)

$$\Gamma^\theta{}_{r\theta} = \frac{1}{r}, \quad \Gamma^\theta{}_{\phi\phi} = -\sin\theta\cos\theta$$

$$\Gamma^\phi{}_{r\phi} = \frac{1}{r}, \quad \Gamma^\phi{}_{\theta\phi} = \frac{\cos\theta}{\sin\theta}.$$

The geodesic equation is given by

$$\frac{d^2x^\rho}{d\lambda^2} + \Gamma^\rho{}_{\mu\nu}\frac{dx^\mu}{d\lambda}\frac{dx^\nu}{d\lambda} = 0. \tag{2.21}$$

Explicitly we have

$$\frac{d^2x^0}{d\lambda^2} + \frac{R_s}{r(r - R_s)}\frac{dx^0}{d\lambda}\frac{dr}{d\lambda} = 0. \tag{2.22}$$

$$\frac{d^2r}{d\lambda^2} + \frac{R_s(r - R_s)}{2r^3}\left(\frac{dx^0}{d\lambda}\right)^2 - \frac{R_s}{2r(r - R_s)}\left(\frac{dr}{d\lambda}\right)^2$$
$$- (r - R_s)\left[\left(\frac{d\theta}{d\lambda}\right)^2 + \sin^2\theta\left(\frac{d\phi}{d\lambda}\right)^2\right] = 0. \tag{2.23}$$

$$\frac{d^2\theta}{d\lambda^2} + \frac{2}{r}\frac{dr}{d\lambda}\frac{d\theta}{d\lambda} - \sin\theta\cos\theta\left(\frac{d\phi}{d\lambda}\right)^2 = 0. \tag{2.24}$$

$$\frac{d^2\phi}{d\lambda^2} + \frac{2}{r}\frac{dr}{d\lambda}\frac{d\phi}{d\lambda} + 2\frac{\cos\theta}{\sin\theta}\frac{d\theta}{d\lambda}\frac{d\phi}{d\lambda} = 0. \tag{2.25}$$

The Schwarzschild metric is obviously invariant under time translations and space rotations. There will therefore be four corresponding Killing vectors K_μ and four conserved quantities given by

$$Q = K_\mu\frac{dx^\mu}{d\lambda}. \tag{2.26}$$

The motion of a particle under a central force of gravity in flat spacetime has invariance under time translation, which leads to conservation of energy and invariance under rotations, which leads to conservation of angular momentum. The angular momentum is a vector in three dimensions with a length (one component) and a direction (two angles). Conservation of the direction means that the motion happens in a plane. In other words, we can choose $\theta = \pi/2$.

In Schwarzschild spacetime the same symmetries are still present and therefore the four Killing vectors K_μ must be associated with time translation and rotations and the four conserved quantities Q are the energy and the angular momentum. The two Killing vectors associated with the conservation of the direction of the angular momentum lead precisely, as in the flat case, to a motion in the plane, viz

$$\theta = \frac{\pi}{2}. \tag{2.27}$$

The metric is independent of x^0 and ϕ and hence the corresponding Killing vectors are

$$K^\mu = (\partial_{x^0})^\mu = \delta^\mu_0 = (1, 0, 0, 0), \quad K_\mu = g_{\mu 0} = \left(-\left(1 - \frac{R_s}{r}\right), 0, 0, 0\right). \tag{2.28}$$

$$R^\mu = (\partial_\phi)^\mu = \delta^\mu_\phi = (0, 0, 0, 1), \quad R_\mu = g_{\mu\phi} = (0, 0, 0, r^2 \sin^2 \theta). \tag{2.29}$$

The corresponding conserved quantities are the energy and the magnitude of the angular momentum given by

$$E = -K_\mu \frac{dx^\mu}{d\lambda} = \left(1 - \frac{R_s}{r}\right)\frac{dx^0}{d\lambda}. \tag{2.30}$$

$$L = R_\mu \frac{dx^\mu}{d\lambda} = r^2 \sin^2 \theta \frac{d\phi}{d\lambda}. \tag{2.31}$$

The minus sign in the energy is consistent with the definition $p_\sigma = p_\mu K^\mu = cp_\mu(\partial_{(\sigma)})^\mu$. Furthermore, E is actually the energy per unit mass for a massive particle, whereas for a massless particles it is indeed the energy since the momentum of a massless particle is identified with its 4-vector velocity. A similar remark applies to the angular momentum. Note that E should be thought of as the total energy including gravitational energy which is the quantity that really needs to be conserved. In other words, E is different from the kinetic energy $-p_a v^a$, which is the energy measured by an observer whose velocity is v^a. Note also that the conservation of angular momentum is precisely Kepler's second law.

There is an extra conserved quantity along the geodesic given by

$$\varepsilon = -g_{\mu\nu} \frac{dx^\mu}{d\lambda} \frac{dx^\nu}{d\lambda}. \tag{2.32}$$

We compute

$$\begin{aligned}
\frac{d\varepsilon}{d\lambda} &= -\frac{dg_{\mu\nu}}{d\lambda} \cdot \frac{dx^\mu}{d\lambda} \frac{dx^\nu}{d\lambda} - 2g_{\mu\nu} \frac{d^2x^\mu}{d\lambda^2} \frac{dx^\nu}{d\lambda} \\
&= -\frac{dg_{\mu\nu}}{d\lambda} \cdot \frac{dx^\mu}{d\lambda} \frac{dx^\nu}{d\lambda} + 2g_{\mu\nu} \Gamma^\mu{}_{\alpha\beta} \frac{dx^\alpha}{d\lambda} \frac{dx^\beta}{d\lambda} \frac{dx^\nu}{d\lambda} \\
&= -\frac{dx^\alpha}{d\lambda} \frac{dx^\beta}{d\lambda} \frac{dx^\rho}{d\lambda} [\partial_\rho g_{\alpha\beta} - 2\Gamma^\mu{}_{\alpha\beta} g_{\mu\rho}] \\
&= -\frac{dx^\alpha}{d\lambda} \frac{dx^\beta}{d\lambda} \frac{dx^\rho}{d\lambda} [\partial_\rho g_{\alpha\beta} - \Gamma^\mu{}_{\alpha\rho} g_{\mu\beta} - \Gamma^\mu{}_{\rho\beta} g_{\mu\alpha}] \\
&= -\frac{dx^\alpha}{d\lambda} \frac{dx^\beta}{d\lambda} \frac{dx^\rho}{d\lambda} \nabla_\rho g_{\alpha\beta} \\
&= 0.
\end{aligned} \tag{2.33}$$

We clearly need to take

$$\varepsilon = c^2, \quad \text{massive particle.} \tag{2.34}$$

$$\varepsilon = 0, \quad \text{massless particle.} \tag{2.35}$$

The above conserved quantity reads explicitly

$$\varepsilon = \frac{E^2}{1 - \dfrac{R_s}{r}} - \frac{1}{1 - \dfrac{R_s}{r}} \left(\frac{dr}{d\lambda}\right)^2 - \frac{L^2}{r^2}. \tag{2.36}$$

Equivalently

$$E^2 - \left(\frac{dr}{d\lambda}\right)^2 - \left(1 - \frac{R_s}{r}\right)\left(\frac{L^2}{r^2} + \varepsilon\right) = 0. \tag{2.37}$$

We also rewrite this as

$$\frac{1}{2}\left(\frac{dr}{d\lambda}\right)^2 + V(r) = \mathcal{E}. \tag{2.38}$$

$$\mathcal{E} = \frac{1}{2}(E^2 - \varepsilon). \tag{2.39}$$

$$
\begin{aligned}
V(r) &= \frac{1}{2}\left(1 - \frac{R_s}{r}\right)\left(\frac{L^2}{r^2} + \varepsilon\right) - \frac{\varepsilon}{2} \\
&= -\frac{\varepsilon GM}{c^2 r} + \frac{L^2}{2r^2} - \frac{GML^2}{c^2 r^3}.
\end{aligned}
\tag{2.40}
$$

This is the equation of a particle with unit mass and energy $\mathcal{E}$ in a potential $V(r)$. In this potential only the last term is new compared to Newtonian gravity. Clearly when $r \longrightarrow 0$ this potential will go to $-\infty$, whereas if the last term is absent (the case of Newtonian gravity) the potential will go to $+\infty$ when $r \longrightarrow 0$. See figure 2.1.

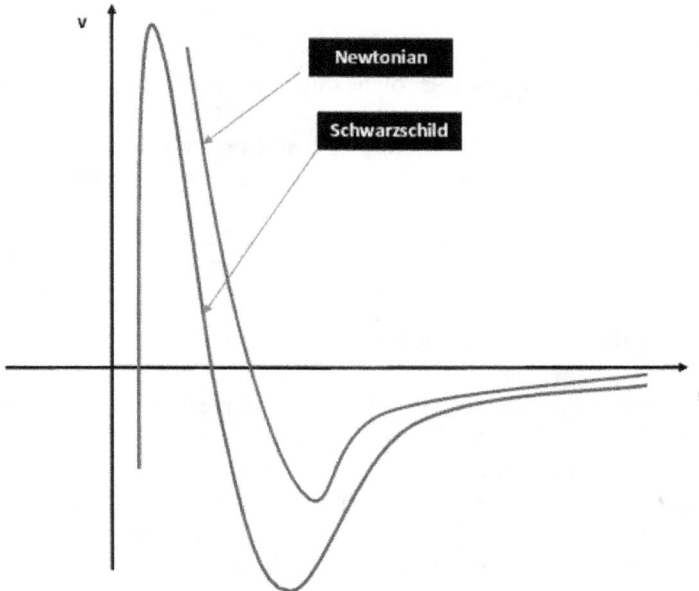

Figure 2.1. Newton and Schwarzschild problems.

The potential $V(r)$ is different for different values of L. It has one maximum and one minimum if $cL/GM > \sqrt{12}$. Indeed we have

$$\frac{dV(r)}{dr} = 0 \Leftrightarrow \varepsilon r^2 - \frac{c^2 L^2}{GM} r + 3L^2 = 0. \tag{2.41}$$

For massive particles the stable (minimum) and unstable (maximum) orbits are located at

$$r_{\text{max}} = \frac{L^2 - \sqrt{L^4 - \dfrac{12 G^2 M^2 L^2}{c^2}}}{2GM}, \quad r_{\text{min}} = \frac{L^2 + \sqrt{L^4 - \dfrac{12 G^2 M^2 L^2}{c^2}}}{2GM}. \tag{2.42}$$

Both orbits are circular. See figure 2.2.

In the limit $L \longrightarrow \infty$ we obtain

$$r_{\text{max}} = \frac{3GM}{c^2}, \quad r_{\text{min}} = \frac{L^2}{GM}. \tag{2.43}$$

The stable circular orbit becomes farther away, whereas the unstable circular orbit approaches $3GM/c^2$.

In the limit of small L, the two orbits coincide when

$$L^4 - \frac{12 G^2 M^2 L^2}{c^2} = 0 \Leftrightarrow L = \sqrt{12} \frac{GM}{c}. \tag{2.44}$$

At which point

$$r_{\text{max}} = r_{\text{min}} = \frac{L^2}{2GM} = \frac{6GM}{c^2}. \tag{2.45}$$

This is the smallest radius possible of a stable circular orbit in a Schwarzschild spacetime.

For massless particles ($\varepsilon = 0$) there is a solution at $r = 3GM/c^2$. This corresponds always to an unstable circular orbit. We have then the following criterion

$$\text{stable circular orbits:} \quad r > \frac{6GM}{c^2}. \tag{2.46}$$

$$\text{unstable circular orbits:} \quad \frac{3GM}{c^3} < r < \frac{6GM}{c^2}. \tag{2.47}$$

These are of course all geodesics, i.e. orbits corresponding to free fall in a gravitational field. There are also bound non-circular orbits which oscillate around the stable circular orbit. For example, if a test particle starts from a point $r_{\text{max}} < r_2 < r_{\text{min}}$ at which $\mathcal{E} = V(r_2) < 0$, it will move in the potential until it hits the potential at a point $r_1 > r_{\text{min}}$ at which $\mathcal{E} = V(r_1)$ where it bounces back. The corresponding bound precessing orbit is shown in figure 2.3.

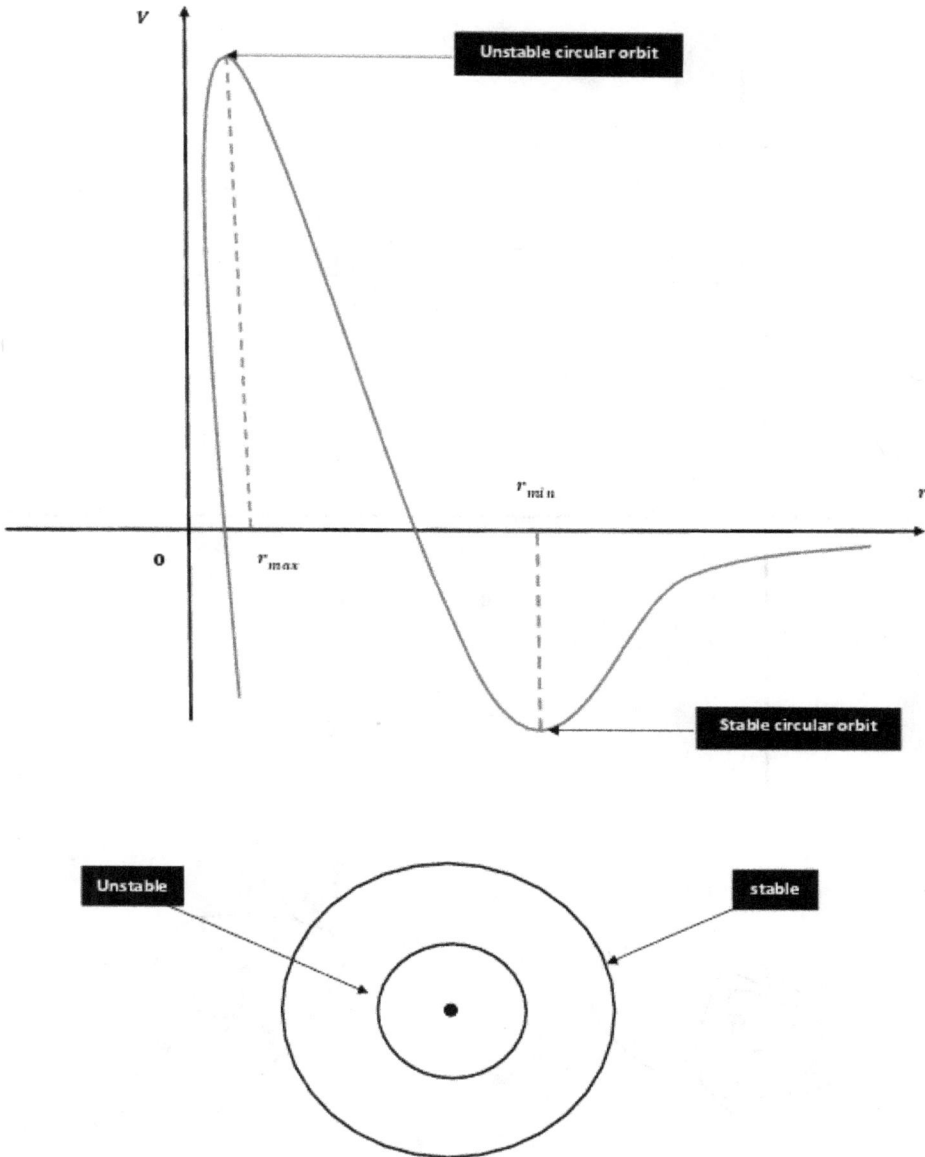

Figure 2.2. Circular orbits.

There exist also scattering orbits. If a test particle comes from infinity with energy $\mathcal{E} > 0$ then it will move in the potential and may hit the wall of the potential at $r_{\max} < r_2 < r_{\min}$ for which $\mathcal{E} = V(r_2) > 0$. If it does not hit the wall of the potential (the energy $\mathcal{E}$ is sufficiently large) then the particle will plunge into the center of the potential at $r = 0$. See figure 2.4.

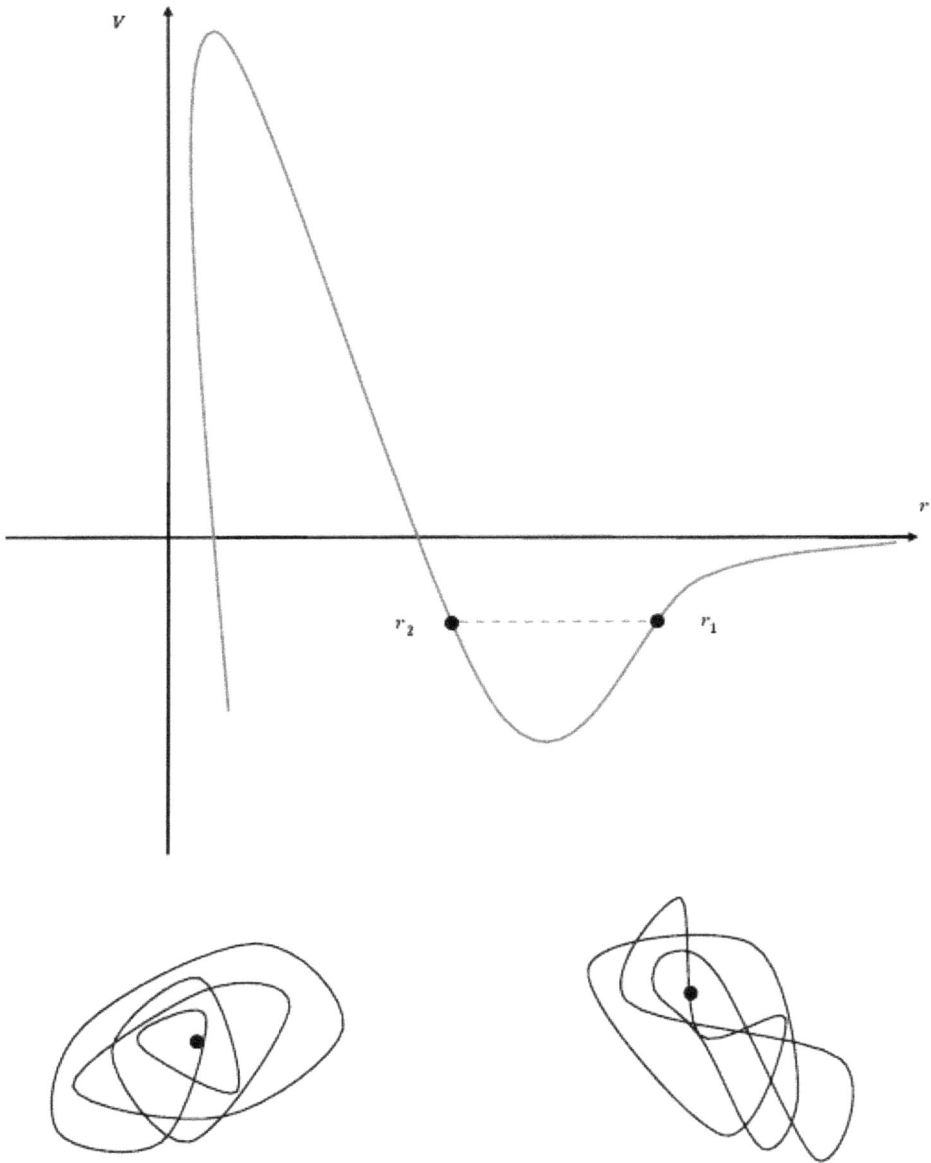

Figure 2.3. Bound orbits.

In contrast to Newtonian gravity these orbits do not correspond to conic section as we will show next.

2.1.3 Precession of perihelia and gravitational redshift

Precession of perihelia: The equation for the conservation of angular momentum reads

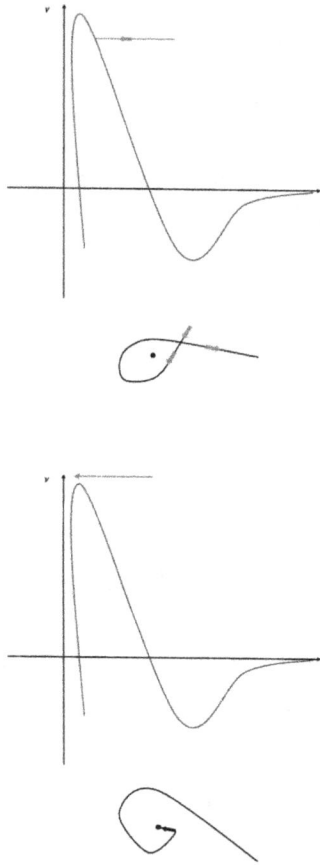

Figure 2.4. Scattering orbits.

$$L = r^2 \frac{d\phi}{d\lambda}.$$ (2.48)

Together with the radial equation

$$\frac{1}{2}\left(\frac{dr}{d\lambda}\right)^2 + V(r) = \mathcal{E}.$$ (2.49)

We have for a massive particle the equation

$$\left(\frac{dr}{d\phi}\right)^2 + \frac{c^2 r^4}{L^2} - \frac{2GMr^3}{L^2} + r^2 - \frac{2GMr}{c^2} = \frac{r^4 E^2}{L^2}.$$ (2.50)

In the case of Newtonian gravity equation (2.41) for a massive particle gives $r = L^2/GM$. This is the radius of a circular orbit in Newtonian gravity. We perform the change of variable

$$x = \frac{L^2}{GMr}.$$ (2.51)

The above last differential equation becomes

$$\left(\frac{dx}{d\phi}\right)^2 + \frac{L^2 c^2}{G^2 M^2} - 2x + x^2 - \frac{2G^2 M^2 x^3}{L^2 c^2} = \frac{L^2 E^2}{G^2 M^2}. \tag{2.52}$$

We differentiate this equation with respect to x to get

$$\frac{d^2 x}{d\phi^2} - 1 + x = \frac{3G^2 M^2}{L^2 c^2} x^2. \tag{2.53}$$

We solve this equation in perturbation theory as follows. We write

$$x = x_0 + x_1. \tag{2.54}$$

The zeroth order equation is

$$\frac{d^2 x_0}{d\phi^2} - 1 + x_0 = 0. \tag{2.55}$$

The first order equation is

$$\frac{d^2 x_1}{d\phi^2} + x_1 = \frac{3G^2 M^2}{L^2 c^2} x_0^2. \tag{2.56}$$

The solution to the zeroth order equation is precisely the Newtonian result

$$x_0 = 1 + e \cos \phi. \tag{2.57}$$

This is an ellipse with eccentricity $e = c/a = \sqrt{1 - b^2/a^2}$ with the center of the coordinate system at the focus $(c, 0)$ and ϕ is the angle measured from the major axis[1].

---

[1] The ellipse is the set of points where the sum of the distances r_1 and r_2 from each point on the ellipse to two fixed points (the foci) is a constant equal $2a$. We have then

$$r_1 + r_2 = 2a. \tag{2.58}$$

Let $2c$ be the distance between the two foci F_1 and F_2 and let O be the middle point of the segment $[F_1, F_2]$. The coordinates of each point on the ellipse are x and y with respect to the Cartesian system with O at the origin. Clearly then $r_1 = \sqrt{(c + x)^2 + y^2}$ and $r_2 = \sqrt{(c - x)^2 + y^2}$. The equation of the ellipse becomes

$$\frac{x^2}{a^2} + \frac{y^2}{b^2} = 1 \tag{2.59}$$

The semi-major axis is a and the semi-minor axis is $b = \sqrt{a^2 - c^2}$. We take the focus F_2 as the center of our system of coordinates and we use polar coordinates. Then $x = r \cos \theta - c$ and $y = r \sin \theta$ and hence the equation of the ellipse becomes (with eccentricity $e = c/a$)

$$\frac{a(1 - e^2)}{r} = 1 - e \cos \theta. \tag{2.60}$$

If we had taken the focus F_1 instead as the center of our system of coordinates we would have obtained

$$\frac{a(1 - e^2)}{r} = 1 + e \cos \theta. \tag{2.61}$$

The semi-major axis a is the distance to the farthest point, whereas the semi-minor axis b is the distance to the closest point. In other words, at $\phi = \pi$ we have $x_0 = 1 - e = a(1 - e^2)/(a + c)$ and at $\phi = 0$ we have $x_0 = 1 + e = a(1 - e^2)/(a - c)$. By comparing also the equation of the ellipse $a(1 - e^2)/r = 1 + e \cos \phi$ with the solution for x_0 we obtain the value of the angular momentum

$$L^2 = GMa(1 - e^2). \tag{2.62}$$

The first order equation becomes

$$\frac{d^2 x_1}{d\phi^2} + x_1 = \frac{3G^2 M^2}{L^2 c^2}(1 + e \cos \phi)^2$$
$$= \frac{3G^2 M^2}{L^2 c^2}\left(1 + \frac{e^2}{2} + \frac{e^2}{2}\cos 2\phi + 2e \cos \phi\right). \tag{2.63}$$

Remark that

$$\frac{d^2}{d\phi^2}(\phi \sin \phi) + \phi \sin \phi = 2 \cos \phi$$
$$\frac{d^2}{d\phi^2}(\cos 2\phi) + \cos 2\phi = - 3 \cos 2\phi. \tag{2.64}$$

Then we can write

$$\frac{d^2 y_1}{d\phi^2} + y_1 = \frac{3G^2 M^2}{L^2 c^2}\left(1 + \frac{e^2}{2}\right), \quad y_1 = x_1 - \frac{3G^2 M^2}{L^2 c^2}\left(-\frac{e^2}{6}\cos 2\phi + e\phi \sin \phi\right). \tag{2.65}$$

Define also

$$z_1 = \frac{y_1}{\dfrac{3G^2 M^2}{L^2 c^2}\left(1 + \dfrac{e^2}{2}\right)}. \tag{2.66}$$

The differential equations becomes

$$\frac{d^2 z_1}{d\phi^2} - 1 + z_1 = 0 \tag{2.67}$$

The solution is immediately given by

$$z_1 = 1 + e \cos \phi \Leftrightarrow x_1 = \frac{3G^2 M^2}{L^2 c^2}\left(1 + \frac{e^2}{2}\right)(1 + e \cos \phi)$$
$$+ \frac{3G^2 M^2}{L^2 c^2}\left(-\frac{e^2}{6}\cos 2\phi + e\phi \sin \phi\right). \tag{2.68}$$

The complete solution is

$$
x = \left[1 + \frac{3G^2M^2}{L^2c^2}\left(1 + \frac{e^2}{2}\right)\right](1 + e\cos\phi)
$$
$$
+ \frac{3G^2M^2}{L^2c^2}\left(-\frac{e^2}{6}\cos 2\phi + e\phi\sin\phi\right).
$$
(2.69)

We can rewrite this in the form

$$
x = \left[1 + \frac{3G^2M^2}{L^2c^2}\left(1 + \frac{e^2}{2}\right)\right](1 + e\cos(1-\alpha)\phi) + \frac{3G^2M^2}{L^2c^2}\left(-\frac{e^2}{6}\cos 2\phi\right).
$$
(2.70)

The small number α is given by

$$
\alpha = \frac{3G^2M^2}{L^2c^2}.
$$
(2.71)

The last term in the above solution oscillates around 0 and hence averages to 0 over successive revolutions and as such it is irrelevant to our consideration.

The above result can be interpreted as follows. The orbit is an ellipse but with a period equal $2\pi/(1-\alpha)$ instead of 2π. Thus the perihelion advances in each revolution by the amount

$$
\Delta\phi = 2\pi\alpha = \frac{6\pi G^2M^2}{L^2c^2}.
$$
(2.72)

By using now the value of the angular momentum for a perfect ellipse given by equation (2.62) we get

$$
\Delta\phi = \frac{6\pi GM}{a(1-e^2)c^2}.
$$
(2.73)

In the case of the motion of Mercury around the Sun we can use the values

$$
\frac{GM}{c^2} = 1.48 \times 10^5 \text{ cm}, \quad a = 5.79 \times 10^{12} \text{ cm}, \quad e = 0.2056.
$$
(2.74)

We obtain

$$
\Delta\phi_{\text{Mercury}} = \frac{6\pi GM}{a(1-e^2)c^2} = 5.03 \times 10^{-7} \text{ rad/orbit}.
$$
(2.75)

However, Mercury completes one orbit every 88 days, thus in a century its perihelion will advance by the amount

$$
\Delta\phi_{\text{Mercury}} = \frac{100 \times 365}{88} \times 5.03 \times 10^{-7}\frac{180 \times 3600}{3.14} \text{ arcsecond/century}
$$
$$
= 43.06 \text{ arcsecond/century}.
$$
(2.76)

The total precession of Mercury is around 575 arcseconds per century[2] with a 532 arcseconds per century due to other planets and 43 arcseconds per century due to the curvature of spacetime caused by the Sun[3].

Gravitational redshift: We consider a stationary observer $(U^i = 0)$ in Schwarzschild spacetime. The 4-vector velocity satisfies $g_{\mu\nu} U^\mu U^\nu = -c^2$ and hence

$$U^0 = \frac{c}{\sqrt{1 - \dfrac{2GM}{c^2 r}}}.$$
(2.77)

The energy (per unit mass) of a photon as measured by this observer is

$$\begin{aligned} E_\gamma &= -U_\mu \frac{dx^\mu}{d\lambda} \\ &= c^2 \sqrt{1 - \frac{2GM}{c^2 r}} \frac{dt}{d\lambda} \\ &= \frac{cE}{\sqrt{1 - \dfrac{2GM}{c^2 r}}}. \end{aligned}$$
(2.78)

The E^2 is the conserved energy (per unit mass) of the Schwarzschild metric given by (6.46). Thus a photon emitted with an energy $E_{\gamma 1}$ at a distance r_1 will be observed at a distance $r_2 > r_1$ with an energy $E_{\gamma 2}$ given by

$$\frac{E_{\gamma 2}}{E_{\gamma 1}} = \sqrt{\frac{1 - \dfrac{2GM}{c^2 r_1}}{1 - \dfrac{2GM}{c^2 r_2}}} < 1.$$
(2.79)

Thus the energy $E_{\gamma 2} < E_{\gamma 1}$, i.e. as the photon climbs out of the gravitational field it gets redshifted. In other words the frequency decreases as the strength of the gravitational field decreases or equivalently as the gravitational potential increases. This is the gravitational redshift. In the limit $r \gg 2GM/c^2$ the formula becomes

$$\frac{E_{\gamma 2}}{E_{\gamma 1}} = 1 + \frac{\Phi_1}{c^2} - \frac{\Phi_2}{c^2}, \quad \Phi = -\frac{GM}{r}.$$
(2.80)

2.1.4 Free fall

For a radially (vertically) freely object we have $d\phi/d\lambda = 0$ and thus the angular momentum is 0, viz $L = 0$. The radial equation of motion becomes

$$\left(\frac{dr}{d\lambda}\right)^2 - \frac{2GM}{r} = E^2 - c^2.$$
(2.81)

---

[2] There is a huge amount of precession due to the precession of equinoxes, which is not discussed here.
[3] There is also a minute contribution due to the oblateness of the Sun.

This is essentially the Newtonian equation of motion. The conserved energy is given by

$$E = c\left(1 - \frac{2GM}{c^2 r}\right)\frac{dt}{d\lambda}. \tag{2.82}$$

We also consider the situation in which the particle was initially at rest at $r = r_i$, viz

$$\frac{dr}{d\lambda}\Big|_{r=r_i} = 0. \tag{2.83}$$

This means in particular that

$$E^2 - c^2 = -\frac{2GM}{r_i}. \tag{2.84}$$

The equation of motion becomes

$$\left(\frac{dr}{d\lambda}\right)^2 = \frac{2GM}{r} - \frac{2GM}{r_i}. \tag{2.85}$$

We can identify the affine parameter λ with the proper time for a massive particle. The proper time required to reach the point $r = r_f$ is

$$\tau = \int_0^\tau d\lambda = -(2GM)^{-\frac{1}{2}} \int_{r_i}^{r_f} dr \sqrt{\frac{r r_i}{r_i - r}}. \tag{2.86}$$

The minus sign is due to the fact that in a free fall $dr/d\lambda < 0$. By performing the change of variables $r = r_i(1 + \cos\alpha)/2$ we find the closed result

$$\tau = \sqrt{\frac{r_i^3}{8GM}}(\alpha_f + \sin\alpha_f). \tag{2.87}$$

This is finite when $r \longrightarrow 2GM/c^2$. Thus a freely falling object will cross the Schwarzschild radius in a finite proper time.

We consider now a distant stationary observer hovering at a fixed radial distance r_∞. Their proper time is

$$\tau_\infty = \sqrt{1 - \frac{2GM}{c^2 r_\infty^2}}\, t. \tag{2.88}$$

By using equations (6.53) and (6.54) we can find dr/dt. We get

$$\begin{aligned}
\frac{dr}{dt} &= -E^{\frac{1}{2}}\frac{d\lambda}{dt}\left(E - c\frac{d\lambda}{dt}\right)^{\frac{1}{2}} \\
&= -\frac{c}{E}\left(1 - \frac{2GM}{c^2 r}\right)\left(E^2 - c^2\left(1 - \frac{2GM}{c^2 r}\right)\right)^{\frac{1}{2}}.
\end{aligned} \tag{2.89}$$

Near $r = 2GM/c^2$ we have

$$\frac{dr}{dt} = -\frac{c^3}{2GM}\left(r - \frac{2GM}{c^2}\right). \tag{2.90}$$

The solution is

$$r - \frac{2GM}{c^2} = \exp\left(-\frac{c^3 t}{2GM}\right). \tag{2.91}$$

Thus when $r \longrightarrow 2GM/c^2$ we have $t \longrightarrow \infty$.

We see that with respect to a stationary distant observer at a fixed radial distance r_∞ the elapsed time τ_∞ goes to infinity as $r \longrightarrow 2GM/c^2$. The correct interpretation of this result is to say that the stationary distant observer can never see the particle actually crossing the Schwarzschild radius $r_s = 2GM/c^2$ although the particle does cross the Schwarzschild radius in a finite proper time as seen by an observer falling with the particle.

2.2 Schwarzschild black hole

We go back to the Schwarzschild metric (2.17), viz (we use units in which $c = 1$)

$$ds^2 = -\left(1 - \frac{2GM}{r}\right)dt^2 + \left(1 - \frac{2GM}{r}\right)^{-1}dr^2 + r^2 d\Omega^2. \tag{2.92}$$

For a radial null curve, which corresponds to a photon moving radially in Schwarzschild spacetime, the angles θ and ϕ are constants and $ds^2 = 0$ and thus

$$0 = -\left(1 - \frac{2GM}{r}\right)dt^2 + \left(1 - \frac{2GM}{r}\right)^{-1}dr^2. \tag{2.93}$$

In other words,

$$\frac{dt}{dr} = \pm \frac{1}{1 - \dfrac{2GM}{r}}. \tag{2.94}$$

This represents the slope of the light cone at a radial distance r on a spacetime diagram of the $t - r$ plane. In the limit $r \longrightarrow \infty$ we get ± 1, which is the flat Minkowski spacetime result, whereas as r decreases the slope increases until we get $\pm\infty$ as $r \longrightarrow 2GM$. The light cones close up at $r = 2GM$ (the Schwarzschild radius).

Thus we reach the conclusion that an infalling observer, as seen by us, never crosses the event horizon $r_s = 2Gm$ in the sense that any fixed interval $\Delta \tau_1$ of its proper time will correspond to a longer and longer interval of our time. In other words, the infalling observer will seem us to move slower and slower as it approaches $r_s = 2GM$ but it will never be seen to actually cross the event horizon. This does not mean that the trajectory of the infalling observer will never reach $r_s = 2GM$ because it actually does, however, we need to change the coordinate system to be able to see this.

We integrate the above equation as follows

$$
\begin{aligned}
t &= \pm \int \frac{dr}{1 - \dfrac{2GM}{r}} \\
&= \pm \int dr \pm 2GM \int \frac{dr}{r - 2GM} \\
&= \pm \left(r + 2GM \log\left(\frac{r}{2GM} - 1\right) \right) + \text{constant} \\
&= \pm r_* + \text{constant}.
\end{aligned}
\tag{2.95}
$$

We call r_* the tortoise coordinate which makes sense only for $r > 2GM$. The event horizon $r = 2GM$ corresponds to $r_* \longrightarrow \infty$. We compute $dr_* = rdr/(r - 2GM)$ and as a consequence the Schwarzschild metric becomes

$$
ds^2 = \left(1 - \frac{2GM}{r}\right)(-dt^2 + dr_*^2) + r^2 d\Omega^2.
\tag{2.96}
$$

Next we define $v = t + r_*$ and $u = t - r_*$. Then

$$
ds^2 = -\left(1 - \frac{2GM}{r}\right)dvdu + r^2 d\Omega^2.
\tag{2.97}
$$

For infalling radial null geodesics we have $t = -r_*$ or equivalently $v = \text{constant}$, whereas for outgoing radial null geodesics we have $t = +r_*$ or equivalently $u = \text{constant}$. We will think of v as our new time coordinate, whereas we will change u back to the radial coordinate r via $u = v - 2r_* = v - 2r - 4GM \log(r/(2GM) - 1)$. Thus $du = dv - 2dr/(1 - 2GM/r)$ and as a consequence

$$
ds^2 = -\left(1 - \frac{2GM}{r}\right)dv^2 + 2dvdr + r^2 d\Omega^2.
\tag{2.98}
$$

These are called the Eddington–Finkelstein coordinates. We remark that the determinant of the metric in this system of coordinates is $g = -r^4 \sin^2 \theta$ which is regular at $r = 2GM$, i.e. the metric is invertible and the original singularity at $r = 2GM$ is simply a coordinate singularity characterizing the system of coordinates (t, r, θ, ϕ). In the Eddington–Finkelstein coordinates the radial null curves are given by the condition

$$
\left[\left(1 - \frac{2GM}{r}\right)\frac{dv}{dr} - 2\right]\frac{dv}{dr} = 0.
\tag{2.99}
$$

We have the following solutions:
- $dv/dr = 0$ or equivalently $v = \text{constant}$ which corresponds to an infalling observer.
- $dv/dr \neq 0$ or equivalently $dv/dr = 2/(1 - \frac{2GM}{r})$. For $r > 2GM$ we obtain the solution $v = 2r + 4GM \log(r/2GM - 1) + \text{constant}$, which corresponds

to an outgoing observer since $dv/dr > 0$. This actually corresponds to $u = $ constant.

- $dv/dr \neq 0$ or equivalently $dv/dr = 2/(1 - \frac{2GM}{r})$. For $r < 2GM$ we obtain the solution $v = 2r + 4GM \log(1 - r/2GM) + $ constant, which corresponds to an infalling observer since $dv/dr < 0$.
- For $r = 2GM$ the above equation reduces to $dvdr = 0$. This corresponds to the observer trapped at $r = 2GM$.

The above solutions are drawn in figure 2.5 in the plane $(v - r) - r$, i.e. the time axis (the perpendicular axis) is $v - r$ and not v. Thus for every point in spacetime we have two solutions:

- The points outside the event horizon such as point 1 in figure (2.5): There are two solutions one infalling and one outgoing.

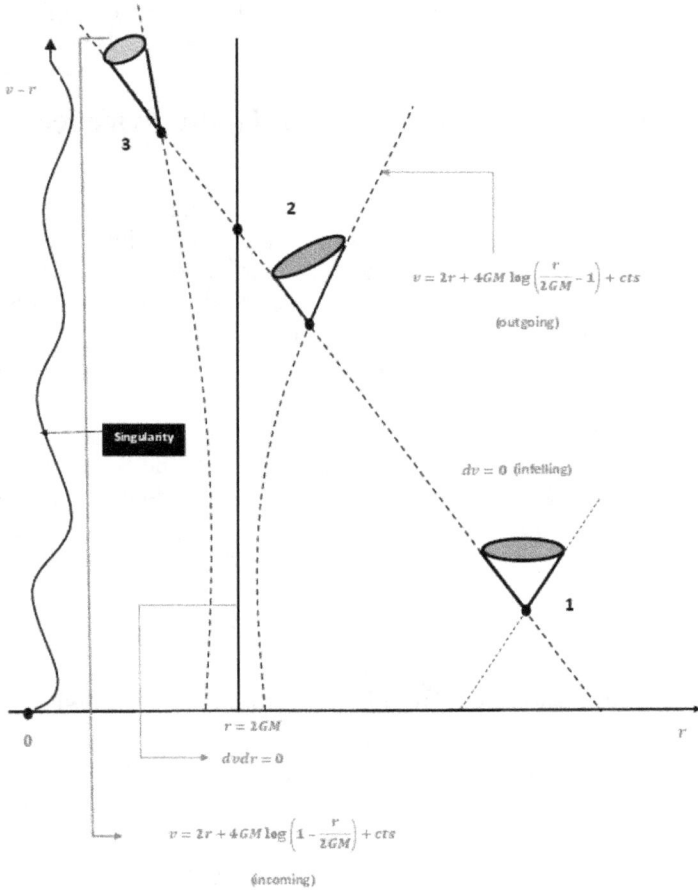

Figure 2.5. The Eddington–Finkelstein coordinates.

- The points inside the event horizon such as point 3 in figure (2.5): There are two solutions both are infalling.
- The points on the event horizon such as point 2 in figure (2.5): There are two solutions one infalling and one trapped.

Several other remarks are of order:
- The light cone at each point of spacetime is determined (bounded) by the two solutions at that point. See figure (2.5).
- The left side of the light cones is always determined by infalling observers.
- The right side of the light cones for $r > 2GM$ is always determined by outgoing observers.
- The right side of the light cones for $r < 2GM$ is always determined by infalling observers.
- The light cone tilt inward as r decreases. For $r < 2GM$ the light cone is sufficiently tilted that no observer can escape the singularity at $r = 0$.
- The horizon $r = 2GM$ is clearly a null surface which consists of observers who can neither fall into the singularity nor escape to infinity (since it is a solution to a null condition which is trapped at $r = 2GM$).

2.3 The Kruskal–Szekres diagram: maximally extended Schwarzschild solution

We have shown explicitly that in the (v, r, θ, ϕ) coordinate system we can cross the horizon at $r = 2GM$ along future directed paths since from the definition $v = t + r_*$ we see that for a fixed v (infalling null radial geodesics) we must have $t = -r_* +$ constant and thus as $r \longrightarrow 2GM$ we must have $t \longrightarrow + \infty$. However, we have also shown that we can cross the horizon at $r = 2GM$ along past-directed paths corresponding to $v = 2r_* +$ constant or equivalently $u =$ constant (outgoing null radial geodesics) and thus as $r \longrightarrow 2GM$ we must have $t \longrightarrow - \infty$. We have also been able to extend the solution to the region $r \leqslant 2GM$.

In the following we will give a maximal extension of the Schwarzschild solution by constructing a coordinate system valid everywhere in Schwarzschild spacetime. We start by rewriting the Schwarzschild metric in the (u, v, θ, ϕ) coordinate system as

$$ds^2 = -\left(1 - \frac{2GM}{r}\right)dv\,du + r^2 d\Omega^2. \tag{2.100}$$

The radial coordinate r should be given in terms of u and v by solving the equations

$$\frac{1}{2}(v - u) = r + 2GM \log\left(\frac{r}{2GM} - 1\right). \tag{2.101}$$

The event horizon $r = 2GM$ is now either at $v = -\infty$ or $u = +\infty$. The coordinates of the event horizon can be pulled to finite values by defining new coordinates u' and v' as

$$v' = \exp\left(\frac{v}{4GM}\right)$$
$$= \sqrt{\frac{r}{2GM} - 1} \, \exp\left(\frac{r+t}{4GM}\right). \tag{2.102}$$

$$u' = -\exp\left(-\frac{u}{4GM}\right)$$
$$= -\sqrt{\frac{r}{2GM} - 1} \, \exp\left(\frac{r-t}{4GM}\right). \tag{2.103}$$

The Schwarzschild metric becomes

$$ds^2 = -\frac{32G^3M^3}{r} \exp\left(-\frac{r}{2GM}\right) dv'du' + r^2 d\Omega^2. \tag{2.104}$$

It is clear that the coordinates u and v are null coordinates since the vectors $\partial/\partial u$ and $\partial/\partial v$ are tangent to light cones and hence they are null vectors. As a consequence, u' and v' are null coordinates. However, we prefer to work with a single time-like coordinate while we prefer the other coordinate to be space-like. We introduce therefore new coordinates T and R defined for $r > 2GM$ by

$$T = \frac{1}{2}(v' + u') = \sqrt{\frac{r}{2GM} - 1} \, \exp\left(\frac{r}{4GM}\right) \sinh \frac{t}{4GM}. \tag{2.105}$$

$$R = \frac{1}{2}(v' - u') = \sqrt{\frac{r}{2GM} - 1} \, \exp\left(\frac{r}{4GM}\right) \cosh \frac{t}{4GM}. \tag{2.106}$$

Clearly T is time-like while R is space-like. This can be confirmed by computing the metric. This is given by

$$ds^2 = \frac{32G^3M^3}{r} \exp\left(-\frac{r}{2GM}\right)(-dT^2 + dR^2) + r^2 d\Omega^2. \tag{2.107}$$

We see that T is always time-like while R is always space-like since the sign of the components of the metric never get reversed.

We remark that

$$T^2 - R^2 = v'u'$$
$$= -\exp \frac{v-u}{4GM}$$
$$= -\exp \frac{r + 2GM \log\left(\frac{r}{2GM} - 1\right)}{2GM} \tag{2.108}$$
$$= \left(1 - \frac{r}{2GM}\right) \exp \frac{r}{2GM}.$$

The radial coordinate r is determined implicitly in terms of T and R from this equation, i.e. equation (6.84). The coordinates (T, R, θ, ϕ) are called Kruskal–Szekres coordinates. Remarks are now in order

- The radial null curves in this system of coordinates are given by

$$T = \pm R + \text{constant}. \tag{2.109}$$

- The horizon defined by $r \longrightarrow 2GM$ is seen to appear at $T^2 - R^2 \longrightarrow 0$, i.e. at (6.85) in the new coordinate system. This shows in an elegant way that the event horizon is a null surface.
- The surfaces of constant r are given from (6.84) by $T^2 - R^2 = \text{constant}$ which are hyperbolae in the R–T plane.
- For $r > 2GM$ the surfaces of constant t are given by $T/R = \tanh t/4GM = \text{constant}$ which are straight lines through the origin. In the limit $t \longrightarrow \pm \infty$ we have $T/R \longrightarrow \pm 1$ which are precisley the horizon $r = 2GM$.
- For $r < 2GM$ we have

$$T = \frac{1}{2}(v' + u') = \sqrt{1 - \frac{r}{2GM}} \exp\left(\frac{r}{4GM}\right) \cosh \frac{t}{4GM}. \tag{2.110}$$

$$R = \frac{1}{2}(v' - u') = \sqrt{1 - \frac{r}{2GM}} \exp\left(\frac{r}{4GM}\right) \sinh \frac{t}{4GM}. \tag{2.111}$$

The metric and the condition determining r implicitly in terms of T and R do not change form in the (T, R, θ, ϕ) system of coordinates and thus the radial null curves, the horizon as well as the surfaces of constant r are given by the same equation as before.

- For $r < 2GM$ the surfaces of constant t are given by $T/R = 1/\tanh t/4GM = \text{constant}$, which are straight lines through the origin.
- It is clear that the allowed range for R and T is (analytic continuation from the region $T^2 - R^2 < 0$ $(r > 2GM)$ to the first singularity which occurs in the region $T^2 - R^2 < 1$ $(r < 2GM)$)

$$-\infty \leqslant R \leqslant +\infty, \quad T^2 - R^2 \leqslant 1. \tag{2.112}$$

A Kruskal–Szekres diagram is shown in figure 2.6. Every point in this diagram is actually a 2D sphere since we are suppressing θ and ϕ and drawing only R and T. The Kruskal–Szekres diagram gives the maximal extension of the Schwarzschild solution. In some sense it represents the entire Schwarzschild spacetime. It can be divided into four regions:

- Region 1: Exterior of black hole with $r > 2GM$ $(R > 0$ and $T^2 - R^2 < 0)$. Clearly future directed time-like (null) worldlines will lead to region 2,

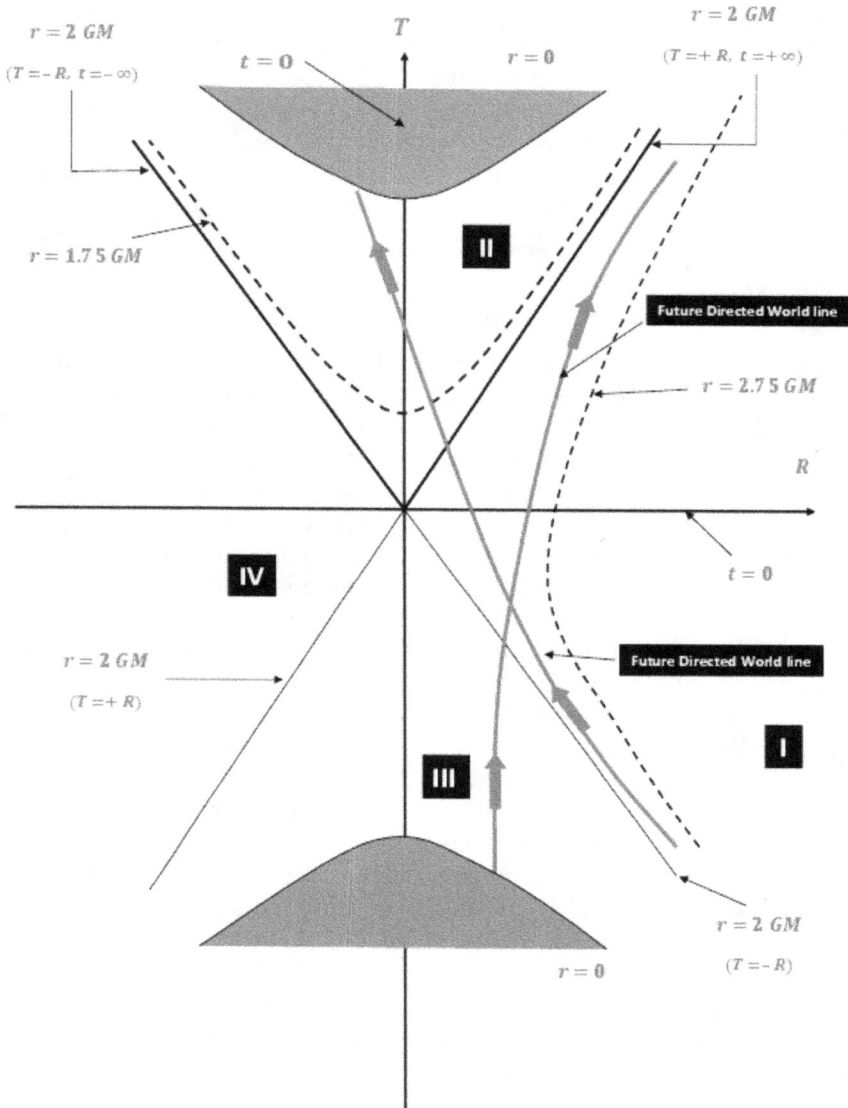

Figure 2.6. The Kruskal–Szekres diagram.

whereas past-directed time-like (null) worldlines can reach it from region 4. Regions 1 and 3 are connected by space-like geodesics.

- Region 2: Inside of black hole with $r < 2GM$ ($T > 0, 0 < T^2 - R^2 < 1$). Any future directed path in this region will hit the singularity. In this region r becomes time-like (while t becomes space-like) and thus we cannot stop moving in the direction of decreasing r in the same way that we cannot stop time progression in region 1.

- Region 3: Parallel exterior region with $r > 2GM$ ($R < 0$, $T^2 - R^2 < 0$). This is another asymptotically flat region of spacetime which we cannot access along future- or past-directed paths.
- Region 4: Inside the white hole with $r < 2GM$ ($T < 0$, $0 < T^2 - R^2 < 1$). The white hole is the time reverse of the black hole. This corresponds to a singularity in the past at which the Universe originated. This is a part of spacetime from which observers can escape to reach us while we cannot go there.

2.4 Various theorems and results

The various theorems and results quoted in this section require a much more careful and detailed analysis more extensive than we are able to do at this stage.

- **Birkhoff's theorem:** The Schwarzschild solution is the only spherically symmetric solution of general relativity in vacuum.

  This is to be compared with Coulomb potential, which is the only spherically symmetric solution of Maxwell's equations in vacuum.

- **No-hair theorem (example):** General relativity coupled to Maxwell's equations admits a small number of stationary asymptotically flat black hole solutions which are non-singular outside the event horizon and which are characterized by a limited number of parameters given by the mass, the charge (electric and magnetic) and the angular momentum.

  In contrast with the above result there exists in general relativity an infinite number of planet solutions and each solution is generically characterized by an infinite number of parameters.

- **Event horizon:** Black holes are characterized by their event horizons. A horizon is a boundary line between two regions of spacetime. Region I consists of all points of spacetime which are connected to infinity by time-like geodesics, whereas region II consists of all spacetime points which are not connected to infinity by time-like geodesics, i.e. observers cannot reach infinity starting from these points. The boundary between regions I and II, which is the event horizon, is a light-like (null) hypersurface.

  The event horizon can be defined as the set of points where the light cones are tilted over (in an appropriate coordinate system). In the Schwarzschild solution the event horizon occurs at $r = 2GM$ which is a null surface, although $r = $ constant is a time-like surface for large r.

  In a general stationary metric we can choose a coordinate system where $\partial_t g_{\mu\nu} = 0$ and on hypersurfaces $t = $ constant the coordinates will resemble spherical polar coordinates (r, θ, ϕ) sufficiently far away. Thus hypersurfaces $r = $ constant are time-like with the topology $S^2 \times R$ as $r \longrightarrow \infty$. It is obvious that $\partial_\mu r$ is a normal one-form to these hypersurfaces with norm

$$g^{rr} = g^{\mu\nu}\partial_\mu r \partial_\nu r. \tag{2.113}$$

If the time-like hypersurfaces $r = $ constant become null at some $r = r_H$ then we will get an event horizon at $r = r_H$ since any time-like geodesic crossing to the region $r < r_H$ will not be able to escape back to infinity. For $r > r_H$ we have clearly $g^{rr} > 0$, whereas for $r < r_H$ we have $g^{rr} < 0$. The event horizon is defined by the condition

$$g^{rr}(r_H) = 0. \tag{2.114}$$

- **Trapped surfaces:** In general relativity, singularities are generic and they are hidden behind event horizons. As shown by Hawking and Penrose, singularities are inevitable if gravitational collapse reaches a point of no return, i.e. the appearance of a trapped surface.

  Let us consider a 2-sphere in Minkowski spacetime. We consider then null rays emanating from the sphere inward or outward. The rays emanating outward describe growing spheres, whereas the rays emanating inward describe shrinking spheres. Consider now a 2-sphere in Schwarzschild spacetime with $r < 2GM$. In this case the rays emanating outward and inward will correspond to shrinking spheres (r is time-like). This is called a trapped surface.

  A trapped surface is a compact space-like 2D surface with the property that outward light rays are in fact moving inward.
- **Singularity theorem (example):** A trapped surface in a manifold M with a generic metric $g_{\mu\nu}$ (which is a solution of Einstein's equation satisfying the strong energy condition) can only be a closed time-like curve or a singularity.
- **Cosmic censorship conjecture:** In general relativity, singularities are hidden behind event horizons. More precisely, naked singularities cannot appear in the gravitational collapse of a non-singular state in an asymptotically flat spacetime which fulfills the dominant energy condition.
- **Hawking's area theorem:** In general relativity, black holes cannot shrink but they can grow in size. Clearly the size of the black hole is measured by the area of the event horizon.

  Hawking's area theorem can be stated as follows. The area of a future event horizon in an asymptotically flat spacetime is always increasing provided the cosmic censorship conjecture and the weak energy condition hold.
- **Stokes's theorem:** Next we recall stokes's theorem

$$\int_\Sigma d\omega = \int_{\partial\Sigma} \omega. \tag{2.115}$$

Explicitly this reads

$$\int_\Sigma d^n x \sqrt{|g|} \; \nabla_\mu V^\mu = \int_{\partial\Sigma} d^{n-1} y \sqrt{|\gamma|} \; \sigma_\mu V^\mu. \tag{2.116}$$

The unit vector σ^μ is normal to the boundary $\partial\Sigma$. In the case that Σ is the whole space, the boundary $\partial\Sigma$ is the 2-sphere at infinity and thus σ^μ is given, in an appropriate system of coordinates, by the components $(0, 1, 0, 0)$.
- **Energy in GR:** The concept of conserved total energy in general relativity is not straightforward.

For a stationary asymptotically flat spacetime with a time-like Killing vector field K^μ we can define a conserved energy–momentum current J_T^μ by

$$J_T^\mu = K_\nu T^{\mu\nu}. \tag{2.117}$$

Let Σ be a space-like hypersurface with a unit normal vector n^μ and an induced metric γ_{ij}. By integrating the component of J_T^μ along the normal n^μ over the surface Σ we get an energy, viz

$$E_T = \int_\Sigma d^3x \sqrt{\gamma} \ n_\mu J_T^\mu. \tag{2.118}$$

This definition is, however, inadequate since it gives zero energy in the case of Schwarzschild spacetime.

Let us consider instead the following current

$$J_R^\mu = K_\nu R^{\mu\nu}$$
$$= 8\pi G K_\nu \left(T^{\mu\nu} - \frac{1}{2} g^{\mu\nu} T \right). \tag{2.119}$$

We compute now

$$\nabla_\mu \ J_R^\mu = K_\nu \ \nabla_\mu \ R^{\mu\nu}. \tag{2.120}$$

By using now the contracted Bianchi identity $\nabla_\mu \ G^{\mu\nu} = \nabla_\mu \ (R^{\mu\nu} - g^{\mu\nu} R/2) = 0$ or equivalently $\nabla_\mu \ R^{\mu\nu} = \nabla^\nu \ R/2$ we get

$$\nabla_\mu \ J_R^\mu = \frac{1}{2} K_\nu \ \nabla^\nu \ R. \tag{2.121}$$

The derivative of the scalar curvature along a Killing vector must vanish and as a consequence J_R^μ is conserved. The corresponding energy is defined by

$$E_R = \frac{1}{4\pi G} \int_\Sigma d^3x \sqrt{\gamma} \ n_\mu J_R^\mu. \tag{2.122}$$

The normalization is chosen for later convenience. The Killing vector K^μ satisfies among other things $\nabla_\nu \ \nabla^\mu K^\nu = R^{\mu\nu} K_\nu$ and hence the vector J_R^μ is actually a total derivative, viz

$$J_R^\mu = \nabla_\nu \ \nabla^\mu \ K^\nu. \tag{2.123}$$

The energy E_R becomes

$$E_R = \frac{1}{4\pi G} \int_\Sigma d^3x \sqrt{\gamma} \ n_\mu \ \nabla_\nu \ \nabla^\mu \ K^\nu$$
$$= \frac{1}{4\pi G} \int_\Sigma d^3x \sqrt{\gamma} \ \nabla_\nu \ (n_\mu \ \nabla^\mu \ K^\nu) \tag{2.124}$$
$$- \frac{1}{4\pi G} \int_\Sigma d^3x \sqrt{\gamma} \ \nabla_\nu \ n_\mu . \nabla^\mu \ K^\nu.$$

In the second term we can clearly replace $\nabla_\nu\, n_\mu$ with $(\nabla_\nu n_\mu - \nabla_\mu\, n_\nu)/2 = (\partial_\nu n_\mu - \partial_\mu n_\nu)/2$. The surface Σ is space-like and thus the unit vector n^μ is time-like. For example, Σ can be the whole of space and thus n^μ must be given, in an appropriate system of coordinates, by the components $(1, 0, 0, 0)$. In this system of coordinates the second term vanishes. The above equation reduces to

$$E_R = \frac{1}{4\pi G} \int_\Sigma d^3x \sqrt{\gamma}\ \ \nabla_\nu\ (n_\mu\ \nabla^\mu\ K^\nu). \tag{2.125}$$

By using stokes's theorem we get the result

$$E_R = \frac{1}{4\pi G} \int_{\partial\Sigma} d^2x \sqrt{\gamma^{(2)}}\ \ \sigma_\nu (n_\mu\ \nabla^\mu\ K^\nu). \tag{2.126}$$

This is the Komar integral, which defines the total energy of the stationary spacetime. For Schwarzschild spacetime we can check that $E_R = M$. The Komar energy agrees with the ADM (Arnowitt, Deser, Misner) energy which is obtained from a Hamiltonian formulation of general relativity and which is associated with invariance under time translations.

2.5 Reissner–Nordström (charged) black hole

2.5.1 Maxwell's equations and charges in GR

Maxwell's equations in flat spacetime are given by

$$\partial_\mu F^{\mu\nu} = -J^\nu. \tag{2.127}$$

$$\partial_\mu F_{\nu\lambda} + \partial_\lambda F_{\mu\nu} + \partial_\nu F_{\lambda\mu} = 0. \tag{2.128}$$

Maxwell's equations in curved spacetime can be obtained from the above equations using the principle of minimal coupling, which consists in making the replacements $\eta_{\mu\nu} \longrightarrow g_{\mu\nu}$ and $\partial_\mu \longrightarrow D_\mu$ where D_ν is the covariant derivative associated with the metric $g_{\mu\nu}$. The homogeneous equations do not change under these substitutions since the extra corrections coming from the Christoffel symbols cancel by virtue of the antisymmetry under permutations of μ, ν and λ. This also means that the field strength tensor $F_{\mu\nu}$ in curved spacetime is still given by the same formula as in the flat case, viz

$$F_{\mu\nu} = \partial_\mu A_\nu - \partial_\nu A_\mu. \tag{2.129}$$

The inhomogeneous Maxwell's equation in curved spacetime is given by

$$D_\mu F^{\mu\nu} = -J^\nu. \tag{2.130}$$

We compute

$$\begin{aligned} D_\mu F^{\mu\nu} &= \partial_\mu F^{\mu\nu} + \Gamma^\mu{}_{\mu\alpha} F^{\alpha\nu} \\ &= \partial_\mu F^{\mu\nu} + \frac{1}{2} g^{\mu\rho} \partial_\alpha g_{\mu\rho} F^{\alpha\nu}. \end{aligned} \tag{2.131}$$

Let $g = \det g_{\mu\nu}$ and let e_i be the eigenvalues of the matrix $g_{\mu\nu}$. We have the result

$$\frac{\partial \sqrt{-g}}{\sqrt{-g}} = \frac{1}{2}\frac{\partial g}{g} = \frac{1}{2}\sum_i \frac{\partial e_i}{e_i} = \frac{1}{2}g^{\mu\rho}\partial g_{\mu\rho}. \tag{2.132}$$

Thus

$$D_\mu F^{\mu\nu} = \partial_\mu F^{\mu\nu} + \frac{\partial_\alpha \sqrt{-g}}{\sqrt{-g}}F^{\alpha\nu}. \tag{2.133}$$

Using this result we can put the inhomogeneous Maxwell's equation in the equivalent form

$$\partial_\mu(\sqrt{-g}\,F^{\mu\nu}) = -\sqrt{-g}\,J^\nu. \tag{2.134}$$

The law of conservation of charge in curved spacetime is now obvious given by

$$\partial_\mu(\sqrt{-g}\,J^\mu) = 0. \tag{2.135}$$

This is equivalent to the form

$$D_\mu J^\mu = 0. \tag{2.136}$$

The energy–momentum tensor of electromagnetism is given by

$$T_{\mu\nu} = F_{\mu\alpha}F_\nu{}^\alpha - \frac{1}{4}g_{\mu\nu}F_{\alpha\beta}F^{\alpha\beta} + g_{\mu\nu}J_\alpha A^\alpha. \tag{2.137}$$

We define the electric and magnetic fields by $F_{0i} = E_i$ and $F_{ij} = \varepsilon_{ijk}B_k$ with $\varepsilon_{123} = -1$.

The amount of electric charge passing through a space-like hypersurface Σ with unit normal vector n^μ is given by the integral

$$\begin{aligned}
Q &= -\int_\Sigma d^3x\sqrt{\gamma}\,n_\mu J^\mu \\
&= -\int_\Sigma d^3x\sqrt{\gamma}\,n_\mu D_\nu F^{\mu\nu}.
\end{aligned} \tag{2.138}$$

The metric γ_{ij} is the induced metric on the surface Σ. By using Stokes's theorem we obtain

$$Q = -\int_{\partial\Sigma} d^2x\sqrt{\gamma^{(2)}}\,n_\mu \sigma_\nu F^{\mu\nu}. \tag{2.139}$$

The unit vector σ^μ is normal to the boundary $\partial\Sigma$.

The magnetic charge P can be defined similarly by considering instead the dual field strength tensor $*F^{\mu\nu} = \varepsilon^{\mu\nu\alpha\beta}F_{\alpha\beta}/2$.

2.5.2 Reissner–Nordström solution

We are interested in finding a spherically symmetric solution of the Einstein–Maxwell equations with some mass M, some electric charge Q and some magnetic charge P, i.e. we want to find the gravitational field around a star of mass M, electric charge Q and magnetic charge P.

We start from the metric

$$ds^2 = -A(r)dt^2 + B(r)dr^2 + r^2(d\theta^2 + \sin^2\theta d\phi^2). \tag{2.140}$$

We compute immediately $\sqrt{-g} = \sqrt{AB}\, r^2 \sin^2\theta$. The components of the Ricci tensor in this metric are given by (with $A = e^{2\alpha}$, $B = e^{2\beta}$)

$$R_{00} = \left(\partial_r^2\alpha + (\partial_r\alpha)^2 - \partial_r\beta\partial_r\alpha + \frac{2}{r}\partial_r\alpha\right)e^{2(\alpha-\beta)}$$

$$R_{rr} = -\partial_r^2\alpha - (\partial_r\alpha)^2 + \partial_r\beta\partial_r\alpha + \frac{2}{r}\partial_r\beta \tag{2.141}$$

$$R_{\theta\theta} = e^{-2\beta}(r\partial_r\beta - r\partial_r\alpha - 1) + 1$$

$$R_{\phi\phi} = \sin^2\theta\,[e^{-2\beta}(r\partial_r\beta - r\partial_r\alpha - 1) + 1].$$

We also need to provide an ansatz for the electromagnetic field. By spherical symmetry the most general electromagnetic field configuration corresponds to a radial electric field and a radial magnetic field. For simplicity we will only consider a radial electric field which is also static, viz

$$E_r = f(r), \quad E_\theta = E_\phi = 0, \quad B_r = B_\theta = B_\phi = 0. \tag{2.142}$$

We will also choose the current J^μ to be zero outside the star where we are interested in finding a solution. We compute $F^{0r} = -f(r)/AB$ while all other components are 0. The only non-trivial component of the inhomogeneous Maxwell's equation is $\partial_r(\sqrt{-g}\,F^{r0}) = 0$ and hence

$$\partial_r\left(\frac{r^2 f(r)}{\sqrt{AB}}\right) = 0 \Leftrightarrow f(r) = \frac{Q\sqrt{AB}}{4\pi r^2}. \tag{2.143}$$

The constant of integration Q will play the role of the electric charge since it is expected that A and B approach 1 when $r \longrightarrow \infty$. The homogeneous Maxwell's equation is satisfied since the only non-zero component of $F^{\mu\nu}$, i.e. F^{0r} is clearly of the form $-\partial^r A^0$ for some potential A^0 while the other components of the vector potential (A^r, A^θ and A^ϕ) are 0.

We have therefore shown that the above electrostatic ansatz solves Maxwell's equations. We are now ready to compute the energy–momentum tensor in this configuration. We compute

$$T_{\mu\nu} = \frac{f^2(r)}{AB}\left(\frac{1}{A}g_{\mu 0}g_{\nu 0} - \frac{1}{B}g_{\mu r}g_{\nu r} + \frac{1}{2}g_{\mu\nu}\right)$$

$$= \frac{f^2(r)}{2AB}\mathrm{diag}(A, -B, r^2, r^2\sin^2\theta). \tag{2.144}$$

Also

$$T_\mu{}^\nu = g^{\nu\lambda}T_{\mu\lambda}$$
$$= \frac{f^2(r)}{2AB}\mathrm{diag}(-1, -1, +1, +1). \tag{2.145}$$

The trace of the energy–momentum is therefore traceless as it should be for the electromagnetic field. Thus Einstein's equation takes the form

$$R_{\mu\nu} = 8\pi G T_{\mu\nu}. \tag{2.146}$$

We find three independent equations given by

$$\left(\partial_r^2\alpha + (\partial_r\alpha)^2 - \partial_r\beta\partial_r\alpha + \frac{2}{r}\partial_r\alpha\right)A = 4\pi G f^2. \tag{2.147}$$

$$\left(-\partial_r^2\alpha - (\partial_r\alpha)^2 + \partial_r\beta\partial_r\alpha + \frac{2}{r}\partial_r\beta\right)A = -4\pi G f^2. \tag{2.148}$$

$$e^{-2\beta}(r\partial_r\beta - r\partial_r\alpha - 1) + 1 = 4\pi G f^2\frac{r^2}{AB}. \tag{2.149}$$

From the first two equations (2.147) and (2.148) we deduce

$$\partial_r(\alpha + \beta) = 0. \tag{2.150}$$

In other words

$$\alpha = -\beta + c \Leftrightarrow B = \frac{c'}{A}. \tag{2.151}$$

c and c' are constants of integration. By substituting this solution in the third equation (2.149) we obtain

$$\partial_r\left(\frac{r}{B}\right) = 1 - GQ^2\frac{1}{4\pi r^2} \Leftrightarrow \frac{1}{B} = 1 + \frac{GQ^2}{4\pi r^2} + \frac{b}{r}. \tag{2.152}$$

In other words,

$$A = c' + \frac{GQ^2c'}{4\pi r^2} + \frac{bc'}{r}. \tag{2.153}$$

The first equation (2.147) is equivalent to

$$\partial_r^2 A + \frac{2}{r}\partial_r A = 8\pi G f^2. \tag{2.154}$$

By substituting the solution (2.153) back in (2.154) we get $c' = 1$. In other words we must have

$$B = \frac{1}{A}, \quad A = 1 + \frac{GQ^2}{4\pi r^2} + \frac{bc'}{r}. \tag{2.155}$$

Similarly to the Schwarzschild solution, we can now invoke the Newtonian limit to set $bc' = -2GM$. We get then the solution

$$A = 1 - \frac{2GM}{r} + \frac{GQ^2}{4\pi r^2}. \tag{2.156}$$

If we also assume a radial magnetic field generated by a magnetic charge P inside the star we obtain the more general metric

$$ds^2 = -\Delta(r)dt^2 + \Delta^{-1}(r)dr^2 + r^2(d\theta^2 + \sin^2\theta d\phi^2). \tag{2.157}$$

$$\Delta = 1 - \frac{2GM}{r} + \frac{G(Q^2 + P^2)}{4\pi r^2}. \tag{2.158}$$

This is the Reissner–Nordström solution. The event horizon is located at $r = r_H$ where

$$\Delta(r_H) = 0 \Leftrightarrow r^2 - 2GMr + \frac{G(Q^2 + P^2)}{4\pi} = 0. \tag{2.159}$$

We should then consider the discriminant

$$\delta = 4G^2M^2 - \frac{G(Q^2 + P^2)}{\pi}. \tag{2.160}$$

There are three possible cases:
- The case $GM^2 < (Q^2 + P^2)/4\pi$. There is a naked singularity at $r = 0$. The coordinate r is always space-like while the coordinate t is always time-like. There is no event horizon. An observer can therefore travel to the singularity and return back. However, the singularity is repulsive. More precisely, a time-like geodesic does not intersect the singularity. Instead it approaches $r = 0$ then it reverses its motion and drives away.

    This solution is in fact unphysical since the condition $GM^2 < (Q^2 + P^2)/4\pi$ means that the total energy is less than the sum of two of its components, which is impossible.
- The case $GM^2 > (Q^2 + P^2)/4\pi$. There are two horizons at

$$r_{\pm} = GM \pm \sqrt{G^2M^2 - \frac{G(Q^2 + P^2)}{4\pi}}. \tag{2.161}$$

These are of course null surfaces. The horizon at $r = r_+$ is similar to the horizon of the Schwarzschild solution. At this point the coordinate r becomes time-like ($\Delta < 0$) and a falling observer will keep going in the direction of decreasing r. At $r = r_-$ the coordinate r becomes space-like again ($\Delta > 0$). Thus the motion in the direction of decreasing r can be reversed, i.e. the singularity at $r = 0$ can be avoided.

The fact that the singularity can be avoided is consistent with the fact that $r = 0$ is a time-like line in the Reissner–Nordström solution as opposed to the singularity $r = 0$ in the Schwarzschild solution which is a space-like surface.

The observer in the region $r < r_-$ can therefore move either towards the singularity at $r = 0$ or towards the null surface $r = r_-$. After passing $r = r_-$ the coordinate r becomes time-like once more and the observer in this case can only move in the direction of increasing r until it emerges from the black hole at $r = r_+$.

- The case $GM^2 = (Q^2 + P^2)/4\pi$ (extremal RN black holes). There is a single horizon at $r = GM$. In this case the coordinate r is always space-like except at $r = GM$ where it is null. Thus the singularity can also be avoided in this case.

2.5.3 Extremal Reissner–Nordström black hole

The metric at $GM^2 = (Q^2 + P^2)/4\pi$ takes the form

$$ds^2 = -\left(1 - \frac{GM}{r}\right)^2 dt^2 + \left(1 - \frac{GM}{r}\right)^{-2} dr^2 + r^2(d\theta^2 + \sin^2\theta d\phi^2). \qquad (2.162)$$

We define the new coordinate $\rho = r - GM$ and the function $H(\rho) = 1 + GM/\rho$. The metric becomes

$$ds^2 = -H^{-2}(\rho)dt^2 + H^2(\rho)(d\rho^2 + \rho^2(d\theta^2 + \sin^2\theta d\phi^2)). \qquad (2.163)$$

Equivalently

$$ds^2 = -H^{-2}(\vec{x})dt^2 + H^2(\vec{x})d\vec{x}^2, \quad H(\vec{x}) = 1 + \frac{GM}{|\vec{x}|}. \qquad (2.164)$$

For simplicity let us consider only a static electric field which is given by $E_r = F_{0r} = Q/4\pi r^2$. From the extremal condition we have $Q^2 = 4\pi GM^2$. For electrostatic fields we have $F_{0r} = -\partial_r A_0$ and the rest are zero. Then it is not difficult to show that

$$A_0 = \frac{Q}{4\pi r} = \frac{1}{\sqrt{4\pi G}}\frac{GM}{\rho + GM}, \quad A_i = 0. \qquad (2.165)$$

Equivalently

$$\sqrt{4\pi G}\,A_0 = 1 - \frac{1}{H(\rho)}, \quad A_i = 0. \qquad (2.166)$$

The metric (2.164) together with the gauge field configuration (2.166) with an arbitrary function $H(\vec{x})$ still solves the Einstein–Maxwell's equations provided $H(\vec{x})$ satisfies the Laplace equation

$$\vec{\nabla}^2 H = 0. \qquad (2.167)$$

The general solution is given by

$$H(\vec{x}) = 1 + \sum_{i=1}^{N} \frac{GM_i}{|\vec{x} - \vec{x_i}|}. \qquad (2.168)$$

This describes a system of N extremal RN black holes located at $\vec{x_i}$ with masses M_i and charges $Q_i^2 = 4\pi G M_i^2$.

2.6 Kerr spacetime

2.6.1 Kerr (rotating) and Kerr–Newman (rotating and charged) black holes

- The Schwarzschild black holes and the Reissner–Nordström black holes are spherically symmetric. Any spherically symmetric vacuum solution of Einstein's equations possesses a time-like Killing vector and thus is stationary. In a stationary metric we can choose coordinates (t, x^1, x^2, x^3) where the killing vector is ∂_t, the metric components are all independent of the time coordinate t and the metric is of the form

$$ds^2 = g_{00}(x)dt^2 + 2g_{0i}(x)dtdx^i + g_{ij}(x)dx^idx^j. \qquad (2.169)$$

  This stationary metric becomes static if the time-like Killing vector ∂_t is also orthogonal to a family of hypersurfaces. In the coordinates (t, x^1, x^2, x^3) the Killing vector ∂_t is orthogonal to the hypersurfaces $t = $ constant and equivalently a stationary metric becomes static if $g_{0i} = 0$.
- In contrast, the Kerr and the Kerr–Newman black holes are not spherically symmetric and are not static but they are stationary. A Kerr black hole is a vacuum solution of Einstein's equations which describes a rotating black hole and thus is characterized by mass and angular momentum, whereas the Kerr–Newman black hole is a charged Kerr black hole and thus is characterized by mass, angular momentum and electric and magnetic charges. The rotation clearly breaks spherical symmetry and makes the black holes not static. However, since the black hole rotates in the same way at all times, it is still stationary. The Kerr and Kerr–Newman metrics must therefore be of the form

$$ds^2 = g_{00}(x)dt^2 + 2g_{0i}(x)dtdx^i + g_{ij}(x)dx^idx^j. \qquad (2.170)$$

- The Kerr metric must be clearly axial symmetric around the axis fixed by the angular momentum. This will correspond to a second Killing vector ∂_ϕ.
- In summary, the metric components, in a properly adapted system of coordinates, will not depend on the time coordinate t (stationary solution) but also will not depend on the angle ϕ (axial symmetry). Furthermore, if we denote the two coordinates t and ϕ by x^a and the other two coordinates by y^i the metric takes then the form

$$ds^2 = g_{ab}(y)dx^adx^b + g_{ij}(y)dx^idx^j. \qquad (2.171)$$

- In the so-called Boyer–Lindquist coordinates (t, r, θ, ϕ) the components of the Kerr metric are found [4] to be given by

$$g_{tt} = -\left(1 - \frac{2GMr}{\rho^2}\right), \quad \rho^2 = r^2 + a^2\cos^2\theta. \qquad (2.172)$$

$$g_{t\phi} = -\frac{2GMar \sin^2 \theta}{\rho^2}. \tag{2.173}$$

$$g_{rr} = \frac{\rho^2}{\Delta}, \quad \Delta = r^2 - 2GMr + a^2. \tag{2.174}$$

$$g_{\theta\theta} = \rho^2, \quad g_{\phi\phi} = \frac{\sin^2 \theta}{\rho^2}[(r^2 + a^2)^2 - a^2\Delta \sin^2 \theta]. \tag{2.175}$$

This solution is characterized by the two numbers M and a. The mass of the Kerr black hole is precisely M, whereas the angular momentum of the black hole is $J = aM$.

- In the limit $a \longrightarrow 0$ (no rotation) we obtain the Schwarzschild solution

$$g_{tt} = -\left(1 - \frac{2GM}{r}\right), \quad g_{rr} = \left(1 - \frac{2GM}{r}\right)^{-1},$$
$$g_{\theta\theta} = r^2, \quad g_{\phi\phi} = r^2 \sin^2 \theta. \tag{2.176}$$

- In the limit $M \longrightarrow 0$ we obtain the solution

$$g_{tt} = -1, \quad g_{rr} = \frac{r^2 + a^2 \cos^2 \theta}{r^2 + a^2},$$
$$g_{\theta\theta} = r^2 + a^2 \cos^2 \theta, \quad g_{\phi\phi} = (r^2 + a^2)\sin^2 \theta. \tag{2.177}$$

A solution with no mass and no rotation must correspond to flat Minkowski spacetime. Indeed the coordinates r, θ and ϕ are nothing but ellipsoidal coordinates in flat space. The corresponding Cartesian coordinates are

$$x = \sqrt{r^2 + a^2} \sin \theta \cos \phi, \quad y = \sqrt{r^2 + a^2} \sin \theta \sin \phi, \quad z = r \cos \theta. \tag{2.178}$$

- The Kerr–Newman black hole is a generalization of the Kerr black hole which includes also electric and magnetic charges and an electromagnetic field. The electric and magnetic charges can be included via the replacement

$$2GMr \longrightarrow 2GMr - G(Q^2 + P^2). \tag{2.179}$$

The electromagnetic field is given by

$$A_t = \frac{Qr - Pa \cos \theta}{\rho^2}, \quad A_\phi = \frac{-Qar \sin^2 \theta + P(r^2 + a^2)\cos \theta}{\rho^2}. \tag{2.180}$$

2.6.2 Killing horizons

In Schwarzschild spacetime the Killing vector $K = \partial_t$ becomes null at the event horizon. We say that the event horizon (which is a null surface) is the Killing horizon of the Killing vector $K = \partial_t$. In general the Killing horizon of a Killing vector χ^μ is a null hypersurface Σ along which the Killing vector χ^μ becomes null. An important result concerning Killing horizons is as follows:

- Every event horizon in a stationary, asymptotically flat spacetime is a Killing horizon for some Killing vector χ^μ.

In the case that the spacetime is stationary and static the Killing vector is precisely $K = \partial_\mu$. In the case that the spacetime is stationary and axial symmetric then the event horizon is a Killing horizon where the Killing vector is a combination of the Killing vector $R = \partial_t$ and the Killing vector $R = \partial_\phi$ associated with axial symmetry. These results are purely geometrical. In the general case of a stationary spacetime then Einstein's equations together with appropriate assumptions on the matter content will also yield the result that every event horizon is a Killing horizon for some Killing vector which is either stationary or axial symmetric.

2.6.3 Surface gravity

Every Killing horizon is associated with an acceleration called the surface gravity. Let Σ be a killing horizon for the Killing vector χ^μ. We know that $\chi^\mu \chi_\mu$ is zero on the Killing horizon and thus $\nabla_\nu(\chi^\mu \chi_\mu) = 2\chi_\mu \nabla_\nu \chi^\mu$ must be normal to the Killing horizon in the sense that it is orthogonal to any vector tangent to the horizon. The normal to the Killing horizon is however unique, given by χ^μ and as a consequence we must have

$$\chi_\mu \nabla_\nu \chi^\mu = -\kappa \chi_\nu. \tag{2.181}$$

This means in particular that the Killing vector χ^μ is a non-affinely parametrized geodesic on the Killing horizon. The coefficient κ is precisely the surface gravity. Since the Killing vector ξ^μ is hypersurface orthogonal we have by the Frobenius's theorem the result

$$\chi_{[\mu}\nabla_\nu\chi_{\sigma]} = 0. \tag{2.182}$$

We compute

$$\nabla^\mu\chi^\nu\chi_{[\mu}\nabla_\nu\chi_{\sigma]} = 2\kappa^2\chi_\sigma + 2\chi_\sigma \nabla^\mu \chi^\nu \nabla_\mu \chi_\nu + 2 \nabla^\mu \chi^\nu(\nabla_\sigma(\chi_\mu\chi_\nu) - \chi_\mu \nabla_\sigma \chi_\nu)$$
$$= 4\kappa^2\chi_\sigma + 2\chi_\sigma \nabla^\mu \chi^\nu \nabla_\mu \chi_\nu. \tag{2.183}$$

We get immediately the surface gravity

$$\kappa^2 = -\frac{1}{2} \nabla^\mu \chi^\nu \nabla_\mu \chi_\nu. \tag{2.184}$$

In a static and asymptotically flat spacetime we have $\chi = K$ where $K = \partial_t$, whereas in a stationary and asymptotically flat spacetime we have $\chi = K + \Omega_H R$ where $R = \partial_\phi$. In both cases, fixing the normalization of K as $K^\mu K_\mu = -1$ at infinity will fix the normalization of χ and as a consequence fixes the surface gravity of any Killing horizon uniquely.

In a static and asymptotically flat spacetime a more physical definition of surface gravity can be given. The surface gravity is the acceleration of a static observer on the horizon as seen by a static observer at infinity. A static observer is an observer whose 4-vector velocity U^μ is proportional to the Killing vector K^μ. By normalizing U^μ as $U^\mu U_\mu = -1$ we have

$$U^\mu = \frac{K^\mu}{\sqrt{-K^\mu K_\mu}}. \tag{2.185}$$

A static observer does not necessarily follow a geodesic. Its acceleration is defined by

$$A^\mu = U^\nu \, \nabla_\nu \, U^\mu. \tag{2.186}$$

We define the redshift factor V by

$$V = \sqrt{-K^\mu K_\mu}. \tag{2.187}$$

We compute

$$
\begin{aligned}
A^\mu &= -\frac{1}{V^3} K^\sigma K^\mu \, \nabla_\sigma \, V + \frac{U_\sigma}{V} \, \nabla^\sigma \, K^\mu \\
&= \frac{1}{V^4} K^\sigma K^\mu K^\alpha \, \nabla_\sigma \, K_\alpha - \frac{U_\sigma}{V} \, \nabla^\mu \, K^\sigma \\
&= -\frac{U_\sigma}{V} \, \nabla^\mu \, K^\sigma \\
&= - \, \nabla^\mu \left(\frac{U_\sigma}{V} K^\sigma \right) + \nabla^\mu \left(\frac{U_\sigma}{V} \right) K^\sigma \\
&= \frac{1}{V} \, \nabla^\mu \, U_\sigma K^\sigma + \nabla^\mu \left(\frac{1}{V} \right) U_\sigma K^\sigma \\
&= \nabla^\mu \ln V.
\end{aligned}
\tag{2.188}
$$

The magnitude of the acceleration is

$$A = \frac{\sqrt{\nabla_\mu V \, \nabla^\mu \, V}}{V}. \tag{2.189}$$

The redshift factor V goes obviously to 0 at the Killing Horizon and hence A goes to infinity. The surface gravity is given precisely by the product VA, viz

$$\kappa = VA = \sqrt{\nabla_\mu V \, \nabla^\mu \, V}. \tag{2.190}$$

This agrees with the original definition (2.184) as one can explicitly check. For a Schwarzschild black hole we compute

$$\kappa = \frac{1}{4GM}. \tag{2.191}$$

2.6.4 Event horizons, ergosphere and singularity

- The event horizons occur at $r = r_H$ where $g^{rr}(r_H) = 0$. Since $g^{rr} = \Delta/\rho^2$ we obtain the equation

$$r^2 - 2GMr + a^2 = 0. \tag{2.192}$$

The discriminant is $\delta = 4(G^2M^2 - a^2)$. As in the case of Reissner–Nordström solution there are three possibilities. We focus only on the more physically interesting case of $G^2M^2 > a^2$. In this case there are two solutions

$$r_\pm = GM \pm \sqrt{G^2M^2 - a^2}. \tag{2.193}$$

These two solutions correspond to two event horizons which are both null surfaces. Since the Kerr solution is stationary and not static the event horizons are not Killing horizons for the Killing vector $K = \partial_t$. In fact, the event horizons for the Kerr solutions are Killing horizons for the linear combination of the time translation Killing vector $K = \partial_t$ and the rotational Killing vector $R = \partial_\phi$, which is given by

$$\chi^\mu = K^\mu + \Omega_H R^\mu. \tag{2.194}$$

We can check that this vector becomes null at the outer event horizon r_+. We check this explicitly as follows. First we compute

$$K^\mu = \partial_t^\mu = \delta_t^\mu = (1, 0, 0, 0) \Leftrightarrow K_\mu = g_{\mu t} = \left(-\left(1 - \frac{2GMr}{\rho^2} \right), 0, 0, 0 \right). \tag{2.195}$$

$$R^\mu = \partial_\phi^\mu = \delta_\phi^\mu = (0, 0, 0, 1) \Leftrightarrow R_\mu = g_{\mu\phi}$$
$$= \left(0, 0, 0, \frac{\sin^2\theta}{\rho^2}[(r^2 + a^2)^2 - a^2\Delta\sin^2\theta] \right). \tag{2.196}$$

Then

$$K^\mu K_\mu = -\frac{1}{\rho^2}(\Delta - a^2\sin^2\theta). \tag{2.197}$$

$$R^\mu R_\mu = \frac{\sin^2\theta}{\rho^2}[(r^2 + a^2)^2 - a^2\Delta\sin^2\theta]. \tag{2.198}$$

$$R^\mu K_\mu = g_{\phi t} = -\frac{2GMar\sin^2\theta}{\rho^2}. \tag{2.199}$$

Thus

$$\chi^\mu\chi_\mu = -\frac{1}{\rho^2}(\Delta - a^2\sin^2\theta) + \Omega_H^2\frac{\sin^2\theta}{\rho^2}[(r^2 + a^2)^2 - a^2\Delta\sin^2\theta]$$
$$- \Omega_H\frac{4GMar\sin^2\theta}{\rho^2}. \tag{2.200}$$

At the outer event horizon $r = r_+$ we have $\Delta = 0$ and thus

$$\chi^\mu \chi_\mu = \frac{\sin^2 \theta}{\rho^2} \left[(r_+^2 + a^2)\Omega_H - a \right]^2. \tag{2.201}$$

This is zero for

$$\Omega_H = \frac{a}{r_+^2 + a^2}. \tag{2.202}$$

As it turns out, Ω_H is the angular velocity of the event horizon $r = r_+$, which is defined as the angular velocity of a particle at the event horizon $r = r_+$.

• Let us consider again the Killing vector $K = \partial_t$. We have

$$K^\mu K_\mu = -\frac{1}{\rho^2}(\Delta - a^2 \sin^2 \theta). \tag{2.203}$$

At $r = r_+$ we have $K^\mu K_\mu = a^2 \sin^2 \theta / \rho^2 \geqslant 0$ and hence this vector is space-like at the outer horizon except at $\theta = 0$ (north pole) and $\theta = \pi$ (south pole) where it becomes null.

The so-called stationary limit surface or ergosurface is defined as the set of points where $K^\mu K_\mu = 0$. This is given by

$$\Delta = a^2 \sin^2 \theta \Leftrightarrow (r - GM)^2 = G^2 M^2 - a^2 \cos^2 \theta. \tag{2.204}$$

The outer event horizon is given by

$$\Delta = 0 \Leftrightarrow (r_+ - GM)^2 = G^2 M^2 - a^2. \tag{2.205}$$

The region between the stationary limit surface and the outer event horizon is called the ergosphere. Inside the ergosphere the Killing vector K^μ is space-like and thus observers cannot remain stationary. In fact they must move in the direction of the rotation of the black hole but they can still move towards the event horizon or away from it.

• The naked singularity in Kerr spacetime occurs at $\rho = 0$. Since $\rho^2 = r^2 + a^2 \cos^2 \theta$ we get the conditions

$$r = 0, \quad \theta = \frac{\pi}{2}. \tag{2.206}$$

To exhibit what these conditions correspond to we substitute them in equation (2.178), which is valid in the limit $M \longrightarrow 0$. We obtain immediately $x^2 + y^2 = a^2$, which is a ring. This ring singularity is, of course, only a coordinate singularity in the limit $M \longrightarrow 0$. For $M \neq 0$ the ring singularity is indeed a true or naked singularity as one can explicitly check. The rotation has therefore softened the naked singularity at $r = 0$ of the Schwarzschild solution but spreading it over a ring.

• A sketch of the Kerr black hole is shown in figure 2.7.

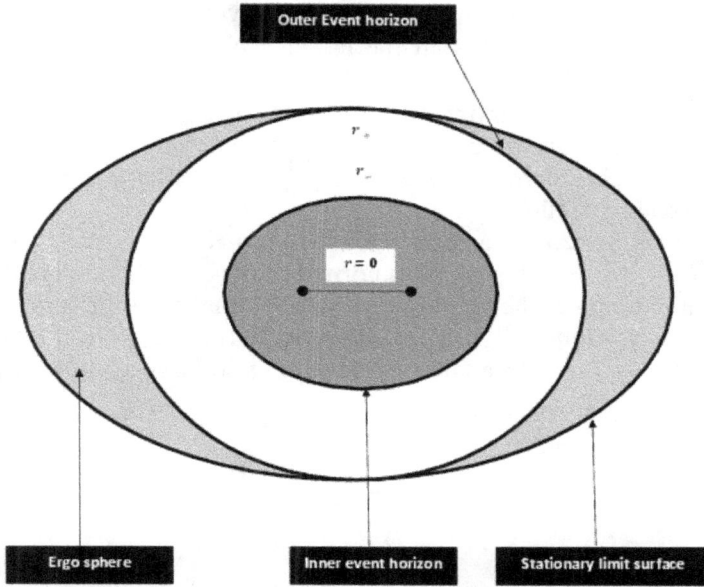

Figure 2.7. Kerr black hole.

2.6.5 Penrose process

The conserved energy of a massive particle with mass m in a Kerr spacetime is given by

$$E = -K_\mu p^\mu = -g_{tt}K^t p^t - g_{t\phi}K^t p^\phi$$
$$= m\left(1 - \frac{2GMr}{\rho^2}\right)\frac{dt}{d\tau} + \frac{2GmMar\sin^2\theta}{\rho^2}\frac{d\phi}{d\tau}. \qquad (2.207)$$

The angular momentum of the particle is given by

$$L = R_\mu p^\mu = g_{\phi\phi}R^\phi p^\phi + g_{\phi t}R^\phi p^t$$
$$= \frac{m\sin^2\theta}{\rho^2}[(r^2+a^2)^2 - a^2\Delta\sin^2\theta]\frac{d\phi}{d\tau} - \frac{2GmMar\sin^2\theta}{\rho^2}\frac{dt}{d\tau}. \qquad (2.208)$$

The minus sign in the definition of the energy guarantees positivity since both K^μ and p^μ are time-like vectors at infinity and as such their scalar product is negative. Inside the ergosphere the Killing vector K^μ becomes space-like and thus it is possible to have particles for which $E = -K_\mu p^\mu < 0$.

We imagine an object starting outside the ergosphere with energy $E^{(0)}$ and momentum $p^{(0)}$ and falling into the black hole. The energy $E^{(0)} = -K^\mu p_\mu^{(0)}$ is positive and conserved along the geodesic. Once the object enters the ergosphere it splits into two with momenta $p^{(1)}$ and $p^{(2)}$. The object with momentum $p^{(1)}$ is allowed to escape back to infinity while the object with momentum $p^{(2)}$ falls into the black hole.

We have the momentum and energy conservations $p^{(0)} = p^{(1)} + p^{(2)}$ and $E^{(0)} = E^{(1)} + E^{(2)}$. It is possible that the infalling object with momentum $p^{(2)}$ has negative energy $E^{(2)}$ and as a consequence $E^{(0)}$ will be less than $E^{(1)}$. In other words, the escaping object can have more energy than the original infalling object. This so-called Penrose process allows us therefore to extract energy from the black hole, which actually happens by decreasing its angular momentum. This process can be made more explicit as follows.

The outer event horizon of a Kerr black hole is a Killing horizon for the Killing vector $\chi^{\mu} = K^{\mu} + \Omega_H R^{\mu}$. This vector is normal to the event horizon and it is future-pointing, i.e. it determines the forward direction in time. Thus the statement that the particle with momentum $p^{(2)}$ crosses the event horizon moving forward in time means that $-p^{(2)\mu}\chi_{\mu} \geqslant 0$. The analog statement in a static spacetime is that particles with positive energy move forward in time, i.e. $E = -p^{(2)\mu}K_{\mu} \geqslant 0$. The condition $-p^{(2)\mu}\chi_{\mu} \geqslant 0$ is equivalent to

$$L^{(2)} \leqslant \frac{E^{(2)}}{\Omega_H} < 0. \tag{2.209}$$

Since $E^{(2)}$ is assumed to be negative and Ω_H is positive the angular momentum $L^{(2)}$ is negative and hence the particle with momentum $p^{(2)}$ is actually moving against the rotation of the black hole. After the particle with momentum $p^{(1)}$ escapes to infinity and the particle with momentum $p^{(2)}$ falls into the black hole the mass and the angular momentum of the Kerr black hole change (decrease) by the amounts

$$\Delta M = E^{(2)}, \quad \Delta J = L^{(2)}. \tag{2.210}$$

The bound $L^{(2)} \leqslant E^{(2)}/\Omega_H$ becomes

$$\Delta J \leqslant \frac{\Delta M}{\Omega_H}. \tag{2.211}$$

Thus, extracting energy from the black hole (or equivalently decreasing its mass) is achieved by decreasing its angular momentum, i.e. by making the infalling particle carry angular momentum opposite to the rotation of the black hole.

In the limit when the particle with momentum $p^{(2)}$ becomes null tangent to the event horizon we get the ideal process $\Delta J = \Delta M/\Omega_H$.

2.7 Black hole thermodynamics

Let us start this section by calculating the area of the outer event horizon $r = r_+$ of a Kerr black hole. Recall first that

$$r_+ = GM + \sqrt{G^2M^2 - a^2}. \tag{2.212}$$

We need the induced metric γ_{ij} on the outer event horizon. Since the outer event horizon is defined by $r = r_+$ the coordinates on the outer event horizon are θ and ϕ.

We set therefore $r = r_+$ ($\Delta = 0$), $dr = 0$ and $dt = 0$ in the Kerr metric. We obtain the metric

$$
\begin{aligned}
ds^2 \mid_{r=r_+} &= \gamma_{ij} dx^i dx^j \\
&= g_{\theta\theta} d\theta^2 + g_{\phi\phi} d\phi^2 \\
&= (r_+^2 + a^2 \cos^2 \theta) d\theta^2 + \frac{(r_+^2 + a^2)^2 \sin^2 \theta}{r_+^2 + a^2 \cos^2 \theta} d\phi^2.
\end{aligned}
\tag{2.213}
$$

The area of the horizon can be constructed from the induced metric as follows

$$
\begin{aligned}
A &= \int \sqrt{|\det \gamma|} \, d\theta d\phi \\
&= \int (r_+^2 + a^2) \sin \theta d\theta d\phi \\
&= 4\pi (r_+^2 + a^2) \\
&= 8\pi G^2 \left(M^2 + \sqrt{M^4 - \frac{M^2 a^2}{G^2}} \right) \\
&= 8\pi G^2 \left(M^2 + \sqrt{M^4 - \frac{J^2}{G^2}} \right).
\end{aligned}
\tag{2.214}
$$

The area is related to the so-called irreducible mass M_{irr}^2 by

$$
\begin{aligned}
M_{\text{irr}}^2 &= \frac{A}{16\pi G^2} \\
&= \frac{1}{2} \left(M^2 + \sqrt{M^4 - \frac{J^2}{G^2}} \right).
\end{aligned}
\tag{2.215}
$$

The area (or equivalently the irreducible mass) depends on the two parameters characterizing the Kerr black hole, namely its mass and its angular momentum. From the other hand we know that the mass and the angular momentum of the Kerr black hole decrease in the Penrose process. Thus the area changes in the Penrose process as follows

$$
\begin{aligned}
\Delta A &= \frac{8\pi G}{\sqrt{G^2 M^2 - a^2}} [2GMr_+ \Delta M - a \Delta J] \\
&= \frac{8\pi G}{\sqrt{G^2 M^2 - a^2}} \left[(r_+^2 + a^2) \Delta M - a \Delta J \right] \\
&= \frac{8\pi G (r_+^2 + a^2)}{\sqrt{G^2 M^2 - a^2}} [\Delta M - \Omega_H \Delta J] \\
&= \frac{8\pi G a}{\Omega_H \sqrt{G^2 M^2 - a^2}} [\Delta M - \Omega_H \Delta J].
\end{aligned}
\tag{2.216}
$$

This is equivalent to

$$\Delta M_{\text{irr}}^2 = \frac{a}{2G\sqrt{G^2M^2 - a^2}} \left[\frac{\Delta M}{\Omega_H} - \Delta J \right] \Leftrightarrow \Delta M_{\text{irr}}$$

$$= \frac{a}{4GM_{\text{irr}}\sqrt{G^2M^2 - a^2}} \left[\frac{\Delta M}{\Omega_H} - \Delta J \right]. \tag{2.217}$$

However, we have already found that in the Penrose process we must have $\Delta J \leqslant \Delta M / \Omega_H$. This leads immediately to

$$\Delta M_{\text{irr}} \geqslant 0. \tag{2.218}$$

The irreducible mass cannot decrease. From this result we deduce immediately that

$$\Delta A \geqslant 0. \tag{2.219}$$

This is the second law of black hole thermodynamics or the area theorem which states that the area of the event horizon is always non-decreasing. The area in black hole thermodynamics plays the role of entropy in thermodynamics.

We can use equation (2.215) to express the mass of the Kerr black hole in terms of the irreducible mass M_{irr} and the angular momentum J. We find

$$M^2 = M_{\text{irr}}^2 + \frac{J^2}{4G^2M_{\text{irr}}^2}$$

$$= \frac{A}{16\pi G^2} + \frac{4\pi J^2}{A}. \tag{2.220}$$

Now we imagine a Penrose process which is reversible, i.e. we reduce the angular momentum of the black hole from J_i to J_f such that $\Delta A = 0$ (clearly $\Delta A > 0$ is not a reversible process simply because the reverse process violates the area theorem). Then

$$M_i^2 - M_f^2 = \frac{4\pi}{A}(J_i^2 - J_f^2). \tag{2.221}$$

If we consider $J_f = 0$ then we obtain

$$M_i^2 - M_f^2 = \frac{4\pi}{A}J_i^2 \Leftrightarrow M_f^2 = \frac{A}{16\pi G^2} = M_{\text{irr}}^2. \tag{2.222}$$

In other words if we reduce the angular momentum of the Kerr black hole to zero, i.e. until the black hole stops rotating, then its mass will reduce to a minimum value given precisely by M_{irr}. This is why this is called the irreducible mass. In fact, M_{irr} is the mass of the resulting Schwarzschild black hole. The maximum energy we can therefore extract from a Kerr black hole via a Penrose process is $M - M_{\text{irr}}$. We have

$$E_{\max} = M - M_{\text{irr}} = M - \frac{1}{\sqrt{2}}\sqrt{M^2 + \sqrt{M^4 - \frac{J^2}{G^2}}}. \tag{2.223}$$

The irreducible mass is minimum at $M^2 = J/G$ or equivalently $GM = a$ (which is the case of extremal Kerr black hole) and as a consequence E_{max} is maximum for $GM = a$. At this point

$$E_{max} = M - M_{irr} = M - \frac{1}{\sqrt{2}}M = 0.29M. \qquad (2.224)$$

We can therefore extract at most 29% of the original mass of Kerr black hole via the Penrose process.

The first law of black hole thermodynamics is essentially given by equation (2.216). This result can be rewritten as

$$\Delta M = \frac{\kappa}{8\pi G}\Delta A + \Omega_H \Delta J. \qquad (2.225)$$

The constant κ is called the surface gravity of the Kerr black hole and it is given by

$$\begin{aligned}
\kappa &= \frac{\Omega_H \sqrt{G^2 M^2 - a^2}}{a} \\
&= \frac{\sqrt{G^2 M^2 - a^2}}{r_+^2 + a^2} \\
&= \frac{\sqrt{G^2 M^2 - a^2}}{2GM(GM + \sqrt{G^2 M^2 - a^2})}.
\end{aligned} \qquad (2.226)$$

The above first law of black hole thermodynamics is similar to the first law of thermodynamics $dU = TdS - pdV$ with the most important identifications

$$\begin{aligned}
U &\leftrightarrow M \\
S &\leftrightarrow \frac{A}{4G} \\
T &\leftrightarrow \frac{\kappa}{2\pi}.
\end{aligned} \qquad (2.227)$$

The quantity $\kappa \Delta A/(8\pi G)$ is heat energy while $\Omega_H \Delta J$ is the work done on the black by throwing particles into it.

The zeroth law of black hole thermodynamics states that surface gravity is constant on the horizon. Again this is the analog of the zeroth law of thermodynamics which states that temperature is constant throughout a system in thermal equilibrium.

2.8 Exercises

Exercise 1: Show explicitly in the case of Schwarzschild black hole that $r = R_s$ is a coordinate singularity and that $r = 0$ is an essential singularity.

Exercise 2:
- The strong energy condition is given by $T_{\mu\nu} t^\mu t^\nu \geqslant T^\lambda{}_\lambda t^\sigma t_\sigma / 2$ for any time-like vector t^μ. Show that this is equivalent to $\rho + P \geqslant 0$ and $\rho + 3P \geqslant 0$.
- The dominant energy condition is given by $T_{\mu\nu} t^\mu t^\nu \geqslant 0$ and $T_{\mu\nu} T^\nu{}_\lambda t^\mu t^\lambda \leqslant 0$ for any time-like vector t^μ. Show that these are equivalent to $\rho \geqslant |P|$.
- The weak energy condition is given by $T_{\mu\nu} t^\mu t^\nu \geqslant 0$ for any time-like vector t^μ. Show that these are equivalent to $\rho \geqslant 0$ and $\rho + P \geqslant 0$.

Exercise 3: Show that for a Schwarzschild black hole the Hawking's area theorem implies that the mass of the black hole can only increase.

Exercise 4: Verify that J_T^μ defined by $J_T^\mu = K_\nu T^{\mu\nu}$ is conserved by using the fact that the energy–momentum tensor is conserved ($\nabla_\mu T^{\mu\nu} = 0$) and the fact that K^μ is a Killing vector ($\nabla_\mu K_\nu + \nabla_\nu K_\mu = 0$).

Exercise 5: Show that $\nabla_\nu \nabla^\mu K^\nu = R^{\mu\nu} K_\nu$ and that the derivative of the scalar curvature along a Killing vector must vanish.

Exercise 6: Show that the total energy of a stationary spacetime given by the Koma integral

$$E_R = \frac{1}{4\pi G} \int_{\partial\Sigma} d^2 x \sqrt{\gamma^{(2)}} \; \sigma_\nu (n_\mu \nabla^\mu K^\nu) \tag{2.228}$$

gives $E_R = M$ for Schwarzschild spacetime.

Exercise 7:
- Show that the homogeneous Maxwell's equations take the same form in curved spacetime. What do you conclude?
- Show that the energy–momentum tensor of the electromagnetic field is given by

$$T_{\mu\nu} = F_{\mu\alpha} F_\nu{}^\alpha - \frac{1}{4} g_{\mu\nu} F_{\alpha\beta} F^{\alpha\beta} + g_{\mu\nu} J_\alpha A^\alpha. \tag{2.229}$$

Exercise 8: Verify directly equation (2.157).

Exercise 9: Show that the metric (2.164) together with the gauge field configuration (2.166) with an arbitrary function $H(\vec{x})$ still solves the Einstein–Maxwell's equations provided $H(\vec{x})$ satisfies the Laplace equation

$$\vec{\nabla}^2 H = 0. \tag{2.230}$$

Exercise 10: Show result (2.178) explicitly.

Exercise 11: Show by using Frobenius's theorem that $\chi_{[\mu}\nabla_\nu\chi_{\sigma]} = 0$.

Exercise 12: Verify that the definition (2.190) of the surface gravity agrees with (2.184). Show that for a Schwarzschild black hole the surface gravity is given by $\kappa = 1/4GM$.

Exercise 13: Show that the angular velocity of the event horizon $r = r_+$ is given by

$$\Omega_H = \frac{a}{r_+^2 + a^2}. \tag{2.231}$$

Compute this velocity directly by computing the angular velocity of a photon emitted in the ϕ direction at some r in the equatorial plane $\theta = \pi/2$ in a Kerr black hole.

Exercise 14: Show that the ring singularity $x^2 + y^2 = a^2$ of the Kerr black hole is a true or naked singularity for $M \neq 0$, whereas it is only a coordinate singularity in the limit $M \longrightarrow 0$. Show, for example, that $R_{\mu\nu\alpha\beta}R^{\mu\nu\alpha\beta}$ diverges at $\rho = 0$.

References

[1] Carroll S M 2004 *Spacetime and Geometry: An Introduction to General Relativity* (San Francisco, CA: Addison-Wesley) p 513
[2] Wald R M 1984 *General Relativity* (Chicago, IL: University of Chicago Press) p 491
[3] 't Hooft G 2009 Introduction to The Theory of Black Holes https://webspace.science.uu.nl/~hooft101/lectures/blackholes/BH_lecturenotes.pdf
[4] Kerr R P 1963 Gravitational field of a spinning mass as an example of algebraically special metrics *Phys. Rev. Lett.* **11** 237–238

IOP Publishing

Lectures on General Relativity, Cosmology and Quantum Black Holes (Second Edition)

Badis Ydri

Chapter 3

Cosmology I: the expanding Universe

The modern science of cosmology is based on three basic observational results:
- The Universe, on very large scales, is homogeneous and isotropic.
- The Universe is expanding.
- The Universe is composed of: matter, radiation, dark matter and dark energy.

Some standard references for this chapter are [1–5]. In particular, we follow in the discussion of the various cosmological models and distances the elegant presentation of Carroll [1], while in the solution of the Friedmann equation as a potential problem we follow Hartle [2].

3.1 Homogeneity and isotropy

The Universe is expected to look exactly the same from every point in it. This is the content of the so-called Copernican principle. On the other hand, the Universe appears perfectly isotropic to us on Earth. Isotropy is the property that at every point in spacetime all spatial directions look the same, i.e. there are no preferred directions in space. The isotropy of the observed Universe is inferred from the cosmic microwave background (CMB) radiation, which is the most distant electromagnetic radiation originating at the time of decoupling, and which is observed at around 3 K, which is found to be isotropic to at least one part in a thousand by various experiments such as COBE, WMAP and PLANK.

The nine years of results of the Wilkinson Microwave Anisotropy Probe (WMAP) for the temperature distribution across the whole sky are shown in figure 3.1. See also [6]. The microwave background is very homogeneous in temperature with a mean of 2.7 K and relative variations from the mean of the order of 5×10^{-5} K. The temperature variations are presented through different colors with the 'red' being hotter (2.7281 K) while the 'blue' is colder 2.7279 K than

doi:10.1088/978-0-7503-5824-8ch3

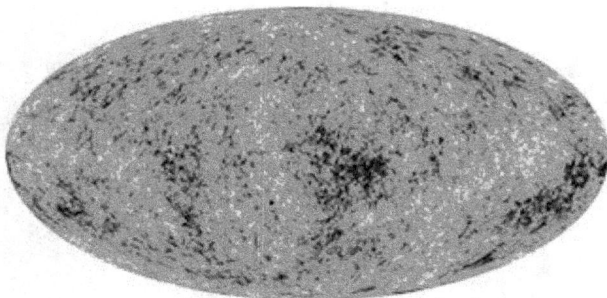

Figure 3.1. The all-sky map of the CMB. Credit: NASA WMAP Science Team.

the average. These fluctuations about isotropy are extremely important since they will lead, in the theory of inflation, by means gravitational interactions, to structure formation.

The Copernican principle together with the observed isotropy means in particular that the Universe on very large scales must look homogeneous and isotropic. Homogeneity is the property that all points of space look the same at every instant of time. This is the content of the so-called cosmological principle. Homogeneity is verified directly by constructing three-dimensional maps of the distribution of galaxies such as the 2-Degree-Field Galaxy Redshift survey (2dFGRS) and the Sloan Digital Sky survey (SDSS). See for example [7]. A slice through the SDSS 3-dimensional map of the distribution of galaxies with the Earth at the center is shown in figure 3.2.

3.2 Expansion and distances

3.2.1 Hubble law

The most fundamental fact about the Universe is its expansion. This can be characterized by the so-called scale factor $a(t)$. At the present time t_0 we set $a(t_0) = 1$. At earlier times, when the Universe was much smaller, the value of $a(t)$ was much smaller.

Spacetime can be viewed as a grid of points where the so-called comoving distance between the points remains constant with the expansion, since it is associated with the coordinates chosen on the grid, while the physical distance evolves with the expansion of the Universe linearly with the scale factor and the comoving distance, viz

$$\text{distance}_{\text{physical}} = a(t) \times \text{distance}_{\text{comoving}}. \tag{3.1}$$

In an expanding universe, galaxies are moving away from each other. Thus galaxies must be receding from us. Now, we know from the Doppler effect that the wavelength of sound or light emitted from a receding source is stretched out in the sense that the observed wavelength is larger than the emitted wavelength. Thus

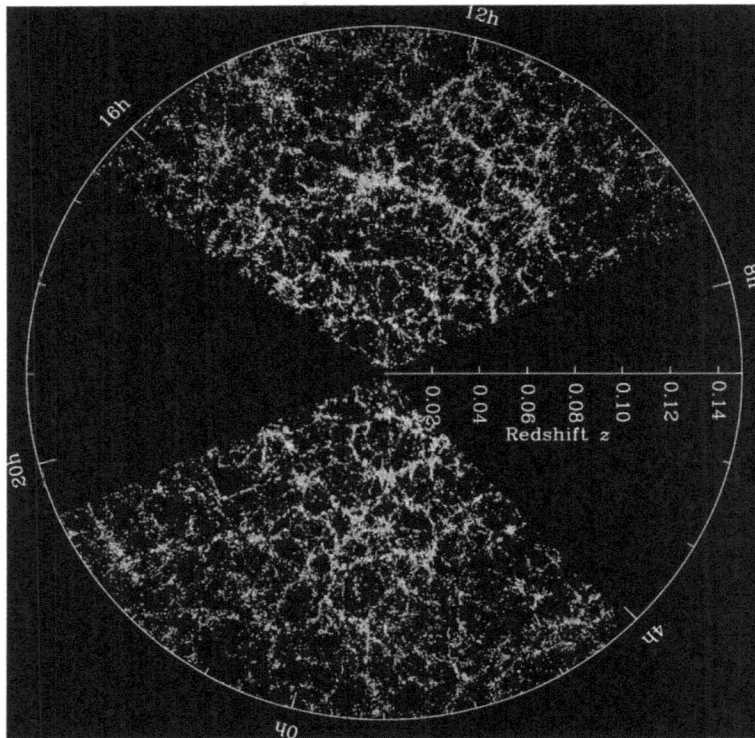

Figure 3.2. The Sloan Digital Sky Survey. Credit: M Blanton and the Sloan Digital Sky Survey.

the spectra of galaxies, since they are receding from us, must be redshifted. This can be characterized by the so-called redshift z defined by

$$1 + z = \frac{\lambda_{\text{obs}}}{\lambda_{\text{emit}}} \geqslant 1 \Leftrightarrow z = \frac{\Delta\lambda}{\lambda}. \tag{3.2}$$

For low redshifts $z \longrightarrow 0$, i.e. for sufficiently close galaxies with receding velocities much smaller than the speed of light, the standard Doppler formula must hold, viz

$$z = \frac{\Delta\lambda}{\lambda} \simeq \frac{v}{c}. \tag{3.3}$$

This allows us to determine the expansion velocities of galaxies by measuring the redshifts of absorption and emission lines. This was done originally by Hubble in 1929. He found a linear relation between the velocity v of recession and the distance d given by

$$v = H_0 d. \tag{3.4}$$

This is the celebrated Hubble law exhibited in figure 3.3. The constant H_0 is the Hubble constant given by the value [8]

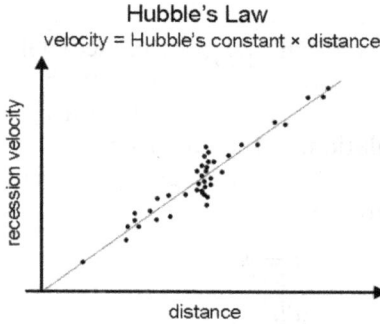

Figure 3.3. The Hubble law. Credit: NASA/Goddard Space Flgiht Center.

$$H_0 = 72 \pm 7 \text{ (km s}^{-1}\text{) Mpc}^{-1}. \tag{3.5}$$

The Mpc is megapersec which is the standard unit of distances in cosmology. We have

$$1 \text{ parsec(pc)} = 3.08 \times 10^{18}\text{cm} = 3.26 \text{ light-year}. \tag{3.6}$$

The Hubble law can also be seen as follows. Starting from the formula relating the physical distance to the comoving distance $d = ax$, and assuming no comoving motion $\dot{x} = 0$, we can show immediately that the relative velocity $v = \dot{d}$ is given by

$$v = Hd, \quad H = \frac{\dot{a}}{a}. \tag{3.7}$$

The Hubble constant sets essentially the age of the Universe by keeping a constant velocity. We get the estimate

$$t_H = \frac{1}{H_0} \sim 14 \text{ billion years}. \tag{3.8}$$

This is believed to be the time of the initial singularity known as the big bang where density, temperature and curvature were infinite.

3.2.2 Cosmic distances from standard candles

It is illuminating to start by noting the following distances:
- The distance to the edge of the observable Universe is 14 Gpc.
- The size of the largest structures in the Universe is around 100 Mpc.
- The distance to the nearest large cluster, the Virgo cluster which contains several thousands galaxies, is 20 Mpc.
- The distance to a typical galaxy in the Local Group which contains 30 galaxies is 50–1000 kpc. For example, Andromeda is 725 kpc away.
- The distance to the center of the Milky Way is 10 kpc.
- The distance to the nearest star is 1 pc.
- The distance to the Sun is 5 μpc.

But the fundamental question that one must immediately pose, given the immense expanses of the Universe, how do we come up with these numbers?

- **Triangulation:** We start with distances to nearby stars which can be determined using triangulation. The angular position of the star is observed from two points on the orbit of Earth giving two angles α and β, and as a consequence, the parallax p is given by

$$2p = \pi - \alpha - \beta. \tag{3.9}$$

For nearby stars the parallax p is a sufficiently small angle and thus the distance d to the star is given by (with a the semi-major axis of Earth's orbit)

$$d = \frac{a}{p}. \tag{3.10}$$

This method was used, by the Hipparchos satellite, to determine the distances to around 120 000 stars in the solar neighborhood.

- **Standard candles:** Most cosmological distances are obtained using the measurements of apparent luminosity of objects of supposedly known intrinsic luminosity. Standard candles are objects, such as stars and supernovae, whose intrinsic luminosity are determined from one of their physical properties, such as color or period, which itself is determined independently. Thus a standard candle is a source with known intrinsic luminosity.

  The intrinsic or absolute luminosity L, which is the energy emitted per unit time, of a star is related to its distance d, determined from triangulation, and to the flux l by the equation

$$L = l. \, 4\pi d^2. \tag{3.11}$$

The flux l is the apparent brightness or luminosity which is the energy received per unit time per unit area. By measuring the flux l and the distance d we can calculate the absolute luminosity L.

  Now, if all stars with a certain physical property, for example a certain blue color, and for which the distances can be determined by triangulation, turn out to have the same intrinsic luminosity, these stars will constitute standard candles. In other words, all blue color stars will be assumed to have the following luminosity:

$$L_{\text{blue color stars}} = l. \, 4\pi d^2_{\text{triangulation}}. \tag{3.12}$$

This means that for stars farther away with the same blue color, for which triangulation does not work, their distances can be determined by the above formula (3.11) assuming the same intrinsic luminosity (3.12) and only requiring the determination of their flux l at Earth, viz

$$d_{\text{blue color stars}} = \sqrt{\frac{L_{\text{blue color stars}}}{4\pi l}}. \tag{3.13}$$

Some of the standard candles are:
- **Main-sequence stars:** These are stars that still burn hydrogen at their cores producing helium through nuclear fusion. They obey a characteristic relation between absolute luminosity and color which both depend on the mass. For example, the luminosity is maximum for blue stars and minimum for red stars. The position of a star along the main sequence is essentially determined by its mass. This is summarized in a so-called Hertzsprung–Russell diagram which plots the intrinsic or absolute luminosity against its color index. An example is shown in figure 3.4.

  All main-sequence stars are in hydrostatic equilibrium since the outward thermal pressure from the hot core is exactly balanced by the inward pressure of gravitational collapse. The main-sequence stars with mass less than $0.23 M_O$ will evolve into white dwarfs, whereas those with mass less than $10 M_O$ will evolve into red giants. Those main-sequence stars with more mass will either gravitationlly collapse into black holes or explode into supernova.

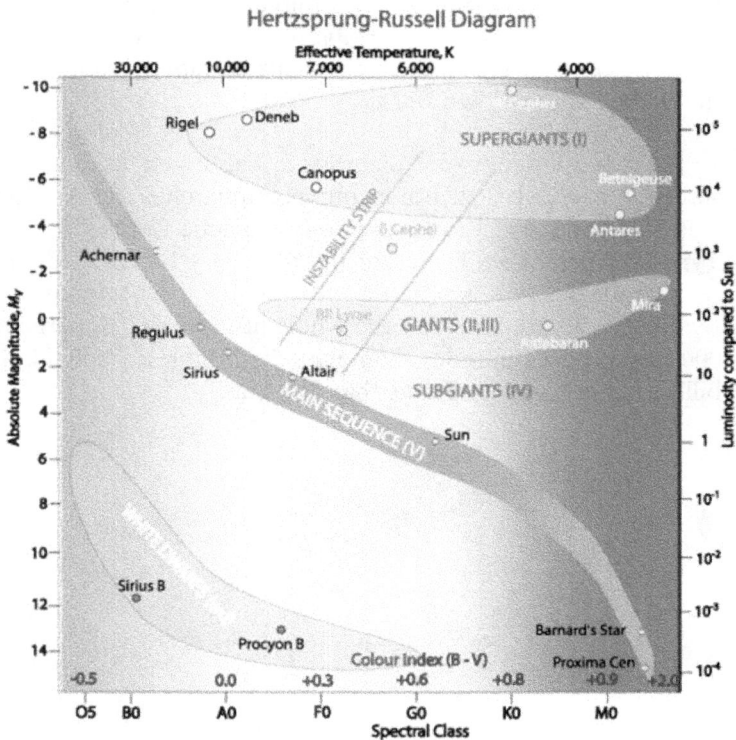

Figure 3.4. The Hertzsprung–Russell diagram including the instability strip. Reprinted with permission from CSIRO. Copyright CSIRO Australia.

The HR diagram of main-sequence stars is calibrated using triangulation: The absolute luminosity, for a given color, is measured by measuring the apparent luminosity and the distance from triangulation and then using the inverse square law (3.11).

By determining the luminosity class of a star, i.e. whether or not it is a main-sequence star, and determining its position on the HR diagram, i.e. its color, we can determine its absolute luminosity. This allows us to calculate its distance from us by measuring its apparent luminosity and using the inverse square law (3.11).

– **Cepheid variable stars:** These are massive, bright, yellow stars which arise in a post main-sequence phase of evolution with luminosity of up to 1000–10 000 times greater than that of the Sun. These stars are also pulsating, i.e. they grow and shrink in size with periods between 3 and 50 days. They are named after the δ Cephei star in the constellation Cepheus which is the first star of this kind. These stars lie in the so-called instability strip of the HR diagram.

The established strong correlation between the luminosity and the period of pulsation allows us to use Cepheid stars as standard candles. By determining the variability of a given Cepheid star, we can determine its absolute luminosity by determining its position on the period–luminosity diagram such as in figure 3.5. From this we can determine its distance from us by determining its apparent luminosity and using the inverse square law (3.11).

The period–luminosity diagram is calibrated using main-sequence stars and triangulation. For example, Hipparchos satellite had provided true parallaxes for a good sample of Galactic Cepheids.

– **Type Ia supernovae:** These are the only very far away discrete objects within galaxies that can be resolved due to their brightness which can rival even the brightness of the whole host galaxy. Supernovae are 100 000 times more luminous than even the brightest Cepheid, and several billion times more luminous than the Sun.

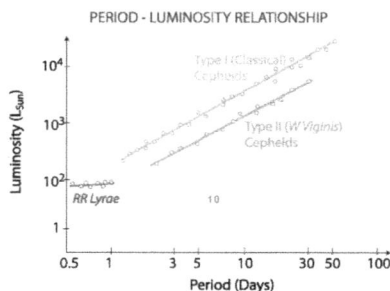

Figure 3.5. The period–luminosity relationship in pulsating Cepheid stars. Reprinted with permission from http://www.atnf.csiro.au/outreach/education/senior/astrophysics/. Copyright CSIRO Australia.

A type Ia supernova occurs when a white dwarf star in a binary system accretes sufficient matter from its companion until its mass reaches the Chandrasekhar limit, which is the maximum possible mass that can be supported by electron degeneracy pressure. The white dwarf then becomes unstable and explodes. These explosions are infrequent and even in a large galaxy only one supernova per century occurs on average.

The exploding white dwarf star in a supernova has always a mass close to the Chandrasekhar limit of 1.4 M_O and as a consequence all supernovae are basically the same, i.e. they have the same absolute luminosity. This absolute luminosity can be calculated by observing supernovae which occur in galaxies whose distances were determined using Cepheid stars. Then we can use this absolute luminosity to measure distances to even farther galaxies, for which Cepheid stars are not available, by observing supernovae in those galaxies and determining their apparent luminosities and using the inverse square law (3.11).

- **Cosmic distance ladder:** Triangulation and the standard candles discussed above: main-sequence stars, Cepheid variable stars and type Ia supernovae provide a cosmic distance ladder.

3.3 Matter, radiation, and vacuum

- **Matter:** This is in the form of stars, gas and dust held together by gravitational forces in bound states called galaxies. The iconic Hubble deep field image, which covers a tiny portion of the sky 1/30th the diameter of the full Moon, is perhaps the most conclusive piece of evidence that galaxies are the most important structures in the Universe. See figure 3.6. The observed

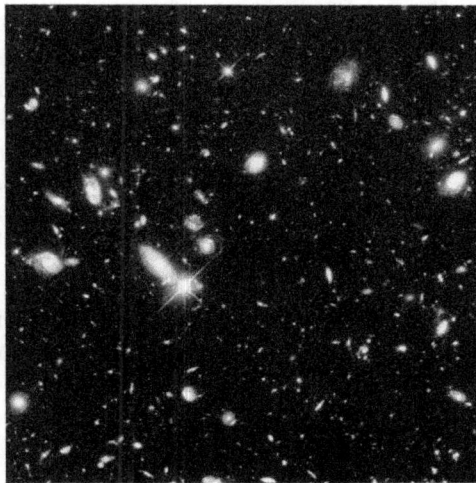

Figure 3.6. The Hubble deep field. Source. R Williams (STScI), HDF-S Team, NASA.

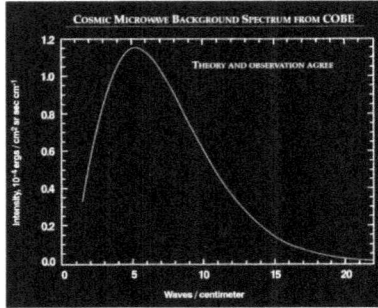

Figure 3.7. The black-body spectrum. Credit: NASA / COBE Science Team.

Universe may contain 10^{11} galaxies, each one contains around 10^{11} stars with a total mass of $10^{12} M_O$. The density of this visible matter is roughly given by

$$\rho_{\text{visible}} = 10^{-31} \text{g/cm}^{-3}. \tag{3.14}$$

- **Radiation**[1]: This consists of zero-mass particles such as photons, gravitons (gravitational waves) and in many circumstances (neutrinos) which are not obviously bound by gravitational forces. The most important example of radiation observed in the Universe is the CMB radiation with density given by

$$\rho_{\text{radiation}} = 10^{-34} g/\text{cm}^{-3}. \tag{3.15}$$

This is much smaller than the observed matter density since we are in a matter-dominated phase in the evolution of the Universe. This CMB radiation is an electromagnetic radiation left over from the hot big bang, and corresponds to a black-body spectrum with a temperature of $T = 2.725 \pm 0.001$ K [9]. See figure 3.7.

- **Dark matter:** This is the most important form of matter in the Universe in the sense that most mass in the Universe is not luminous (the visible matter) but dark, although its effect can still be seen from its gravitational effect.

    It is customary to dynamically measure the mass of a given galaxy by using Kepler's third law:

$$GM(r) = v^2(r)r. \tag{3.16}$$

In the above equation we have implicitly assumed spherical symmetry, $v(r)$ is the orbital (rotational) velocity of the galaxy at a distance r from the center, and $M(r)$ is the mass inside r. The plot of $v(r)$ as a function of the distance r is known as the rotation curve of the galaxy.

---

[1] Strictly speaking radiation should be included with matter.

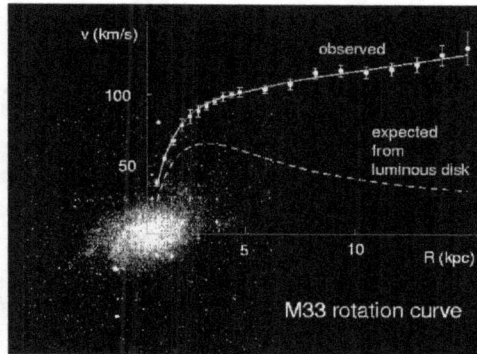

Figure 3.8. The galaxy rotation curve. This M33 rotation curve HI.gif image has been obtained by the author from the Wikimedia website https://commons.wikimedia.org/wiki/File:M33rotation_curve_HI.gif, where it is stated to have been released into the public domain. It is included within this book on that basis.

Applying this law to spiral galaxies, which are disks of stars and dust rotating about a central nucleus, taking r the radius of the galaxy, i.e. the radius within which much of the light emitted by the galaxy is emitted, one finds precisely the mass density $\rho_{\text{visible}} = 10^{-31} g/cm^{-3}$ quoted above. This is the luminous mass density since it is associated with the emission of light.

Optical observations are obviously limited due to the interstellar dust which does not allow the penetration of light waves. However, this problem does not arise when making radio measurements of atomic hydrogen. More precisely, neutral hydrogen (HI) atoms, which are abundant and ubiquitous in low density regions of the interstellar medium, are detectable in the 21 cm hyperfine line. This transition results from the magnetic interaction between the quantized electron and proton spins when the relative spins change from parallel to antiparallel.

Observations of the 21 cm line from neutral hydrogen regions in spiral galaxies can therefore be used to measure the speed of rotation of objects [10]. More precisely, since objects in galaxies are moving, they are Doppler shifted and the receiver can determine their velocities by comparing the observed wavelengths to the standard wavelength of 21 cm. By extending to distances beyond the point where light emitted from the Galaxy effectively ceases, one finds the behavior, shown in figure 3.8, which is given by

$$v \sim \text{constant.} \tag{3.17}$$

We would have expected that outside the radius of the galaxy, with the luminous matter providing the only mass, the velocity should have behaved as

$$v \sim 1/r^{1/2}. \tag{3.18}$$

The result (3.17) indicates that even in the outer region of the galaxies the mass behaves as

$$M(r) \sim r. \tag{3.19}$$

In other words, the mass always grows with r. We conclude that spiral galaxies, and in fact most other galaxies, contain dark, i.e. invisible, matter which permeates the galaxy and extends into the galaxy's halo with a density of at least 3–10 times the mass density of the visible matter, viz

$$\rho_{\text{halo}} = (3 - 10) \times \rho_{\text{visible}}. \qquad (3.20)$$

This form of matter is expected to be: (1) mostly nonbaryonic, (2) cold, i.e. non-relativistic during most of the Universe's history, so that structure formation is not suppressed, and (3) very weakly interacting since they are hard to detect. The most important candidate for dark matter is WIMPs (weakly interacting massive particles) such as the neutralinos, which are the lightest of the additional stable particles predicted by supersymmetry with mass around 100 GeV.

• **Dark energy:** This is speculated to be the energy of empty space, i.e. vacuum energy, and is the dominant component in the Universe: around 70%. The best candidate for dark energy is usually identified with the cosmological constant.

3.4 Flat universe

The simplest isotropic and homogeneous spacetime is the one in which the line element is given by

$$\begin{aligned} ds^2 &= -dt^2 + a^2(t)(dx^2 + dy^2 + dz^2) \\ &= -dt^2 + a^2(t)(dr^2 + r^2(d\theta^2 + \sin^2\theta d\phi^2)). \end{aligned} \qquad (3.21)$$

The function $a(t)$ is the scale factor. This is a flat universe. The homogeneity, isotropy and flatness are properties of the space and not spacetime.

The coordinate or comoving distance between any two points is given by

$$d_{\text{comoving}} = \sqrt{\Delta x^2 + \Delta y^2 + \Delta z^2}. \qquad (3.22)$$

This is for example the distance between any pair of galaxies. This distance is constant in time which can be seen as follows. Since we will view the distribution of galaxies as a smoothed out cosmological fluid, and thus a given galaxy is a particle in this fluid with coordinates x^i, the velocity dx^i/dt of the galaxy must vanish, otherwise it will provide a preferred direction contradicting the isotropy property. On the other hand, the physical distance between any two points depends on time and is given by

$$d_{\text{physical}}(t) = a(t)d_{\text{comoving}}. \qquad (3.23)$$

Clearly, if $a(t)$ increases with time then the physical distance $d_{\text{physical}}(t)$ must increase with time which is what happens in an expanding universe.

The energy of a particle moving in this spacetime will change similarly to the way that the energy of a particle moving in a time-dependent potential will change. For a photon this change in energy is precisely the cosmological redshift. The worldline of the photon satisfies

$$ds^2 = 0. \tag{3.24}$$

By assuming that we are at the origin of the spherical coordinates r, θ and ϕ, and that the photon is emitted in a galaxy a comoving distance $r = R$ away with a frequency ω_e at time t_e, and is received here at time $t = t_0$ with frequency ω_0, the worldline of the photon is therefore the radial null geodesics

$$ds^2 = -dt^2 + a^2(t)dr^2 = 0. \tag{3.25}$$

Integration yields immediately

$$R = \int_{t_e}^{t_0} \frac{dt}{a(t)}. \tag{3.26}$$

For a photon emitted at time $t_e + \delta t_e$ and observed at time $t_0 + \delta t_0$ we will have instead

$$R = \int_{t_e + \delta t_e}^{t_0 + \delta t_0} \frac{dt}{a(t)}. \tag{3.27}$$

Thus we get

$$\frac{\delta t_0}{a(t_0)} = \frac{\delta t_e}{a(t_e)}. \tag{3.28}$$

In particular, if δt_e is the period of the emitted light, i.e. $\delta t_e = 1/\nu_e$, the period of the observed light will be different given by $\delta t_0 = 1/\nu_0$. The relation between ν_e and ν_0 defines the redshift z through

$$1 + z = \frac{\lambda_0}{\lambda_e} = \frac{\nu_e}{\nu_0} = \frac{a(t_0)}{a(t_e)}. \tag{3.29}$$

This can be rewritten as

$$z = \frac{\Delta\lambda}{\lambda} = \frac{a(t_0) - a(t_e)}{a(t_e)} = \frac{\dot{a}(t_e)}{a(t_e)}(t_e - t_0) + \cdots \tag{3.30}$$

The physical distance d is related to the comoving distance R by $d = a(t_0)R$. By assuming that R is small we have from $ds^2 = 0$ the result

$$t_e - t_0 = \int_0^R a(t)dr = a(t_0)R + O(R^2). \tag{3.31}$$

Thus

$$z = \frac{\Delta\lambda}{\lambda} = \frac{\dot{a}(t_0)}{a(t_0)}d + \cdots \tag{3.32}$$

This is Hubble law. The Hubble constant is

$$H_0 = \frac{\dot{a}(t_0)}{a(t_0)}. \tag{3.33}$$

The Hubble time t_H and the Hubble distance d_H are defined by

$$t_H = \frac{1}{H_0}, \quad d_H = c t_H. \tag{3.34}$$

The line element (3.21) is called the flat Robertson–Walker metric, and when the scale factor $a(t)$ is specified via Einstein's equations, it is called the flat Friedman–Robertson–Walker metric. The time evolution of the scale factor $a(t)$ is controled by the Friedman equation

$$\frac{\dot{a}^2}{a^2} = \frac{8\pi G\rho}{3}. \tag{3.35}$$

For a detailed derivation starting from Einstein equations see next sections 3.8 and 3.9. ρ is the total mass–energy density. At the present time this equation gives the relation between the Hubble constant H_0 and the critical mass density ρ_c given by

$$H_0^2 \equiv \frac{\dot{a}^2(t_0)}{a^2(t_0)} = \frac{8\pi G\rho(t_0)}{3} \Rightarrow \rho(t_0) = \frac{3H_0^2}{8\pi G} \equiv \rho_c. \tag{3.36}$$

We can choose, without any loss of generality, $a(t_0) = 1$. We have the following numerical values

$$H_0 = 100h \text{ (km s}^{-1}) \text{ Mpc}^{-1}, \quad t_H = 9.78h^{-1} \text{ Gyr},$$
$$d_H = 2998h^{-1} \text{ Mpc}, \quad \rho_c = 1.88 \times 10^{-29}h^2 \text{ g cm}^{-3}. \tag{3.37}$$

The matter, radiation and vacuum contributions to the critical mass density are given by the fractions

$$\Omega_M = \frac{\rho_M(t_0)}{\rho_c}, \quad \Omega_R = \frac{\rho_R(t_0)}{\rho_c}, \quad \Omega_V = \frac{\rho_V(t_0)}{\rho_c}. \tag{3.38}$$

Obviously

$$1 = \Omega_M + \Omega_R + \Omega_V. \tag{3.39}$$

The generalization of this equation to $t \neq t_0$ is given by

$$\rho(a) = \rho_c \left(\frac{\Omega_M}{a^3} + \frac{\Omega_R}{a^4} + \Omega_V \right). \tag{3.40}$$

By using this last equation in the Friedmann equation we get the equivalent equation

$$\frac{1}{2H_0^2}\dot{a}^2 + V_{\text{eff}}(a) = 0, \quad V_{\text{eff}}(a) = -\frac{1}{2}\left(\frac{\Omega_M}{a} + \frac{\Omega_R}{a^2} + a^2\Omega_V \right). \tag{3.41}$$

This is effectively the equation of motion of a zero-energy particle moving in one dimension under the influence of the potential $V_{\text{eff}}(a)$. The three possible distinct solutions are:

- **Matter-dominated Universe:** In this case $\Omega_M = 1$, $\Omega_R = \Omega_V = 0$ and thus

$$V_{\text{eff}}(a) = -\frac{1}{2a} \Rightarrow \frac{1}{2H_0^2}\dot{a}^2 - \frac{1}{2a} = 0 \Rightarrow a = \left(\frac{t}{t_0}\right)^{2/3}, \quad t_0 = \frac{2}{3H_0}. \tag{3.42}$$

- **Radiation-dominated Universe:** In this case $\Omega_R = 1$, $\Omega_M = \Omega_V = 0$ and thus

$$V_{\text{eff}}(a) = -\frac{1}{2a^2} \Rightarrow \frac{1}{2H_0^2}\dot{a}^2 - \frac{1}{2a^2} = 0 \Rightarrow a = \left(\frac{t}{t_0}\right)^{1/2}, \quad t_0 = \frac{1}{2H_0}. \tag{3.43}$$

In this case, as well as in the matter-dominated case, the Universe starts at $t = 0$ with $a = 0$ and thus $\rho = \infty$, and then expands forever. This physical singularity is what we mean by the Big Bang. Here the expansion is decelerating since the potentials $-1/2a$ and $-1/2a^2$ increase without limit from $-\infty$ to 0, as a increases from 0 to ∞, and thus corresponds to kinetic energies $1/2a$ and $1/2a^2$ which decrease without limit from $+\infty$ to 0 over the same range of a.

- **Vacuum-dominated Universe:** In this case $\Omega_V = 1$, $\Omega_M = \Omega_R = 0$ and thus

$$V_{\text{eff}}(a) = -\frac{a^2}{2} \Rightarrow \frac{1}{2H_0^2}\dot{a}^2 - \frac{a^2}{2} = 0 \Rightarrow a = \exp(H_0(t - t_0)), \quad H_0^2 = c^4\frac{\Lambda}{3}. \tag{3.44}$$

In this case the Hubble constant is truely a constant for all times.

In the actual evolution of the Universe the three effects are present. The addition of the vacuum energy results typically in a maximum in the potential $V_{\text{eff}}(a)$ when plotted as a function of a. Thus the Universe is initially in a decelerating expansion phase consisting of radiation-dominated and matter-dominated regions, then it becomes a vacuum-dominated with an accelerating expansion. This is because beyond the maximum the potential becomes a decreasing function of a and as a consequence the kinetic energy is an increasing function of a.

3.5 Closed and open universes

There are two more possible Friedman–Robertson–Walker universes, besides the flat case, which are isotropic and homogeneous. These are the closed universe given by a 3-sphere and the open universe given by a 3-hyperboloid. The spacetime metric in the three cases is given by

$$ds^2 = -dt^2 + a^2(t)dl^2. \tag{3.45}$$

The spatial metric in the flat case can be rewritten as (with $\chi \equiv r$)

$$dl^2 = d\chi^2 + \chi^2(d\theta^2 + \sin^2\theta d\phi^2). \tag{3.46}$$

Now we discuss the other two cases.

Closed FRW universe: A 3-sphere can be embedded in R^4 in the usual way by

$$X^2 + Y^2 + Z^2 + W^2 = 1. \tag{3.47}$$

We introduce spherical coordinates $0 \leqslant \theta \leqslant \pi$, $0 \leqslant \phi \leqslant 2\pi$ and $0 \leqslant \chi \leqslant \pi$ by

$$X = \sin\chi \sin\theta \cos\phi, \quad Y = \sin\chi \sin\theta \sin\phi, \quad Z = \sin\chi \cos\theta, \quad W = \cos\chi. \tag{3.48}$$

The line element on the 3-sphere is given by

$$\begin{aligned} dl^2 &= (dX^2 + dY^2 + dZ^2 + dW^2)_{S^3} \\ &= d\chi^2 + \sin^2\chi(d\theta^2 + \sin^2\theta d\phi^2). \end{aligned} \tag{3.49}$$

This is a closed space with finite volume and without boundary. The comoving volume is given by

$$\begin{aligned} dV &= \int \sqrt{\det g}\, d^4X \\ &= \int_0^{2\pi} d\phi \int_0^{\pi} d\theta \sin\theta \int_0^{\pi} d\chi \sin^2\chi \\ &= 2\pi^2. \end{aligned} \tag{3.50}$$

The physical volume is of course given by $dV(t) = a^3(t)dV$.

Open FRW universe: A 3-hyperboloid is a 3-surface in a Minkowski spacetime M^4 analogous to a 3-sphere in R^4. It is embedded in M^4 by the relation

$$X^2 + Y^2 + Z^2 - T^2 = -1. \tag{3.51}$$

We introduce hyperbolic coordinates $0 \leqslant \theta \leqslant \pi$, $0 \leqslant \phi \leqslant 2\pi$ and $0 \leqslant \chi \leqslant \infty$ by

$$\begin{aligned} X &= \sinh\chi \sin\theta \cos\phi, \quad Y = \sinh\chi \sin\theta \sin\phi, \\ Z &= \sinh\chi \cos\theta, \quad T = \cosh\chi. \end{aligned} \tag{3.52}$$

The line element on this 3-surface is given by

$$\begin{aligned} dl^2 &= (dX^2 + dY^2 + dZ^2 - dT^2)_{H^3} \\ &= d\chi^2 + \sinh^2\chi(d\theta^2 + \sin^2\theta d\phi^2). \end{aligned} \tag{3.53}$$

This is an open space with infinite volume.

The three metrics (3.46), (3.49) and (3.53) can be rewritten collectively as

$$dl^2 = \frac{dr^2}{1 - kr^2} + r^2(d\theta^2 + \sin^2\theta d\phi^2). \tag{3.54}$$

The variable r and the parameter k, called the spatial curvature, are given by

$$r = \sin\chi, \quad k = +1: \quad \text{closed}. \tag{3.55}$$

$$r = \chi, \quad k = 0: \quad \text{flat.} \tag{3.56}$$

$$r = \sinh \chi, \quad k = -1: \quad \text{open.} \tag{3.57}$$

The metric of spacetime is thus given by

$$ds^2 = -dt^2 + a^2(t) \left[\frac{dr^2}{1 - kr^2} + r^2(d\theta^2 + \sin^2 \theta d\phi^2) \right]. \tag{3.58}$$

Thus the open and closed cases are characterized by a non-zero spatial curvature. As before, the scale factor must be given by the Friedman equation. This is given by

$$\frac{\dot{a}^2}{a^2} = \frac{8\pi G\rho}{3} - \frac{kc^2}{a^2}. \tag{3.59}$$

At $t = t_0$ we get

$$H_0^2 = \frac{8\pi G\rho(t_0)}{3} - \frac{kc^2}{a^2(t_0)} \Rightarrow \rho(t_0) - \rho_c = \frac{3kc^2}{8\pi Ga^2(t_0)}. \tag{3.60}$$

The critical density is of course defined by

$$\rho_c = \frac{3H_0^2}{8\pi G}. \tag{3.61}$$

Thus for a closed universe the spacetime is positively curved and as a consequence the current energy density is larger than the critical density, i.e. $\Omega = \rho(t_0)/\rho_c > 1$, whereas for an open universe the spacetime is negatively curved and as a consequence the current energy density is smaller than the critical density, i.e. $\Omega = \rho(t_0)/\rho_c < 1$. Only for a flat universe the current energy density is equal to the critical density, i.e. $\Omega = \rho(t_0)/\rho_c = 1$. The above equation can also be rewritten as

$$\Omega = 1 + \frac{kc^2}{H_0^2 a^2(t_0)}. \tag{3.62}$$

Equivalently

$$\Omega_M + \Omega_R + \Omega_V + \Omega_C = 1 \Rightarrow \Omega_C = 1 - \Omega_M - \Omega_R - \Omega_V. \tag{3.63}$$

The density parameter Ω_C associated with the spatial curvature is defined by

$$\Omega_C = -\frac{kc^2}{H_0^2 a^2(t_0)}. \tag{3.64}$$

We use now the formula

$$\rho(t) = \rho_c \Omega(t)$$

$$= \rho_c \left(\frac{\Omega_M}{\tilde{a}^3(t)} + \frac{\Omega_R}{\tilde{a}^4(t)} + \Omega_V \right), \quad \tilde{a}(t) = \frac{a(t)}{a(t_0)}. \tag{3.65}$$

The Friedman equation can then be put in the form (with $\tilde{t} = t/t_H = H_0 t$)

$$\frac{1}{2}\left(\frac{d\tilde{a}}{d\tilde{t}}\right)^2 + V_{\text{eff}}(\tilde{a}) = \frac{\Omega_C}{2}. \tag{3.66}$$

$$V_{\text{eff}}(\tilde{a}) = -\frac{1}{2}\left(\frac{\Omega_M}{\tilde{a}} + \frac{\Omega_R}{\tilde{a}^2} + \tilde{a}^2\Omega_V\right). \tag{3.67}$$

We need to solve (3.63), (3.66) and (3.67). This is a generalization of the potential problem (3.41) corresponding to the flat FRW model to the generic curved FRW models. This is effectively the equation of motion of a particle moving in one dimension under the influence of the potential $V_{\text{eff}}(\tilde{a})$ with energy $\Omega_C/2$. There are therefore four independent cosmological parameters Ω_M, Ω_R, Ω_V and H_0. The solution of the above equation determines the scale factor $a(t)$ as well as the present age t_0.

There are two general features worthy of mention here:

- **Open and flat:** In this case $\Omega \leqslant 1$ and thus $\Omega_C = 1 - \Omega \geqslant 0$. From the other hand, $V_{\text{eff}} < 0$. Thus V_{eff} is strictly below the line $\Omega_C/2$. In other words, there are no turning points where 'the total energy' $\Omega_C/2$ becomes equal to the 'potential energy' V_{eff}, i.e. 'the kinetic energy' $\dot{a}^2/2$ never vanishes and thus we never have $\dot{a} = 0$. The Universe starts from a Big Bang singularity at $a = 0$ and keeps expanding forever.
- **Closed:** In this case $\Omega > 1$ and thus $\Omega_C = 1 - \Omega < 0$. There are here two scenarios:
    - The potential is strictly below the line $\Omega_C/2$ and thus there are no turning points. The Universe starts from a Big Bang singularity at $a = 0$ and keeps expanding forever.
    - The potential intersects the line $\Omega_C/2$. There are two turning points given by the intersection points. We have two possibilities depending on where $\tilde{a} = 1$ is located below the smaller turning point or above the larger turning point.
        * $\tilde{a} = 1$ is below the smaller turning point. The Universe starts from a Big Bang singularity at $a = 0$, expands to a maximum radius corresponding to the smaller turning point, then recollapses to a big crunch singularity at $a = 0$.
        * $\tilde{a} = 1$ is above the larger turning point. The Universe collapses from a larger value of a, it bounces when it hits the largest turning point and then re-expands forever. There is no singularity in this case. This case is ruled out by current observations.

For an FRW universe dominated by matter and vacuum like ours the above possibilities are sketched in the plane of the least certain cosmological parameters Ω_M and Ω_V in figure 3.9. Flat FRW models are on the line $\Omega_V = 1 - \Omega_M$. Open models lie below this line while closed models lie above it. The solid lines are from

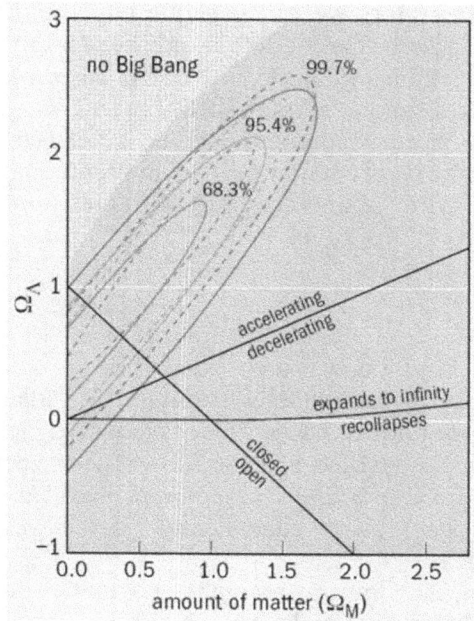

Figure 3.9. The Friedmann–Lemaître–Robertson–Walker models in the $\Omega_M - \Omega_V$ plane. Reproduced from [13], copyright IOP Publishing. All rights reserved.

data of the High-Z Supernova Search [11] while the dashed lines are from data of the Supernova Cosmology Project [12].

3.6 Aspects of the early Universe

The most central property of the Universe is expansion. The evidence for the expansion of the Universe comes from three main sets of observations. Firstly, light from distant galaxies is shifted towards the red which can be accounted for by the expansion of the Universe. Secondly, the observed abundance of light elements can be calculated from Big Bang nucleosynthesis (BBN). Thirdly, the CMB radiation can be interpreted as the afterglow of a hot early Universe. The temperature of the Universe at any instant of time t is inversely proportional to the scale factor $a(t)$, viz

$$T \propto \frac{1}{a(t)}. \tag{3.68}$$

The early Universe is obviously radiation-dominated because of the relativistic energies involved. During this era the temperature is related to time by

$$\frac{t}{1 \text{ s}} = \left(\frac{10^{10} \text{ K}}{T} \right)^2. \tag{3.69}$$

In particle physics accelerators we can generate temperatures up to $T = 10^{15}$ K which means that we can probe the conditions of the early Universe down to 10^{-10} s. From 10^{-10} s to today the history of the Universe is based on well-understood and well-tested physics. For example at 1 s the BBN begins where light nuclei start to form, and at 10^4 years matter–radiation equality is reached where the density of photons drops below that of matter. After matter–radiation equality, which corresponds to a scale factor of about $a = 10^{-4}$, the relation between temperature and time changes to

$$t \propto \frac{1}{T^{3/2}}. \tag{3.70}$$

The Universe after the Big Bang was a hot and dense plasma of photons, electrons and protons which was very opaque to electromagnetic radiation. As the Universe expanded it cooled down until it reached the stage where the formation of neutral hydrogen was energetically favored and the ratio of free electrons and protons to neutral hydrogen decreased to 1/10000. This event is called recombination and it occurred at around $T \simeq 0.3$ eV or equivalently 378 000 years ago, which corresponds to a scale factor $a = 1/1200$.

After recombination, the Universe becomes fully matter-dominated, and shortly after recombination, photons decouple from matter and as a consequence the mean free path of photons approaches infinity. In other words after photon decoupling the Universe becomes effectively transparent. These photons are seen today as the CMB radiation. The decoupling period is also called the surface of last scattering.

3.7 Concordance model

From a combination of CMB and large-scale structure (LSS) observations we deduce that the Universe is spatially flat and is composed of 4% ordinary mater, 23% dark matter and 73% dark energy (vaccum energy or cosmological constant Λ), i.e.

$$\Omega_k \sim 0. \tag{3.71}$$

$$\Omega_M \sim 0.04, \quad \Omega_{DM} \sim 0.23, \quad \Omega_\Lambda \sim 0.73. \tag{3.72}$$

3.8 Friedmann–Lemaître–Robertson–Walker metric

The Universe on very large scales is homogeneous and isotropic. This is the cosmological principle.

A spatially isotropic spacetime is a manifold in which there exists a congruence of time-like curves representing observers with tangents u^a such that given any two unit spatial tangent vectors s_1^a and s_2^a at a point p, orthogonal to u^a, there exists an isometry of the metric g_{ab} which rotates s_1^a into s_2^a while leaving p and u^a fixed. The fact that we can rotate s_1^a into s_2^a means that there is no preferred direction in space.

On the other hand, a spacetime is spatially homogeneous if there exists a foliation of spacetime, i.e. a one-parameter family of space-like hypersurfaces Σ_t foliating spacetime such that any two points p, $q \in \Sigma_t$ can be connected by an isometry of the metric g_{ab}. The surfaces of homogeneity Σ_t are orthogonal to the isotropic observers with tangents u^a and they must be unique. In flat spacetime the isotropic observers and the surfaces of homogeneity are not unique.

A manifold can be homogeneous but not isotropic such as $R \times S^2$ or it can be isotropic around a point but not homogeneous such as the cone around its vertex. However, a spacetime which is isotropic everywhere must be also homogeneous, and a spacetime which is isotropic at a point and homogeneous must be isotropic everywhere.

The requirement of spatial isotropy and homogeneity of spacetime means that there exists a foliation of spacetime consisting of 3D maximally symmetric spatial slices Σ_t. The Universe is therefore given by the manifold $R \times \Sigma$ with metric

$$ds^2 = -c^2 dt^2 + R^2(t) d\vec{u}^2. \tag{3.73}$$

The metric on Σ is given by

$$d\sigma^2 = d\vec{u}^2 = \gamma_{ij} du^i du^j. \tag{3.74}$$

The scale factor $R(t)$ gives the volume of the spatial slice Σ at the instant of time t. The coordinates t, u^1, u^2 and u^3 are called comoving coordinates. An observer whose spatial coordinates u^i remain fixed is a comoving observer. Obviously, the Universe can look isotropic only with respect to a comoving observer. It is obvious that the relative distance between particles at fixed spatial coordinates grows with time t as $R(t)$. These particles draw worldlines in spacetime which are said to be comoving. Similarly, a comoving volume is a region of space which expands along with its boundaries defined by fixed spatial coordinates with the expansion of the Universe.

A maximally symmetric metric is certainly a spherically symmetric metric. Recall that the metric $d\vec{x}^2 = dx^2 + dy^2 + dz^2$ of the flat 3D space in spherical coordinates is $d\vec{x}^2 = dr^2 + r^2 d\Omega^2$ where $d\Omega^2 = d\theta^2 + \sin^2\theta d\phi^2$. A general 3D metric with spherical symmetry is therefore necessarily of the form

$$d\vec{u}^2 = e^{2\beta(r)} dr^2 + r^2 d\Omega^2. \tag{3.75}$$

The Christoffel symbols are computed to be given by

$$\Gamma^r_{rr} = \partial_r\beta, \quad \Gamma^r_{\theta\theta} = -re^{-2\beta(r)}, \quad \Gamma^r_{\phi\phi} = -r\sin^2\theta e^{-2\beta(r)}, \quad \Gamma^r_{r\theta} = \Gamma^r_{r\phi} = \Gamma^r_{\theta\phi} = 0. \tag{3.76}$$

$$\Gamma^\theta_{r\theta} = \frac{1}{r}, \quad \Gamma^\theta_{\phi\phi} = -\sin\theta\cos\theta, \quad \Gamma^\theta_{rr} = \Gamma^\theta_{r\phi} = \Gamma^\theta_{\theta\theta} = \Gamma^\theta_{\theta\phi} = 0. \tag{3.77}$$

$$\Gamma^\phi_{r\phi} = \frac{1}{r}, \quad \Gamma^\phi_{\theta\phi} = \frac{\cos\theta}{\sin\theta}, \quad \Gamma^\phi_{rr} = \Gamma^\phi_{r\theta} = \Gamma^\phi_{\theta\theta} = \Gamma^\phi_{\phi\phi} = 0. \tag{3.78}$$

The Ricci tensor is then given by

$$R_{rr} = \frac{2}{r}\partial_r\beta. \tag{3.79}$$

$$R_{r\theta} = 0, \quad R_{r\phi} = 0. \tag{3.80}$$

$$R_{\theta\theta} = 1 + e^{-2\beta}(r\partial_r\beta - 1). \tag{3.81}$$

$$R_{\theta\phi} = 0. \tag{3.82}$$

$$R_{\phi\phi} = \sin^2\theta[1 + e^{-2\beta}(r\partial_r\beta - 1)]. \tag{3.83}$$

The above spatial metric is a maximally symmetric metric. Hence, we know that the 3D Riemann curvature tensor must be of the form

$$R^{(3)}_{ijkl} = \frac{R^{(3)}}{3(3-1)}(\gamma_{ik}\gamma_{jl} - \gamma_{il}\gamma_{jk}). \tag{3.84}$$

In other words, the Ricci tensor is actually given by

$$R^{(3)}_{ik} = (R^{(3)})_{ijk}{}^{j} = R^{(3)}_{ijkl}\gamma^{lj} = \frac{R^{(3)}}{3}\gamma_{ik}. \tag{3.85}$$

By comparison we get the two independent equations (with $k = R^{(3)}/6$)

$$2ke^{2\beta} = \frac{2}{r}\partial_r\beta. \tag{3.86}$$

$$2kr^2 = 1 + e^{-2\beta}(r\partial_r\beta - 1). \tag{3.87}$$

From the first equation we determine that the solution must be such that $\exp(-2\beta) = -kr^2 +$ constant, whereas from the second equation we determine that constant=1. We then get the solution

$$\beta = -\frac{1}{2}\ln(1 - kr^2). \tag{3.88}$$

The spatial metric becomes

$$d\vec{u}^2 = \frac{dr^2}{1 - kr^2} + r^2 d\Omega^2. \tag{3.89}$$

The constant k is proportional to the scalar curvature, which can be positive, negative or 0. It also obviously sets the size of the spatial slices. Without any loss of generality we can normalize it such that $k = +1, 0, -1$ since any other scale can be absorbed into the scale factor $R(t)$ which multiplies the length $|d\vec{u}|$ in the formula for ds^2.

We introduce a new radial coordinate χ by the formula

$$d\chi = \frac{dr}{\sqrt{1 - kr^2}}. \tag{3.90}$$

By integrating both sides we obtain

$$\begin{aligned}
r &= \sin\chi, \quad k = +1 \\
r &= \chi, \quad k = 0 \\
r &= \sinh\chi, \quad k = -1.
\end{aligned} \tag{3.91}$$

Thus the metric becomes

$$\begin{aligned}
d\vec{u}^2 &= d\chi^2 + \sin^2\chi\, d\Omega^2, \quad k = +1 \\
d\vec{u}^2 &= d\chi^2 + \chi^2 d\Omega^2, \quad k = 0 \\
d\vec{u}^2 &= d\chi^2 + \sinh^2\chi\, d\Omega^2, \quad k = -1.
\end{aligned} \tag{3.92}$$

The physical interpretation of this result is as follows:
- The case $k = +1$ corresponds to a constant positive curvature on the manifold Σ and is called closed. We recognize the metric $d\vec{u}^2 = d\chi^2 + \sin^2\chi\, d\Omega^2$ to be that of a 3-sphere, i.e. $\Sigma = S^3$. This is obviously a closed sphere in the same sense that the 2-sphere S^2 is closed.
- The case $k = 0$ corresponds to 0 curvature on the manifold Σ and as such is naturally called flat. In this case the metric $d\vec{u}^2 = d\chi^2 + \chi^2 d\Omega^2$ corresponds to the flat 3D Euclidean space, i.e. $\Gamma = R^3$.
- The case $k = -1$ corresponds to a constant negative curvature on the manifold Σ and is called open. We recognize the metric $d\vec{u}^2 = d\chi^2 + \sinh^2\chi\, d\Omega^2$ to be that of a 3D hyperboloid, i.e. $\Sigma = H^3$. This is an open space.

The so-called Robertson–Walker metric on a spatially homogeneous and spatially isotropic spacetime is therefore given by

$$ds^2 = -c^2 dt^2 + R^2(t)\left[\frac{dr^2}{1 - kr^2} + r^2 d\Omega^2\right]. \tag{3.93}$$

3.9 Friedmann equations

3.9.1 The first Friedmann equation

The scale factor $R(t)$ has units of distance and thus r is actually dimensionless. We reinstate a dimensionful radius ρ by $\rho = R_0 r$. The scale factor becomes dimensionless given by $a(t) = R(t)/R_0$, whereas the curvature becomes dimensionful $\kappa = k/R_0^2$. The Robertson–Walker metric becomes

$$ds^2 = -c^2 dt^2 + a^2(t)\left[\frac{d\rho^2}{1 - \kappa\rho^2} + \rho^2 d\Omega^2\right]. \tag{3.94}$$

The non-zero components of the metric are $g_{00} = -1$, $g_{\rho\rho} = a^2/(1 - \kappa\rho^2)$, $g_{\theta\theta} = a^2\rho^2$, $g_{\phi\phi} = a^2\rho^2 \sin^2\theta$. We compute now the non-zero Christoffel symbols

$$\Gamma^0_{\rho\rho} = \frac{a\dot{a}}{c(1 - \kappa\rho^2)}, \quad \Gamma^0_{\theta\theta} = \frac{a\dot{a}\rho^2}{c}, \quad \Gamma^0_{\phi\phi} = \frac{a\dot{a}\rho^2 \sin^2\theta}{c}. \tag{3.95}$$

$$\Gamma^\rho_{0\rho} = \frac{\dot{a}}{ca}, \quad \Gamma^\rho_{\rho\rho} = \frac{\kappa\rho}{1 - \kappa\rho^2}, \quad \Gamma^\rho_{\theta\theta} = -\rho(1 - \kappa\rho^2),$$

$$\Gamma^\rho_{\phi\phi} = -\rho(1 - \kappa\rho^2)\sin^2\theta. \tag{3.96}$$

$$\Gamma^\theta_{0\theta} = \frac{\dot{a}}{ca}, \quad \Gamma^\theta_{\rho\theta} = \frac{1}{\rho}, \quad \Gamma^\theta_{\phi\phi} = -\sin\theta\cos\theta. \tag{3.97}$$

$$\Gamma^\phi_{0\phi} = \frac{\dot{a}}{ca}, \quad \Gamma^\phi_{\rho\phi} = \frac{1}{\rho}, \quad \Gamma^\phi_{\theta\phi} = \frac{\cos\theta}{\sin\theta}. \tag{3.98}$$

The non-zero components of the Ricci tensor are

$$R_{00} = -\frac{3}{c^2}\frac{\ddot{a}}{a}. \tag{3.99}$$

$$R_{\rho\rho} = \frac{1}{c^2(1 - \kappa\rho^2)}(a\ddot{a} + 2\dot{a}^2 + 2\kappa c^2). \tag{3.100}$$

$$R_{\theta\theta} = \frac{\rho^2}{c^2}(a\ddot{a} + 2\dot{a}^2 + 2\kappa c^2), \quad R_{\phi\phi} = \frac{\rho^2 \sin^2\theta}{c^2}(a\ddot{a} + 2\dot{a}^2 + 2\kappa c^2). \tag{3.101}$$

The scalar curvature is therefore given by

$$R = g^{\mu\nu}R_{\mu\nu} = \frac{6}{c^2}\left(\frac{\ddot{a}}{a} + \left(\frac{\dot{a}}{a}\right)^2 + \frac{\kappa c^2}{a^2}\right). \tag{3.102}$$

Einstein's equations are

$$R_{\mu\nu} = \frac{8\pi G}{c^4}\left(T_{\mu\nu} - \frac{1}{2}g_{\mu\nu}T\right). \tag{3.103}$$

The stress–energy–momentum tensor

$$T^{\mu\nu} = \left(\rho + \frac{P}{c^2}\right)U^\mu U^\nu + Pg^{\mu\nu}. \tag{3.104}$$

The fluid is obviously at rest in comoving coordinates. In other words, $U^\mu = (c, 0, 0, 0)$ and hence

$$T^{\mu\nu} = \text{diag}(\rho c^2, Pg^{11}, Pg^{22}, Pg^{33}) \Rightarrow T_\mu^{\ \lambda} = \text{diag}(-\rho c^2, P, P, P). \tag{3.105}$$

Thus $T = T_\mu{}^\mu = -\rho c^2 + 3P$. The $\mu = 0$, $\nu = 0$ component of Einstein's equations is

$$R_{00} = \frac{8\pi G}{c^4}\left(T_{00} + \frac{1}{2}T\right) \Rightarrow -3\frac{\ddot{a}}{a} = 4\pi G\left(\rho + 3\frac{P}{c^2}\right). \qquad (3.106)$$

The $\mu = \rho$, $\nu = \rho$ component of Einstein's equations is

$$R_{\rho\rho} = \frac{8\pi G}{c^4}\left(T_{\rho\rho} - \frac{1}{2}g_{\rho\rho}T\right) \Rightarrow a\ddot{a} + 2\dot{a}^2 + 2\kappa c^2 = 4\pi G\left(\rho - \frac{P}{c^2}\right)a^2. \qquad (3.107)$$

There are no other independent equations. The Einstein's equation (3.106) is known as the second Friedmann equation. This is given by

$$\frac{\ddot{a}}{a} = -\frac{4\pi G}{3}\left(\rho + 3\frac{P}{c^2}\right). \qquad (3.108)$$

Using this result in Einstein's equation (3.107) yields immediately the first Friedmann equation. This is given by

$$\frac{\dot{a}^2}{a^2} = \frac{8\pi G\rho}{3} - \frac{\kappa c^2}{a^2}. \qquad (3.109)$$

In most cases, in which we know how ρ depends on a, the first Friedmann equation is sufficient to solve for the problem.

3.9.2 Cosmological parameters

We introduce the following cosmological parameters:
- **The Hubble parameter H**: This is given by

$$H = \frac{\dot{a}}{a}. \qquad (3.110)$$

This provides the rate of expansion. At present time we have

$$H_0 = 100h \ \text{km} \ \text{sec}^{-1} \ \text{Mpc}^{-1}. \qquad (3.111)$$

The dimensionless Hubble parameter h is around 0.7 ± 0.1. The megaparsec Mpc is 3.09×10^{24} cm.
- **The density parameter Ω and the critical density ρ_c**: These are defined by

$$\Omega = \frac{8\pi G}{3H^2}\rho = \frac{\rho}{\rho_c}. \qquad (3.112)$$

$$\rho_c = \frac{3H^2}{8\pi G}. \qquad (3.113)$$

- **The deceleration parameter q**: This provides the rate of change of the rate of the expansion of the Universe. This is defined by

$$q = -\frac{a\ddot{a}}{\dot{a}^2}. \tag{3.114}$$

Using the first two parameters in the first Friedmann equation we obtain

$$\frac{\rho - \rho_c}{\rho_c} = \Omega - 1 = \frac{\kappa c^2}{H^2 a^2}. \tag{3.115}$$

We get immediately the behavior

$$\textbf{closed universe}: \kappa > 0 \leftrightarrow \Omega > 1 \leftrightarrow \rho > \rho_c. \tag{3.116}$$

$$\textbf{flat universe}: \kappa = 0 \leftrightarrow \Omega = 1 \leftrightarrow \rho = \rho_c. \tag{3.117}$$

$$\textbf{open universe}: \kappa < 0 \leftrightarrow \Omega < 1 \leftrightarrow \rho < \rho_c. \tag{3.118}$$

3.9.3 Energy conservation

Let us now consider the conservation law $\nabla_\mu \ T^\mu_\nu = \partial_\mu T^\mu_\nu + \Gamma^\mu_{\mu\alpha} T^\alpha_\nu - \Gamma^\alpha_{\mu\nu} T^\mu_\alpha = 0$. In the comoving coordinates we have $T_\mu{}^\nu = \mathrm{diag}(-\rho c^2, P, P, P)$. The $\nu = 0$ component of the conservation law is

$$-c\dot{\rho} - \frac{3\dot{a}}{ca}(\rho c^2 + P) = 0. \tag{3.119}$$

In cosmology the pressure P and the rest mass density ρ are generally related by the equation of state

$$P = w\rho c^2. \tag{3.120}$$

The conservation of energy becomes

$$\frac{\dot{\rho}}{\rho} = -3(1 + w)\frac{\dot{a}}{a}. \tag{3.121}$$

For constant w the solution is of the form

$$\rho \propto a^{-3(1+w)}. \tag{3.122}$$

There are three cases of interest:
- **The matter-dominated Universe:** Matter (also called dust) is a set of collisionless non-relativistic particles which have zero pressure. For example, stars and galaxies may be considered as dust since pressure can be neglected to a very good accuracy. Since $P_M = 0$ we have $w = 0$ and as a consequence

$$\rho_M \propto a^{-3}. \tag{3.123}$$

This can be seen also as follows. The energy density for dust comes entirely from the rest mass of the particles. The mass density is $\rho = nm$ where n is the number density which is inversely proportional to the volume. Hence, the mass density must go as the inverse of a^3 which is the physical volume of a comoving region.

- **The radiation-dominated Universe:** Radiation consists of photons (obviously) but also includes any particles with speeds close to the speed of light. For an electromagnetic field we can show that the stress–energy-tensor satisfies $T_\mu^\mu = 0$. However, the stress–energy–momentum tensor of a perfect fluid satisfies $T_\mu{}^\mu = -\rho c^2 + 3P$. Thus for radiation we must have the equation of state $P_R = \rho_R c^2/3$ and as a consequence $w = 1/3$ and hence

$$\rho_R \propto a^{-4}. \tag{3.124}$$

In this case the energy of each photon will redshift away as $1/a$ (see below) as the Universe expands which is the extra factor that multiplies the original factor $1/a^3$ coming from number density.

- **The vacuum-dominated Universe:** The vacuum energy is a perfect fluid with equation of state $P_V = -\rho_V$, i.e. $w = -1$ and hence

$$\rho_V \propto a^0. \tag{3.125}$$

The vacuum energy is an unchanging form of energy in any physical volume which does not redshift.

The null dominant energy condition allows for densities which satisfy the requirements $\rho \geqslant 0$, $\rho \geqslant |P|/c^2$ or $\rho \leqslant 0$, $P = -c^2\rho < 0$, thus in particular allowing the vacuum energy to be either positive or negative, and as a consequence we must have in all the above discussed cases $|w| \leqslant 1$.

In general matter, radiation and vacuum can contribute simultaneously to the evolution of the Universe. Let us simply assume that all densities evolve as power laws, viz

$$\rho_i = \rho_{i0} a^{-n_i} \Leftrightarrow w_i = \frac{n_i}{3} - 1. \tag{3.126}$$

The first Friedmann equation can then be put in the form

$$
\begin{aligned}
H^2 &= \frac{8\pi G}{3} \sum_i \rho_i - \frac{\kappa c^2}{a^2} \\
&= \frac{8\pi G}{3} \sum_{i,C} \rho_i.
\end{aligned}
\tag{3.127}
$$

In the above equation the spatial curvature is thought of as giving another contribution to the rest mass density given by

$$\rho_C = -\frac{3}{8\pi G} \frac{\kappa c^2}{a^2}. \tag{3.128}$$

This rest mass density corresponds to the values $w_C = -1/3$ and $n_C = 2$. The total density parameter Ω is defined by $\Omega = \sum_i 8\pi G \rho_i / 3H^2$. By analogy the density parameter of the spatial curvature is given by

$$\Omega_C = \frac{8\pi G \rho_C}{3H^2} = -\frac{\kappa c^2}{H^2 a^2}. \tag{3.129}$$

The first Friedmann equation becomes the identity

$$\sum_{i,C} \Omega_i = 1 \Leftrightarrow \Omega_C = 1 - \Omega = 1 - \Omega_M - \Omega_R - \Omega_V. \tag{3.130}$$

The rest mass densities of matter and radiation are always positive, whereas the rest mass densities corresponding to vacuum and curvature can be either positive or negative.

The Hubble parameter is the rate of expansion of the Universe. The derivative of the Hubble parameter is

$$\begin{aligned}
\dot{H} &= \frac{\ddot{a}}{a} - \left(\frac{\dot{a}}{a}\right)^2 \\
&= -\frac{4\pi G}{3} \sum_i \rho_i (1 + 3w_i) - \frac{8\pi G}{3} \sum_i \rho_i + \frac{\kappa c^2}{a^2} \\
&= -4\pi G \sum_i \rho_i (1 + w_i) + \frac{\kappa c^2}{a^2} \\
&= -4\pi G \sum_{i,C} \rho_i (1 + w_i).
\end{aligned} \tag{3.131}$$

This is effectively the second Friedmann equation. In terms of the deceleration parameter this reads

$$\frac{\dot{H}}{H^2} = -1 - q. \tag{3.132}$$

An open or flat universe $\rho_C \geqslant 0$ ($\kappa \leqslant 0$) with $\rho_i > 0$ will never contract as long as $\sum_{i,C} \rho_i \neq 0$ since $H^2 \propto \sum_{i,C} \rho_i$ from the first Friedmann equation (7.186). On the other hand, we have $|w_i| \leqslant 1$, and thus we deduce from the second Friedmann equation (7.188) the condition $\dot{H} \leqslant 0$ which indicates that the expansion of the Universe decelerates.

For a flat universe dominated by a single component w_i we can show that the deceleration parameter is given by

$$q_i = \frac{1}{2}(1 + 3w_i). \tag{3.133}$$

This is positive and thus the expansion is accelerating for a matter-dominated universe ($w_i = 0$), whereas it is negative and thus the expansion is decelerating for a vacuum-dominated universe ($w_i = -1$). The current cosmological data strongly favors the second possibility.

3.10 Examples of scale factors

3.10.1 Matter and radiation-dominated universes

From observation we know that the Universe was radiation-dominated at early times, whereas it is matter-dominated at the current epoch. Let us then consider a single kind of rest mass density $\rho \propto a^{-n}$. The Friedmann equation gives therefore $\dot{a} \propto a^{1-n/2}$. The solution behaves as

$$a \sim t^{\frac{2}{n}}. \tag{3.134}$$

For a flat (since $\rho_C = 0$) universe dominated by matter we have $\Omega = \Omega_M = 1$ and $n = 3$. In this case

$$a \sim t^{\frac{2}{3}}, \text{ Matter-dominated Universe.} \tag{3.135}$$

This is also known as the Einstein–de Sitter Universe. For a flat Universe dominated by radiation we have $\Omega = \Omega_R = 1$ and $n = 4$ and hence $a \sim t^{\frac{1}{2}}$.

$$a \sim t^{\frac{1}{2}}, \text{ Radiation-dominated Universe.} \tag{3.136}$$

These solutions exhibit a singularity at $a = 0$ known as the Big Bang. Indeed the rest mass density diverges as $a \longrightarrow 0$. At this regime general relativity breaks down and quantum gravity takes over. The so-called cosmological singularity theorems show that any universe with $\rho > 0$ and $p \geqslant 0$ must start at a singularity.

3.10.2 Vacuum-dominated universes

For a flat universe dominated by vacuum energy we have $H = $ constant since $\rho_\Lambda = $ constant and hence $a = \exp(Ht)$. The universe expands exponentially. The metric reads explicitly $ds^2 = -c^2 dt^2 + \exp(Ht)(dx^2 + dy^2 + dz^2)$. This is the maximally symmetric spacetime known as de Sitter spacetime. Indeed, the corresponding Riemann curvature tensor has the characteristic form of a maximally symmetric spacetime in 4D. Since de Sitter spacetime has a positive scalar curvature, whereas this space has zero curvature, the coordinates (t, x, y, z) must only cover part of the de Sitter spacetime. Indeed, we can show that these coordinates are incomplete in the past.

From observation $\Omega_R \ll \Omega_{M, V, C}$. We will therefore neglect the effect of radiation and set $\Omega = \Omega_M + \Omega_V$. The curvature is $\Omega_C = 1 - \Omega_M - \Omega_V$. Recall that $\Omega_C \propto 1/a^2$, $\Omega_M \propto 1/a^3$ and $\Omega_V \propto 1/a^0$. Thus in the limit $a \longrightarrow 0$ (the past), matter dominates and spacetime approaches Einstein–de Sitter spacetime, whereas in the limt $a \longrightarrow \infty$ (the future), vacuum dominates and spacetime approaches de Sitter spacetime.

3.10.3 Milne universe

For an empty space with spatial curvature we have

$$H^2 = -\frac{\kappa c^2}{a^2}. \tag{3.137}$$

The curvature must be negative. This corresponds to the so-called Milne universe with a rest mass density $\rho_C \propto a^{-2}$, i.e. $n = 2$. Hence the Milne universe expands linearly, viz

$$a \sim t, \quad \text{Milne Universe}. \tag{3.138}$$

The Milne universe can only be Minkowski spacetime in a certain incomplete coordinate system which can be checked by showing that its Riemann curvature tensor is actually 0. In fact, the Milne universe is the interior of the future light cone of some fixed event in spacetime foliated by hyperboloids which have negative scalar curvature.

3.10.4 The static universe

A static universe satisfies $\dot{a} = \ddot{a} = 0$. The Friedmann equations become

$$\frac{\kappa c^2}{a^2} = \frac{8\pi G}{3} \sum_i \rho_i, \quad \sum_i \left(\rho_i + 3\frac{P}{c^2} \right) = 0. \tag{3.139}$$

Again by neglecting radiation the second equation leads to

$$\rho_M + \rho_V = -\frac{3}{c^2}(P_M + P_V) = 3\rho_V \Rightarrow \rho_M = 2\rho_V. \tag{3.140}$$

The first equation gives the scalar curvature

$$\kappa = \frac{4\pi G \rho_M a^2}{c^2}. \tag{3.141}$$

3.10.5 Expansion versus recollapse

Recall that $H = \dot{a}/a$. Thus if $H > 0$ the Universe is expanding while if $H < 0$ the Universe is collapsing. The point a_* at which the Universe goes from expansion to collapse corresponds to $H = 0$. By using the Friedmann equation this gives the condition

$$\rho_{M0}a_*^{-3} + \rho_{V0} + \rho_{C0}a_*^{-2} = 0. \tag{3.142}$$

Recall also that $\Omega_{C0} = 1 - \Omega_{M0} - \Omega_{V0}$ and $\Omega_i \propto \rho_i/H^2$. By dividing the above equation on H_0^2 we get

$$\begin{aligned} &\Omega_{M0}a_*^{-3} + \Omega_{V0} + (1 - \Omega_{M0} - \Omega_{V0})a_*^{-2} \\ &= 0 \Rightarrow \Omega_{V0}a_*^3 + (1 - \Omega_{M0} - \Omega_{V0})a_* + \Omega_{M0} = 0. \end{aligned} \tag{3.143}$$

First we consider $\Omega_{V0} = 0$. For open and flat universes we have $\Omega_0 = \Omega_{M0} \leqslant 1$ and thus the above equation has no solution. In other words, open and flat universes keep expanding. For a closed universe $\Omega_0 = \Omega_{M0} > 1$ and the above equation admits a solution a_* and as a consequence the closed universe will recollapse.

For $\Omega_{V0} > 0$ the situation is more complicated. For $0 \leqslant \Omega_{M0} \leqslant 1$ the Universe will always expand, whereas for $\Omega_{M0} > 1$ the Universe will always expand only if Ω_Λ is further bounded from below as

$$\Omega_{V0} \geqslant \hat{\Omega}_{V0} = 4\Omega_{M0}\cos^3\left[\frac{1}{3}\cos^{-1}\left(\frac{1 - \Omega_{M0}}{\Omega_{M0}}\right) + \frac{4\pi}{3}\right]. \tag{3.144}$$

This means in particular that the Universe with sufficiently large Ω_{M0} can recollapse for $0 < \Omega_{V0} < \hat{\Omega}_{V0}$. Thus, a sufficiently large Ω_M can halt the expansion before Ω_V becomes dominant.

Note also from the above solution that the Universe will always recollapse for $\Omega_{V0} < 0$. Indeed, the effect of $\Omega_{V0} < 0$ is to cause deceleration and recollapse.

3.11 Redshift, distances and age

3.11.1 Redshift in a flat universe

Let us consider the metric

$$ds^2 = -c^2 dt^2 + a^2(t)[dx^2 + dy^2 + dz^2]. \tag{3.145}$$

Thus space at each fixed instant of time t is the Euclidean 3D space R^3. The Universe described by this metric is expanding in the sense that the volume of the 3D spatial slice, which is given by the so-called scale factor $a(t)$, is a function of time. The above metric is also rewritten as

$$g_{00} = -1, \quad g_{ij} = a^2(t)\delta_{ij}, \quad g_{0i} = 0. \tag{3.146}$$

It is obvious that the relative distance between particles at fixed spatial coordinates grows with time t as $a(t)$. These particles draw worldlines in spacetime which are said to be comoving. Similarly, a comoving volume is a region of space which expands along with its boundaries defined by fixed spatial coordinates with the expansion of the Universe.

We recall the formula of the Christoffel symbols

$$\Gamma^\lambda_{\mu\nu} = \frac{1}{2}g^{\lambda\rho}(\partial_\mu g_{\nu\rho} + \partial_\nu g_{\mu\rho} - \partial_\rho g_{\mu\nu}). \tag{3.147}$$

We compute

$$\Gamma^0_{\mu\nu} = -\frac{1}{2}(\partial_\mu g_{\nu 0} + \partial_\nu g_{\mu 0} - \partial_0 g_{\mu\nu}) \Rightarrow \Gamma^0_{00} = \Gamma^0_{0i} = \Gamma^0_{i0} = 0, \quad \Gamma^0_{ij} = \frac{a\dot{a}}{c}\delta_{ij}. \tag{3.148}$$

$$\Gamma^i_{\mu\nu} = \frac{1}{2a^2}(\partial_\mu g_{\nu i} + \partial_\nu g_{\mu i} - \partial_i g_{\mu\nu}) \Rightarrow \Gamma^i_{00} = \Gamma^i_{jk} = 0, \quad \Gamma^i_{0j} = \frac{\dot{a}}{ac}\delta_{ij}. \tag{3.149}$$

The geodesic equation reads

$$\frac{d^2x^\lambda}{d\tau^2} + \Gamma^\lambda_{\mu\nu}\frac{dx^\mu}{d\tau}\frac{dx^\nu}{d\tau} = 0. \tag{3.150}$$

In particular

$$\frac{d^2x^0}{d\lambda^2} + \Gamma^0_{ij}\frac{dx^i}{d\lambda}\frac{dx^j}{d\lambda} = 0 \Rightarrow \frac{d^2t}{d\lambda^2} + \frac{a\dot{a}}{c^2}\left(\frac{d\vec{x}}{d\lambda}\right)^2 = 0. \tag{3.151}$$

For null geodesics (which are paths followed by massless particles such as photons) we have $ds^2 = -c^2dt^2 + a^2(t)d\vec{x}^2 = 0$. In other words, we must have along a null geodesic with parameter λ the condition $a^2(t)d\vec{x}^2/d\lambda^2 = c^2dt^2/d\lambda^2$. We get then the equation

$$\frac{d^2t}{d\lambda^2} + \frac{\dot{a}}{a}\left(\frac{dt}{d\lambda}\right)^2 = 0. \tag{3.152}$$

The solution is immediately given by (with ω_0 a constant)

$$\frac{dt}{d\lambda} = \frac{\omega_0}{c^2a}. \tag{3.153}$$

The energy of the photon as measured by an observer whose velocity is U^μ is given by $E = -p^\mu U_\mu$ where p^μ is the 4-vector energy–momentum of the photon. A comoving observer is an observer with fixed spatial coordinates and thus $U^\mu = (U^0, 0, 0, 0)$. Since $g_{\mu\nu}U^\mu U^\nu = -c^2$ we must have $U^0 = \sqrt{-c^2/g_{00}} = c$. Furthermore, we choose the parameter λ along the null geodesic such that the 4-vector energy–momentum of the photon is $p^\mu = dx^\mu/d\lambda$. We get then

$$\begin{aligned} E &= -g_{\mu\nu}p^\mu U^\nu \\ &= p^0 U^0 \\ &= \frac{dx^0}{d\lambda}c \\ &= \frac{\omega_0}{a}. \end{aligned} \tag{3.154}$$

Thus if a photon is emitted with energy E_1 at a scale factor a_1 and then observed with energy E_2 at a scale factor a_2 we must have

$$\frac{E_1}{E_2} = \frac{a_2}{a_1}. \tag{3.155}$$

In terms of wavelengths this reads

$$\frac{\lambda_2}{\lambda_1} = \frac{a_2}{a_1}. \tag{3.156}$$

This is the phenomena of cosmological redshift: In an expanding universe we have $a_2 > a_1$ and as a consequence we must have $\lambda_2 > \lambda_1$, i.e. the wavelength of the photon grows with time. The amount of redshift is

$$z = \frac{E_1 - E_2}{E_2} = \frac{a_2}{a_1} - 1. \tag{3.157}$$

This effect allows us to measure the change in the scale factor between distant galaxies (where the photons are emitted) and here (where the photons are observed). Also, it can be used to infer the distance between us and distant galaxies. Indeed, a greater redshift means a greater distance. For example z close to 0 means that there was not sufficient time for the universe to expand because the emitter and observer are very close to each other.

The scale factor $a(t)$ as a function of time might be of the form

$$a(t) = t^q, \quad 0 < q < 1. \tag{3.158}$$

In the limit $t \longrightarrow 0$ we have $a(t) \longrightarrow 0$. In fact the time $t = 0$ is a true singularity of this geometry, which represents a Big Bang event, and hence it must be excluded. The physical range of t is

$$0 < t < \infty. \tag{3.159}$$

The light cones of this curved spacetime are defined by the null paths $ds^2 = -c^2 dt^2 + a^2(t) d\vec{x}^2 = 0$. In $1 + 1$ dimensions this reads

$$ds^2 = -c^2 dt^2 + a^2(t) dx^2 = 0 \Rightarrow \frac{dx}{dt} = \pm ct^{-q}. \tag{3.160}$$

The solution is

$$t = \left(\pm \frac{1-q}{c}(x - x_0) \right)^{\frac{1}{1-q}}. \tag{3.161}$$

These are the light cones of our expanding universe. Compare with the light cones of a flat Minkowski universe obtained by setting $q = 0$ in this formula. These light cones are tangent to the singularity at $t = 0$. As a consequence, the light cones in this curved geometry of any two points do not necessarily need to intersect in the past as opposed to the flat Minkowski universe where the light cones of any two points intersect both in the past and in the future.

3.11.2 Cosmological redshift

Recall that a Killing vector is any vector which satisfies the Killing equation $\nabla_\mu K_\nu + \nabla_\nu K_\mu = 0$. This Killing vector generates an isometry of the metric which is associated with the conservation of the momentum $p_\mu K^\mu$ along the geodesic whose tangent vector is p^μ.

In an FLRW universe there could be no Killing vector along time-like geodesic and thus no concept of conserved energy. However, we can define Killing tensor along a time-like geodesic. We introduce the tensor

$$K_{\mu\nu} = a^2(t)\left(g_{\mu\nu} + \frac{U_\mu U_\nu}{c^2}\right). \tag{3.162}$$

We have

$$\begin{aligned}
\nabla_{(\sigma K_{\mu\nu})} &= \nabla_\sigma\ K_{\mu\nu} + \nabla_\mu\ K_{\sigma\nu} + \nabla_\nu\ K_{\mu\sigma} \\
&= \partial_\sigma K_{\mu\nu} + \partial_\mu K_{\sigma\nu} + \partial_\nu K_{\mu\sigma} - 2\Gamma^\rho_{\sigma\mu}K_{\rho\nu} - 2\Gamma^\rho_{\sigma\nu}K_{\mu\rho} - 2\Gamma^\rho_{\mu\nu}K_{\sigma\rho}.
\end{aligned} \tag{3.163}$$

Since U^μ is the 4-vector velocity of comoving observers in the FLRW universe we have $U^\mu = (c, 0, 0, 0)$ and $U_\mu = (-c, 0, 0, 0)$ and as a consequence $K_{\mu\nu} = a^4\text{diag}(0, 1/(1 - \kappa\rho^2), \rho^2, \rho^2\sin^2\theta)$. In other words, $K_{ij} = a^2 g_{ij} = a^4\gamma_{ij}$, $K_{0i} = K_{00} = 0$. The first set of nontrivial components of $\nabla_{(\sigma K_{\mu\nu})}$ are

$$\begin{aligned}
\nabla_{(0 K_{ij})} = \nabla_{(i K_{0j})} = \nabla_{(j K_{i0})} &= \nabla_{0 K_{ij}} \\
&= \partial_0 K_{ij} - 2\Gamma^k_{0i}K_{kj} - 2\Gamma^k_{0j}K_{ik}.
\end{aligned} \tag{3.164}$$

By using the result $\Gamma^k_{0i} = \dot{a}\delta^k_i/ca$ we get

$$\begin{aligned}
\nabla_{(0 K_{ij})} &= \partial_0 K_{ij} - 4\frac{\dot{a}a}{c}g_{ij} \\
&= \partial_0 K_{ij} - \frac{d}{dx^0}a^4 \cdot \gamma_{ij} \\
&= 0.
\end{aligned} \tag{3.165}$$

The other set of nontrivial components of $\nabla_{(\sigma K_{\mu\nu})}$ are

$$\begin{aligned}
\nabla_{(i K_{jk})} &= \nabla_i\ K_{jk} + \nabla_j\ K_{ik} + \nabla_k\ K_{ij} \\
&= a^4(\nabla_i\gamma_{jk} + \nabla_j\ \gamma_{ik} + \nabla_k\ \gamma_{ij}) \\
&= 0.
\end{aligned} \tag{3.166}$$

In the last step we have used the fact that the 3D metric γ_{ij} is covariantly constant, which can be verified directly.

We conclude therefore that the tensor $K_{\mu\nu}$ is a Killing tensor and hence $K^2 = K_{\mu\nu}V^\mu V^\nu$ where $V^\mu = dx^\mu/d\tau$ is the 4-vector velocity of a particle is conserved along its geodesic. We have two cases to consider:

- Massive particles: In this case $V^\mu V_\mu = -c^2$ and thus $(V^0)^2 = c^2 + g_{ij}V^i V^j = c^2 + \vec{V}^2$. But since $U_\mu V^\mu = -cV^0$ we have

$$\begin{aligned}
K^2 &= K_{\mu\nu}V^\mu V^\nu \\
&= a^2(V^\mu V_\mu + \frac{(U_\mu V^\mu)^2}{c^2}) \\
&= a^2\vec{V}^2.
\end{aligned} \tag{3.167}$$

We get then the result

$$|\vec{V}| = \frac{K}{a}. \tag{3.168}$$

In other words, particles slow down with respect to comoving coordinates as the universe expands. This is equivalent to the statement that the universe cools down as it expands.

- Massless particles: In this case $V^\mu V_\mu = 0$ and hence

$$
\begin{aligned}
K^2 &= K_{\mu\nu} V^\mu V^\nu \\
&= a^2 \left(V^\mu V_\mu + \frac{(U_\mu V^\mu)^2}{c^2} \right) \\
&= \frac{a^2}{c^2} (U_\mu V^\mu)^2.
\end{aligned} \tag{3.169}
$$

We get now the result

$$|U_\mu V^\mu| = \frac{cK}{a}. \tag{3.170}$$

However, recall that the energy E of the photon as measured with respect to the comoving observer whose 4-vector velocity is U^μ is given by $E = -p^\mu U_\mu$. But the 4-vector energy–momentum of the photon is given by $p^\mu = V^\mu$. Hence we obtain

$$E = \frac{cK}{a}. \tag{3.171}$$

An emitted photon with energy E_{em} will be observed with a lower energy E_{ob} as the universe expands, viz

$$\frac{E_{\text{em}}}{E_{\text{ob}}} = \frac{a_{\text{ob}}}{a_{\text{em}}} > 1. \tag{3.172}$$

We define the redshift

$$z_{\text{em}} = \frac{E_{\text{em}} - E_{\text{ob}}}{E_{\text{ob}}}. \tag{3.173}$$

This means that

$$a_{\text{em}} = \frac{a_{\text{ob}}}{1 + z_{\text{em}}}. \tag{3.174}$$

Recall that $a(t) = R(t)/R_0$. Thus if we are observing the photon today we must have $a_{\text{ob}}(t) = 1$ or equivalently $R_{\text{ob}}(t) = R_0$. We get then

$$a_{\text{em}} = \frac{1}{1 + z_{\text{em}}}. \tag{3.175}$$

The redshift is a direct measure of the scale factor at the time of emission.

3.11.3 Comoving and instantaneous physical distances

Note that the above described redshift is due to the expansion of the Universe and not to the relative velocity between the observer and emitter and thus it is not the same as the Doppler effect. However, over distances which are much smaller than the Hubble radius $1/H_0$ and the radius of spatial curvature $1/\sqrt{\kappa}$ we can view the expansion of the Universe as galaxies moving apart and as a consequence the redshift can be thought of as a Doppler effect. The redshift can therefore be thought of as a relative velocity. We stress that this picture is only an approximation which is valid at sufficiently small distances.

The distance d from us to a given galaxy can be taken to be the instantaneous physical distance d_p. Recall the metric of the FLRW universe given by

$$ds^2 = -c^2dt^2 + R_0^2 a^2(t)(d\chi^2 + S_k^2(\chi)d\Omega^2). \tag{3.176}$$

$$\begin{aligned} S_k(\chi) &= \sin\chi, \quad k = +1 \\ S_k(\chi) &= \chi, \quad k = 0 \\ S_k(\chi) &= \sinh\chi, \quad k = -1. \end{aligned} \tag{3.177}$$

Clearly the instantaneous physical distance d_p from us at $\chi = 0$ to a galaxy which lies on a sphere centered on us of radius χ is

$$d_p = R_0 a(t)\chi. \tag{3.178}$$

The radial coordinate χ is constant since we are assuming that us and the galaxy are perfectly comoving. The relative velocity (which we can define only within the approximation that the redshift is a Doppler effect) is therefore

$$v = \dot{d}_p = R_0\dot{a}\chi = \frac{\dot{a}}{a}d_p = Hd_p. \tag{3.179}$$

At the present time this law reads

$$v = H_0 d_p. \tag{3.180}$$

This is the famous Lemaître–Hubble law: galaxies which are not very far from us move away from us with a recess velocity which is linearly proportional to their distance.

The instantaneous physical distance d_p is obviously not a measurable quantity since measurement relates to events on our past-light cone, whereas d_p relates to events on our current spatial hypersurface.

3.11.4 Luminosity distance

The luminosity distance is the distance inferred from comparing the proper luminosity to the observed brightness if we were in a flat and non-expanding universe. Recall that the luminosity L of a source is the amount of energy emitted per unit time. This is the proper or intrinsic luminosity of the source. We will assume that the source radiates equally in all directions. In flat space the flux of the source as measured by an observer a distance d away is the amount of energy per unit time per unit area, given by $F = L/4\pi d^2$. This is the apparent brightness at the location of the observer. We write this result as

$$\frac{F}{L} = \frac{1}{4\pi d^2}. \tag{3.181}$$

Now in the FLRW universe the flux will be diluted by two effects. First the energy of each photon will be redshifted by the factor $1/a = 1 + z$ due to the expansion of the universe. In other words the luminosity L must be changed as $L \longrightarrow (1 + z)L$. In a comoving system light will travel a distance $|\vec{du}| = cdt/(R_0 a)$ during a time dt. Hence two photons emitted a time δt apart will be observed a time $(1 + z)\delta t$ apart. The flux F must therefore be changed as $F \longrightarrow F/(1 + z)$. Hence in the FLRW universe we must have

$$\frac{F}{L} = \frac{1}{(1 + z)^2 A}. \tag{3.182}$$

The area A of a sphere of radius χ in the comoving system of coordinates is from the FLRW metric

$$A = 4\pi R_0^2 a^2(t) S_k^2(\chi) = 4\pi R_0^2 S_k^2(\chi). \tag{3.183}$$

Again we used the fact that at the current epoch $a(t) = 1$. The luminosity distance d_L is the analog of d and thus it must be defined by

$$d_L^2 = \frac{L}{4\pi F} \Rightarrow d_L = (1 + z)R_0 S_k(\chi). \tag{3.184}$$

Next, on a null radial geodesic we have $-c^2 dt^2 + a^2(t)R_0^2 d\chi^2 = 0$ and thus we obtain (by using $dt = da/(aH)$ and remembering that at the emitter position $\chi' = 0$ and $a = a(t)$, whereas at our position $\chi' = \chi$ and $a = 1$)

$$\int_0^\chi d\chi' = \frac{c}{R_0} \int_{t_a}^{t_1} \frac{dt'}{a(t')} = \frac{c}{R_0} \int_a^1 \frac{da'}{a'^2 H(a')}. \tag{3.185}$$

We convert to redshift by the formula $a = 1/(1 + z')$. We get

$$\chi = \frac{c}{R_0} \int_0^z \frac{dz'}{H(z')}. \tag{3.186}$$

The Friedmann equation is

$$\begin{aligned}
H^2 &= \frac{8\pi G}{3}\sum_{i,c}\rho_i \\
&= \frac{8\pi G}{3}\sum_{i,c}\rho_{i0} a^{-n_i} \\
&= \frac{8\pi G}{3}\sum_{i,c}\rho_{i0}(1 + z')^{n_i} \\
&= H_0^2 \sum_{i,c}\Omega_{i0}(1 + z')^{n_i} \\
&= H_0^2 E^2(z').
\end{aligned} \tag{3.187}$$

Thus

$$H(z') = H_0 E(z'), \ E(z') = \left[\sum_{i,c} \Omega_{i0}(1 + z')^{n_i} \right]^{\frac{1}{2}}. \tag{3.188}$$

Hence

$$\chi = \frac{c}{R_0 H_0} \int_0^z \frac{dz'}{E(z')}. \tag{3.189}$$

The luminosity distance becomes

$$d_L = (1 + z) R_0 S_k \left(\frac{c}{R_0 H_0} \int_0^z \frac{dz'}{E(z')} \right). \tag{3.190}$$

Recall that the curvature density is $\Omega_c = -\kappa c^2/(H^2 a^2) = -k^2 c^2/(H^2 R^2(t))$. Thus

$$\Omega_{c0} = -\frac{kc^2}{H_0^2 R_0^2} \Rightarrow R_0 = \frac{c}{H_0} \sqrt{-\frac{k}{\Omega_{c0}}} = \frac{c}{H_0} \sqrt{\frac{1}{|\Omega_{c0}|}}. \tag{3.191}$$

The above formula works for $k = \pm 1$. This formula will also lead to the correct result for $k = 0$ as we will now show. Thus for $k = \pm 1$ we have

$$\frac{c}{R_0 H_0} = \sqrt{|\Omega_{c0}|}. \tag{3.192}$$

In other words,

$$d_L = (1 + z) \frac{c}{H_0} \frac{1}{\sqrt{|\Omega_{c0}|}} S_k \left(\sqrt{|\Omega_{c0}|} \int_0^z \frac{dz'}{E(z')} \right). \tag{3.193}$$

For $k = 0$, the curvature density Ω_{c0} vanishes but it cancels exactly in this last formula for d_L and we get therefore the correct answer which can be checked by comparing with the original formula (3.190).

The above formula allows us to compute the distance to any source at redshift z given H_0 and Ω_{i0} which are the Hubble parameter and the density parameters at our epoch. Conversely, given the distance d_L at various values of the redshift we can extract H_0 and Ω_{i0}.

3.11.5 Other distances

Proper motion distance: This is the distance inferred from the proper and observable motion of the source. This is given by

$$d_M = \frac{u}{\dot{\theta}}. \tag{3.194}$$

The u is the proper transverse velocity, whereas $\dot{\theta}$ is the observed angular velocity. We can check that

$$d_M = \frac{d_L}{1 + z}. \tag{3.195}$$

The angular diameter distance: This is the distance inferred from the proper and observed size of the source. This is given by

$$d_A = \frac{S}{\theta}. \tag{3.196}$$

The S is the proper size of the source and θ is the observed angular diameter. We can check that

$$d_A = \frac{d_L}{(1 + z)^2}. \tag{3.197}$$

3.11.6 Age of the Universe

Let t_0 be the age of the Universe today and let t_* be the age of the Universe when the photon was emitted. The difference $t_0 - t_*$ is called lookback time. This is given by

$$
\begin{aligned}
t_0 - t_* &= \int_{t_*}^{t_0} dt \\
&= \int_{a_*}^{1} \frac{da}{aH(a)} \\
&= \frac{1}{H_0} \int_0^{z_*} \frac{dz'}{(1 + z')E(z')}.
\end{aligned}
\tag{3.198}
$$

For a flat ($k = 0$) matter-dominated ($\rho \simeq \rho_M = \rho_{M0}a^{-3}$) universe we have $\Omega_{M0} \simeq 1$ and hence $E(z') = \sqrt{\Omega_{M0}(1 + z')^3 + \cdots} = (1 + z')^{3/2}$. Thus

$$
\begin{aligned}
t_0 - t_* &= \frac{1}{H_0} \int_0^{z_*} \frac{dz'}{(1 + z')^{\frac{5}{2}}} \\
&= \frac{2}{3H_0}\left[1 - \frac{1}{(1 + z_*)^{\frac{3}{2}}} \right].
\end{aligned}
\tag{3.199}
$$

By allowing $t_* \longrightarrow 0$ we get the actual age of the Universe. This is equivalent to $z_* \longrightarrow \infty$ since a photon emitted at the time of the Big Bang will be infinitely redshifted, i.e. unobservable. We get then

$$t_0 = \frac{2}{3H_0}. \tag{3.200}$$

3.12 Exercises

Exercise 1: Explain how did we make the measurement of the velocity away from the luminous core of the M33 galaxy in the galaxy rotation curve (figure 3.8).

Exercise 2: Explain the cause behind the scattered data points (around the Virgo cluster) in the Hubble diagram (figure 3.3).

Exercise 3: Show that generalization of the equation $\Omega_M + \Omega_R + \Omega_V = 1$ to $t \neq t_0$ is given by

$$\rho(a) = \rho_c\left(\frac{\Omega_M}{a^3} + \frac{\Omega_R}{a^4} + \Omega_V\right). \tag{3.201}$$

Employ energy conservation.

Solution 3: By employing the principle of local conservation as expressed by the first law of thermodynamics we have

$$dE = -PdV.$$

Thus the change in the total energy in a volume V, containing a fixed number of particles and a pressure P, due to any change dV in the volume is equal to the work done on it. The heat flow in any direction is zero because of isotropy. Alternatively, because of homogeneity the temperature T depends only on time and thus no place is hotter or colder than any other.

The volume dV is the physical volume and thus it is related to the time-independent comoving volume $dV_{comoving} = dxdydz$ by $dV = a^3(t)dV_{comoving}$. On the other hand, the energy E is given in terms of the density ρ by $E = \rho dV$. The first law of thermodynamics becomes

$$\frac{d}{dt}(\rho a^3(t)) = -P\frac{d}{dt}(a^3(t)).$$

We have the following three possibilities:

- **Matter-dominated Universe:** In this case galaxies are approximated by a pressureless dust and thus $P_M = 0$. Also, in this case all the energy comes from the rest mass since kinetic motion is neglected. We then get

$$\frac{d}{dt}(\rho_M a^3(t)) = 0 \Rightarrow \rho_M(t) = \rho_M(t_0)\frac{a^3(t_0)}{a^3(t)}.$$

- **Radiation-dominated Universe:** In this case $P_R = \rho_R/3$ (see below for a proof). Thus

$$\frac{d}{dt}(\rho_R a^3(t)) = -\frac{1}{3}\rho_R\frac{d}{dt}(a^3(t)) \Rightarrow \rho_R(t) = \rho_R(t_0)\frac{a^4(t_0)}{a^4(t)}.$$

It is not difficult to check that radiation dominates matter when the scale factor satisfies $a(t) \leqslant a(t_0)/1000$, i.e. when the Universe was 1/1000 of its present size. Thus over most of the Universe's history matter-dominated radiation.

- **Vacuum-dominated Universe:** In this case $P_V = -\rho_V$. Thus

$$\frac{d}{dt}(\rho_V a^3(t)) = \rho_V \frac{d}{dt}(a^3(t)) \Rightarrow \rho_V(t) = \rho_V(t_0)\frac{a^0(t_0)}{a^0(t)}.$$

In other words, ρ_V is always a constant and thus, unlike matter and radiation, it does not decay away with the expansion of the universe. In particular, the future of any perpetually expanding universe will be dominated by vacuum energy. In the case of a cosmological constant we write

$$\rho_V = \frac{c^4}{8\pi G}\Lambda.$$

We compute immediately

$$\Omega_M(t) = \frac{\rho_M(t)}{\rho_c} = \frac{\Omega_M}{a^3}.$$

$$\Omega_R(t) = \frac{\rho_R(t)}{\rho_c} = \frac{\Omega_R}{a^4}.$$

$$\Omega_V(t) = \frac{\rho_V(t)}{\rho_c} = \frac{\Omega_V}{a^0}.$$

The total mass–energy density is given by $\rho(t) = \rho_c \Omega(t) = \rho_c \Omega_M(t) + \Omega_R(t) + \Omega_V(t)$ or equivalently

$$\rho(a) = \rho_c\left(\frac{\Omega_M}{a^3} + \frac{\Omega_R}{a^4} + \Omega_V\right).$$

Exercise 4: Give an estimation of the age of the Universe in a matter-dominated world.

Solution 4: In a matter-dominated universe the scale factor is $a = (t/t_0)^{2/3}$ where $t_0 = 2/3H_0$. The age of the universe is given in terms of the Hubble time by the relation

$$t_0 = \frac{2}{3H_0} = \frac{2}{3}t_H.$$

This gives around 9 Gyr which is not correct since there are stars as old as 12 Gyr in our own Galaxy.

Exercise 5: The size of the Universe may be given in terms of the Hubble distance d_H. A more accurate measure is given by the cosmological horizon, which is the largest radius $r_{\text{horizon}}(t)$ from which a signal could have reached the observer at t

since the Big Bang. Calculate the physical distance to the horizon and compare with the current estimate.

Solution 5: A more accurate measure of the size of the Universe now is given in terms of the conformal time η defined as follows

$$d\eta = \frac{dt}{a(t)}.$$

In the $\eta - r$ spacetime diagram, radial geodesics are the $45°$ lines. In this diagram the Big Bang is the line $\eta = 0$, while our worldline may be chosen to be the line $r = 0$. At any conformal instant η only signals from points inside the past-light cone can be received. To each conformal time η corresponds an instant t given through the equation

$$\eta = \int_0^t \frac{dt'}{a(t')}.$$

We have assumed that the Big Bang occurs at $t = 0$. Since $ds^2 = a^2(t)(-d\eta^2 + dr^2) = 0$, the largest radius $r_{\text{horizon}}(t)$ from which a signal could have reached the observer at t since the big bang is given by

$$r_{\text{horizon}}(t) = \eta = \int_0^t \frac{dt'}{a(t')}.$$

The 3D surface in spacetime with radius $r_{\text{horizon}}(t)$ is called the cosmological horizon. This radius $r_{\text{horizon}}(t)$ and as a consequence the cosmological horizon grows with time and thus a larger region becomes visible as time goes on. The physical distance to the horizon is obviously given by

$$d_{\text{horizon}}(t) = a(t)r_{\text{horizon}}(t) = a(t)\int_0^t \frac{dt'}{a(t')}.$$

The physical radius at the current epoch in a matter-dominated universe is

$$d_{\text{horizon}}(t) = t_0^{2/3}\int_0^{t_0} t^{-2/3}dt = 2t_H = 8\text{Gpc}.$$

This is different from the currently best measured age of 14 Gpc.

Exercise 6:
- The Robertson–Walker metric is

$$ds^2 = -c^2dt^2 + a^2(t)\left[\frac{d\rho^2}{1 - \kappa\rho^2} + \rho^2d\Omega^2\right]. \tag{3.202}$$

     – Compute the non-zero Christoffel symbols.
     – Compute the Ricci tensor.
     – Compute the Ricci scalar.

- In comoving coordinates the stress–energy–momentum tensor is given by a perfect fluid, i.e. of the form

$$T_\mu{}^\lambda = \mathrm{diag}(-\rho c^2,\ P,\ P,\ P). \tag{3.203}$$

Derive Einsetin's equations in this case.

  Hint: There are only two equations in this case known as Friedmann equations.

Exercise 7: The Friedmann equation can be put in the form

$$\frac{1}{2H_0^2}\left(\frac{da}{dt}\right)^2 + V_{\mathrm{eff}}(a) = \frac{\Omega_C}{2}. \tag{3.204}$$

The effective potential is

$$V_{\mathrm{eff}}(a) = -\frac{1}{2}\left(\frac{\Omega_M}{a} + \frac{\Omega_R}{a^2} + a^2\Omega_V\right). \tag{3.205}$$

The curvature density is

$$\Omega_C = -\frac{k}{H_0^2}. \tag{3.206}$$

The density parameters satisfy

$$\Omega_M + \Omega_R + \Omega_V + \Omega_C = 1. \tag{3.207}$$

- Show that for a matter-dominated universe the Friedmann equation can be put in the form

$$\eta = \frac{1}{H_0}\int\frac{da}{\sqrt{(1-\Omega_M)a^2 + a\Omega_M}}. \tag{3.208}$$

  η is the conformal time defined by $dt = a\,d\eta$.
- For a closed universe ($k = +1$, $\Omega_C < 0$, $\Omega_M > 1$) perform the change of variable

$$a \longrightarrow \theta:\ a = \frac{\Omega_M}{2(\Omega_M - 1)}(1 - \cos\theta), \tag{3.209}$$

  to integrate the Friedmann equation explicitly to find a as a function of the conformal time η.
- Find t as a function of the conformal time.
- What are the properties of this solution, i.e. draw $a(\eta)$ as a function of η. Determine the total age and maximum spatial volume of this universe.

- Show that the solution for an open universe ($k = -1$, $\Omega_C > 0$, $\Omega_M < 1$) is of the form

$$a(\eta) = \frac{\Omega_M H_0^2}{2}(\cosh \eta - 1). \tag{3.210}$$

- Determine the flat solution ($k = 0$, $\Omega_C = 0$, $\Omega_M = 1$) by taking the limit $\Omega_M \longrightarrow 1$ in the closed and open solutions.
  Hint: for consistency treat η as small.

Solution 7:
- For a matter-dominated universe: $\Omega_R = \Omega_V = 0$, $\Omega_M + \Omega_C = 1$, $V_{\text{eff}} = -\Omega_M/2a$, $\Omega_C = 1 - \Omega_M$. The Friedmann equation becomes

$$\frac{1}{2H_0^2}\left(\frac{da}{dt}\right)^2 - \frac{1}{2a}\Omega_M = \frac{1}{2}(1 - \Omega_M) \Rightarrow \frac{da}{d\eta} = H_0\sqrt{a\Omega_M + a^2(1 - \Omega_M)}.$$

Thus

$$\eta = \frac{1}{H_0}\int \frac{da}{\sqrt{a\Omega_M + a^2(1 - \Omega_M)}}.$$

- For a closed universe $k = +1$ and thus $\Omega_C = -1/H_0^2 < 0$, i.e. $\Omega_M > 1$. We consider the change of variable

$$a = \frac{\Omega_M}{2(\Omega_M - 1)}(1 - \cos\theta) \Rightarrow da = \frac{\Omega_M}{2(\Omega_M - 1)}(d\theta \sin\theta).$$

We find

$$\eta = \frac{1}{H_0\sqrt{\Omega_M - 1}}\int d\theta = \theta.$$

The solution is therefore

$$a = \frac{\Omega_M}{2(\Omega_M - 1)}(1 - \cos\eta).$$

- We have $dt = ad\eta$. Thus

$$t = \int ad\eta$$

$$= \frac{\Omega_M}{2(\Omega_M - 1)}(\eta - \sin\eta).$$

- The universe starts at a singularity (**Big Bang**) at $\eta = 0$ with $a = 0$, it expands to a maximum size at $\eta = \pi$, then contracts to zero size at $\eta = 2\pi$ (big crunch).

The age is

$$T = \frac{\Omega_M}{2(\Omega_M - 1)}(2\pi - \sin 2\pi) = \frac{\Omega_M \pi}{H_0(\Omega_M - 1)^{3/2}}.$$

The maximum volume is

$$
\begin{aligned}
V &= a^3(\pi) \int \sqrt{\det g}\, d^3x \\
&= a^3(\pi).\, 2\pi^2 \\
&= 2\pi^2 H_0^3 \left(\frac{\Omega_M}{\sqrt{\Omega_M - 1}}\right)^3.
\end{aligned}
$$

- For an open universe $k = -1$ and thus $\Omega_C = 1/H_0^2 > 0$, i.e. $\Omega_M < 1$. We consider the change of variable

$$a = \frac{\Omega_M}{2(\Omega_M - 1)}(\cosh\theta - 1).$$

The solution is

$$a = \frac{\Omega_M}{2(\Omega_M - 1)}(\cosh\eta - 1).$$

- The flat universe is $k = 0$, $\Omega_C = 0$ and $\Omega_M = 1$.
  - The closed universe with η small:

$$a = \frac{H_0^2 \eta^2}{4}.$$

However,

$$t = \frac{H_0^2 \eta^3}{12}.$$

Hence

$$a = \left(\frac{t}{t_0}\right)^{2/3}, \quad \frac{1}{t_0^{2/3}} = \frac{H_0^2}{4}(12H_0^2)^{2/3}.$$

  - The open universe with η small:

$$a = \frac{H_0^2 \eta^2}{4}.$$

The rest is the same as in the case of the closed universe.

References

[1] Carroll S M 2004 *Spacetime and Geometry: An Introduction to General Relativity* (San Francisco, CA: Addison-Wesley) p 513

[2] Hartle J 2014 *Gravity: An Introduction to Einsteinas General Relativity* (London: Pearson)

[3] Weinberg S 2008 *Cosmology* (Oxford: Oxford University Press) p 593

[4] Mukhanov V 2005 *Physical Foundations of Cosmology* (Cambridge: Cambridge University Press) p 421

[5] Dodelson S 2003 *Modern Cosmology* (Amsterdam: Academic) p 440

[6] Benett C L[WMAP Collaboration] *et al* 2013 Nine-year Wilkinson microwave anisotropy probe (WMAP) observations: final maps and results *Astrophys. J. Suppl.* **208** 20

[7] Ahn C P[SDSS Collaboration] *et al* 2014 The tenth data release of the Sloan digital sky survey: first spectroscopic data from the SDSS-III Apache point observatory galactic evolution experiment *J. Astrophys. J. Suppl.* **211** 17

[8] Freedman W L[HST Collaboration] *et al* 2001 Final results from the Hubble Space Telescope key project to measure the Hubble constant *Astrophys. J.* **553** 47

[9] Mather J C *et al* 1994 Measurement of the cosmic microwave background spectrum by the COBE FIRAS instrument *Astrophys. J.* **420** 439

[10] Corbelli E and Salucci P 2000 The extended rotation curve and the dark matter halo of M33 *Mon. Not. Roy. Astron. Soc.* **311** 441

[11] Riess A G[Supernova Search Team] *et al* 1998 Observational evidence from supernovae for an accelerating universe and a cosmological constant *Astron. J.* **116** 1009

[12] Perlmutter S[Supernova Cosmology Project Collaboration] *et al* 1999 Measurements of Omega and Lambda from 42 high redshift supernovae *Astrophys. J.* **517** 565

[13] Crease R P 2007 Dark energy *Phys. World* **20** 19–22

IOP Publishing

Lectures on General Relativity, Cosmology and Quantum Black Holes (Second Edition)

Badis Ydri

Chapter 4

Cosmology II: the inflationary Universe

The phenomenology of inflation is discussed in great detail in [1, 2]. The problem of gauge fixing and cosmological perturbations can be found in [3] but also in [4–6]. The central question of deriving the inflationary dynamics from first principles, i.e. from Einstein's equations, and then deriving their corresponding predictions, such as power spectra and cosmic microwave background (CMB) anisotropies, is addressed in great detail in [7]. In preparing this chapter we also found [8–14] very useful.

4.1 Cosmological puzzles

The isotropy and homogeneity of the Universe and its spatial flatness are two properties which are highly non-generic and, as such, they can only arise from very special set of initial conditions which is a very unsatisfactory state of affairs. Inflation is a dynamical mechanism which allows us to get around this problem by permitting the Universe to evolve to the state of isotropy/homogeneity and spatial flatness from a wide range of initial conditions.

Another problem solved by inflation is the relics problem. Relics refer to magnetic monopoles, domain walls and supersymmetric particles which are assumed to have been produced during the early Universe yet they are not seen in observations.

As it turns out, inflation does also provide the mechanism for the formation of large-scale structures in the Universe starting from minute quantum fluctuations in the early Universe.

4.1.1 Homogeneity/horizon problem

The metric of the Friedmann–Lemaître–Robertson–Walker (FLRW) universe can be put in the form

$$ds^2 = -dt^2 + R^2(t)(d\chi^2 + S_k^2(\chi)d\Omega^2). \tag{4.1}$$

doi:10.1088/978-0-7503-5824-8ch4

$$S_k(\chi) = \sin\chi, \quad k = +1$$
$$S_k(\chi) = \chi, \quad k = 0 \qquad\qquad (4.2)$$
$$S_k(\chi) = \sinh\chi, \quad k = -1.$$

We introduce the conformal time τ by

$$\tau = \int \frac{dt}{R(t)}. \qquad\qquad (4.3)$$

The FLRW metric becomes

$$ds^2 = R^2(\tau)\left[-d\tau^2 + d\chi^2 + S_k^2(\chi)d\Omega^2\right]$$
$$\equiv R^2(\tau)[-d\tau^2 + d\vec{\chi}^2]. \qquad\qquad (4.4)$$

The motion of photons in the FLRW universe is given by null geodesics $ds^2 = 0$. In an isotropic universe it is sufficient to consider only radial motion. The condition $ds^2 = 0$ is then equivalent to $d\tau = d\chi$. The maximum comoving distance a photon can travel since the initial singularity at $t = t_i$ ($R(t_i) = 0$) is

$$\chi_{\rm hor}(t) \equiv \tau - \tau_i = \int_{t_i}^{t} \frac{dt_1}{R(t_1)}. \qquad\qquad (4.5)$$

The is called the particle horizon. Indeed, particles separated by distances larger than $\chi_{\rm hor}$ could never have been in causal contact. The physical size of the particle horizon is

$$d_{\rm hor} = R\chi_{\rm hor}. \qquad\qquad (4.6)$$

The existence of particle horizons is at the heart of the so-called horizon problem, i.e. of the problem of why the Universe is isotropic and homogeneous.

On the other hand, the comoving Hubble radius $1/aH$ is such that particles separated by distances larger than $1/aH$ cannot communicate with each other, i.e. $1/aH$ is a measure of how far things can travel in the Universe. The comoving Hubble radius $1/aH$ increases during standard expansion and as a consequence we can show that

$$d_{\rm hor} \sim d_H. \qquad\qquad (4.7)$$

The so-called Hubble distance d_H is defined simply as the inverse of the Hubble parameter H. This is the source of the horizon problem. Inflation solves this problem by making $d_{\rm hor} \gg d_H$, i.e. by making $1/aH$ decreases dramatically instead of increasing. After the end of inflation normal expansion returns but the Universe remains within a smooth thermalized patch. This means in particular that the portion of the Universe visible before inflation is far larger than the portion of the Universe visible after inflation. See figure 4.1.

Let us put this important point in different words. The CMB radiation consists of photons from the epochs of recombination and photon decoupling. The CMB radiation comes uniformly from every direction of the sky. The physical distance at

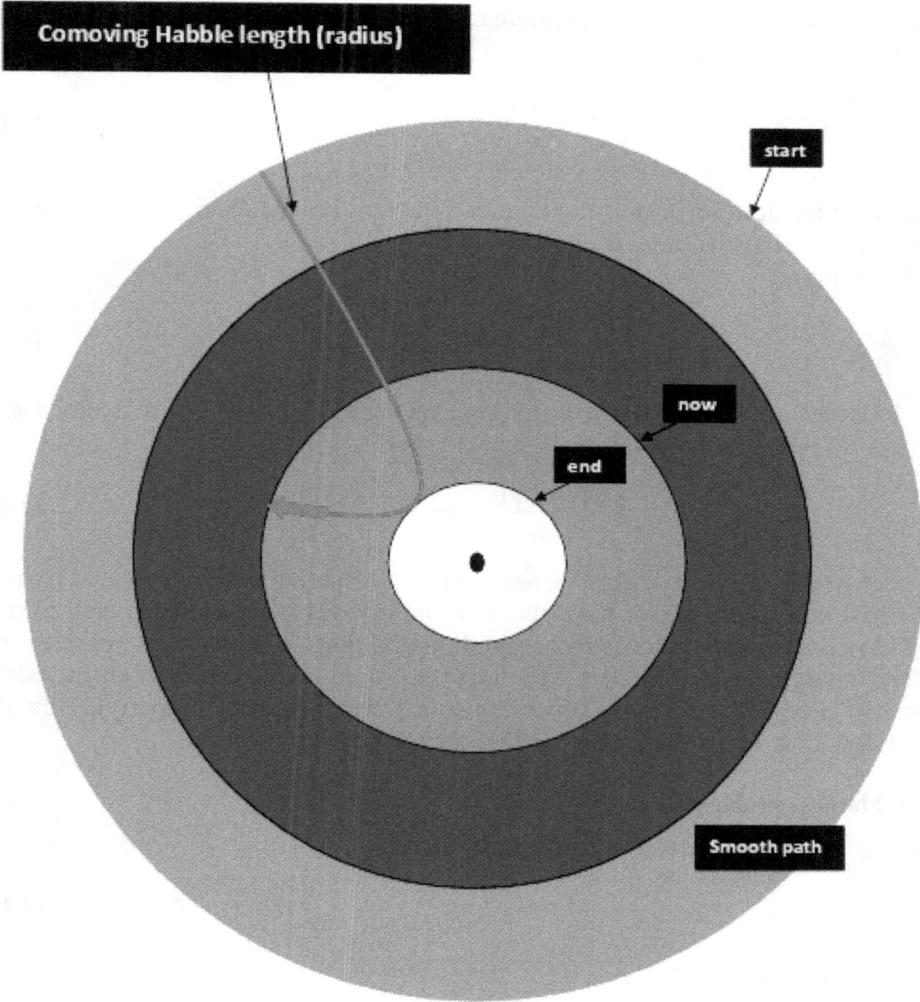

Figure 4.1. The resolution of the horizon problem.

the time of emission t_e of the source of the CMB radiation as measured from an observer on Earth making an observation at time t_0 is given by

$$\Delta d(t_e) = a(t_e) \int_{t_e}^{t_0} \frac{dt_1}{a(t_1)} \tag{4.8}$$
$$= 3t_e^{2/3}(t_0^{1/3} - t_e^{1/3}): \quad \text{MD}.$$

The physical distance between sources of CMB radiation coming from opposite directions of the sky at the time of emission is therefore given by

$$2\Delta d(t_e) = 6t_e^{2/3}(t_0^{1/3} - t_e^{1/3}): \quad \text{MD}. \tag{4.9}$$

At the time of emission t_e the maximum distance a photon had traveled since the Big Bang is

$$d_{\text{hor}}(t_e) = a(t_e) \int_0^{t_e} \frac{dt_1}{a(t_1)} \tag{4.10}$$

$$= 3t_e: \quad \text{MD}.$$

This is the particle horizon at the time of emission, i.e. the maximum distance that light can travel at the time of emission.

We compute

$$\frac{2\Delta d(t_e)}{d_{\text{hor}}(t_e)} = 2(a(t_e)^{-1/2} - 1). \tag{4.11}$$

By looking at CMB, we are looking at the Universe at a scale factor $a_{\text{CMB}} \equiv a(t_e) = 1/1200$. Thus

$$\frac{2\Delta d(t_e)}{d_{\text{hor}}(t_e)} \simeq 67.28. \tag{4.12}$$

In other words, $2\Delta d(t_e) > d_{\text{hor}}(t_e)$. See figure 4.2. The two widely separated parts of the CMB considered above have therefore non-overlapping horizons and as such they have no causal contact at recombination (the time of emission t_e), yet these two widely separated parts of the CMB have the same temperature to an incredible degree of precision (this is the observed isotropy/homogeneity property). How did they know how to do that? This is precisely the horizon problem.

4.1.2 Flatness problem

The first Friedmann equation can be rewritten as

$$\Omega - 1 = \frac{\kappa}{a^2 H^2}. \tag{4.13}$$

The density parameter is

$$\Omega = \frac{\rho}{\rho_c}. \tag{4.14}$$

The critical density is

$$\rho_c = \frac{3}{8\pi G} H^2. \tag{4.15}$$

We know that $1/(a^2 H^2) = a^{1+3w}/H_0^2$ and thus as the Universe expands the quantity $\Omega - 1$ increases, i.e. Ω moves away from 1. The value $\Omega = 1$ is therefore a repulsive (unstable) fixed point since any deviation from this value will tend to increase with time. Indeed, we compute (with $g = \Omega - 1$)

$$a\frac{dg}{da} = (1 + 3w)g. \tag{4.16}$$

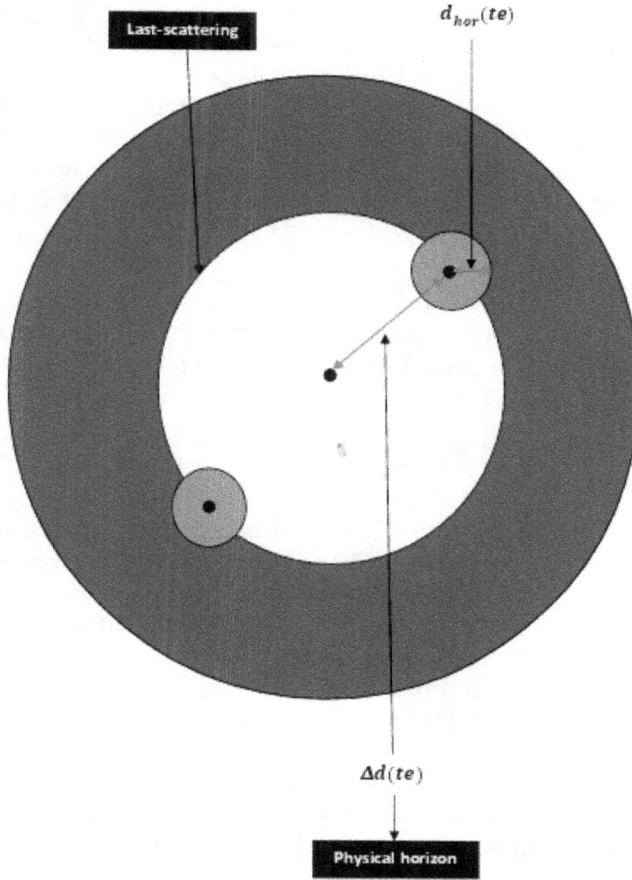

Figure 4.2. The horizon problem.

By assuming the strong energy condition we have $\rho + P \geqslant 0$ and $\rho + 3P \geqslant 0$, i.e. $1 + 3w > 0$. The value $\Omega = 1$ is then clearly a repulsive fixed point since $d\Omega/d \ln a > 0$.

As a consequence, the value $\Omega \sim 1$ observed today can only be obtained if the value of Ω in the early Universe is fine-tuned to be extremely close to 1. This is the flatness problem.

4.2 Elements of inflation

4.2.1 Solving the flatness and horizon problems

The second Friedmann equation can be put into the form

$$\frac{\ddot{a}}{a} = -\frac{4\pi G}{3}(\rho + 3P). \tag{4.17}$$

For matter satisfying the strong energy condition, i.e. $\rho + 2P \geqslant 0$, we have $\ddot{a} < 0$. Inflation is any epoch with $\ddot{a} > 0$. We explain this further below.

We have shown that the first Friedmann equation can be put into the form $|\Omega - 1| = \kappa/(a^2 H^2)$. The problem with the hot Big Bang model (big bang without inflation) is simply that aH is always decreasing so that Ω is always flowing away from 1. Indeed, in a universe filled with a fluid with an equation of state $w = P/\rho$ with the strong energy condition $1 + 3w > 0$ the comoving Hubble radius is given by

$$\frac{1}{aH} \propto a^{\frac{1+3w}{2}}. \qquad (4.18)$$

Thus $\ddot{a} = d(aH)/dt$ is always negative. Inflation is the hypothesis that during the early Universe there was a period of accelerated expansion $\ddot{a} > 0$. We write this condition as

$$\ddot{a} = \frac{d(aH)}{dt} > 0 \Leftrightarrow P < -\frac{\rho}{3}. \qquad (4.19)$$

Thus the comoving Hubble length $1/(aH)$ is decreasing during inflation, whereas in any other epoch it will be increasing. This behavior holds in a vacuum-dominated flat universe ($P = -\rho$, $\rho \propto a^0$, $a(t) \propto \exp(Ht)$). However, inflation can only be a phenomena of the early Universe and thus must terminate quickly in order for the hot Big Bang theory to proceed normally.

Inflation solves the flatness problem by construction since in the first Friedmann equation $|\Omega - 1| = \kappa/(a^2 H^2)$ the right-hand side decreases rapidly during inflation and is thus driving Ω towards 1 (towards flatness) quite fast. Another way of putting it using the first Friedmann equation in the form $H^2 = 8\pi G\rho/3 - \kappa/a^2$ is as follows. In a vacuum-dominated (for example) universe the mass density $\rho \propto a^0$ grows very fast with respect to the spatial curvature term $-\kappa/a^2$ and hence the universe becomes flatter very quickly.

The horizon problem is also solved by inflation. Recall that this problem arises from the fact that the physical horizon length $d_{\text{hor}}(t_e)$ grows more rapidly with the scale factor (in a matter-dominated or radiation-dominated universe) than the physical distance $2\Delta d(t_e)$ between any two comoving objects. We need therefore to reverse this situation so that

$$\Delta d(t_e) \ll d_{\text{hor}}(t_e). \qquad (4.20)$$

Or equivalently

$$\int_0^{t_e} \frac{dt_1}{a(t_1)} \gg \int_{t_e}^{t_0} \frac{dt_1}{a(t_1)}. \qquad (4.21)$$

This means in particular that we want a situation where photons can travel much further before recombination/decoupling than they can afterwards. Equivalently, if the Hubble radius is decreasing then the strong energy condition is viloated and as a consequence the Big Bang singularity is pushed to infinite negative conformal time since

$$\tau = \int \frac{dt}{a(t)} \propto \frac{2}{1+3w} a^{\frac{1+3w}{2}}. \tag{4.22}$$

In other words, there is much more conformal time between the initial Big Bang singularity and decoupling with inflation.

In a universe with a period of inflation, the comoving Hubble length $1/(aH)$ is decreasing during inflation. Thus if we start with a large Hubble length then a sufficiently large and smooth patch within the Hubble length can form by ordinary causal interactions. Inflation will cause this Hubble length to shrink enormously to within the smooth patch and after inflation comes to an end the Hubble length starts increasing again but remains within the smooth patch. See figure 4.1.

This can also be stated as follows. All comoving scales k^{-1} which are relevant today were larger than the Hubble radius until $a = 10^{-5}$ (start of inflation). At earlier times these scales were within the Hubble radius and thus were causally connected, whereas at recent times these scales re-entered again within the Hubble radius. See figure 4.3.

The observable Universe is therefore one causal patch of a much larger unobservable Universe. In other words, there are parts of the Universe which cannot communicate with us yet but they will eventually come into view as the cosmological horizon moves out and which will appear to us no different from any other region of space we have already seen since they are within the smooth patch. This explains homogeneity or the horizon problem. However, there are possibly other parts of the Universe outside the smooth patch which are different from the observable Universe.

4.2.2 Inflation

Inflation can be driven by a field called inflaton. This is a scalar field coupled to gravity with dynamics given by the usual action

$$S_\phi = \int d^4x \sqrt{-\det g} \left[-\frac{1}{2} g^{\mu\nu} \nabla_\mu \phi \nabla_\nu \phi - V(\phi) \right]. \tag{4.23}$$

The equations of motion read

$$\frac{\delta S_\phi}{\delta \phi} \equiv \nabla_\nu (g^{\mu\nu} \nabla_\mu \phi) - \frac{\delta V}{\delta \phi}$$

$$= \frac{1}{\sqrt{-\det g}} \partial_\mu (\sqrt{-\det g} \ \partial^\mu \phi) - \frac{\delta V}{\delta \phi} \tag{4.24}$$

$$= \partial_\mu \partial^\mu \phi + \frac{1}{2} g^{\alpha\beta} \partial_\mu g_{\alpha\beta} \partial^\mu \phi - \frac{\delta V}{\delta \phi}$$

$$= 0.$$

For a homogeneous field $\phi \equiv \phi(t, \vec{x}) = \phi(t)$ we obtain

$$\partial_0 \partial^0 \phi + \frac{1}{2} g^{\alpha\beta} \partial_0 g_{\alpha\beta} \partial^0 \phi - \frac{\delta V}{\delta \phi} = 0. \tag{4.25}$$

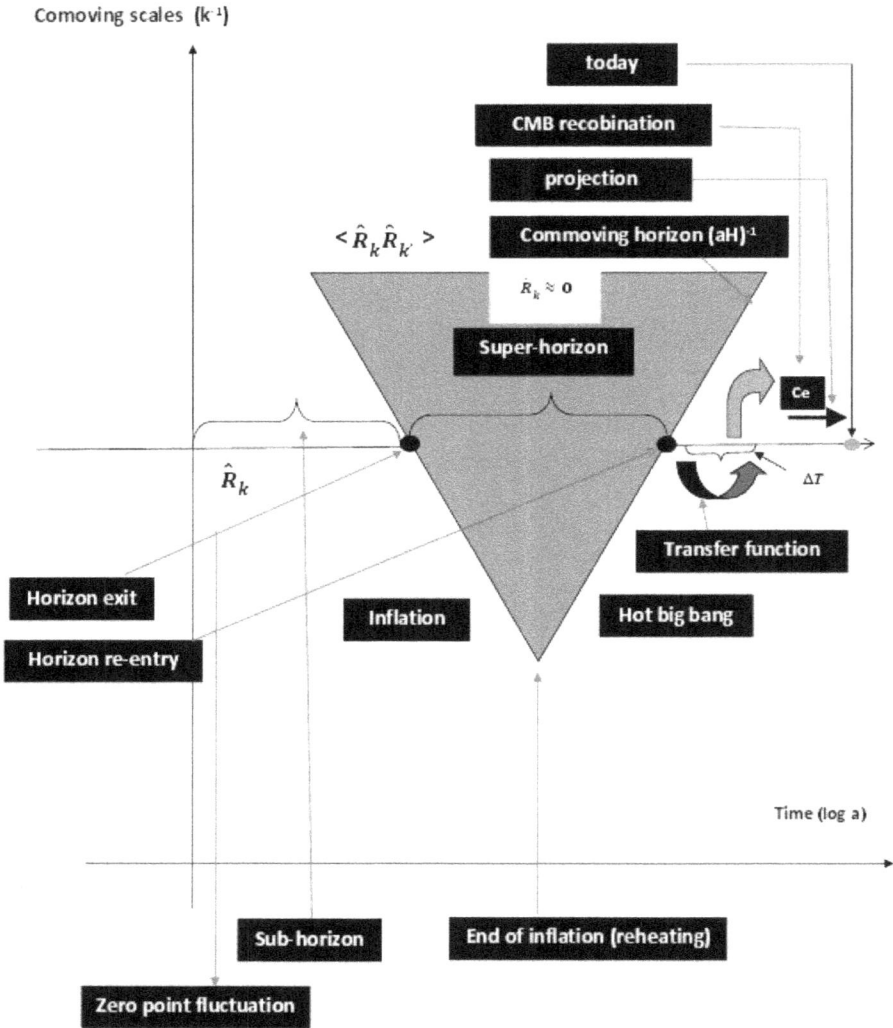

Figure 4.3. The resolution of the horizon problem.

In the Ropertson–Walker metric we obtain

$$\ddot{\phi} + 3H\dot{\phi} + \frac{\delta V}{\delta \phi} = 0. \tag{4.26}$$

The corresponding stress–energy–momentum tensor is calculated to be given by

$$T^{(\phi)}_{\mu\nu} = \nabla_\mu \phi \nabla_\nu \phi - \frac{1}{2}g_{\mu\nu}g^{\rho\sigma} \nabla_\rho \phi \nabla_\sigma \phi - g_{\mu\nu}V(\phi). \tag{4.27}$$

Explicit calculation shows that this stress–energy–momentum tensor is of the form of the stress–energy–momentum tensor of a perfect fluid $T_\mu{}^\nu = (-\rho_\phi, P_\phi, P_\phi, P_\phi)$ with

$$\rho_\phi = \frac{1}{2}\dot\phi^2 + V, \quad P_\phi = \frac{1}{2}\dot\phi^2 - V. \tag{4.28}$$

The equation of state is therefore

$$w_\phi = \frac{P_\phi}{\rho_\phi} = \frac{\frac{1}{2}\dot\phi^2 - V}{\frac{1}{2}\dot\phi^2 + V}. \tag{4.29}$$

We can have accelerated expansion $w_\phi < -1/3$ if the potential dominates over the kinetic energy. In other words we will have inflation whenever the potential dominates. The first Friedmann equation in this case reads (assuming also flatness)

$$H^2 = \frac{8\pi G}{3}\left(\frac{1}{2}\dot\phi^2 + V\right). \tag{4.30}$$

The second Friedmann equation reads

$$\frac{\ddot a}{a} = -\frac{8\pi G}{3}(\dot\phi^2 - V) \tag{4.31}$$

$$= H^2(1 - \epsilon).$$

The so-called slow-roll parameter is given by

$$\epsilon = 4\pi G\frac{\dot\phi^2}{H^2} \tag{4.32}$$

$$= \frac{3}{2}(1 + w_\phi).$$

This can also be expressed as

$$\epsilon = -\frac{\dot H}{H^2}. \tag{4.33}$$

Let us introduce $N = \ln a$, i.e. $dN = H dt$. Then we can show that

$$\epsilon = -\frac{d \ln H}{dN}. \tag{4.34}$$

Inflation corresponds to $\epsilon < 1$. In the so-called de Sitter limit $P_\phi \longrightarrow -\rho_\phi$ ($w_\phi \longrightarrow -1, \epsilon \longrightarrow 0$) we observe that the kinetic energy can be neglected compared to the potential energy, i.e. $\dot\phi^2 \ll V$. We have then

$$\epsilon \ll 1 \Leftrightarrow \dot\phi^2 \ll V. \tag{4.35}$$

This condition means that the scalar field is moving very slowly because for example the potential is flat.

In order to maintain accelerated expansion for a sufficient long time we require also that the second derivative of ϕ is small enough, viz $|\ddot{\phi}| \ll |3H\dot{\phi}|$ and $|\ddot{\phi}| \ll |\delta V/\delta\phi|$. This second condition means that the field keeps moving slowly over a wide range of its values and hence the term slowly rolling. We compute

$$\frac{1}{2\epsilon}\frac{d\epsilon}{dt} = \frac{\ddot{\phi}}{\dot{\phi}} - \frac{\dot{H}}{H} \Rightarrow \frac{1}{2\epsilon}\frac{d\epsilon}{dN} = \frac{\ddot{\phi}}{H\dot{\phi}} + \epsilon. \tag{4.36}$$

We introduce a second slow-roll parameter by

$$\eta = -\frac{\ddot{\phi}}{\dot{\phi}H}$$
$$= \epsilon - \frac{1}{2\epsilon}\frac{d\epsilon}{dN}. \tag{4.37}$$

It is clear that sustained acceleration is equivalent to the condition $\eta \ll 1$. In other words

$$\eta \ll 1 \Leftrightarrow |\ddot{\phi}| \ll |3H\dot{\phi}|, \quad |\ddot{\phi}| \ll |\frac{\delta V}{\delta\phi}|. \tag{4.38}$$

The equations of motion in the slow-roll regime are

$$H^2 \simeq \frac{8\pi G}{3}V, \quad \dot{\phi} = -\frac{\delta V/\delta\phi}{3H}. \tag{4.39}$$

Since ϕ is almost constant during slow-roll we can assume that $H^2 \simeq$ constant in this regime and hence $a(t) \simeq \exp(Ht)$. This is de Sitter spacetime.

Instead of the Hubble slow-roll parameters ϵ and η we can work with the potential slow-roll parameters ϵ_V and η_V defined as follows. The first slow-roll parameter ϵ is equivalent to

$$\epsilon \simeq \frac{3}{2}\frac{\dot{\phi}^2}{V}$$
$$\simeq \frac{1}{6H^2}\frac{(\delta V/\delta\phi)^2}{V}$$
$$\simeq \frac{1}{16\pi G}\frac{(\delta V/\delta\phi)^2}{V^2} \tag{4.40}$$
$$\simeq \epsilon_V.$$

We compute

$$\frac{1}{2\epsilon}\frac{d\epsilon}{dt} \simeq \frac{\dot{\phi}}{\delta V/\delta\phi}\left(\frac{\delta^2 V}{\delta\phi^2} - \frac{1}{V}\left(\frac{\delta V}{\delta\phi}\right)^2\right) \Leftrightarrow \frac{1}{2\epsilon}\frac{d\epsilon}{dN} \simeq \frac{\dot{\phi}}{H\delta V/\delta\phi}\left(\frac{\delta^2 V}{\delta\phi^2} - \frac{1}{V}\left(\frac{\delta V}{\delta\phi}\right)^2\right). \tag{4.41}$$

The second slow-roll parameter η is therefore equivalent to

$$\eta \simeq \frac{1}{8\pi G} \frac{\delta^2 V/\delta\phi^2}{V} - \epsilon_V \qquad (4.42)$$
$$\simeq \eta_V - \epsilon_V.$$

The slow-roll conditions ϵ, $|\eta| \ll 1$ are equivalent to

$$\epsilon_V, |\eta_V| \ll 1. \qquad (4.43)$$

These are obviously conditions on the shape of the inflationary potential. The first (inflation) states that the slope of the potential is small, whereas the second (prolonged inflation) states that the curvature of the potential is small. These are necessary conditions for the slow-roll state but they are not sufficient. For example, a potential could be very flat but the velocity of the field is very large.

The amount of inflation is defined by the logarithm of the expansion or equivalently the number of e-foldings N given by

$$N = \ln \frac{a(t_{\text{end}})}{a(t_{\text{initial}})} = \int_{t_i}^{t_e} H dt$$
$$= \int_{\phi(t_i)}^{\phi(t_e)} \frac{H}{\dot{\phi}} d\phi$$
$$= 8\pi G \int_{\phi(t_i)}^{\phi(t_e)} \frac{V}{\delta V/\delta\phi} d\phi$$
$$= \sqrt{8\pi G} \int_{\phi(t_i)}^{\phi(t_e)} \frac{1}{\sqrt{2\epsilon_V}} d\phi. \qquad (4.44)$$

In order to solve the horizon and the flatness problems we need a minimum amount of inflation of at least 60 e-foldings which is equivalent to an expansion by a factor of 10^{30}.

Inflation ends at the value of the field ϕ_{end} where the kinetic energy becomes comparable to the potential energy. This is the value where the slow-roll conditions break down, viz $\epsilon(\phi_{\text{end}}) = 1$, $\epsilon_V(\phi_{\text{end}}) \simeq 1$. After inflation ends, the scalar field starts oscillating around the minimum of the potential and then decays into conventional matter. This is the process of reheating which is followed by the usual hot Big Bang theory. See figure 4.4.

The most simple and interesting models of inflation involve (1) a single rolling scalar field, and (2) a potential V which satisfies the slow-roll conditions in some regions and possesses a minimum with zero potential where inflation terminates. Some of these models are

$$V = \lambda_p \phi^p, \quad \text{chaotic inflation.} \qquad (4.45)$$

$$V = V_0(1 + \cos\frac{\phi}{f}), \quad \text{natural inflation.} \qquad (4.46)$$

$$V = V_0 \exp(\alpha\phi), \quad \text{power-law inflation.} \qquad (4.47)$$

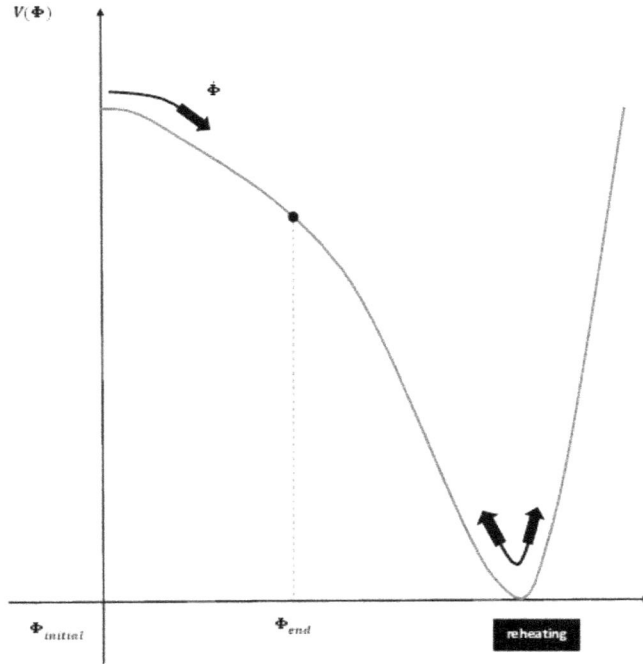

Figure 4.4. The end of inflation: reheating.

4.2.3 Amount of inflation

As pointed out above, in order to solve the horizon and the flatness problems we need a minimum amount of inflation of at least 60 e-foldings which is equivalent to an expansion by a factor of 10^{30}. A clean argument is found in [8].

We imagine a universe in which inflation starts at t_i with a scale factor $a(t_i)$ and ends at t_f with a scale factor $a(t_f)$. The current time is t_0 with a scale factor $a(t_0)$. During inflation we can assume that H is constant (de Sitter spacetime) and as a consequence the (vacuum) mass density ρ_V is constant. We assume for simplicity that between the end of inflation and the current moment the Universe is radiation-dominated with a density ρ_R behaving as $1/a^4$. We assume that at t_f the vacuum density is fully converted into radiation, viz $\rho_R(t_f) = \rho_V$. We recall that the density parameter Ω_C associated with curvature is given by

$$\Omega_C = \frac{\rho_C}{\rho_c} = -\frac{\kappa}{a^2 H^2} = -\frac{\kappa}{\dot{a}^2}. \tag{4.48}$$

During inflation the expansion is accelerating, since gravity is acting as repulsive due to the dominance of the vacuum energy, and thus $\dot{a}$ increases and hence Ω_C decreases. Thus $\Omega_C \ll 1$ today at t_0 can be easily explained with more inflation. On the other hand, if inflation is preceded with a long phase of deceleration in which gravity acts in the usual way as attractive, then $\dot{a}$ at t_i must be very small and hence $\Omega_C \gg 1$ at the beginning of inflation. This case also would only require more inflation to explain. Thus it is sufficient to assume that $\Omega_C \sim 1$ at t_0 and t_i. This means that at t_0 and t_i ρ_C is equal to the critical density ρ_c. In other words, at t_0 and t_i we have nothing but curvature, viz

$$\rho_V(t_i) = \rho_C(t_i), \quad \rho_R(t_0) = \rho_C(t_0). \tag{4.49}$$

We compute now

$$
\begin{aligned}
\frac{\rho_C(a_0)}{\rho_C(a_i)} &= \left(\frac{a_i}{a_0}\right)^2 \\
&= \frac{\rho_R(a_0)}{\rho_V(a_i)} \\
&= \frac{\rho_R(a_0)}{\rho_V(a_f)} \\
&= \frac{\rho_R(a_0)}{\rho_R(a_f)} \\
&= \left(\frac{a_f}{a_0}\right)^4 .
\end{aligned}
\tag{4.50}
$$

The solution is thus

$$\frac{a_f}{a_i} = \sqrt{\frac{a_0}{a_i}} = \frac{a_0}{a_f}. \tag{4.51}$$

In general we obtain

$$\frac{a_f}{a_i} \geqslant \frac{a_0}{a_f}. \tag{4.52}$$

In terms of the e-fold number $N = \ln a$ this reads

$$N_f - N_i \geqslant N_0 - N_f. \tag{4.53}$$

The amount of inflation is precisely $\Delta N = N_f - N_i$. Thus we have more expansion during inflation than after the end of inflation. Although there is no upper bound on the amount of inflation, there is a lower bound given by

$$
\begin{aligned}
\Delta N_{\min} &= N_0 - N_f \\
&= \ln \frac{a(t_0)}{a(t_f)} \\
&= \frac{1}{4} \ln \frac{\rho_R(a_f)}{\rho_R(a_0)} .
\end{aligned}
\tag{4.54}
$$

As we will show later the energy scale of inflation is $\rho_R(a_f) \sim 10^{-12}\rho_{\rm pl}$. Also, we have already seen that the energy density contained in radiation is $\rho_{\rm rad} = 10^{-34} {\rm g/cm}^3 = 10^{-127}\rho_{\rm pl}$ where $\rho_{\rm pl} = 10^{93} {\rm g/cm}^3$. Hence the minimum amount of inflation is

$$\Delta N_{\min} = \frac{1}{4} \ln \frac{10^{-12}}{10^{-127}} \sim 66. \tag{4.55}$$

From this approach we can also get the other important estimation

$$1 + \frac{\bar{P}_i}{\bar{\rho}_i} < \frac{1}{99} \sim 10^{-2}. \tag{4.56}$$

4.2.4 End of inflation: reheating and scalar-matter-dominated epoch

We start by summarizing the main results of the previous section. The main equations are the equation of motion of the inflaton scalar field and the Friedmann equation given, respectively, by

$$\ddot{\phi} + 3H\dot{\phi} + \frac{\partial V}{\partial \phi} = 0. \tag{4.57}$$

$$H^2 = \frac{8\pi G}{3} \left(\frac{1}{2} \dot{\phi}^2 + V \right). \tag{4.58}$$

The slow-roll approximation is given by the conditions

$$\epsilon \ll 1, \quad \eta \ll 1 \Leftrightarrow \dot{\phi}^2 \ll V, \quad |\ddot{\phi}| \ll |3H\dot{\phi}|, \quad |\ddot{\phi}| \ll \left| \frac{\delta V}{\delta \phi} \right|. \tag{4.59}$$

Equivalently

$$\epsilon_V \ll 1, \quad \eta_V \ll 1 \Leftrightarrow \frac{(\delta V / \delta \phi)^2}{V^2} \ll 1, \quad \frac{\delta^2 V / \delta \phi^2}{V} \ll 1. \tag{4.60}$$

The equations of motion during slow-roll are:

$$H^2 \simeq \frac{8\pi G}{3} V, \quad \dot{\phi} = -\frac{\delta V / \delta \phi}{3H}. \tag{4.61}$$

These two equations can be combined to give the equation of motion

$$\frac{d \ln a}{d\phi} = -8\pi G \frac{V}{\frac{\partial V}{\partial \phi}}. \tag{4.62}$$

The solution is

$$a(\phi) = a_i \exp\left(-8\pi G \int_{\phi_i}^{\phi} \frac{V}{\frac{\partial V}{\partial \phi}} d\phi \right). \tag{4.63}$$

For a power-law potential $V = \lambda\phi^n/n$ the slow-roll conditions are equivalent to $\phi \gg 1$ and the above solution is given by

$$a(\phi) = a_i \exp\left(-\frac{4\pi G}{n}(\phi^2 - \phi_i^2)\right). \tag{4.64}$$

Let us consider a quadratic potential $V = m^2\phi^2/2$ at the end of inflation. By combining the Friedmann equation and the equation of motion of the scalar field we obtain a differential equation for $\dot{\phi}$ as a function of ϕ given by

$$\frac{d\dot{\phi}}{d\phi} = -\frac{m^2\phi + \sqrt{12\pi G(\dot{\phi}^2 + m^2\phi^2)}}{\dot{\phi}}. \tag{4.65}$$

In slow-roll we have $d\dot{\phi}/d\phi \sim 0$, $\dot{\phi} = $ constant and $\phi \gg 1$. A solution is given by

$$\dot{\phi} = -\frac{m}{\sqrt{12\pi G}}, \quad \phi = -\frac{m}{\sqrt{12\pi G}}(t_f - t). \tag{4.66}$$

Since $|\phi| \gg 1$ during inflation we must have $mt \gg 1$. The pressure is given by

$$P = -\rho + \dot{\phi}^2 = -\rho + \frac{m^2}{12\pi G}. \tag{4.67}$$

When the scalar field drops to its Planck value $\phi \sim 1/\sqrt{12\pi G}$ we observe that the energy density drops to $m^2/12\pi G$ and hence the pressure vanishes. Inflation is then over. Thus inflation ends when the scalar field becomes of order 1 in Planck units. The duration of inflation is

$$\Delta t = t_f - t_i = -\frac{\sqrt{12\pi G}}{m}\Delta\phi = \sqrt{12\pi G}\frac{\phi_i}{m}. \tag{4.68}$$

By substituting the above solution into the Friedmann equation we get

$$H = \frac{1}{3}m^2(t_f - t) = -\sqrt{\frac{4\pi G}{3}}m\phi. \tag{4.69}$$

$$a(t) = a(t_f)\exp\left(-\frac{m^2}{6}(t_f - t)^2\right) \tag{4.70}$$

We get immediately

$$\frac{a_f}{a_i} = \exp\left(2\pi G\phi_i^2\right). \tag{4.71}$$

Thus in order to get 75 e-folds we must start with a value of the scalar field which is four times the Planck value, viz $\phi_i \sim 4/\sqrt{G}$.

We can also show that the scale factor behaves as

$$a = t^{2/3}\left(1 + O\left(\frac{1}{(m^2t^2)}\right)\right). \tag{4.72}$$

We get therefore a graceful exit into a matter-dominated phase. In summary, if the mass is sufficiently small compared to the Planck mass the inflationary phase will last sufficiently long and is followed by a matter-dominated phase. Furthermore, a quadratic potential gives rise naturally to a post-inflationary matter-dominated universe consisting of heavy scalar particles which will eventually be converted into photons, baryons and leptons (reheating). The usual radiation-dominated, matter-dominated and vacuum-dominated phases follow after reheating.

Other power-law potentials will also give oscillatory stages with scale factors behaving as $a \sim t^p$. For example, a quartic potential will give an oscillating scalar field with the scale factor of a radiation-dominated universe, viz $a \sim t^{1/2}$.

4.3 Perfect fluid revisited

Let ρ be the mass density of a perfect fluid, P its pressure, S its entropy per unit mass and $\vec{u}$ its flow velocity vector, i.e. the velocity of an element of fluid at a point $\vec{x}$ at a time t. The equation of state of the perfect fluid allows us to determine the pressure in terms of the mass density ρ and the entropy S, viz

$$P = P(\rho, S). \tag{4.73}$$

The state of the prefect fluid is therefore completely determined by the mass density ρ, the entropy per unit mass S and the flow velocity vector $\vec{u}$. In the absence of dissipation the entropy is conserved, i.e.

$$\frac{dS}{dt} = \frac{\partial S}{\partial t} + (\vec{u} \cdot \vec{\nabla})S \tag{4.74}$$
$$= 0.$$

The mass M contained in a volume V is given by

$$M = \int_V dV \rho. \tag{4.75}$$

The rate of change of the mass contained in V is obviously given by

$$\frac{dM}{dt} = \int_V dV \frac{\partial \rho}{\partial t}. \tag{4.76}$$

This rate of change is also obviously given by the mass flowing through the surface Σ which encloses the volume V. Since the amount of mass flowing per unit area per unit time is $\vec{J} = \rho \vec{u}$ the rate of change dM/dt can be rewritten as

$$\frac{dM}{dt} = -\oint_\Sigma d\vec{\sigma} \vec{J} = -\int_V dV \vec{\nabla} (\rho \vec{u}). \tag{4.77}$$

We get therefore the continuity equation

$$\frac{\partial \rho}{\partial t} + \vec{\nabla} (\rho \vec{u}) = 0. \tag{4.78}$$

The Newtonian gravitational potential Φ generated by the mass density ρ is given by the Poisson equation

$$\nabla\Phi = 4\pi G\rho. \tag{4.79}$$

The force exerted by this potential Φ on a mass ΔM is given by Newton's law of gravitation, viz

$$\vec{F}_{\mathrm{gr}} = -\Delta M\,\vec{\nabla}\,\Phi. \tag{4.80}$$

The other force acting on ΔM is due to the pressure P of the perfect fluid and is given by

$$\begin{aligned} \vec{F}_{\mathrm{pr}} &= -\oint_{\Delta\Sigma} P d\vec{\sigma} \\ &= -\oint_{\Delta V} \vec{\nabla}\,P dV \\ &\simeq -\vec{\nabla}\,P\Delta V. \end{aligned} \tag{4.81}$$

Newton's second law then reads

$$-\Delta M\,\vec{\nabla}\,\Phi - \vec{\nabla}\,P\Delta V = \Delta M\vec{g}. \tag{4.82}$$

However,

$$\vec{g} = \frac{d\vec{u}}{dt} = \frac{\partial\vec{u}}{\partial t} + (\vec{u}\,\vec{\nabla})\vec{u}. \tag{4.83}$$

By identification we obtain Euler's equation

$$\frac{\partial\vec{u}}{\partial t} + (\vec{u}\,\vec{\nabla})\vec{u} + \frac{\vec{\nabla}P}{\rho} + \vec{\nabla}\,\Phi = 0. \tag{4.84}$$

We have seven equations: equation of state (4.73), conservation of the entropy (4.74), continuity equation (4.78), Poisson's equation (4.79) and three Euler's equations (4.84) for seven unknowns: ρ, P, S, $\vec{u}$ and Φ.

Linearization of these equations around an expanding homogeneous and isotropic universe with mass density $\rho_0 = \rho_0(t)$ and flow velocity vector $\vec{u_0}$ obeying the Hubble law, i.e. $\vec{u_0} = H(t)\vec{x}$, leads to a Newtonian theory of gravitational instabilities. This topic is discussed at length in [7]. In the following we will concentrate instead on the corresponding general relativistic theory following mostly [3, 7].

4.4 Cosmological perturbations

4.4.1 Metric perturbations

The Universe is isotropic and homogeneous and spatially flat with a gravitational field described by the Robertson–Walker metric. This is the punchline so far.

However, this is just an approximation which neglects the most obvious fact we observe directly around us which is the presence of structure: galaxies, stars and us. All departures from homogeneity and isotropy will be assumed to be small given by weak first order fluctuations. The perturbed metric is given by

$$g_{\mu\nu} = \bar{g}_{\mu\nu} + \delta g_{\mu\nu}. \tag{4.85}$$

$$d\bar{s}^2 = \bar{g}_{\mu\nu}dx^\mu dx^\nu = -dt^2 + a^2(t)dx^i dx^i. \tag{4.86}$$

The inverse is given by

$$g^{\mu\nu} = \bar{g}^{\mu\nu} - \delta g^{\mu\nu}, \quad \delta g^{\mu\nu} = \bar{g}^{\mu\alpha}\bar{g}^{\nu\beta}\delta g_{\alpha\beta}. \tag{4.87}$$

The metric is a symmetric (0, 2) tensor containing 10 degrees of freedom. The 00 component is a scalar under spatial rotations, the $0i$ (or equivalently $i0$) components form a vector under spatial rotations and the ij components form a rank 2 tensor under spatial rotations. We introduce two scalars Φ and Ψ, a vector B_i and a traceless rank 2 tensor E_{ij} under the action of spatial rotations by the relations [3]

$$\begin{aligned} \delta g_{00} &= -2\Phi \\ \delta g_{0i} &= aB_i \\ \delta g_{ij} &= a^2[2E_{ij} - 2\Psi\delta_{ij}]. \end{aligned} \tag{4.88}$$

The metric takes the form

$$ds^2 = -(1 + 2\Phi)dt^2 + 2aB_i dt dx^i + a^2[2E_{ij} + (1 - 2\Psi)\delta_{ij}]dx^i dx^j. \tag{4.89}$$

$\Psi\delta_{ij}$ and E_{ij} form together a rank 2 tensor under spatial rotations where Ψ is precisely its trace. We call Ψ the spatial curvature perturbation, E_{ij} spatial shear tensor, B_i the shift and Φ the lapse.

We can decompose any 3-vector such as B_i into a divergenceless 3-vector S_i satisfying $\partial^i S_i = 0$, and a total derivative $\partial_i B$ as follows

$$B_i = -S_i + \partial_i B. \tag{4.90}$$

This is Helmholtz's decomposition. Similarly, we can Hodge decompose any symmetric traceless 3-tensor such as E_{ij} into a divergenceless symmetric traceless 3-tensor h_{ij} satisfying $\partial^i h_{ij} = 0$ and $h_i^i = 0$, and a divergenceless 3-vector F_i satisfying $\partial^i F_i = 0$, and a scalar E as follows

$$E_{ij} = \partial_i\partial_j E + \partial_i F_j + \partial_j F_i + \frac{1}{2}h_{ij}. \tag{4.91}$$

We get then the metric

$$\begin{aligned} ds^2 = &-(1 + 2\Phi)dt^2 + 2a(\partial_i B - S_i)dt dx^i \\ &+ a^2[h_{ij} + 2\partial_i\partial_j E + 2\partial_i F_j + 2\partial_j F_i + (1 - 2\Psi)\delta_{ij}]dx^i dx^j. \end{aligned} \tag{4.92}$$

There are four scalar degrees of freedom Φ, B, E and Ψ, four vector degrees of freedom contained in S_i and F_i which satisfy two constraints, and two tensor degrees of freedom contained in h_{ij} which satisfies four constraints. Thus the total number of degrees of freedom is $4 + 4 + 2 = 10$, which is precisely the correct number of degrees of freedom contained in the perturbed metric.

As we will see, scalars lead to, or are induced by, density fluctuations, and since they can suffer from gravitational instabilities they can also lead to structure formation. On the other hand, tensors lead to gravitational waves and as such they are absent in the Newtonian theory. Further, it can be shown that the vector perturbations S_i and F_i, which are related to the rotational motions of the fluid, decay as $1/a^2$ with the expansion of the Universe, which holds true already in Newtonian theory, and thus they do not play an important role in cosmology. The scalar, vector and tensor perturbations evolve independently of each other at the linear order and as such they can be treated separately.

In the following we will mostly neglect vector perturbations for simplicity. The metric will then read

$$ds^2 = -(1 + 2\Phi)dt^2 + 2a\partial_i B dt dx^i + a^2[h_{ij} + 2\partial_i\partial_j E + (1 - 2\Psi)\delta_{ij}]dx^i dx^j. \tag{4.93}$$

4.4.2 Gauge transformations

We recall that coordinate transformations are given by

$$x^\mu \longrightarrow x'^\mu = x^\mu + \epsilon^\mu(x). \tag{4.94}$$

$$g_{\mu\nu}(x) \longrightarrow g'_{\mu\nu}(x') = \frac{\partial x^\alpha}{\partial x'^\mu}\frac{\partial x^\beta}{\partial x'^\nu}g_{\alpha\beta}(x). \tag{4.95}$$

Explicitly we have

$$
\begin{aligned}
g'_{\mu\nu}(x') &= \left[\delta^\alpha_\mu - \frac{\partial \epsilon^\alpha}{\partial x'^\mu}\right]\left[\delta^\beta_\nu - \frac{\partial \epsilon^\beta}{\partial x'^\nu}\right]g_{\alpha\beta}(x) \\
&= g_{\mu\nu}(x) - \frac{\partial \epsilon^\alpha}{\partial x'^\nu}g_{\mu\beta}(x) - \frac{\partial \epsilon^\alpha}{\partial x'^\mu}g_{\alpha\nu}(x) \\
&= g_{\mu\nu}(x) - \frac{\partial \epsilon^\alpha}{\partial x^\nu}g_{\mu\alpha}(x) - \frac{\partial \epsilon^\alpha}{\partial x^\mu}g_{\alpha\nu}(x).
\end{aligned}
\tag{4.96}
$$

We will reinterpret these coordinate transformations as gauge transformations where all change is encoded only in the field perturbations, viz

$$\delta g_{\mu\nu}(x) \longrightarrow \delta g'_{\mu\nu}(x) = \delta g_{\mu\nu}(x) + \Delta\delta g_{\mu\nu}(x). \tag{4.97}$$

Equivalently

$$
\begin{aligned}
\Delta\delta g_{\mu\nu}(x) &= g'_{\mu\nu}(x) - g_{\mu\nu}(x) \\
&= \delta g'_{\mu\nu}(x) - \delta g_{\mu\nu}(x).
\end{aligned}
\tag{4.98}
$$

We compute

$$\Delta \delta g_{\mu\nu}(x) = g'_{\mu\nu}(x' - \epsilon) - g_{\mu\nu}(x)$$

$$= g'_{\mu\nu}(x') - \epsilon^\lambda \frac{\partial}{\partial x^\lambda} g_{\mu\nu}(x) - g_{\mu\nu}(x) \qquad (4.99)$$

$$= -\frac{\partial \epsilon^\beta}{\partial x^\nu} \bar{g}_{\mu\beta}(x) - \frac{\partial \epsilon^\alpha}{\partial x^\mu} \bar{g}_{\alpha\nu}(x) - \epsilon^\lambda \frac{\partial}{\partial x^\lambda} \bar{g}_{\mu\nu}(x).$$

Explicitly we have

$$\Delta \delta g_{ij}(x) = -\frac{\partial \epsilon_i}{\partial x^j} - \frac{\partial \epsilon_j}{\partial x^i} + 2a\dot{a}\epsilon_0 \delta_{ij}. \qquad (4.100)$$

$$\Delta \delta g_{i0}(x) = -\frac{\partial \epsilon_i}{\partial t} - \frac{\partial \epsilon_0}{\partial x^i} + 2\frac{\dot{a}}{a}\epsilon_i. \qquad (4.101)$$

$$\Delta \delta g_{00}(x) = -2\frac{\partial \epsilon_0}{\partial t}. \qquad (4.102)$$

Let us consider the metric (4.92) with the vector and tensor perturbations set to zero. We have then

$$ds^2 = -(1 + 2\Phi)dt^2 + 2a\partial_i B dt dx^i + a^2[2\partial_i\partial_j E + (1 - 2\Psi)\delta_{ij}]dx^i dx^j. \qquad (4.103)$$

The coordinate transformations (4.94) are given explicitly by $x^0 \longrightarrow x'^0 = x^0 + \epsilon^0$, $x^i \longrightarrow x'^i = x^i + \epsilon^i$. We will write $\epsilon^0 = \alpha$. The vector ϵ^i can be decomposed as $\epsilon_i = a^2\partial_i\beta + \epsilon_i^V$ where $\partial^i\epsilon_i^V = 0$. As before, we will neglect the vector contribution coming from ϵ_i^V since it will only contribute to the gauge transformations of the vector perturbations which we have dropped. By setting $\epsilon_i^V = 0$ the coordinate transformations (4.94) take now the form

$$t \longrightarrow t' = t + \alpha, \quad x^i \longrightarrow x'^i = x^i + \partial_i\beta. \qquad (4.104)$$

The corresponding gauge transformations are (with $\bar{g}_{ij} = a^2\delta_{ij}$, $\bar{g}_{i0} = 0$, $\bar{g}_{00} = -1$ and $H = \dot{a}/a$)

$$\Delta\Phi = -\frac{\partial\alpha}{\partial t}$$

$$\Delta B = a^{-1}\alpha - a\frac{\partial\beta}{\partial t} \qquad (4.105)$$

$$\Delta E = -\beta$$

$$\Delta\Psi = H\alpha.$$

This depends only on two functions α and β. Thus by choosing α and β appropriately we can make any two of the four scalar perturbations E, B, Φ and Ψ vanish. In other words, the space of the physical scalar perturbations is two-dimensional. This space is spanned by the two gauge-invariant linear combinations Φ_B and Ψ_B known as Bardeen potentials which are defined by

$$\Phi_B = \Phi - \frac{d}{dt}[a(a\partial_t E - B)] \tag{4.106}$$

$$\Psi_B = \Psi + \dot{a}[a\partial_t E - B].$$

Indeed, we compute

$$\Delta\Phi_B = \Delta\Phi - \frac{d}{dt}[a(a\partial_t \Delta E - \Delta B)] = 0 \tag{4.107}$$

$$\Delta\Psi_B = \Delta\Psi + \dot{a}[a\partial_t \Delta E - \Delta B] = 0.$$

Let us now consider the metric (4.92) with the scalar and vector perturbations set to zero. We obtain

$$ds^2 = -dt^2 + a^2[\delta_{ij} + h_{ij}]dx^i dx^j. \tag{4.108}$$

It is obvious from the above discussion that the tensor h_{ij} is invariant under gauge transformations, viz

$$\Delta h_{ij} = 0 \tag{4.109}$$

Some of the used and most useful gauge choices are as follows:

- **Longitudinal, conformal-Newtonian gauge:** We can choose β so that $E = 0$ and then choose α so that $B = 0$. These are unique choices which fix the gauge uniquely. This gauge is therefore given by

$$E = B = 0. \tag{4.110}$$

  The metric becomes

$$ds^2 = -(1 + 2\Phi)dt^2 + a^2(1 - 2\Psi)\delta_{ij}dx^i dx^j. \tag{4.111}$$

- **Synchronous gauge:** We can choose β so that $B = 0$ and then choose α so that $\Phi = 0$. This gauge is therefore given by

$$B = \Phi = 0. \tag{4.112}$$

  The metric becomes

$$ds^2 = -dt^2 + a^2[2\partial_i\partial_j E + (1 - 2\Psi)\delta_{ij}]dx^i dx^j. \tag{4.113}$$

The synchronous gauge does not fix the gauge completely. Indeed, we can check immediately that the choice $B = \Phi = 0$ remains intact under the gauge transformations

$$\alpha(t, x^i) = f_1(x^i), \quad \beta(t, x^i) = f_1(x^i)\int_{-\infty}^{t} \frac{dt'}{a^2(t')} + f_2(x^i), \tag{4.114}$$

for any functions $f_i(x^i)$.

4.4.3 Linearized Einstein equations

We want to linearize the Einstein equations

$$G_\nu^\mu = R_\nu^\mu - \frac{1}{2}Rg_\nu^\mu = 8\pi GT_\nu^\mu, \tag{4.115}$$

around the perturbed metric (4.85). We have already computed the components of the unperturbed Ricci tensor. We recall (with the prime denoting differentiation with respect to the conformal time, $d\eta = adt$ and $\mathcal{H} = a'/a$)

$$\bar{R}_0^0 = 3\frac{\ddot{a}}{a} = 3\frac{\mathcal{H}'}{a^2}. \tag{4.116}$$

$$\bar{R}_j^i = \frac{\delta_j^i}{a^2}(a\ddot{a} + 2\dot{a}^2) = \frac{\delta_j^i}{a^2}(\mathcal{H}'^2 + 2\mathcal{H}^2). \tag{4.117}$$

$$\bar{R}_0^i = 0. \tag{4.118}$$

The background stress–energy–momentum must also be diagonal by the background Einstein equations, viz

$$\bar{T}_0^0 \neq 0, \quad \bar{T}_0^i = 0, \quad \bar{T}_j^i \propto \delta_j^i. \tag{4.119}$$

The linearized Einstein's equations are of the form

$$\delta G_\nu^\mu = 8\pi G\delta T_\nu^\mu. \tag{4.120}$$

Both δG_ν^μ and δT_ν^μ are not gauge invariant. Indeed, under the gauge transformations (4.94) the tensors $\delta X_\nu^\mu = \delta G_\nu^\mu$, δT_ν^μ will transform as second rank tensors similarly to δg_ν^μ, i.e. as (4.97) with

$$\Delta\delta X_{\mu\nu}(x) = X'_{\mu\nu}(x' - \epsilon) - X_{\mu\nu}(x)$$
$$= -\frac{\partial\epsilon^\beta}{\partial x^\nu}\bar{X}_{\mu\beta}(x) - \frac{\partial\epsilon^\alpha}{\partial x^\mu}\bar{X}_{\alpha\nu}(x) - \epsilon^\lambda\frac{\partial}{\partial x^\lambda}\bar{X}_{\mu\nu}(x). \tag{4.121}$$

More explicitly we have

$$\Delta\delta X_{ij}(x) = -2\partial_i\partial_j\beta.\frac{\bar{X}_{kk}}{3} - \alpha\partial_t\bar{X}_{ij}. \tag{4.122}$$

$$\Delta\delta X_{i0}(x) = -\partial_t\partial_i\beta.\frac{\bar{X}_{kk}}{3} - \partial_i\alpha\bar{X}_{00}. \tag{4.123}$$

$$\Delta\delta X_{00}(x) = -2\partial_t\alpha\bar{X}_{00} - \alpha\partial_t\bar{X}_{00}. \tag{4.124}$$

We can construct gauge invariant quantities as follows. We observe that

$$\Delta I = -\alpha, \quad I = a(a\partial_t E - B) \text{ and } \Delta E = -\beta. \tag{4.125}$$

Thus the following combinations are gauge invariant:

$$\Delta\delta\hat{X}_{ij} = 0, \quad \delta\hat{X}_{ij} = \delta X_{ij} - 2\partial_i\partial_j E. \frac{\bar{X}_{kk}}{3} - I\partial_t\bar{X}_{ij}. \tag{4.126}$$

$$\Delta\delta\hat{X}_{i0} = 0, \quad \delta\hat{X}_{i0} = \delta X_{i0} - \partial_i\left(\partial_t E. \frac{\bar{X}_{kk}}{3} + I\bar{X}_{00}\right). \tag{4.127}$$

$$\Delta\delta\hat{X}_{00} = 0, \quad \delta\hat{X}_{00} = \delta X_{00} - 2\bar{X}_{00}\partial_t I - I\partial_t\bar{X}_{00}. \tag{4.128}$$

We use now the result

$$\begin{aligned}\Delta\delta X_\nu^\alpha &= \bar{g}^{\alpha\mu}\Delta\delta X_{\mu\nu} - \Delta\delta g^{\alpha\mu}. \bar{X}_{\mu\nu}\\ &= \bar{g}^{\alpha\mu}\Delta\delta X_{\mu\nu} + (\partial^\mu\epsilon^\alpha + \partial^\alpha\epsilon^\mu + \bar{g}^{\alpha\rho}\bar{g}^{\mu\sigma}\epsilon^\lambda\partial_\lambda\bar{g}_{\rho\sigma})\bar{X}_{\mu\nu}.\end{aligned} \tag{4.129}$$

We get now the gauge invariant observables

$$\Delta\delta\hat{X}_0^0 = 0, \quad \delta\hat{X}_0^0 = \delta X_0^0 - I\partial_t\bar{X}_0^0. \tag{4.130}$$

$$\Delta\delta\hat{X}_i^0 = 0, \quad \delta\hat{X}_i^0 = \delta X_i^0 - \partial_i I(\bar{X}_0^0 - \frac{1}{3}\bar{X}_k^k). \tag{4.131}$$

$$\Delta\delta\hat{X}_j^i = 0, \quad \delta\hat{X}_j^i = \delta X_j^i - I\partial_t\bar{X}_j^i. \tag{4.132}$$

We can then write the linearized Einstein equations in a gauge invariant way as follows

$$\delta\hat{G}_\nu^\mu = 8\pi G\delta\hat{T}_\nu^\mu. \tag{4.133}$$

4.4.4 Explicit calculation of $\delta\hat{G}_\nu^\mu$

We start from

$$ds^2 = -dt^2 + a^2\delta_{ij}dx^i dx^j = a^2[-d\eta^2 + \delta_{ij}dx^i dx^j]. \tag{4.134}$$

From here on the subscript 0 indicates conformal time. We compute

$$\Gamma_{00}^0 = \frac{\dot{a}}{a}, \quad \Gamma_{ij}^0 = \frac{\dot{a}}{a}\delta_{ij}, \quad \Gamma_{0j}^i = \frac{\dot{a}}{a}\delta_{ij}. \tag{4.135}$$

$$\bar{R}_{00} = -3\partial_0\left(\frac{\dot{a}}{a}\right), \quad \bar{R}_{ij} = \left(\partial_0\left(\frac{\dot{a}}{a}\right) + 2\frac{\dot{a}^2}{a^2}\right)\delta_{ij}, \quad \bar{R} = \frac{6}{a^2}(\partial_0\left(\frac{\dot{a}}{a}\right) + \frac{\dot{a}^2}{a^2}). \tag{4.136}$$

Now we have the perturbations

$$\delta R_{\mu\nu} = \partial_\alpha\delta\Gamma_{\mu\nu}^\alpha - \partial_\mu\delta\Gamma_{\alpha\nu}^\alpha + \delta\Gamma_{\mu\nu}^\beta\bar{\Gamma}_{\alpha\beta}^\alpha + \bar{\Gamma}_{\mu\nu}^\beta\delta\Gamma_{\alpha\beta}^\alpha - \delta\Gamma_{\alpha\nu}^\beta\bar{\Gamma}_{\mu\beta}^\alpha - \bar{\Gamma}_{\alpha\nu}^\beta\delta\Gamma_{\mu\beta}^\alpha. \tag{4.137}$$

And[1]

$$\delta\Gamma^\rho_{\mu\nu} = \frac{1}{2}\bar{g}^{\rho\sigma}(\partial_\mu\delta g_{\nu\sigma} + \partial_\nu\delta g_{\mu\sigma} - \partial_\sigma\delta g_{\mu\nu}) - \frac{1}{2}\bar{g}^{\rho\sigma}\bar{g}^{\sigma\beta}\delta g_{\alpha\beta}(\partial_\mu\bar{g}_{\nu\sigma} + \partial_\nu\bar{g}_{\mu\sigma} - \partial_\sigma\bar{g}_{\mu\nu}). \quad (4.138)$$

$$\delta G_{\mu\nu} = \delta R_{\mu\nu} - \frac{1}{2}\bar{g}_{\mu\nu}\delta R - \frac{3}{a^2}(\partial_0\left(\frac{\dot{a}}{a}\right) + \frac{\dot{a}^2}{a^2})\delta g_{\mu\nu}. \quad (4.139)$$

We will work in the conformal-Newtonian gauge $E = B = 0$ in which the metric takes the form

$$ds^2 = -(1 + 2\Phi)dt^2 + a^2(1 - 2\Psi)\delta_{ij}dx^idx^j = a^2[-(1 + 2\Phi)d\eta^2 + (1 - 2\Psi)\delta_{ij}dx^idx^j]. \quad (4.140)$$

We have

$$\delta g_{00} = -2a^2\Phi, \quad \delta g_{ij} = -2a^2\Psi\delta_{ij}. \quad (4.141)$$

$$\delta g^{00} = -\frac{2}{a^2}\Phi, \quad \delta g^{ij} = -\frac{2}{a^2}\Psi\delta_{ij}. \quad (4.142)$$

We compute immediately

$$\delta\Gamma^0_{\mu\nu} = -\frac{1}{2a^2}(\partial_\mu\delta g_{\nu 0} + \partial_\nu\delta g_{\mu 0} - \partial_0\delta g_{\mu\nu}) - \frac{1}{2a^4}\delta g_{00}(\partial_\mu\bar{g}_{\nu 0} + \partial_\nu\bar{g}_{\mu 0} - \partial_0\bar{g}_{\mu\nu}). \quad (4.143)$$

$$\delta\Gamma^i_{\mu\nu} = \frac{1}{2a^2}(\partial_\mu\delta g_{\nu i} + \partial_\nu\delta g_{\mu i} - \partial_i\delta g_{\mu\nu}) - \frac{1}{2a^4}\delta g_{ik}(\partial_\mu\bar{g}_{\nu k} + \partial_\nu\bar{g}_{\mu k} - \partial_k\bar{g}_{\mu\nu}). \quad (4.144)$$

Step 1: We set $a = 1$. In this case we compute

$$\delta\Gamma^0_{00} = \partial_0\Phi, \quad \delta\Gamma^0_{0i} = \partial_i\Phi, \quad \delta\Gamma^0_{ij} = -\partial_0\Psi\delta_{ij} \quad (4.145)$$

$$\delta\Gamma^i_{00} = \partial_i\Phi, \quad \delta\Gamma^i_{0j} = -\partial_0\Psi\delta_{ij}, \quad \delta\Gamma^i_{jl} = \partial_i\Psi\delta_{jl} - \partial_j\Psi\delta_{li} - \partial_l\Psi\delta_{ij}. \quad (4.146)$$

Thus

$$\delta R_{00} = \partial_i^2\Phi + 3\partial_0^2\Psi. \quad (4.147)$$

$$\delta R_{0i} = 2\partial_0\partial_i\Psi. \quad (4.148)$$

$$\delta R_{ij} = \delta_{ij}\left(-\partial_0^2\Psi + \partial_k^2\Psi\right) - \partial_i\partial_j(\Phi - \Psi). \quad (4.149)$$

And

$$\delta R = -\delta R_{00} + \delta R_{ii}. \quad (4.150)$$

Thus

$$\delta G_{00} = \frac{1}{2}(\delta R_{00} + \delta R_{ii}) = 2\partial_i^2\Psi. \quad (4.151)$$

---

[1] The minus sign in the second term is due to our 'bad' definition: $-\delta g^{\mu\nu} = g^{\mu\nu} - \bar{g}^{\mu\nu}$.

$$\delta G_{0i} = \delta R_{0i} = 2\partial_i \partial_0 \Psi. \tag{4.152}$$

$$\delta G_{ij} = \delta R_{ij} + \frac{1}{2}\delta_{ij}(\delta R_{00} - \delta R_{kk}) = 2\delta_{ij}\left(\partial_0^2 \Psi + \frac{1}{2}\partial_i^2(\Phi - \Psi)\right) - \partial_i \partial_j(\Phi - \Psi). \tag{4.153}$$

Step 2: The next step we perform the conformal transformation

$$g_{\mu\nu} \longrightarrow \tilde{g}_{\mu\nu} = Fg_{\mu\nu}, \quad F = a^2. \tag{4.154}$$

Under this transformation we have

$$\Gamma^\rho_{\mu\nu} \longrightarrow \tilde{\Gamma}^\rho_{\mu\nu} = \Gamma^\rho_{\mu\nu} + \frac{1}{2}(\partial_\mu \ln F. \, g_\nu^\rho + \partial_\nu \ln F. \, g_\mu^\rho - \partial^\rho \ln F. \, g_{\mu\nu}). \tag{4.155}$$

Also (by using also the fact that the metric is covariantly constant)

$$R_{\mu\nu} \longrightarrow \tilde{R}_{\mu\nu} = R_{\mu\nu} - \frac{1}{F}\nabla_\mu\,\nabla_\nu\,F - \frac{1}{2F}g_{\mu\nu}\nabla_\alpha\,\nabla^\alpha\,F + \frac{3}{2F^2}\nabla_\mu\,F\nabla_\nu\,F. \tag{4.156}$$

Thus

$$R \longrightarrow \tilde{R} = \frac{1}{F}R - \frac{3}{F^2}\nabla_\mu\,\nabla^\mu\,F + \frac{3}{2F^3}\nabla_\mu\,F\nabla^\mu\,F. \tag{4.157}$$

$$G_{\mu\nu} \longrightarrow \tilde{G}_{\mu\nu} = G_{\mu\nu} - \frac{1}{F}\nabla_\mu\,\nabla_\nu\,F + \frac{1}{F}g_{\mu\nu}\nabla_\alpha\,\nabla^\alpha\,F + \frac{3}{2F^2}\nabla_\mu\,F\nabla_\nu\,F - \frac{3}{4F^2}g_{\mu\nu}\nabla_\alpha\,F\nabla^\alpha\,F. \tag{4.158}$$

For our case we need

$$\delta\left(-\frac{1}{F}\nabla_\mu\,\nabla_\nu\,F\right) = 2\frac{\dot{a}}{a}\delta\Gamma^0_{\mu\nu}. \tag{4.159}$$

$$\delta\left(\frac{1}{F}g_{\mu\nu}\nabla_\alpha\,\nabla^\alpha\,F\right) = \delta\left(\frac{1}{F}g_{\mu\nu}g^{\alpha\beta}\nabla_\alpha\,\partial_\beta F\right)$$
$$= \delta\left(\frac{1}{F}g_{\mu\nu}g^{\alpha\beta}\left(\partial_\alpha\partial_\beta F - \Gamma^\rho_{\alpha\beta}\partial_\rho F\right)\right) \tag{4.160}$$
$$= \left(-2\frac{\ddot{a}}{a} - 2\frac{\dot{a}^2}{a^2}\right)\delta g_{\mu\nu} + \left(4\frac{\ddot{a}}{a}\Phi + 4\frac{\dot{a}^2}{a^2}\Phi + 2\frac{\dot{a}}{a}\partial_0\Phi + 6\frac{\dot{a}}{a}\partial_0\Psi\right)\tilde{g}_{\mu\nu}.$$

$$\delta\left(-\frac{3}{4F^2}g_{\mu\nu}\nabla_\alpha\,F\nabla^\alpha\,F\right) = 3\frac{\dot{a}^2}{a^2}\delta g_{\mu\nu} - 6\frac{\dot{a}^2}{a^2}\Phi\tilde{g}_{\mu\nu}. \tag{4.161}$$

$$\delta\tilde{G}_{\mu\nu} = \delta G_{\mu\nu} + 2\frac{\dot{a}}{a}\delta\Gamma^0_{\mu\nu} + \left(-2\frac{\ddot{a}}{a} + \frac{\dot{a}^2}{a^2}\right)\delta g_{\mu\nu} + \left(4\frac{\ddot{a}}{a}\Phi - 2\frac{\dot{a}^2}{a^2}\Phi + 2\frac{\dot{a}}{a}\partial_0\Phi + 6\frac{\dot{a}}{a}\partial_0\Psi\right)\tilde{g}_{\mu\nu}. \tag{4.162}$$

Explicitly we have

$$\delta\tilde{G}_{00} = \delta G_{00} - 6\frac{\dot{a}}{a}\partial_0\Psi. \tag{4.163}$$

$$\delta\tilde{G}_{0i} = \delta G_{0i} + 2\frac{\dot{a}}{a}\partial_i\Phi. \tag{4.164}$$

$$\delta\tilde{G}_{ij} = \delta G_{ij} + \left(4\frac{\dot{a}}{a}\partial_0\Psi + 2\frac{\dot{a}}{a}\partial_0\Phi + \left(4\frac{\ddot{a}}{a} - 2\frac{\dot{a}^2}{a^2}\right)\Phi + \left(4\frac{\ddot{a}}{a} - 2\frac{\dot{a}^2}{a^2}\right)\Psi\right)\delta_{ij}. \tag{4.165}$$

We rewrite these as

$$-a^2\delta\tilde{G}^0_0 = \delta G_{00} - 6\frac{\dot{a}^2}{a^2}\Phi = 2\partial_i^2\Psi - 6\frac{\dot{a}}{a}\partial_0\Psi - 6\frac{\dot{a}^2}{a^2}\Phi. \tag{4.166}$$

$$-a^2\delta\tilde{G}^0_i = \delta\tilde{G}_{0i} = 2\partial_i\left(\partial_0\Psi + \frac{\dot{a}}{a}\Phi\right). \tag{4.167}$$

$$a^2\delta\tilde{G}^0_0 = \delta G_{ij} + \left(-4\frac{\ddot{a}}{a} + 2\frac{\dot{a}^2}{a^2}\right)\Psi\delta_{ij}$$

$$= 2\delta_{ij}\left(\partial_0^2\Psi + \frac{1}{2}\partial_i^2(\Phi - \Psi)\right) - \partial_i\partial_j(\Phi - \Psi) + \left(4\frac{\dot{a}}{a}\partial_0\Psi + 2\frac{\dot{a}}{a}\partial_0\Phi + \left(4\frac{\ddot{a}}{a} - 2\frac{\dot{a}^2}{a^2}\right)\Phi\right)\delta_{ij}$$

$$= 2\delta_{ij}\left(\partial_0^2\Psi + \frac{1}{2}\partial_i^2(\Phi - \Psi) + 2\frac{\dot{a}}{a}\partial_0\Psi + \frac{\dot{a}}{a}\partial_0\Phi + \left(2\partial_0\left(\frac{\dot{a}}{a}\right) + \frac{\dot{a}^2}{a^2}\right)\Phi\right) - \partial_i\partial_j(\Phi - \Psi). \tag{4.168}$$

The linearized Einstein's equations are therefore given by

$$\partial_i^2\Psi - 3\frac{\dot{a}}{a}\partial_0\Psi - 3\frac{\dot{a}^2}{a^2}\Phi = -4\pi Ga^2\delta T^0_0. \tag{4.169}$$

$$\partial_i\left(\partial_0\Psi + \frac{\dot{a}}{a}\Phi\right) = -4\pi Ga^2\delta T^0_i. \tag{4.170}$$

$$\delta_{ij}\left(\partial_0^2\Psi + \frac{1}{2}\partial_i^2(\Phi - \Psi) + 2\frac{\dot{a}}{a}\partial_0\Psi + \frac{\dot{a}}{a}\partial_0\Phi + \left(2\partial_0\left(\frac{\dot{a}}{a}\right) + \frac{\dot{a}^2}{a^2}\right)\Phi\right) - \frac{1}{2}\partial_i\partial_j(\Phi - \Psi) = 4\pi Ga^2\delta T^i_j. \tag{4.171}$$

The gauge invariant objects are obtained by replacing Φ and Ψ by the Bardeen potentials Φ_B and Ψ_B, respectively.

4.4.5 Matter perturbations

Now we discuss matter perturbations. The stress–energy–momentum tensor $T^{\mu\nu}$ of a perfect fluid is given by

$$T^{\mu\nu} = (\rho + P)U^\mu U^\nu + Pg^{\mu\nu}, \quad g_{\mu\nu}U^\mu U^\nu = -1. \tag{4.172}$$

Again we will work with the conformal time denoted by 0 for simplicity. The unperturbed velocity satisfies $\bar{g}_{\mu\nu}\bar{U}^\mu\bar{U}^\nu = -1$ and thus $\bar{U}^\mu = (1/a, 0, 0, 0)$. We compute then from $2\bar{g}_{\mu\nu}\bar{U}^\mu\delta U^\nu + \delta g_{\mu\nu}\bar{U}^\mu\bar{U}^\nu = 0$ the result $\delta U^0 = \delta U_0/a^2 = \delta g_{00}/2a^3$ while δU^i is an independent dynamical variable. We have then $U^0 = (1 - \Phi)/a$, $U_0 = -a - a\Phi$ and $U^i = g^{i\mu}U_\mu \Rightarrow \delta U^i = \delta U_i/a^2 - B_i/a$. We will use the notation $\delta U_i = a v_i$ and thus $\delta U^i = (v_i - B_i)/a$. The first order perturbation of the stress–energy–momentum tensor is

$$\delta T_{\mu\nu} = (\delta\rho + \delta P)\bar{U}_\mu\bar{U}_\nu + (\bar{\rho} + \bar{P})\delta U_\mu\bar{U}_\nu + (\bar{\rho} + \bar{P})\bar{U}_\mu\delta U_\nu + \delta P\bar{g}_{\mu\nu} + \bar{P}\delta g_{\mu\nu}. \quad (4.173)$$

Explicitly we have (using $\delta g_\mu{}^\nu = 0$)

$$\delta T_{00} = a^2(\delta\rho + 2\Phi\bar{\rho}) \Leftrightarrow \delta T_0{}^0 = -\delta\rho. \quad (4.174)$$

$$\delta T_{i0} = -a^2(\bar{\rho} + \bar{P})v_i + a^2\bar{P}B_i \Leftrightarrow \delta T_i^0 = (\bar{\rho} + \bar{P})v_i, \quad \delta T_0^i = -(\bar{\rho} + \bar{P})(v_i - B_i). \quad (4.175)$$

$$\delta T_{ij} = a^2\delta_{ij}\delta P + \bar{P}\delta g_{ij} \Leftrightarrow \delta T_i^j = \delta_{ij}\delta P. \quad (4.176)$$

There is an extra contribution to the stress–energy–momentum tensor $T_{\mu\nu}$ which is the anisotropic stress tensor $\Sigma_{\mu\nu}$ which vanishes in the unperturbed theory. This tensor is therefore a first order perturbation which is constrained to satisfy $\Sigma_{\mu\nu}U^\nu = 0$ and $\Sigma_\mu{}^\mu = 0$ and as a consequence $\Sigma_{00} = \Sigma_{i0} = 0$ and $\Sigma_i{}^i = 0$. The anisotropic stress tensor is therefore a traceless symmetric 3-tensor Σ_{ij}. In other words, we need to change equation (4.176) as follows

$$\delta T_{ij} = a^2\delta_{ij}\delta P + \bar{P}\delta g_{ij} + \Sigma_{ij} \Leftrightarrow \delta T_i^j = \delta_i^j\delta P + \Sigma_i{}^j. \quad (4.177)$$

It is obvious that the tensor $\delta T_{\mu\nu}$ must transform under gauge transformations in the same way as the tensor $\delta g_{\mu\nu}$. These have been already computed in (4.130), (4.131) and (4.131). In conformal time we need to make the replacements $\bar{X}_0^0 \longrightarrow \bar{X}_0^0, \bar{X}_i^0 \longrightarrow \bar{X}_i^0/a, \bar{X}_j^i \longrightarrow \bar{X}_j^i$ where 0 stands now for conformal time. The gauge invariant quantities are given by

$$\Delta\delta\hat{X}_0^0 = 0, \quad \delta\hat{X}_0^0 = \delta X_0^0 + (B - E')(\bar{X}_0^0)'. \quad (4.178)$$

$$\Delta\delta\hat{X}_i^0 = 0, \quad \delta\hat{X}_i^0 = \delta X_i^0 + \partial_i(B - E')(\bar{X}_0^0 - \frac{1}{3}\bar{X}_k^k). \quad (4.179)$$

$$\Delta\delta\hat{X}_j^i = 0, \quad \delta\hat{X}_j^i = \delta X_j^i + (B - E')(\bar{X}_j^i)'. \quad (4.180)$$

The gauge invariant linearized Einstein's equations becomes given by

$$\partial_i^2\Psi_B - 3\mathcal{H}\Psi_B' - 3\mathcal{H}^2\Phi_B = 4\pi G a^2\delta\hat{\rho}. \quad (4.181)$$

$$\partial_i\left(\Psi_B' + \mathcal{H}\Phi_B\right) = -4\pi G a^2(\bar{\rho} + \bar{P})\frac{\delta\hat{U}_i}{a}. \quad (4.182)$$

$$\delta_{ij}\left(\Psi_B'' + \frac{1}{2}\partial_i^2(\Phi_B - \Psi_B) + 2\mathcal{H}\Psi_B' + \mathcal{H}\Phi_B' + (2\mathcal{H}' + \mathcal{H}^2)\Phi_B\right) - \frac{1}{2}\partial_i\partial_j(\Phi_B - \Psi_B) = 4\pi G a^2\delta\hat{P}\delta_{ij}. \quad (4.183)$$

$$\delta\hat{T}_0^0 = -\delta\hat{\rho} = -\delta\rho - \bar{\rho}'(B - E'). \quad (4.184)$$

$$\delta \hat{T}_i^0 = (\bar{\rho} + \bar{P})\left(\frac{\delta U_i}{a} - \partial_i(B - E')\right) = (\bar{\rho} + \bar{P})\frac{\delta \hat{U}_i}{a}. \tag{4.185}$$

$$\delta \hat{T}_j^i = \delta_{ij}(\delta P + \bar{P}(B - E')) = \delta_{ij}\delta \hat{P}. \tag{4.186}$$

In the above second equation $\delta \hat{U}_i$ is the gauge invariant velocity perturbation. As before, only the parallel part of this velocity, which is of the form $a^2 \partial_i \gamma$ for some scalar function γ, will contribute to scalar perturbation. Remember that we are neglecting vector perturbations throughout.

4.5 Matter–radiation equality

We recall the two Friedmann equations and the energy conservation law

$$\mathcal{H}^2 = \frac{8\pi G a^2}{3}\bar{\rho}, \quad \mathcal{H}^2 - \mathcal{H}' = 4\pi G a^2(\bar{\rho} + \bar{P}), \quad \bar{\rho}' = -3\mathcal{H}(\bar{\rho} + \bar{P}). \tag{4.187}$$

By combining the two Friedmann equations as $\mathcal{H}^2 - (\mathcal{H}^2 - \mathcal{H}')$ we obtain

$$a'' = \frac{4\pi G a^3}{3}(\bar{\rho} - 3\bar{P}). \tag{4.188}$$

We have already shown that the density of radiation falls off as $1/a^4$, whereas the density of matter falls off as $1/a^3$ and thus in a universe filled with matter and radiation we have the energy density

$$\bar{\rho} = \bar{\rho}_m + \bar{\rho}_r$$
$$= \frac{\bar{\rho}_{eq}}{2}\left(\frac{a_{eq}^3}{a^3} + \frac{a_{eq}^4}{a^4}\right). \tag{4.189}$$

The ρ_{eq} is the energy density at the time of equality η_{eq} at which matter and radiation densities become equal and $a_{eq} = a(\eta_{eq})$ is the corresponding scale factor. We also recall that the pressure of matter (dust) is 0, whereas the pressure of radiation is $\bar{P}_r = \bar{\rho}_r/3$ and thus

$$\bar{P} = \bar{P}_m + \bar{P}_r = \frac{\bar{\rho}_r}{3}. \tag{4.190}$$

By using these last two equations in the Friedmann equation (4.188) we obtain

$$a'' = \frac{2\pi G a_{eq}^3 \bar{\rho}_{eq}}{3} = 2C_0. \tag{4.191}$$

The solution is immediately given by

$$a = C_0 \eta^2 + C_1 \eta + C_2. \tag{4.192}$$

We find $C_2 = 0$ from the boundary condition $a(0) = 0$. By substituting this solution in the Friedmann equation $\mathcal{H}^2 = 8\pi G a^2 \bar{\rho}/3$ or equivalently

$$a'^2 = \frac{4\pi G}{3}\bar{\rho}_{eq}(a_{eq}^3 a + a_{eq}^4),$$

(4.193)

we obtain $C_1 = \sqrt{4C_0 a_{eq}}$. The scale factor is therefore given by

$$a = a_{eq}\left(\frac{\eta^2}{\eta_*^2} + 2\frac{\eta}{\eta_*}\right).$$

(4.194)

The time η_* is related to the time of equality η_{eq} by $\eta_{eq} = (\sqrt{2} - 1)\eta_*$. In the radiation-dominated universe corresponding to $\eta \ll \eta_{eq}$ we have $a \propto \eta$, whereas in the matter-dominated universe corresponding to $\eta \gg \eta_{eq}$ we have $a \propto \eta^2$.

4.6 Hydrodynamical adiabatic scalar perturbations

The Einstein's equation (4.183) for $i \neq j$ gives $\partial_i\partial_j(\Phi - \Psi) = 0$. The only solutions consistent with Φ and Ψ being perturbations are $\Phi = \Psi$. The remaining Einstein's equations simplify therefore to

$$\partial_i^2\Phi_B - 3\mathcal{H}\Phi'_B - 3\mathcal{H}^2\Phi_B = 4\pi Ga^2\delta\hat{\rho}.$$

(4.195)

$$\partial_i(a\Phi_B)' = -4\pi Ga^2(\bar{\rho} + \bar{P})\delta\hat{U}_i.$$

(4.196)

$$\Phi_B'' + 3\mathcal{H}\Phi_B' + (2\mathcal{H}' + \mathcal{H}^2)\Phi_B = 4\pi Ga^2\delta\hat{P}.$$

(4.197)

The first equation is the generalization of Poisson's equation for the Newtonian gravitational potential which is identified here with the Bardeen potential Φ_B. Recall that the sub-Hubble or sub-horizon scales correspond to comoving Fourier scales k^{-1} such that $k > \mathcal{H} = aH$. The second and third terms in (4.195) can be rewritten as $-3\mathcal{H}(a\Phi_B)'/a$, i.e. they are suppressed by a factor $1/\mathcal{H}$ on sub-Hubble scales and thus can be neglected compared to the first term. Equation (4.195) reduces therefore to the usual Poisson's equation for the Newtonian gravitational potential in this limit. The combination $(a\Phi_B)'$ is precisely the velocity potential which is given by equation (4.196).

Now we will split the pressure perturbation into an adiabatic (curvature) piece and an entropy (isocurvature) piece as follows

$$\delta\hat{P} = c_s^2\delta\hat{\rho} + \tau\delta S.$$

(4.198)

The first component $\delta\hat{P} = c_s^2\delta\hat{\rho}$ is the adiabatic perturbation and it corresponds to fluctuations in the energy density and thus induces inhomogeneities in the spatial curvature. The second component $\delta\hat{P} = \tau\delta S$ is the entropy perturbation and it corresponds to fluctuations in the form of the local equation of state of the system, i.e. fluctuations in the relative number densities of the different particle types present in the system. The two perturbations are orthogonal since any other perturbation can be written as a linear combination of the two. The coefficients c_s^2 and τ are given by

$$c_s^2 = \left(\frac{\partial P}{\partial \rho}\right)_S, \quad \tau = \left(\frac{\partial p}{\partial S}\right)_\rho. \tag{4.199}$$

In particular, c_s^2 is the speed of sound as we now show. We combine the two Einstein's equations (4.195) and (4.197) as follows

$$c_s^2\left(\partial_i^2\Phi_B - 3\mathcal{H}\Phi_B' - 3\mathcal{H}^2\Phi_B\right) - \left(\Phi_B'' + 3\mathcal{H}\Phi_B' + (2\mathcal{H}' + \mathcal{H}^2)\Phi_B\right) = 4\pi Ga^2(c_s^2\delta\hat{\rho} - \delta\hat{P}). \tag{4.200}$$

We get then the general relativistic Poisson's equation for the Newtonian gravitational potential given by

$$\Phi_B'' + 3\mathcal{H}(1 + c_s^2)\Phi_B' - c_s^2\partial_i^2\Phi_B + \left(2\mathcal{H}' + \mathcal{H}^2(1 + 3c_s^2)\right)\Phi_B = 4\pi Ga^2\tau\delta S. \tag{4.201}$$

Adiabatic perturbations: We will only concentrate here on adiabatic perturbations. The case of entropy perturbations is treated in the excellent book [7].

In this case we set

$$\delta S = 0. \tag{4.202}$$

Equivalently

$$
\begin{aligned}
c_s^2 &= \frac{\delta\hat{P}}{\delta\hat{\rho}} = \left(\frac{\partial P}{\partial \rho}\right)_S \\
&= \left(\frac{\partial\eta}{\partial\rho}\frac{\partial P}{\partial\eta}\right)_S \\
&= \frac{\bar{P}'}{\bar{\rho}'}.
\end{aligned}
\tag{4.203}
$$

The above general relativistic Poisson's equation equation can be simplified by introducing the variable u defined by

$$
\begin{aligned}
u &= \exp\left(\frac{3}{2}\int (1 + c_s^2)\mathcal{H}d\eta\right)\Phi_B \\
&= \exp\left(-\frac{1}{2}\int\left(1 + \frac{\bar{P}'}{\bar{\rho}'}\right)\frac{\bar{\rho}'}{\bar{\rho} + \bar{P}}d\eta\right)\Phi_B \\
&= \frac{1}{\sqrt{\bar{\rho} + \bar{P}}}\Phi_B.
\end{aligned}
\tag{4.204}
$$

We rewrite this as

$$u = \frac{a}{\sqrt{\bar{\rho}}}\theta\Phi_B, \quad \theta = \frac{1}{a\sqrt{1 + \dfrac{\bar{P}}{\bar{\rho}}}}. \tag{4.205}$$

We remark that

$$\theta = \frac{1}{a\sqrt{\frac{2}{3}\left(1 - \frac{\mathcal{H}'}{\mathcal{H}^2}\right)}}.$$

(4.206)

We compute immediately

$$\Phi_B'' + 3\mathcal{H}(1 + c_s^2)\Phi_B' = \exp\left(-\frac{3}{2}\int(1 + c_s^2)\mathcal{H}d\eta\right)\left[u'' - \left[-\frac{3}{2}(1 + c_s^2)\mathcal{H}\right]^2 u + \left[-\frac{3}{2}(1 + c_s^2)\mathcal{H}\right]'u\right].$$

(4.207)

We observe that the friction term cancels exactly. Also

$$\left(2\mathcal{H}' + \mathcal{H}^2(1 + 3c_s^2)\right)\Phi_B = \exp\left(-\frac{3}{2}\int(1 + c_s^2)\mathcal{H}d\eta\right)\left[-\frac{3\mathcal{H}^2}{a^2\theta^2} + 3(1 + c_s^2)\mathcal{H}^2\right]u.$$

(4.208)

We use

$$1 - \frac{\mathcal{H}'}{\mathcal{H}^2} = \frac{3}{2a^2\theta^2}.$$

(4.209)

$$1 + c_s^2 = \frac{1}{a^2\theta^2} + \frac{2}{3} + \frac{2}{3\mathcal{H}}\frac{\theta'}{\theta}.$$

(4.210)

Thus

$$\left(2\mathcal{H}' + \mathcal{H}^2(1 + 3c_s^2)\right)\Phi_B = \exp\left(-\frac{3}{2}\int(1 + c_s^2)\mathcal{H}d\eta\right)\left[2\mathcal{H}^2 + 2\mathcal{H}\frac{\theta'}{\theta}\right]u.$$

(4.211)

After some calculation we get

$$\left(2\mathcal{H}' + \mathcal{H}^2(1 + 3c_s^2) + \left(2\mathcal{H}' + \mathcal{H}^2(1 + 3c_s^2)\right)\right)\Phi_B = \exp\left(-\frac{3}{2}\int(1 + c_s^2)\mathcal{H}d\eta\right)$$
$$\times\left[u'' + \left(-\frac{3\mathcal{H}'}{2a^2\theta^2} - \mathcal{H}' + \mathcal{H}^2 - \frac{9\mathcal{H}^2}{4a^4\theta^4} - \frac{\theta''}{\theta}\right)u\right].$$

(4.212)

After some inspection we get

$$\left(2\mathcal{H}' + \mathcal{H}^2(1 + 3c_s^2) + \left(2\mathcal{H}' + \mathcal{H}^2(1 + 3c_s^2)\right)\right)\Phi_B = \exp\left(-\frac{3}{2}\int(1 + c_s^2)\mathcal{H}d\eta\right)$$
$$\times\left[u'' - \frac{\theta''}{\theta}u\right].$$

(4.213)

Poisson's equation reduces therefore to

$$u'' - c_s^2\partial_i^2 u - \frac{\theta''}{\theta}u = 0.$$

(4.214)

We look for plane wave solutions of the form

$$u = u_{\vec{k}}(\vec{x}, \eta) = \exp(i\vec{k}\vec{x})\chi_{\vec{k}}(\eta).$$

(4.215)

We need to solve

$$\chi_{\vec{k}}'' + (c_s^2 \vec{k}^2 - \frac{\theta''}{\theta})\chi_{\vec{k}} = 0. \tag{4.216}$$

Let us first assume that θ''/θ is a constant, viz

$$\frac{\theta''}{\theta} = \sigma^2. \tag{4.217}$$

The above differential equation becomes

$$\chi_{\vec{k}}'' + \omega_{\vec{k}}^2 \chi_{\vec{k}} = 0, \quad \omega_{\vec{k}} = \sqrt{c_s^2 \vec{k}^2 - \sigma^2}. \tag{4.218}$$

We define the so-called Jeans length by

$$\lambda_J = \frac{2\pi}{k_J}, \quad k_J = \frac{\sigma}{c_s}. \tag{4.219}$$

In other words,

$$\omega_{\vec{k}} = c_s \sqrt{\vec{k}^2 - \vec{k}_J^2}. \tag{4.220}$$

The behavior of the perturbation depends therefore crucially on its spatial size given by the Jeans length. Two interesting limiting cases emerge immediately:

- Large scales corresponding to long-wavelengths where gravity dominates given by $k \ll k_J$, $\lambda \gg \lambda_J$: in this case we get the solutions

$$\chi_{\vec{k}} \sim \exp(\pm|\omega_{\vec{k}}|\eta). \tag{4.221}$$

  The plus sign describes exponentially fast growth of inhomogeneities, whereas the negative sign describes a decaying solution. We have when $k \longrightarrow 0$ the behavior

$$|\omega_{\vec{k}}|\eta \longrightarrow c_s k_J \eta = \frac{\eta}{\eta_{gr}}, \quad \eta_{gr} = \frac{1}{\sigma}. \tag{4.222}$$

  From this we can deduce that gravity is very efficient in amplifying adiabatic perturbations. As an example, if the initial adiabatic perturbation is extremely small, of the order of 10^{-100}, gravity will only need $\eta = 230\eta_{gr}$ to amplify it to order 1. We remark that this limit $k \ll k_J$ corresponds to $c_s k \eta \ll \eta/\eta_{gr} = \lambda/\lambda_J$ where the Jeans length $\lambda_J = c_s t_{gr}$ is the sound communication scale, i.e. the scale over which pressure can react to changes in the energy density due to gravity. Thus this limit can be characterized simply by $c_s k \eta \ll 1$.

- Small scales corresponding to short-wavelengths where gravity is negligible compared to pressure given by $k \gg k_J$, $\lambda \ll \lambda_J$: in this case we get the solutions

$$\chi_{\vec{k}} \sim \exp(\pm i\omega_{\vec{k}}\eta). \tag{4.223}$$

These are sound waves with phase velocity given by

$$c_{\text{phase}} = \frac{\omega_{\vec{k}}}{k} = c_s\sqrt{1 - \frac{k_J^2}{k^2}} \longrightarrow c_s. \tag{4.224}$$

We solve now the differential equation (4.216) more rigorously in these two limiting cases.

Large scales or long wavelengths$(c_s k\eta \ll 1)$: In this case we can neglect the spatial derivative in (4.216). The Bardeen potential in the case that the universe is a mixture of radiation and matter is found to be given by (with $\xi = \eta/\eta_*$)

$$\begin{aligned}
\Phi_B &= \frac{A}{\xi(\xi+2)}\frac{d}{d\xi}\left(\frac{1}{\xi+2}\left(\frac{1}{5}\xi^4 + \xi^3 + \frac{4}{3}\xi^2\right) + \frac{A'}{\xi(\xi+2)}\right) \\
&= \frac{A(\xi+1)}{(\xi+2)^3}\left(\frac{3}{5}\xi^2 + 3\xi + \frac{13}{3} + \frac{1}{\xi+1}\right) + \frac{B(\xi+1)}{\xi^3(\xi+2)^3}.
\end{aligned} \tag{4.225}$$

The A term is the term corresponding to the growth of inhomogeneities, whereas the B term is the decaying mode which we can neglect.

By using the Friedmann equation $\mathcal{H}^2 = 8\pi G a^2 \bar{\rho}/3$ and the Einstein equation (4.195) we obtain an expression for the energy density perturbation given by

$$\frac{\delta\hat{\rho}}{\bar{\rho}} = -2\Phi_B - \frac{2}{\mathcal{H}}\Phi'_B + \frac{2}{3\mathcal{H}^2}\partial_i^2\Phi_B. \tag{4.226}$$

We use the results

$$\mathcal{H} = \frac{2(\xi+1)}{\eta_*\xi(\xi+2)}, \quad \frac{d\Phi_B}{d\xi} = -\frac{4A(\xi+5)}{15(\xi+2)^4}. \tag{4.227}$$

Thus

$$\frac{\delta\hat{\rho}}{\bar{\rho}} = -2\Phi_B + \frac{4A\xi(\xi+5)}{15(\xi+1)(\xi+2)^3} - \frac{\vec{k}^2\eta^2(\xi+2)^2}{6(\xi+1)^2}\Phi_B. \tag{4.228}$$

The last term is of course negligible for long-wavelengths $k \longrightarrow 0$. At early times compared to $\eta_{\text{eq}} \sim \eta_*$ we have $\xi \longrightarrow 0$ and $\Phi_B \longrightarrow 2A/3$, $\delta\hat{\rho}/\bar{\rho} \longrightarrow -4A/3$, whereas at late times compared to $\eta_{\text{eq}} \sim \eta_*$ we have $\xi \longrightarrow \infty$ and $\Phi_B \longrightarrow 3A/5$, $\delta\hat{\rho}/\bar{\rho} \longrightarrow -6A/5$. Thus Φ_B and $\delta\hat{\rho}/\bar{\rho}$ are both constants during radiation-dominated (early times) and matter-dominated (late times) epochs with the amplitude decreasing by a factor of $9/10$ at the time of radiation-matter equality. In the matter-dominated epoch the gravitational potential remains always a constant, whereas the energy density fluctuation starts to increase as η^2 at the time of horizon crossing around $\eta \sim k^{-1}$.

Small scales or short-wavelengths ($c_s k\eta \gg 1$): In this case we can neglect the last term (gravity effect) in (4.216) and the equation reduces to

$$\chi_{\vec{k}}'' + c_s^2 \vec{k}^2 \chi_{\vec{k}} = 0. \tag{4.229}$$

This is a wave equation for sound perturbations with time-dependent amplitude which can be solved explicitly in the Wentzel–Kramers–Brillouin (WKB) approximation for slowly varying speed of sound.

4.7 Quantum cosmological scalar perturbations

4.7.1 Slow-roll revisited

We consider a flat universe filled with a scalar field ϕ with an action

$$S = \int \sqrt{-g}\, d^4x\, \mathcal{P}(X, \phi), \quad X = -\frac{1}{2}g^{\alpha\beta}\, \nabla_\alpha\, \phi\, \nabla_\beta\, \phi. \tag{4.230}$$

A canonical scalar field is given by

$$\mathcal{P}(X, \phi) = X - V(\phi). \tag{4.231}$$

The energy–momentum tensor is defined by

$$\begin{aligned}
T_{\mu\nu} &= -\frac{2}{\sqrt{-g}}\frac{\delta S}{\delta g^{\mu\nu}} \\
&= 2X\frac{\partial\mathcal{P}}{\partial X}u_\mu u_\nu + \mathcal{P}g_{\mu\nu}, \quad u_\mu = -\frac{1}{\sqrt{2X}}\, \nabla_\mu\, \phi.
\end{aligned} \tag{4.232}$$

We observe that $g^{\mu\nu}u_\mu u_\nu = -1$. Since $T_{00} = \rho a^2$ we deduce

$$\rho = 2X\frac{\partial\mathcal{P}}{\partial X} - \mathcal{P}. \tag{4.233}$$

Thus

$$T_{\mu\nu} = (\rho + \mathcal{P})u_\mu u_\nu + \mathcal{P}g_{\mu\nu}. \tag{4.234}$$

In other words, $\mathcal{P}$ plays the role of pressure.

The unperturbed system consists of the usual scale factor $a(\eta)$ and a homogeneous field $\phi_0(\eta)$. The equations of motion of the scale factor are the Friedmann equations

$$\mathcal{H}^2 = \frac{8\pi G a^2}{3}\rho, \quad \mathcal{H}^2 - \mathcal{H}' = 4\pi G a^2(\rho + \mathcal{P}). \tag{4.235}$$

Also, we note the continuity equation

$$\rho' = -3\mathcal{H}(\rho + \mathcal{P}) = \frac{\partial\rho}{\partial\phi}\phi_0' + \frac{\partial\rho}{\partial X}X_0'. \tag{4.236}$$

The equation of motion of a canonical scalar field ϕ is given by

$$
\begin{aligned}
\frac{\delta S}{\delta \phi} &= \frac{1}{\sqrt{-g}} \partial_\alpha \left(\sqrt{-g}\, g^{\alpha\beta} \frac{\partial P}{\partial X} \partial_\beta \phi \right) + \frac{\partial P}{\partial \phi} \\
&= \frac{1}{\sqrt{-g}} \partial_\alpha (\sqrt{-g}\, g^{\alpha\beta} \partial_\beta \phi) - \frac{\partial V}{\partial \phi} \\
&= 0.
\end{aligned}
\tag{4.237}
$$

For the background ϕ_0 this reads explicitly

$$
\phi_0'' + 2\mathcal{H}\phi_0' + a^2 \frac{\partial V}{\partial \phi} = 0, \quad \ddot{\phi}_0 + 3H\dot{\phi}_0 + \frac{\partial V}{\partial \phi} = 0.
\tag{4.238}
$$

We consider scalar perturbation of the form

$$
\phi = \phi_0 + \delta\phi.
\tag{4.239}
$$

The gauge transformation of the scalar perturbation is computed as follows

$$
\begin{aligned}
\Delta\delta\phi &= \phi'(x' - \epsilon) - \phi(x) \\
&= -\epsilon^\lambda \frac{\partial}{\partial x^\lambda} \phi_0 \\
&= -\alpha\dot{\phi}_0.
\end{aligned}
\tag{4.240}
$$

Thus the gauge invariant scalar perturbation is given by

$$
\delta\hat{\phi} = \delta\phi - (E' - B)\phi_0', \quad \Delta\delta\hat{\phi} = 0.
\tag{4.241}
$$

The above scalar perturbation induces scalar metric perturbation of the form

$$
ds^2 = a^2(-(1 + 2\Phi)d\eta^2 + 2a\partial_i B d\eta dx^i + a^2[2\partial_i\partial_j E + (1 - 2\Psi)\delta_{ij}]dx^i dx^j).
\tag{4.242}
$$

Again we will work in the longitudinal (conformal-Newtonian) gauge, viz

$$
ds^2 = a^2(-(1 + 2\Phi)d\eta^2 + (1 - 2\Psi)\delta_{ij}dx^i dx^j).
\tag{4.243}
$$

To linear order, the equation of motion of the scalar field perturbation $\delta\phi$ reads

$$
\partial_\alpha(-a^4 \delta g^{\alpha\beta} \partial_\beta \phi_0 + a^4(\Phi - 3\Psi)\bar{g}^{\alpha\beta} \partial_\beta \phi_0 + a^4 \bar{g}^{\alpha\beta} \partial_\beta \delta\phi) - (\Phi - 3\Psi)\partial_\alpha(a^4 \bar{g}^{\alpha\beta} \partial_\beta \phi_0) - a^4 \delta\phi \frac{\partial^2 V}{\partial\phi^2} = 0.
\tag{4.244}
$$

Or equivalently

$$
\delta\phi'' + 2\mathcal{H}\delta\phi' - (\Phi' + 3\Psi')\phi_0' + 2a^2\Phi \frac{\partial V}{\partial\phi} + a^2\delta\phi \frac{\partial^2 V}{\partial\phi^2} - \partial_i^2 \delta\phi = 0.
\tag{4.245}
$$

The gauge invariant version of this equation is obtained by making the replacements $\delta\phi \longrightarrow \delta\hat{\phi}$, $\Phi \longrightarrow \Phi_B$ and $\Psi \longrightarrow \Psi_B$, viz

$$
\delta\hat{\phi}'' + 2\mathcal{H}\delta\hat{\phi}' - (\Phi_B' + 3\Psi_B')\phi_0' + 2a^2\Phi_B \frac{\partial V}{\partial\phi} + a^2\delta\hat{\phi} \frac{\partial^2 V}{\partial\phi^2} - \partial_i^2 \delta\hat{\phi} = 0.
\tag{4.246}
$$

Small scales or short-wavelengths: This corresponds to wavelengths $\lambda \ll 1/H$ or equivalently wavenumbers $k \gg aH$ where gravity can be neglected. Remember that $1/H$ is the Hubble distance and $1/aH$ is the Hubble length or radius. During inflation since $a \sim \exp(Ht)$ we have $aH = -1/\eta$. Thus this limit corresponds to $k\eta \gg 1$. The last term in the above equation therefore dominates and we end up with a solution of the form $\delta\hat{\phi} \sim \exp(\pm ik\eta)$. By using equations (4.233) and (4.282) (see below) we find that the gravitational potential solves the equations

$$\Psi'_B + \mathcal{H}\Phi_B = 4\pi G\phi'_0\delta\hat{\phi}. \tag{4.247}$$

We must also have $\Phi_B = \Psi_B$ (see below). The gravitational potential therefore oscillates as

$$\Psi_B = \Phi_B \sim \frac{4\pi G}{k}\phi'_0\delta\hat{\phi}. \tag{4.248}$$

The third and fourth terms can therefore be neglected. The fifth term can also be neglected since during inflation $\partial^2 V/\partial\phi^2 \ll V \sim H^2(\eta_V \ll 1)$. Equation (4.246) reduces therefore with $\delta\hat{\phi} = \exp(i\vec{k}\,\vec{x})\delta\hat{\phi}_k$ to

$$\delta\hat{\phi}_k'' + 2\mathcal{H}\delta\hat{\phi}_k' + \vec{k}^2\delta\hat{\phi}_k = 0. \tag{4.249}$$

In terms of $u_k = a\delta\hat{\phi}_k$ this reads

$$u_k'' + \left(\vec{k}^2 - \frac{a''}{a}\right)u_k = 0. \tag{4.250}$$

Since $k\eta \gg 1$ the solution is of the form

$$\delta\hat{\phi}_k \simeq \frac{C_k}{a}\exp(\pm ik\eta). \tag{4.251}$$

We fix the constant of integration C_k by requiring that the initial scalar mode arises as vacuum quantum fluctuation.

The minimal vacuum fluctuations must satisfy the Heisenberg uncertainty principle $\Delta X \Delta P \sim 1$. From (4.230) the action of the perturbation $\delta\phi$ starts as

$$S = \int dt \int dV \left[\frac{1}{2}\dot{\delta\phi}^2 + \cdots\right]. \tag{4.252}$$

Obviously $dV = a^3 d^3x$. Thus in a finite volume $V = L^3$ the canonical field is $X = L^{3/2}\delta\phi$ while the conjugate field is $P = L^{3/2}\dot{\delta\phi}$. For a massless field we have the estimate $P = L^{1/2}\delta\phi$ and as a consequence the Heisenberg uncertainty principle yields $\Delta\delta\varphi = 1/L$. In other words, minimal quantum fluctuations of the scalar perturbation are of the order of $1/L$. However, quantum fluctuations of the Fourier mode $\delta\phi_k$ are related to quantum fluctuations of the scalar perturbations $\delta\phi$ by (see below for a derivation)

$$\delta\phi \sim \delta\phi_k k^{3/2}. \tag{4.253}$$

Since $k \sim a/L$ we conclude that $\delta\phi_k \sim L^{1/2}/a^{3/2}$ or equivalently $\delta\phi_k \sim 1/a\sqrt{k}$. Hence

$$\delta\hat{\phi}_k \simeq \frac{1}{\sqrt{k}\,a}. \tag{4.254}$$

In other words, $C_k = 1/\sqrt{k}$. The evolution of the mode in this region is such that the vacuum spectrum is preserved. We observe that the amplitude of fluctuation is such that

$$\delta_\phi \sim \delta\phi_k k^{3/2} = \frac{k}{a} \gg H. \tag{4.255}$$

Thus every mode will eventually be stretched to very large scales while new modes will be generated. The moment $\eta_k \sim 1/k$ at which the mode k leaves the horizon is called horizon crossing and is defined by

$$\delta_\phi \sim \delta\phi_k k^{3/2} = \frac{k}{a_k} = H_{k \sim Ha}. \tag{4.256}$$

If this mode was classical it will be completely washed out, i.e. become very small, after it is stretched out to galactic scales.

Large scales or long-wavelengths: In the slow-roll approximation the equation of motion (4.238) becomes

$$3H\dot{\phi}_0 + \frac{\partial V}{\partial \phi} = 0. \tag{4.257}$$

The equations of motion (4.246) and (4.247) in terms of the physical time read

$$\ddot{\delta\hat{\phi}} + 3H\dot{\delta\hat{\phi}} - 4\dot{\Phi}_B\dot{\phi}_0 + 2\Phi_B\frac{\partial V}{\partial \phi} + \delta\hat{\phi}\frac{\partial^2 V}{\partial \phi^2} - \frac{1}{a^2}\partial_i^2\delta\hat{\phi} = 0. \tag{4.258}$$

$$\dot{\Phi}_B + H\Phi_B = 4\pi G\dot{\phi}_0\delta\hat{\phi}. \tag{4.259}$$

For long-wavelengths $k \ll aH$ we can drop the Laplacian term. As we will see the terms $\ddot{\delta\hat{\phi}}$ and $\dot{\Phi}_B$ are also negligible in this limit. The equations become

$$3H\dot{\delta\hat{\phi}} + 2\Phi_B\frac{\partial V}{\partial \phi} + \delta\hat{\phi}\frac{\partial^2 V}{\partial \phi^2} = 0. \tag{4.260}$$

$$H\Phi_B = 4\pi G\dot{\phi}_0\delta\hat{\phi}. \tag{4.261}$$

We introduce the variable

$$y = \frac{\delta\hat{\phi}}{\dfrac{\partial V}{\partial \phi}} = -\frac{\delta\hat{\phi}}{3H\dot{\phi}_0}. \tag{4.262}$$

Thus

$$H\Phi_B = 4\pi Gy(-3H\dot{\phi}_0^2) = 4\pi Gy\dot{V}. \tag{4.263}$$

Also (by neglecting $\ddot{\phi}_0$ and $\partial^2 V/\partial\phi^2$ and $\dot{H}$ during inflation)

$$3H\dot{y} + 2\Phi_B = 0. \tag{4.264}$$

By using also $3H^2 = 8\pi GV$ during inflation we have

$$\frac{d}{dt}(yV) = \frac{H}{8\pi G}(3H\dot{y} + 2\Phi_B) = 0. \tag{4.265}$$

The solutions are immediately given by

$$\delta\phi_k = C_k\frac{1}{V}\frac{\partial V}{\partial\phi}. \tag{4.266}$$

$$\Phi_B = \frac{4\pi GC_k}{H}\frac{\dot{V}}{V}$$
$$= -\frac{1}{2}C_k\left(\frac{1}{V}\frac{\partial V}{\partial\phi}\right)^2. \tag{4.267}$$

We fix the constant of integration C_k by comparing with (4.256) at the instant of horizon crossing. We obtain

$$C_k = \frac{k^{-1/2}}{a_k}\left(\frac{V}{\dfrac{\partial V}{\partial\phi}}\right)_{k\sim Ha}. \tag{4.268}$$

The solutions (4.251) and (4.266) are sketched in figure 4.5. After horizon crossing the short-wavelengths modes are stretched to galactic scales in such a way that they do not lose their amplitudes. Remember that inside the horizon gravity is negligible. Thus perturbations which are initially inside the horizon, will eventually exit the horizon, and then start feeling the curvature effects of gravity preserving therefore their amplitudes from decay. We say that the perturbation is frozen after horizon crossing. This is how we get the required amplitude $\Phi \sim 10^{-5}$ on large scales from initial quantum fluctuations.

At the end of inflation the slow-roll condition is violated since $V/(\partial V/\partial\phi)$ becomes of order 1 and the amplitude of fluctuations is

$$\delta_\phi(k)_{t_f} \sim C_k k^{3/2}$$

$$\sim \left(H\frac{V}{\dfrac{\partial V}{\partial\phi}}\right)_{k\sim Ha}$$

$$\sim \left(\frac{V^{3/2}}{\dfrac{\partial V}{\partial\phi}}\right)_{k\sim Ha}. \tag{4.269}$$

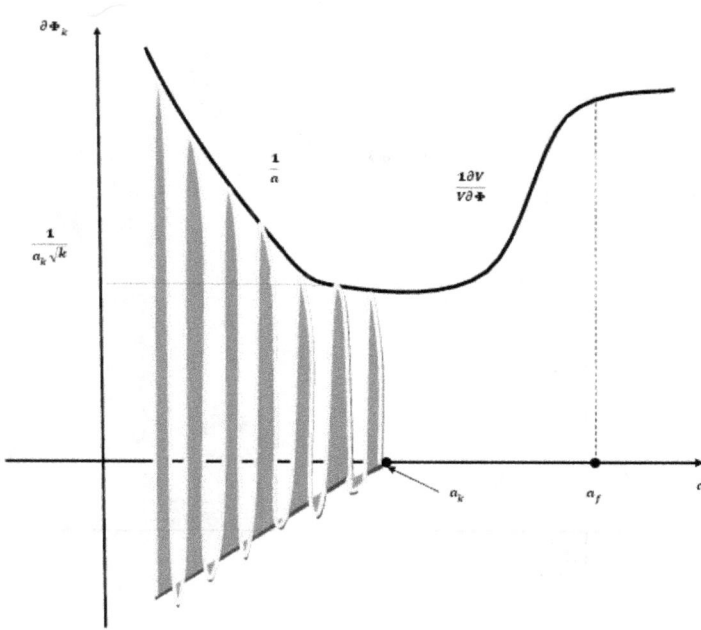

Figure 4.5. The short-wavelength and long-wavelength solutions.

This depends only on quantities evaluated at the moment of horizon crossing. For a power-law potential $V = \lambda\phi^n/n$ we get

$$\delta_\phi(k)_{t_f} \sim \lambda^{1/2}\left(\phi^2_{k\sim Ha}\right)^{\frac{n+2}{4}}. \tag{4.270}$$

By using (4.64) we have

$$\phi^2_{k\sim Ha} \sim \ln\frac{a(t_f)}{a(t_k)}$$

$$\sim \ln\frac{1}{(aH)_k}H_k\,a(t_f) \tag{4.271}$$

$$\sim \ln\lambda_{\mathrm{ph}}H_k.$$

The physical wavelength is $\lambda_{\mathrm{ph}} = a(t_f)/k$. Thus the amplitude of fluctuations at the end of inflation is

$$\delta_\phi(k)_{t_f} \sim \lambda^{1/2}(\ln\lambda_{\mathrm{ph}}H_k)^{\frac{n+2}{4}}. \tag{4.272}$$

We can further make the approximation $H_k \sim H_f$ since curvature scale does not change very much during inflation, which is essentially the defining property of inflation. We get finally the amplitude

$$\delta_\phi(k)_{t_f} \sim \lambda^{1/2}(\ln\lambda_{\mathrm{ph}}H_f)^{\frac{n+2}{4}}. \tag{4.273}$$

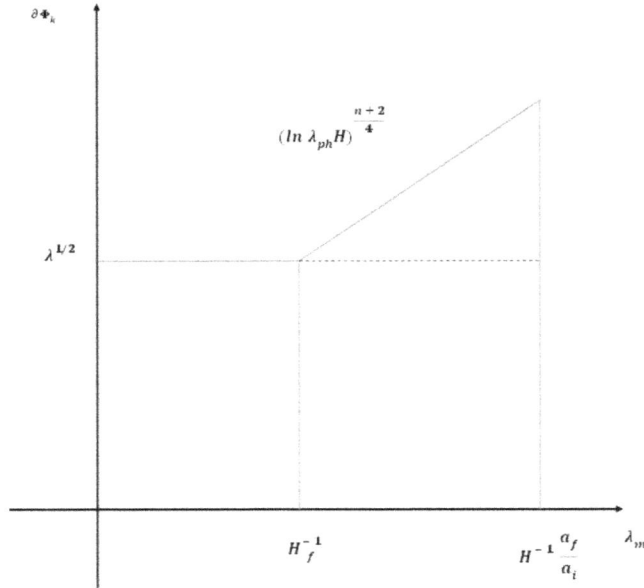

Figure 4.6. The amplitude of fluctuations.

Inside the horizon $\lambda_{\rm ph} < 1/H_f$ or equivalently $k > H_f a_f$, i.e. the logarithm becomes negative and thus we should instead make the replacment $\phi^2_{k \sim Ha} = \phi^2_f$, i.e. the amplitude comes out proportional to $\lambda^{1/2}$ in this regime. This is the flat space result since gravity is neglected inside the horizon. This is sketched in figure 4.6.

For a quadratic potential $V = m^2\phi^2/2$ we get the amplitude $\delta_\phi = m \ln \lambda_{\rm ph} H_f$. Galactic scales correspond to $L = 10^{25}$cm or equivalently $\ln \lambda_{\rm ph} H_f \sim 50$ and thus in order to get an amplitude of the gravitational potential around 10^{-5} the mass of the inflaton scalar field should be around 10^{-6} in Planck units, viz $m = 10^{-6}$. $m_{\rm pl} = 10^{-6}$ $\sqrt{\hbar c}/\sqrt{8\pi G} = 10^{-6}10^{18}$GeV $= 10^{12}$GeV. At the end of inflation the scalar field is around 1 in Planck units, viz $\phi = 1.1/l_{\rm pl} = \sqrt{c^3}/\sqrt{\hbar G}$. The energy density at the end of inflation is therefore $\rho \sim m^2\phi^2 \sim 10^{-12}$. $\rho_{\rm pl}$.

4.7.2 Mukhanov action

Equation (4.246) contains three unknown variables Φ_B, Ψ_B and $\delta\hat\phi$ which should also satisfy Einstein's equations. Thus we need to compute the energy–momentum tensor explicitly. We will drop in the following the subscript B and the hat for ease of notation.

We compute (with $u_0 \mid_{\phi_0} = -a$, $u_i \mid_{\phi_0} = 0$, $X_0 = (\phi'_0)^2/2a^2$, etc)

$$\delta T^i_j = 2\bar{\mathcal{P}}\Psi\delta_{ij} + \frac{1}{a^2}\delta T_{ij}$$
$$= \delta\mathcal{P}\delta_{ij}.$$

(4.274)

Thus from the Einstein's equation with $i \neq j$ we conclude, as before, that $\Phi_B = \Psi_B$. The other two Einstein's equations are therefore sufficient to determine Φ_B and $\delta\hat{\phi}$. We compute then

$$\delta T_i^0 = -\frac{1}{a^2}\delta T_{0i}$$
$$= (\bar{\rho} + \bar{\mathcal{P}})u^0 \mid_{\phi_0} \delta u_i. \tag{4.275}$$

$$\delta T_0^0 = 2\bar{\rho}\Phi - \frac{1}{a^2}\delta T_{00}$$
$$= 2(\bar{\rho} + \bar{\mathcal{P}})\Phi + \frac{2}{a}(\bar{\rho} + \bar{\mathcal{P}})\delta u_0 - \delta\rho \tag{4.276}$$
$$= -\delta\rho.$$

In the above two equations we have used

$$\delta u_\mu = -\frac{a}{\phi_0'}\partial_\mu\delta\phi + (\Phi - \frac{\delta\phi'}{\phi_0'})u_\mu \mid_{\phi_0}. \tag{4.277}$$

Further, we compute

$$-\delta T_0^0 = \delta\rho$$
$$= \frac{\partial\rho}{\partial X}\delta X + \frac{\partial\rho}{\partial\phi}\delta\phi$$
$$= \frac{\partial\rho}{\partial X}\left(-\frac{\Phi\phi_0'^2}{a^2} + \frac{\phi_0'\delta\phi'}{a^2}\right) + \left(-\frac{\partial\rho}{\partial X}\frac{X_0'}{\phi_0'} - \frac{3\mathcal{H}(\bar{\rho} + \bar{\mathcal{P}})}{\phi_0'}\right)\delta\phi$$
$$= \frac{\partial\rho}{\partial X}\left(-\frac{\Phi\phi_0'^2}{a^2} + \frac{\phi_0'\delta\phi'}{a^2} - \frac{X_0'}{\phi_0'}\delta\phi\right) + \left(-\frac{3\mathcal{H}(\bar{\rho} + \bar{\mathcal{P}})}{\phi_0'}\right)\delta\phi$$
$$= \frac{\partial\rho}{\partial X}\left(-\frac{\Phi\phi_0'^2}{a^2} + \frac{\phi_0'\delta\phi'}{a^2} - \frac{\phi_0''}{a^2}\delta\phi + \frac{\phi_0'\mathcal{H}}{a^2}\delta\phi\right) + \left(-\frac{3\mathcal{H}(\bar{\rho} + \bar{\mathcal{P}})}{\phi_0'}\right)\delta\phi \tag{4.278}$$
$$= 2X_0\frac{\partial\rho}{\partial X}\left(-\Phi + \left(\frac{\delta\phi}{\phi_0'}\right)' + \frac{\mathcal{H}}{\phi_0'}\delta\phi\right) + \left(-\frac{3\mathcal{H}(\bar{\rho} + \bar{\mathcal{P}})}{\phi_0'}\right)\delta\phi$$
$$= \frac{\bar{\rho} + \bar{\mathcal{P}}}{c_s^2}\left(-\Phi + \left(\frac{\delta\phi}{\phi_0'}\right)' + \frac{\mathcal{H}}{\phi_0'}\delta\phi\right) + \left(-\frac{3\mathcal{H}(\bar{\rho} + \bar{\mathcal{P}})}{\phi_0'}\right)\delta\phi.$$

In the last equation we have introduced the speed of sound by the relation

$$c_s^2 = \frac{\partial\mathcal{P}}{\partial X}\frac{\partial X}{\partial\rho} = \frac{\rho + \mathcal{P}}{2X}\frac{\partial X}{\partial\rho}. \tag{4.279}$$

Also,

$$\delta T_i^0 = -\frac{\bar{\rho} + \bar{\mathcal{P}}}{\phi_0'}\partial_i \delta\phi. \tag{4.280}$$

The relevant Einstein's equations are now given explicitly by

$$\partial_i^2 \Psi - 3\mathcal{H}(\Psi' + \mathcal{H}\Phi) = 4\pi G a^2 (\bar{\rho} + \bar{\mathcal{P}})\left[\frac{1}{c_s^2}\left(-\Phi + \left(\frac{\delta\phi}{\phi_0'}\right)' + \frac{\mathcal{H}}{\phi_0'}\delta\phi\right) - \frac{3\mathcal{H}}{\phi_0'}\delta\phi\right]. \tag{4.281}$$

$$(\Psi' + \mathcal{H}\Phi) = 4\pi G a^2 (\bar{\rho} + \bar{\mathcal{P}})\frac{\delta\phi}{\phi_0'}. \tag{4.282}$$

By substituting this last equation into the previous one we obtain

$$\partial_i^2 \Psi = \frac{4\pi G a^2}{\mathcal{H} c_s^2}(\bar{\rho} + \bar{\mathcal{P}})\left[\Psi' - 4\pi G a^2 (\bar{\rho} + \bar{\mathcal{P}})\frac{\delta\phi}{\phi_0'} + \mathcal{H}\left(\frac{\delta\phi}{\phi_0'}\right)' + \frac{\mathcal{H}^2}{\phi_0'}\delta\phi\right]$$

$$= \frac{4\pi G a^2}{\mathcal{H} c_s^2}(\bar{\rho} + \bar{\mathcal{P}})\left(\Psi + \mathcal{H}\frac{\delta\phi}{\phi_0'}\right)'. \tag{4.283}$$

Further,

$$\left(a^2 \frac{\Psi}{\mathcal{H}}\right)' = a^2\left(\frac{\Psi'}{\mathcal{H}} + \Psi + 4\pi G a^2 \frac{\bar{\rho} + \bar{\mathcal{P}}}{\mathcal{H}^2}\Psi\right). \tag{4.284}$$

We have then the two equations

$$\partial_i^2 \Psi = \frac{4\pi G a^2}{\mathcal{H} c_s^2}(\bar{\rho} + \bar{\mathcal{P}})\left(\Psi + \mathcal{H}\frac{\delta\phi}{\phi_0'}\right)'. \tag{4.285}$$

$$\left(a^2 \frac{\Psi}{\mathcal{H}}\right)' = \frac{4\pi G a^4}{\mathcal{H}^2}(\bar{\rho} + \bar{\mathcal{P}})\left(\Psi + \mathcal{H}\frac{\delta\phi}{\phi_0'}\right). \tag{4.286}$$

We introduce the variables u and v and the parameters z and θ by

$$u = \frac{\Psi}{4\pi G\sqrt{\bar{\rho} + \bar{\mathcal{P}}}}, \quad v = \sqrt{\frac{\partial\rho}{\partial X}}\,a\left(\delta\phi + \frac{\phi_0'}{\mathcal{H}}\Psi\right). \tag{4.287}$$

$$z = \frac{a^2\sqrt{\bar{\rho} + \bar{\mathcal{P}}}}{c_s \mathcal{H}}, \quad \theta = \frac{1}{c_s z} = \sqrt{\frac{8\pi G}{3}}\frac{1}{a}\frac{1}{\sqrt{1 + \dfrac{\bar{\mathcal{P}}}{\bar{\rho}}}}. \tag{4.288}$$

The Einstein's equations in terms of these new variables take the simpler form

$$\partial_i^2 u = \frac{1}{c_s} z \left(\Psi + \mathcal{H} \frac{\delta\phi}{\phi_0'} \right)'$$

$$= \frac{1}{c_s} z \left(\frac{v}{z} \right)'. \tag{4.289}$$

$$\left(4\pi G \frac{u}{\theta} \right)' = \frac{4\pi G a^4}{\mathcal{H}^2} (\bar{p} + \mathcal{P}) \frac{v}{z} \Rightarrow \left(\frac{u}{\theta} \right)' = c_s \frac{v}{\theta}. \tag{4.290}$$

By substituting one of the equations into the other one we find the second order differential equation

$$u'' - c_s^2 \partial_i^2 u - \frac{\theta''}{\theta} u = 0. \tag{4.291}$$

This is precisely the Poisson equation (4.214). In fact the definitions of u and θ used here for the scalar field are essentially those used in the hydrodynamical fluid. A similar equation for v holds, viz

$$v'' - c_s^2 \partial_i^2 v - \frac{z''}{z} v = 0. \tag{4.292}$$

Since we are interested in quantizing the scalar metric perturbation we will have to quantize the fields u and v. Thus one must start from an appropriate action which gives as equations of motion of the fields u and v precisely the above Poisson equations. This is straightforward and one finds for the field v the action

$$S = \int d\eta d^3x \ \mathcal{L} = \frac{1}{2} \int d\eta d^3x \left(v'^2 + c_s^2 v \partial_i^2 v + \frac{z''}{z} v^2 \right). \tag{4.293}$$

From this result we see that metric scalar perturbations are given by a massless scalar field in a de Sitter spacetime (see chapter 5). This is the most fundamental result in our view and a direct derivation of this action using ADM (Arnowitt, Deser and Misner) formalism, which is a very complex calculation, is included in the next section for completeness. The ADM formalism is reviewed in detail at the end of next chapter.

Small scales or short-wavelengths: For a plane wave perturbation with a wavenumber k such that $c_s^2 k^2 \gg |\theta''/\theta|$ we have the WKB (slowly varying speed of sound c_s) solution

$$u = \frac{C}{\sqrt{c_s}} \exp\left(\pm ik \int c_s d\eta \right). \tag{4.294}$$

The gravitational potential is immediately given

$$\Phi = 4\pi G \dot{\phi_0} C \sqrt{\frac{\frac{\partial P}{\partial x}}{c_s}} \exp\left(\pm ik \int c_s d\eta \right). \tag{4.295}$$

On the other hand, we can determine the perturbation of the scalar field from (4.282). We get

$$\delta\phi = C \frac{1}{\sqrt{c_s \dfrac{\partial P}{\partial x}}} \left(\pm i c_s \frac{k}{a} + H + \cdots \right) \exp\left(\pm i k \int c_s \, d\eta \right). \tag{4.296}$$

The most important observation here is that both the gravitational potential and the scalar perturbation oscillate in this regime. The amplitude of the gravitational potential is proportional to $\dot{\phi}_0$ and thus will grow at the end of inflation while the amplitude of the scalar perturbation decays as $1/a$.

Large scales or long-wavelengths: These are characterized by $c_s^2 k^2 \ll |\theta''/\theta|$. The solution was found in previous sections and it is given by

$$\Phi = A \frac{d}{dt}\left(\frac{1}{a} \int a \, dt \right) = A\left(1 - \frac{H}{a} \int a \, dt \right). \tag{4.297}$$

The perturbation of the scalar field from (4.282) is given by

$$4\pi G a^2 (\bar{\rho} + \bar{\mathcal{P}}) \frac{\delta\phi}{\phi_0'} = \frac{d}{dt}(a\Phi)$$

$$= -A\dot{H} \int a \, dt \tag{4.298}$$

$$= A 4\pi G (\bar{\rho} + \bar{\mathcal{P}}) \int a \, dt.$$

Thus

$$\delta\phi = A\dot{\phi}_0 \frac{1}{a} \int a \, dt. \tag{4.299}$$

During slow-roll inflation we can make the approximation

$$\frac{H}{a} \int a \, dt = 1 - \frac{d}{dt}\left(\frac{1}{H}\right) + \frac{H}{a} \int \frac{da}{H} \frac{d}{dt}\left(\frac{1}{H} \frac{d}{dt}\left(\frac{1}{H} \right) \right)$$

$$\simeq 1 - \frac{d}{dt}\left(\frac{1}{H}\right) + \cdots \tag{4.300}$$

Our results reduce therefore during inflation to

$$\Phi = -\frac{A}{H^2}\dot{H} \sim A\left(\frac{1}{V} \frac{\partial V}{\partial \phi} \right)^2. \tag{4.301}$$

$$\delta\phi = \frac{A\dot{\phi}_0}{H} \sim A \frac{1}{V}\frac{\partial V}{\partial \phi}. \tag{4.302}$$

These are precisely the equations (4.267) and (4.266), respectively. At the end of inflation $V/(\partial V/\partial \phi)$ becomes of order 1 and thus

$$\Phi = A = \left(H\frac{\delta\phi}{\dot{\phi}_0}\right)_{c_s k \sim Ha}. \tag{4.303}$$

We evaluated the different quantities at the instant of horizon crossing.

After inflation the scale factor behaves as $a \propto t^p$. In this case we get the results

$$\Phi = \frac{A}{p+1}. \tag{4.304}$$

$$\delta\phi = \frac{A\dot{\phi}_0}{p+1}t. \tag{4.305}$$

Hence the amplitude of the gravitational field freezes out after inflation. In the radiation-dominated phase corresponding to $p = 1/2$ we get then

$$\Phi = \frac{2A}{3}$$
$$= \frac{2}{3}\left(H\frac{\delta\phi}{\dot{\phi}_0}\right)_{c_s k \sim Ha}. \tag{4.306}$$

Thus the amplitudes at the end of inflation and in the radiation-dominated phase differ only by a numerical coefficient.

4.7.3 Quantization and inflationary spectrum

The canonical momentum is defined by the usual formula

$$\pi = \frac{\partial \mathcal{L}}{\partial v'} = v'. \tag{4.307}$$

In the quantum theory we replace v and π with operators $\hat{v}$ and $\hat{\pi}$ satisfying the equal-time commutation relations given by

$$[\hat{v}(\eta, \vec{x}), \hat{\pi}(\eta, \vec{y})] = i\delta^3(\vec{x}-\vec{y}). \tag{4.308}$$

$$[\hat{v}(\eta, \vec{x}), \hat{v}(\eta, \vec{y})] = [\hat{\pi}(\eta, \vec{x}), \hat{\pi}(\eta, \vec{y})] = 0. \tag{4.309}$$

We expand the field as

$$\hat{v}(\eta, \vec{x}) = \frac{1}{\sqrt{2}}\int\frac{d^3k}{(2\pi)^{3/2}}\left(\hat{a}_k v_k^*(\eta)e^{i\vec{k}\vec{x}} + \hat{a}_k^+ v_k(\eta)e^{-i\vec{k}\vec{x}}\right). \tag{4.310}$$

Thus

$$\hat{\pi}(\eta, \vec{x}) = \frac{1}{\sqrt{2}}\int\frac{d^3k}{(2\pi)^{3/2}}\left(\hat{a}_k v_k^{*'}(\eta)e^{i\vec{k}\vec{x}} + \hat{a}_k^+ v_k'(\eta)e^{-i\vec{k}\vec{x}}\right). \tag{4.311}$$

This field obeys the equation of motion

$$\hat{v}'' - c_s^2 \partial_i^2 \hat{v} - \frac{z''}{z}\hat{v} = 0. \tag{4.312}$$

Equivalently

$$v_k'' + \omega_k^2(\eta)v_k = 0, \quad \omega_k^2(\eta) = c_s^2\vec{k}^2 - \frac{z''}{z}. \tag{4.313}$$

The creation and annihilation operators are expected to satisfy the commutation relations

$$[\hat{a}_k, \hat{a}_p^+] = \delta^3(\vec{k} - \vec{p}). \tag{4.314}$$

$$[\hat{a}_k, \hat{a}_p] = [\hat{a}_k^+, \hat{a}_p^+] = 0. \tag{4.315}$$

We compute then

$$[\hat{v}(\eta, \vec{x}), \hat{\pi}(\eta, \vec{y})] = \frac{1}{2}\int \frac{d^3k}{(2\pi)^3}e^{i\vec{k}\cdot(\vec{x}-\vec{y})}\Big(v_k^*v_k' - v_k v_k^{*'}\Big). \tag{4.316}$$

Thus we must have

$$v_k^*v_k' - v_k v_k^{*'} = 2i. \tag{4.317}$$

This is the condition for v_k to be a positive norm solution (see later for more detail). The negative norm solution is immediately given by v_k^*. Alternatively, the above condition is the Wronskian which expresses the linear independence of these two solutions.

The Hamiltonian is given by

$$\begin{aligned}
\hat{H} &= \int d^3x(\hat{\pi}\hat{v}' - \mathcal{L}) \\
&= \frac{1}{2}\int d^3x(\hat{\pi}^2 - c_s^2\hat{v}\partial_i^2\hat{v} - \frac{z''}{z}\hat{v}^2) \\
&= \int d^3k\Big(E_k(\hat{a}_k\hat{a}_k^+ + \hat{a}_k^+a_k) + F_k\hat{a}_k^+\hat{a}_{-k}^+ + F_k^*\hat{a}_k\hat{a}_{-k}\Big),
\end{aligned} \tag{4.318}$$

where

$$E_k = \frac{1}{2}(|v_k'|^2 + \omega_k^2|v_k|^2), \quad F_k = \frac{1}{2}((v_k')^2 + \omega_k^2(v_k)^2). \tag{4.319}$$

The choice of the vacuum state is a very subtle issue in a curved spacetime (see chapter 5). Here, we will simply define the vacuum state as the state annihilated by all the $\hat{a}_k$, viz

$$\hat{a}_k|0\rangle = 0. \tag{4.320}$$

Then

$$\langle 0|\hat{H}|0\rangle = \int d^3k E_k$$
$$= \frac{1}{2}\int d^3k(|v_k{'}|^2 + \omega_k^2|v_k|^2). \tag{4.321}$$

We consider now the ansatz for v_k given by

$$v_k = r_k \exp(i\alpha_k). \tag{4.322}$$

The Wronskian condition becomes

$$r_k^2 \alpha_k' = 1. \tag{4.323}$$

The energy of the vacuum in this vacuum becomes

$$\langle 0|\hat{H}|0\rangle = \frac{1}{2}\int d^3k(r_k'^2 + \frac{1}{r_k^2} + \omega_k^2 r_k^2). \tag{4.324}$$

This energy is minimized when $r_k'(\eta) = 0$ and $r_k(\eta) = 1/\sqrt{\omega_k(\eta)}$. Thus at a given initial time η_0 the energy in the vacuum $|0\rangle$ is minimum iff

$$v_k(\eta_0) = \frac{1}{\sqrt{\omega_k(\eta_0)}}\exp(i\alpha_k(\eta_0)), \quad v_k'(\eta_0) = i\sqrt{\omega_k(\eta_0)}\exp(i\alpha_k(\eta_0)). \tag{4.325}$$

The phases $\eta(\eta_0)$ can clearly be set to zero. These are the initial conditions for v_k and v_k'. These considerations are well defined for modes with $\omega_k^2 > 0$ or equivalently $c_s^2 k^2 > (z''/z)_{\eta_0}$. This is the sub-horizon or sub-Hubble regime. By allowing c_s to change only adiabatically, the modes $c_s^2 k^2 > (z''/z)_{\eta_0}$ remain not exited and the above minimal fluctuations are well defined. In the case that ω_k is independent of time, the vacuum state $|0\rangle$ coincides precisely with the Minkowski vacuum and minimal fluctuations are obviously well defined.

On the other hand, the super-horizon or super-Hubble modes $c_s^2 k^2 < (z''/z)_{\eta_0}$ cannot be well determined in the same way but fortunately they will be stretched to extreme unobservable distances subsequent to inflation.

We compute now the 2-point function

$$\langle 0|\hat{\Phi}(\eta, \vec{x})\hat{\Phi}(\eta, \vec{y})|0\rangle = (4\pi G)^2(\bar{p} + \bar{P})\langle 0|\hat{u}(\eta, \vec{x})\hat{u}(\eta, \vec{y})|0\rangle. \tag{4.326}$$

The expansion of the field operator $\hat{u}$ is similarly given by

$$\hat{u}(\eta, \vec{x}) = \frac{1}{\sqrt{2}}\int \frac{d^3k}{(2\pi)^{3/2}}\left(\hat{a}_k u_k^*(\eta)e^{i\vec{k}\vec{x}} + \hat{a}_k^+ u_k(\eta)e^{-i\vec{k}\vec{x}}\right). \tag{4.327}$$

Thus

$$\langle 0|\hat{\Phi}(\eta,\,\vec{x})\hat{\Phi}(\eta,\,\vec{y})|0\rangle = \frac{1}{2}(4\pi G)^2(\bar{\rho}+\bar{\mathcal{P}})\left\langle 0\Big|\int\frac{d^3p}{(2\pi)^{3/2}}\Big(\hat{a}_p u_p^*(\eta)e^{i\vec{p}\,\vec{x}}\Big).\int\frac{d^3k}{(2\pi)^{3/2}}\Big(\hat{a}_k^+ u_k(\eta)e^{-i\vec{k}\,\vec{y}}\Big)\Big|0\right\rangle$$

$$= \frac{1}{2}(4\pi G)^2(\bar{\rho}+\bar{\mathcal{P}})\int\frac{d^3k}{(2\pi)^3}|u_k(\eta)|^2 e^{i\vec{k}\,(\vec{x}-\vec{y})} \tag{4.328}$$

$$= 4G^2(\bar{\rho}+\bar{\mathcal{P}})\int k^3|u_k(\eta)|^2\frac{\sin k\,r}{kr}\frac{dk}{k}.$$

In this equation $r = |\vec{x}-\vec{y}|$. This can be related to the variance σ_k^2 of the gravitational potential Φ as follows. First we write the gravitational potential in the form

$$\hat{\Phi}(\eta,\,\vec{x}) = \int\frac{d^3k}{(2\pi)^{3/2}}\hat{\Phi}(\eta,\,\vec{k})e^{i\vec{k}\vec{x}},\quad \hat{\Phi}(\eta,\,\vec{k}) = \frac{4\pi G\sqrt{\bar{\rho}+\bar{\mathcal{P}}}}{\sqrt{2}}\Big(\hat{a}_k u_k^*(\eta) + \hat{a}_{-k}^+ u_{-k}(\eta)\Big). \tag{4.329}$$

Then we have

$$\langle 0|\hat{\Phi}(\eta,\,\vec{x})\hat{\Phi}(\eta,\,\vec{y})|0\rangle = \int\frac{d^3k}{(2\pi)^{3/2}}\int\frac{d^3p}{(2\pi)^{3/2}}\left\langle 0|\hat{\Phi}(\eta,\,\vec{k})\hat{\Phi}(\eta,\,\vec{p})|0\right\rangle e^{i\vec{k}\,\vec{x}}e^{i\vec{p}\,\vec{y}}$$

$$= \int\frac{d^3k}{(2\pi)^{3/2}}\int\frac{d^3p}{(2\pi)^{3/2}}\sigma_k^2\delta^3(\vec{k}+\vec{p})e^{i\vec{k}\,\vec{x}}e^{i\vec{p}\,\vec{y}} \tag{4.330}$$

$$= \int\frac{k^3\sigma_k^2}{2\pi^2}\frac{\sin k\,r}{kr}\frac{dk}{k}.$$

σ_k^2 is precisely the variance of the gravitational potential Φ given by

$$\sigma_k^2 = 8\pi^2 G^2(\bar{\rho}+\bar{\mathcal{P}})|u_k(\eta)|^2. \tag{4.331}$$

The dimensionless variance or power spectrum is defined by

$$\delta_\Phi^2(k) = \frac{k^3\sigma_k^2}{2\pi^2}. \tag{4.332}$$

Short-wavelengths: From the equations of motion $c_s\partial_i^2 u = z(v/z)'$ and $(u/\theta)' = c_s v/\theta$ we find

$$u_k = -\frac{1}{c_s|\vec{k}|^2}(v_k' - \frac{z'}{z}v_k) \Rightarrow u_k(\eta_0)$$

$$= -\frac{i}{\sqrt{c_s}|\vec{k}|^{\frac{3}{2}}}\left[1 - \frac{1}{c_s^2|\vec{k}|^2}\frac{z''}{z}\right]^{\frac{1}{4}} + \frac{z'}{z}\frac{1}{c_s^{\frac{3}{2}}|\vec{k}|^{\frac{5}{2}}}\left[1 - \frac{1}{c_s^2|\vec{k}|^2}\frac{z''}{z}\right]^{-\frac{1}{4}}. \tag{4.333}$$

$$u_k' = c_s v_k - \left(\frac{c_s'}{c_s} + \frac{z'}{z}\right)u_k \Rightarrow u_k'(\eta_0) = \frac{\sqrt{c_s}}{|\vec{k}|^{\frac{1}{2}}}\left[1 - \frac{1}{c_s^2|\vec{k}|^2}\frac{z''}{z}\right]^{-\frac{1}{4}} - \left(\frac{c_s'}{c_s} + \frac{z'}{z}\right)u_k(\eta_0). \tag{4.334}$$

All functions are of course evaluated at the initial time $\eta = \eta_0$. In the relevant regime of short-wavelengths where $c_s^2|\vec{k}|^2 \gg (z''/z)_{\eta_0}$ or equivalently $c_s^2|\vec{k}|^2 \gg (\theta''/\theta)_{\eta_0}$

we can neglect the gravity terms in equations (4.291) and (4.292) and we obtain the initial conditions

$$u_k(\eta_0) = -\frac{i}{\sqrt{c_s}\,|\overrightarrow{k}\,|^{\frac{3}{2}}}.$$

(4.335)

$$u_k'(\eta_0) = \frac{\sqrt{c_s}}{|\overrightarrow{k}\,|^{\frac{1}{2}}}.$$

(4.336)

In this regime, the WKB solution of equation (4.291) is therefore given by

$$u_k(\eta) = -\frac{i}{\sqrt{c_s}\,|\overrightarrow{k}\,|^{\frac{3}{2}}}\,\exp\left(ik\int_{\eta_0}^{\eta}c_s(\eta')d\eta'\right).$$

(4.337)

During inflation $|\dot{H}/H^2| \ll 1$ and thus in this regime θ behaves as $\theta \sim 1/a$ while a behaves as $a \sim -1/\eta H$. Thus $|\theta''/\theta| \sim |\dot{H}/\eta^2 H^2| \ll 1/\eta^2$. The short-wavelengths regime is given by $c_s^2|\overrightarrow{k}\,|^2 \gg (\theta''/\theta)_\eta$ or equivalently, with $c_s \ll 1$ during inflation, by

$$|\eta| \gg \frac{1}{k}\sqrt{|\frac{\dot{H}}{H^2}|}.$$

(4.338)

Remember that at the end of inflation $\dot{H}/H^2$ becomes of order 1. Equivalently short-wavelengths regime is given by $|\eta| \gg 1/c_s k$ which is much larger than the previous estimate. On the other hand, horizon crossing is given by $c_s k|\eta| \sim 1$ and long-wavelengths regime is given by $|\eta| \ll 1/c_s k$. Hence there is a short time interval outside the horizon given by

$$\frac{1}{c_s k} > |\eta| > \frac{1}{k}\sqrt{|\frac{\dot{H}}{H^2}|},$$

(4.339)

in which the solution (4.337) is still valid. Since the above time interval is very narrow the solution (4.337) in this range is effectively a constant, i.e. the gravitational potential freezes at horizon crossing.

In this case the power spectrum is given by

$$\sigma_k^2 = 8\pi^2 G^2(\bar{\rho} + \bar{\mathcal{P}})\frac{1}{c_s k^3} \Rightarrow \delta_\Phi^2(k,\,t) = 4G^2\frac{\bar{\rho} + \bar{\mathcal{P}}}{c_s},\quad c_s k \gg Ha.$$

(4.340)

Long-wavelengths: In this case the solution is given by (4.297), viz

$$u_k = \frac{\Phi_k}{4\pi G\sqrt{\bar{\rho} + \bar{\mathcal{P}}}} = \frac{A_k}{4\pi G\sqrt{\bar{\rho} + \bar{\mathcal{P}}}}\left(1 - \frac{H}{a}\int a\,dt\right).$$

(4.341)

During inflation, and using $\dot{H}/4\pi G = -(\bar{\rho} + \bar{\mathcal{P}})$, we have

$$u_k = \frac{A_k}{4\pi G\sqrt{\bar{\rho} + \bar{\mathcal{P}}}} \frac{d}{dt}\left(\frac{1}{H}\right)$$

$$= A_k \frac{\sqrt{\bar{\rho} + \bar{\mathcal{P}}}}{H^2}.$$

(4.342)

This is constant in the time interval (4.339). This should be compared with the solution (4.337) which holds in the time interval (4.339). Since both η_0 and η_0 are in this short time interval they can be taken both to be equal to the moment of horizon crossing. This allows us to fix A_k as

$$A_k = -\frac{i}{k^{\frac{3}{2}}}\left(\frac{H^2}{\sqrt{c_s(\bar{\rho} + \bar{\mathcal{P}})}}\right)_{c_s k \sim Ha}.$$

(4.343)

In this case the power spectrum is given by

$$\sigma_k^2 = \frac{1}{2k^3}\left(\frac{H^4}{c_s(\bar{\rho} + \bar{\mathcal{P}})}\right)_{c_s k \sim Ha}\left(1 - \frac{H}{a}\int a\,dt\right)^2$$

$$\Rightarrow \delta_\Phi^2(k, t) = \frac{16}{9}G^2\left(\frac{\bar{\rho}}{c_s\left(1 + \frac{\bar{\mathcal{P}}}{\bar{\rho}}\right)}\right)_{c_s k \sim Ha}\left(1 - \frac{H}{a}\int a\,dt\right)^2, \quad (Ha/c_s)_i < k \ll Ha/c_s.$$

(4.344)

This formula gives the time evolution of long-wavelength perturbations even after inflation. After inflation the universe is radiation-dominated (where CMB originated) and hence $a \sim t^{1/2}$. In this case we get the power spectrum

$$\delta_\Phi^2 = \frac{64}{81}G^2\left(\frac{\bar{\rho}}{c_s\left(1 + \frac{\bar{\mathcal{P}}}{\bar{\rho}}\right)}\right)_{c_s k \sim Ha}, \quad (Ha/c_s)_i < k < (Ha/c_s)_f.$$

(4.345)

This results applies for large scales which includes the whole universe. This depends on the energy density $\bar{\rho}$ and the deviation of the equation of state from the vacuum given by $\Delta w = 1 + \bar{\mathcal{P}}/\bar{\rho}$ at the instant of horizon crossing. We know that $\delta_\Phi \sim 10^{-5}$ on galactic scales while Δw is estimated as 10^{-2} thus the energy density at horizon crossing must be of the order of $10^{-12}G^{-2}$, i.e. 10^{-12} of the Planck density. This is the same estimate obtained previously.

The above spectrum depends on the scale slightly. The requirement that inflation must have a graceful exist means that the energy density decreases slowly while the deviation of the equation of state from the vacuum increases slowly at the end of inflation, and as a consequence, the perturbations which cross the horizon earlier have larger amplitudes than those which cross the horizon later. A flat (scale-invariant) spectrum is characterized by a spectral index $n_s = 1$ where n_s is defined through the power law

$$\delta_\Phi^2 \sim k^{n_s - 1}. \tag{4.346}$$

Obviously

$$n_s - 1 = \frac{d \ln \delta_\Phi^2}{d \ln k}. \tag{4.347}$$

On the other hand,

$$\begin{aligned} n_s - 1 &= \frac{1}{H} \frac{\dot{\rho}}{\rho} - \frac{1}{H} \frac{d}{dt} \left(\ln c_s + \ln \left(1 + \frac{\mathcal{P}}{\rho} \right) \right) \\ &= 2 \frac{\dot{H}}{H^2} - \frac{1}{H} \frac{d}{dt} \left(\ln c_s + \ln \left(1 + \frac{\mathcal{P}}{\rho} \right) \right) \\ &= - 3 \left(1 + \frac{\mathcal{P}}{\rho} \right) - \frac{1}{H} \frac{d}{dt} \left(\ln c_s + \ln \left(1 + \frac{\mathcal{P}}{\rho} \right) \right). \end{aligned} \tag{4.348}$$

In the above equation we have used the approximation $d \ln k = d \ln a_k = H dt$. Since all correction terms are negative we have $n_s < 1$ and thus the amplitude increases slightly for small k corresponding to larger scales. We say that the spectrum is red-tilted. This tilt can be traced to the requirement that inflation must have a graceful exit.

An estimation for n_s can be given as follows. Galactic scales cross the horizon at 50 e-folds before the end of inflation. At this time the deviation of the equation of state from the vacuum is around 10^{-2} and the second term in (4.348) is also around 10^{-2} and hence $n_s = 0.96$. This should be compared with the 2013 Planck result $n_s = 0.9603 \pm 0.0073$.

For inflation with a potential V the above formula becomes

$$\begin{aligned} n_s - 1 &= - \frac{1}{8\pi G} \left(\frac{1}{V} \frac{\partial V}{\partial \phi} \right)^2 - \frac{2}{H} \frac{d}{dt} \ln \frac{1}{V} \frac{\partial V}{\partial \phi} \\ &= - \frac{1}{8\pi G} \left(\frac{1}{V} \frac{\partial V}{\partial \phi} \right)^2 + \frac{1}{4\pi G} \left(\frac{1}{V} \frac{\partial^2 V}{\partial \phi^2} - \left(\frac{1}{V} \frac{\partial V}{\partial \phi} \right)^2 \right) \\ &= - \frac{3}{8\pi G} \left(\frac{1}{V} \frac{\partial V}{\partial \phi} \right)^2 + \frac{1}{4\pi G} \frac{1}{V} \frac{\partial^2 V}{\partial \phi^2} \\ &= - 6\epsilon_V + 2\eta_V. \end{aligned} \tag{4.349}$$

4.8 Rederivation of the Mukhanov action

4.8.1 Mukhanov action from ADM

In this section we will rederive the Mukhanov action from the ADM formalism following the method of [11]. The action of interest here is of course

$$S = \frac{1}{2} \int d^4x \sqrt{-\det g} \ [R - g^{\mu\nu} \nabla_\mu \phi \nabla_\nu \phi - 2V(\phi)]. \tag{4.350}$$

By going now through the steps of the famous ADM formalism [15] we can express this action in terms of 3D quantities which is very useful if one is interested in canonical quantization. The ADM formalism starts with the metric put in the form

$$ds^2 = -N^2 dt^2 + \gamma_{ij}(dx^i + N^i dt)(dx^j + N^j dt). \tag{4.351}$$

In other words we slice spacetime into 3D spatial hypersurfaces. Indeed γ_{ij} is the metric on the spatial 3D slices of constant t. The function N and the vector N_i are called lapse function and shift vector. We have

$$g_{00} = \gamma_{ij} N^i N^j - N^2, \quad g_{0j} = \gamma_{ij} N^i, \quad g_{i0} = \gamma_{ij} N^j, \quad g_{ij} = \gamma_{ij}. \tag{4.352}$$

A straightforward calculation shows that

$$\sqrt{-\det g} = N \sqrt{\det \gamma}. \tag{4.353}$$

$$g^{00} = -\frac{1}{N^2}, \quad g^{0j} = \frac{1}{N^2} N^j, \quad g^{i0} = \frac{1}{N^2} N^i, \quad g^{ij} = \gamma^{ij} - \frac{1}{N^2} N^i N^j. \tag{4.354}$$

The variables γ_{ij}, N_i and N contain the same information as the original spacetime metric $g_{\mu\nu}$. As it turns out N and N_i are only Lagrange multipliers.

We get after some calculation the action

$$S = \frac{1}{2} \int d^4x \sqrt{\det \gamma} \ [N R_{(3)} + N^{-1}(E_{ij}E^{ij} - E^2) + N^{-1}(\partial_t \phi - N^i \partial_i \phi)^2 - N\gamma^{ij}\partial_i\phi\partial_j\phi - 2NV]. \tag{4.355}$$

The extrinsic curvature of the 3D spatial slices is $K_{ij} = N^{-1}E_{ij}$ where

$$E_{ij} = \frac{1}{2}(\partial_t \gamma_{ij} - \nabla_i N_j - \nabla_j N_i), \quad E = \gamma^{ij}E_{ij}. \tag{4.356}$$

Recall that

$$\nabla_i N_j = \partial_i N_j - \Gamma^k_{\ ij} N_k, \quad \Gamma^k_{\ ij} = \frac{1}{2}\gamma^{kl}(\partial_i \gamma_{lj} + \partial_j \gamma_{li} - \partial_l \gamma_{ij}). \tag{4.357}$$

By varying the above action with respect to N and N^i we obtain the equations of motion

$$R_{(3)} - N^{-2}(E_{ij}E^{ij} - E^2) - N^{-2}(\partial_t \phi - N^i \partial_i \phi)^2 - \gamma^{ij}\partial_i\phi\partial_j\phi - 2V = 0. \tag{4.358}$$

$$-N^{-1}\partial_i\phi(\partial_t\phi - N^i\partial_i\phi) + \nabla_j \left(N^{-1}(E_i^j - \gamma_i^j E)\right) = 0. \tag{4.359}$$

These are constraints equations for the lapse function N and the shift vector N^i. In the comoving gauge we will choose $\delta\phi = 0$ and hence $\phi = \bar{\phi}$ where the unperturbed configuration $\bar{\phi}$ is uniform. Hence the above equations of motion reduce to

$$R_{(3)} - N^{-2}(E_{ij}E^{ij} - E^2) - N^{-2}(\partial_t\bar{\phi})^2 - 2\bar{V} = 0. \tag{4.360}$$

$$\nabla_j \left(N^{-1}(E_i^j - \gamma_i^j E)\right) = 0. \tag{4.361}$$

In the comoving gauge we also choose

$$\gamma_{ij} = a^2(1 - 2\mathcal{R})\delta_{ij} + h_{ij}, \quad h^i{}_i = \partial^i h_{ij} = 0. \tag{4.362}$$

In most of the following we will set $h = 0$. Then

$$R_{(3)} = \frac{4}{a^2}\vec{\nabla}^2\mathcal{R}. \tag{4.363}$$

We resolve the shift vector N_i into the sum of a total derivative (irrotational scalar) and a divergenceless vector (incompressible vector) as

$$N_i = \partial_i\psi + \tilde{N}_i, \quad \partial^i\tilde{N}_i = 0. \tag{4.364}$$

We also introduce the lapse perturbation α as

$$N = 1 + \alpha. \tag{4.365}$$

We expand ψ, $\tilde{N}_i$ and α in powers of $\mathcal{R}$ as follows

$$\psi = \psi_1 + \psi_2 + \cdots \tag{4.366}$$

$$\alpha = \alpha_1 + \alpha_2 + \cdots \tag{4.367}$$

$$\tilde{N}_i = \tilde{N}_i^{(1)} + \tilde{N}_i^{(2)} + \cdots \tag{4.368}$$

We have

$$\gamma^{ij} = \frac{1}{a^2(1 - 2\mathcal{R})}\delta_{ij}. \tag{4.369}$$

Since we are only going to keep the first order in powers of $\mathcal{R}$ we can approximate E_{ij} by

$$E_{ij} = \frac{1}{2}(\partial_t\gamma_{ij} - \partial_i N_j - \partial_j N_i)$$
$$= a^2 H[1 - 2\mathcal{R} - \frac{\dot{\mathcal{R}}}{H}]\delta_{ij} - \frac{1}{2}(\partial_i N_j + \partial_j N_i). \tag{4.370}$$

Thus we compute

$$E_{ij}E^{ij} = \frac{1}{a^4(1-2\mathcal{R})^2}E_{ij}E_{ij}$$

$$\simeq 3H^2[1 - 2\frac{\dot{\mathcal{R}}}{H}] - \frac{2H}{a^2}\partial_i N_i. \tag{4.371}$$

$$E \simeq 3H\left[1 - \frac{\dot{\mathcal{R}}}{H}\right] - \frac{1}{a^2}\partial_i N_i \Rightarrow E^2 \simeq 9H^2[1 - 2\frac{\dot{\mathcal{R}}}{H}] - \frac{6H}{a^2}\partial_i N_i. \tag{4.372}$$

The constraints become (by using the first Friedmann equation in the form $6H^2 = (\partial_t\bar{\phi})^2 + 2\bar{V}$ and $8\pi G = 1$)

$$\frac{4}{a^2}\vec{\nabla}^2\mathcal{R} - 12H\dot{\mathcal{R}} - \frac{4H}{a^2}\partial_i N_i - 12\alpha H^2 + 2\alpha(\partial_t\bar{\phi})^2 = 0. \tag{4.373}$$

$$\partial_j\left(-2H[1 - \alpha - \frac{\dot{\mathcal{R}}}{H}]\delta_{ij} + \frac{1}{a^2}\partial_k N_k\delta_{ij} - \frac{1}{2a^2}(\partial_i N_j + \partial_j N_i)\right) = 0. \tag{4.374}$$

Equivalently (with $\vec{\nabla}_i^2 = \partial^i\partial_i$)

$$\frac{4}{a^2}\vec{\nabla}^2\mathcal{R} - 12H\dot{\mathcal{R}} - 4H\vec{\nabla}^2\psi_1 - 4\alpha_1\bar{V} = 0. \tag{4.375}$$

$$2H\partial_i(\alpha_1 + \frac{\dot{\mathcal{R}}}{H}) - \frac{1}{2}\vec{\nabla}^2\tilde{N}_i^{(1)} = 0. \tag{4.376}$$

From the second constraint we obtain

$$\alpha_1 = -\frac{\dot{\mathcal{R}}}{H}, \quad \tilde{N}_i^{(1)} = 0. \tag{4.377}$$

The first constraint gives then

$$\vec{\nabla}^2\psi_1 = \frac{\vec{\nabla}^2\mathcal{R}}{a^2H} + \dot{\mathcal{R}}\left(\frac{\bar{V}}{H^2} - 3\right). \tag{4.378}$$

Recall that the slow-roll parameter ϵ is given by

$$\epsilon = \frac{(\partial_t\bar{\phi})^2}{2H^2} = 3 - \frac{\bar{V}}{H^2}. \tag{4.379}$$

Hence we obtain

$$\psi_1 = \frac{\mathcal{R}}{a^2H} - \epsilon(\vec{\nabla}^2)^{-1}\dot{\mathcal{R}}. \tag{4.380}$$

We compute

$$\sqrt{\det\gamma} = a^3\sqrt{\det\gamma_{(3)}}$$
$$= a^3[1 - 3\mathcal{R} + \frac{3}{2}\mathcal{R}^2 + \cdots]. \tag{4.381}$$

$$\mathcal{L} = N\mathcal{R}_{(3)} + N^{-1}(E_{ij}E^{ij} - E^2) + N^{-1}(\partial_t\phi - N^i\partial_i\phi)^2 - N\gamma^{ij}\partial_i\phi\partial_j\phi - 2NV$$
$$= \mathcal{L}_0 + \mathcal{L}_1 + \mathcal{L}_2 + \cdots \tag{4.382}$$

$$\mathcal{L}_0 = (\partial_t\bar\phi)^2 - 2\bar{V} + (E_{ij}E^{ij} - E^2)^{(0)}. \tag{4.383}$$

$$\mathcal{L}_1 = \mathcal{R}_{(3)} - \alpha_1(\partial_t\bar\phi)^2 - 2\alpha_1\bar{V} + (E_{ij}E^{ij} - E^2)^{(1)} - \alpha_1(E_{ij}E^{ij} - E^2)^{(0)}. \tag{4.384}$$

$$\mathcal{L}_2 = \alpha_1\mathcal{R}_{(3)} + (-\alpha_2 + \alpha_1^2)(\partial_t\bar\phi)^2 - 2\alpha_2\bar{V} + (E_{ij}E^{ij} - E^2)^{(2)} - \alpha_1(E_{ij}E^{ij} - E^2)^{(1)}$$
$$+ (-\alpha_2 + \alpha_1^2)(E_{ij}E^{ij} - E^2)^{(0)}. \tag{4.385}$$

A more precise formula for $E_{ij}E^{ij} - E^2$ is

$$E_{ij}E^{ij} - E^2 = -6H^2\left[1 - \frac{2\dot{\mathcal{R}}}{H} + \frac{\dot{\mathcal{R}}^2}{H^2} - \frac{4\mathcal{R}\dot{\mathcal{R}}}{H}\right] + \frac{4H}{a^2}\left[1 - \frac{\dot{\mathcal{R}}}{H} + 2\mathcal{R}\right]\nabla_i N_i$$
$$+ \frac{1}{4a^4}(\nabla_i N_j + \nabla_j N_i)^2 - \frac{1}{a^4}(\nabla_i N_i)^2. \tag{4.386}$$

The last term is already of order 2 and thus we can set $\nabla_i N_j = \partial_i N_j$. By partial integration we can see that this term actually cancels. Further, we compute

$$\nabla_i N_j = \partial_i N_j + \partial_i\mathcal{R}. N_j + \partial_j\mathcal{R}. N_i - \delta_{ij}\partial_k\mathcal{R}. N_k. \tag{4.387}$$

By using the equation of motion $(\partial_t\bar\phi)^2 + 2\bar{V} + (E_{ij}E^{ij} - E^2)^{(0)} = 0$ we obtain

$$\mathcal{L}_0 = -4\bar{V}. \tag{4.388}$$

$$\mathcal{L}_1 = \mathcal{R}_{(3)} + (E_{ij}E^{ij} - E^2)^{(1)}$$
$$= \frac{8}{a^2}\vec{\nabla}^2\mathcal{R} + 12H\dot{\mathcal{R}} - 4\epsilon H\dot{\mathcal{R}}. \tag{4.389}$$

$$\mathcal{L}_2 = \alpha_1\mathcal{R}_{(3)} + \alpha_1^2(\partial_t\bar\phi)^2 + (E_{ij}E^{ij} - E^2)^{(2)} - \alpha_1(E_{ij}E^{ij} - E^2)^{(1)} + \alpha_1^2(E_{ij}E^{ij} - E^2)^{(0)}$$
$$= -\frac{4}{a^2 H}\dot{\mathcal{R}}\vec{\nabla}^2\mathcal{R} + \frac{\dot{\mathcal{R}}^2}{H^2}(\partial_t\bar\phi)^2 + 24H\mathcal{R}\dot{\mathcal{R}} + \frac{12}{a^2}\mathcal{R}\vec{\nabla}^2\mathcal{R} - 12\epsilon H\mathcal{R}\dot{\mathcal{R}}. \tag{4.390}$$

The first term in $\mathcal{L}_1$ is a boundary term by Stokes's theorem and thus it will be neglected, viz

$$\int d^4x\sqrt{\det\gamma}\,\frac{8}{a^2}\vec{\nabla}^2\mathcal{R} = 8\int dt\,a(t)\int d^3x\sqrt{\det\gamma_{(3)}}\,\partial^i\partial_i\mathcal{R}$$
$$= 8\int dt\,a(t)\int d^2x\sqrt{\det\gamma_{(2)}}\,n^i\partial_i\mathcal{R} \tag{4.391}$$
$$= 0.$$

The quadratic contribution coming from $\mathcal{L}_0$ is

$$\left(\frac{3}{2}\mathcal{R}^2\right)\mathcal{L}_0 = -6\mathcal{R}^2\bar{V}. \tag{4.392}$$

The quadratic contribution coming from $\mathcal{L}_1$ is

$$(-3\mathcal{R})\mathcal{L}_1' = -36H\mathcal{R}\dot{\mathcal{R}} + 12\epsilon H\mathcal{R}\dot{\mathcal{R}}. \tag{4.393}$$

Thus

$$\left(\frac{3}{2}\mathcal{R}^2\right)\mathcal{L}_0 + (-3\mathcal{R})\mathcal{L}_1 + \mathcal{L}_2 = -\frac{4}{a^2 H}\dot{\mathcal{R}}\,\vec{\nabla}^2\,\mathcal{R} + \frac{\dot{\mathcal{R}}^2}{H^2}(\partial_t\bar{\phi})^2$$
$$- 12H\mathcal{R}\dot{\mathcal{R}} + \frac{12}{a^2}\mathcal{R}\,\vec{\nabla}^2\,\mathcal{R} - 6\mathcal{R}^2\bar{V}. \tag{4.394}$$

This must be multiplied by a^3. Integration by parts gives

$$\left(\frac{3}{2}\mathcal{R}^2\right)\mathcal{L}_0 + (-3\mathcal{R})\mathcal{L}_1 + \mathcal{L}_2 = -\frac{4}{a^2 H}\dot{\mathcal{R}}\,\vec{\nabla}^2\,\mathcal{R} + \frac{\dot{\mathcal{R}}^2}{H^2}(\partial_t\bar{\phi})^2 + \frac{12}{a^2}\mathcal{R}\,\vec{\nabla}^2\,\mathcal{R}$$
$$= \frac{(\partial_t\bar{\phi})^2}{H^2}\left(\dot{\mathcal{R}}^2 - \frac{1}{a^2}\partial^i\mathcal{R}\partial_i\mathcal{R}\right) + \frac{14}{a^2}\mathcal{R}\,\vec{\nabla}^2\,\mathcal{R}. \tag{4.395}$$

Since we are only keeping quadratic terms in the curvature perturbation $\mathcal{R}$ the last term in the above equation vanishes by (4.391). Indeed we have

$$14\int dt\; a(t)\int d^3x\mathcal{R}\,\vec{\nabla}^2\,\mathcal{R} \simeq -\frac{14}{3}\int dt\; a(t)\int d^3x\sqrt{\det\gamma_{(3)}}\;\partial^i\partial_i\mathcal{R} \tag{4.396}$$
$$= 0$$

We obtain the final action

$$S = \frac{1}{2}\int d^4x a^3\frac{(\partial_t\bar{\phi})^2}{H^2}\left(\dot{\mathcal{R}}^2 - \frac{1}{a^2}\partial^i\mathcal{R}\partial_i\mathcal{R}\right). \tag{4.397}$$

There is also a linear term in $\mathcal{R}$ which we must discuss. This is given by

$$(-3\mathcal{R})\mathcal{L}_0 + (1)\mathcal{L}_1 = \frac{4\bar{V}}{H}(\dot{\mathcal{R}} + 3H\mathcal{R}). \tag{4.398}$$

Again this must be multiplied by a^3. After integration by parts we obtain

$$(-3\mathcal{R})\mathcal{L}_0 + (1)\mathcal{L}_1 = -4\epsilon\mathcal{R}(\bar{V} + 2H\frac{\delta\bar{V}}{\delta\bar{\phi}}). \tag{4.399}$$

This can be neglected in the slow-roll limit $\epsilon \longrightarrow 0$.

The conformal time is defined by $dt = ad\eta$. We introduce also Mukhanov variables

$$v = z\mathcal{R}, \quad z = a\frac{\partial_t\bar{\phi}}{H}. \tag{4.400}$$

Now a really straightforward calculation gives the action

$$S = \frac{1}{2} \int d\eta d^3x \left((v')^2 + \frac{z''}{z} v^2 - \partial^i v \partial_i v \right).$$

(4.401)

4.8.2 Power spectra and tensor perturbations

The equation of motion derived from the above action reads

$$v'' - \partial_i \partial^i v - \frac{z''}{z} v = 0.$$

(4.402)

A solution is given by $u_k = \exp(i\vec{k}\,\vec{x})\chi_k$ (with $\vec{k}\,\vec{x} = k^i x_i$) provided

$$\chi_k'' + (k^2 - \frac{z''}{z})\chi_k = 0.$$

(4.403)

These solutions are positive norm solutions, viz $(u_k, u_l) = \delta_{kl}$ if and only if

$$iV(\chi_k^* \dot{\chi}_k - \chi_k \dot{\chi}_k^*) = 1.$$

(4.404)

The negative norm solutions are u_k^*. As before we will choose $\chi_k = v_k^*/\sqrt{2}$. The field v can then be expanded as

$$v = \int \frac{d^3k}{(2\pi)^3} \frac{1}{\sqrt{2}} \left[a_k v_k^*(\eta) e^{i\vec{k}\,\vec{x}} + a_k^* v_k(\eta) e^{-i\vec{k}\,\vec{x}} \right].$$

(4.405)

In the quantum theory a_k and a_k^+ become operators $\hat{a}_k$ and $\hat{a}_k^+$ satisfying $[\hat{a}_k, \hat{a}_l^+] = V\delta_{kl}$. The field operator is

$$\hat{v} = \int \frac{d^3k}{(2\pi)^3} \frac{1}{\sqrt{2}} \left[\hat{a}_k v_k^*(\eta) e^{i\vec{k}\,\vec{x}} + \hat{a}_k^+ v_k(\eta) e^{-i\vec{k}\,\vec{x}} \right].$$

(4.406)

We are interested in the 2-point function

$$\langle \hat{\mathcal{R}}(x_1)\hat{\mathcal{R}}(x_2) \rangle = \frac{H^2}{a^2(\partial_t\bar{\phi})^2} \langle \hat{v}(t_1)\hat{v}(t_2) \rangle$$

$$= \frac{H^2}{2a^2(\partial_t\bar{\phi})^2} \int \frac{d^3k}{(2\pi)^3} v_k^*(t_1)v_k(t_2).$$

(4.407)

We define the Fourier transform of $\hat{\mathcal{R}}(x)$ by

$$\hat{\mathcal{R}}(x) = \int \frac{d^3k}{(2\pi)^3} \hat{\mathcal{R}}_k(t) e^{i\vec{k}\,\vec{x}}.$$

(4.408)

We define the power spectrum $P_{\mathcal{R}}(k)$ of the curvature perturbation $\mathcal{R}$ by

$$\langle \hat{\mathcal{R}}_{k_1}(t_1)\hat{\mathcal{R}}_{k_2}(t_2) \rangle = (2\pi)^3 \delta^3(k_1 + k_2) P_{\mathcal{R}}(k_1).$$

(4.409)

We compute then

$$\langle \hat{\mathcal{R}}(x_1)\hat{\mathcal{R}}(x_2)\rangle = \int \frac{d^3k}{(2\pi)^3} P_{\mathcal{R}}(k_1) e^{i\vec{k}_1(\vec{x}_1 - \vec{x}_2)}. \tag{4.410}$$

Let us now consider the de Sitter limit $\epsilon \longrightarrow 0$ in which H can be treated as a constant and $a \simeq e^{Ht}$ or equivalently $a \simeq -1/(H\eta)$. We compute

$$\frac{z''}{z} = a\dot{a}\frac{\dot{z}}{z} + a^2\frac{\ddot{z}}{z}. \tag{4.411}$$

$$\dot{z} = \dot{a}\frac{\partial_t\bar{\phi}}{H} - a\frac{\partial_t\bar{\phi}}{H^2}\dot{H} + a\frac{\ddot{a}}{H}. \tag{4.412}$$

In the de Sitter limit case we can make the approximations

$$\dot{z} \simeq \dot{a}\frac{\partial_t\bar{\phi}}{H} \Rightarrow \frac{\dot{z}}{z} \simeq \frac{\dot{a}}{a}, \ \frac{\ddot{z}}{z} \simeq \frac{\ddot{a}}{a}. \tag{4.413}$$

Thus

$$\frac{z''}{z} \simeq \frac{a''}{a}$$
$$\simeq \frac{2}{\eta^2}. \tag{4.414}$$

The equation of motion becomes

$$\chi_k'' + (k^2 - \frac{2}{\eta^2})\chi_k = 0. \tag{4.415}$$

In the limit $\eta \longrightarrow -\infty$ the frequency approaches the flat space result and hence we can choose the vacuum state to be given by the Minkowski vacuum. This is the Bunch–Davies vacuum given by equation (5.128). We have then

$$v_k = -\frac{e^{ik\eta}}{\sqrt{k}}\left(1 + \frac{i}{k\eta}\right). \tag{4.416}$$

We can then compute in the de Sitter limit the real space variance

$$\langle \hat{\mathcal{R}}(x)\hat{\mathcal{R}}(x)\rangle = \int_0^\infty d\ln k \Delta_{\mathcal{R}}^2(k). \tag{4.417}$$

The dimensionless power spectrum $\Delta_{\mathcal{R}}^2(k)$ is given by

$$\Delta_{\mathcal{R}}^2(k) = \frac{H^2}{(2\pi)^2}\frac{H^2}{(\partial_t\bar{\phi})^2}(1 + k^2\eta^2). \tag{4.418}$$

For super-horizon scales ($|k\eta| \ll 1$ or equivalently $k \ll aH$) this dimensionless power spectrum becomes constant. This is precisely the statement that $\mathcal{R}$ remains constant outside the horizon. We may then restrict the calculation to the instant of horizon crossing given by

$$|k\eta_*| = 1 \Leftrightarrow k = a(t_*)H(t_*).$$ (4.419)

The dimensionless power spectrum $\Delta_{\mathcal{R}}^2(k)$ and the power spectrum $P_{\mathcal{R}}(k)$ at horizon crossing are given respectively by[2]

$$\Delta_{\mathcal{R}}^2(k) = \frac{H_*^2}{2\pi^2} \frac{H_*^2}{(\partial_t\bar{\phi})_*^2}.$$ (4.420)

$$P_{\mathcal{R}}(k) = \frac{2\pi^2}{k^3} \Delta_{\mathcal{R}}^2(k)$$
$$= \frac{H_*^2}{k^3} \frac{H_*^2}{(\partial_t\bar{\phi})_*^2}.$$ (4.421)

In summary the primordial power spectrum of comoving curvature perturbation $\mathcal{R}$ at horizon crossing is found to be given by

$$P_{\mathcal{R}}(k) = \frac{H_*^2}{k^3} \frac{H_*^2}{(\partial_t\bar{\phi})_*^2} \Leftrightarrow \Delta_{\mathcal{R}}^2(k) = \frac{H_*^2}{2\pi^2} \frac{H_*^2}{(\partial_t\bar{\phi})_*^2}.$$ (4.422)

This is the scalar power spectrum, viz

$$P_s(k) \equiv P_{\mathcal{R}}(k) \Leftrightarrow \Delta_s^2(k) \equiv \Delta_{\mathcal{R}}^2(k) = \frac{1}{4\pi^2} \frac{H^2}{M_{pl}^2} \frac{1}{\epsilon} |_{k=aH}.$$ (4.423)

As seen from the gauge fixing condition (4.362) there are extra degrees of freedom (two polarizations) encoded in the symmetric traceless and divergenceless tensor h_{ij} which we have not considered at all until now. These degrees of freedom correspond to gravitational waves. In order to determine the primordial power spectrum $\Delta_t^2(k)$ of the tensor perturbation h at horizon crossing $k = aH$ we go back to (4.362) and set $\mathcal{R} = 0$ and then go through some very similar calculations to those which led to $\Delta_s^2(k)$. We find at the end the result

$$\Delta_t^2(k) \equiv 2\Delta_h^2(k) = \frac{2}{\pi^2} \frac{H^2}{M_{pl}^2} |_{k=aH}.$$ (4.424)

The scalar-to-tensor ratio is defined by

$$r \equiv \frac{\Delta_t^2(k)}{\Delta_s^2(k)}$$
$$= 8\epsilon_*.$$ (4.425)

---

[2] These two formulas differ by a factor of 1/2 compared with reference [1].

The scale-dependence of the power spectrum $\Delta_s^2(k)$ can be given by the so-called spectral index n_s defined by

$$n_s = 1 + \frac{d \ln \Delta_s^2}{d \ln k}. \tag{4.426}$$

Obviously scale invariance corresponds to $n_s = 1$. We may approximate $\Delta_s^2(k)$ by a power law as follows

$$\Delta_s^2(k) = A_s(k_*)\left(\frac{k}{k_*}\right)^{n_s(k_*)-1+\frac{1}{2}\alpha_s(k_*)\ln \frac{k}{k_*}}, \quad \alpha_s(k) = \frac{dn_s}{d \ln k}. \tag{4.427}$$

Similarly we define

$$n_t = \frac{d \ln \Delta_s^2}{d \ln k}. \tag{4.428}$$

In terms of the Hubble slow-roll parameters ϵ and η the indices n_s and n_t are given by

$$n_s = 1 + 2\eta_* - 4\epsilon_*. \tag{4.429}$$

$$n_t = -2\epsilon_*. \tag{4.430}$$

In the slow-roll limit with $m^2\phi^2$ potential we obtain the predictions

$$n_s = 0.96, \quad r = 0.05. \tag{4.431}$$

Let us summarize our results so far. During inflation the comoving horizon $1/(aH)$ decreases while after inflation it increases. In this inflationary universe fluctuation is created quantum mechanically on all scales with a spectrum of wavenumbers k. The comoving scales k^{-1} are constant during and after inflation. The physically relevant fluctuations are created at sub-horizon scales $k > aH$. Any given fluctuation with a wavenumber k starts thus inside the horizon and at some point it will exit the horizon (during inflation) and then it will re-enter again the horizon at a later time (after inflation during the hot **Big Bang**). All fluctuations after they exit the horizon (corresponding to super-horizon scales $k < aH$) are frozen until they re-enter the horizon in the sense that they are not affected by and they cannot affect the physics inside the horizon. This is the statement that the curvature perturbation $\mathcal{R}$ is constant outside the horizon which allows us to concentrate on the value of $\mathcal{R}$ at the time of exit (crossing) since that value will not change until re-entry. This is the main result of this chapter. This is summarized in figure 4.7.

4.8.3 CMB temperature anisotropies

The remaining question we would like to discuss is how to relate the power spectrum P_s to CMB temperature anisotropies. The CMB temperature fluctuations $\Delta T(\hat{n})$ relative to the background temperature $T = 2.7K$ are given by

$$\frac{\Delta T(\vec{n})}{T} = \sum_{lm} a_{lm} \hat{Y}_{lm}(\hat{n}). \tag{4.432}$$

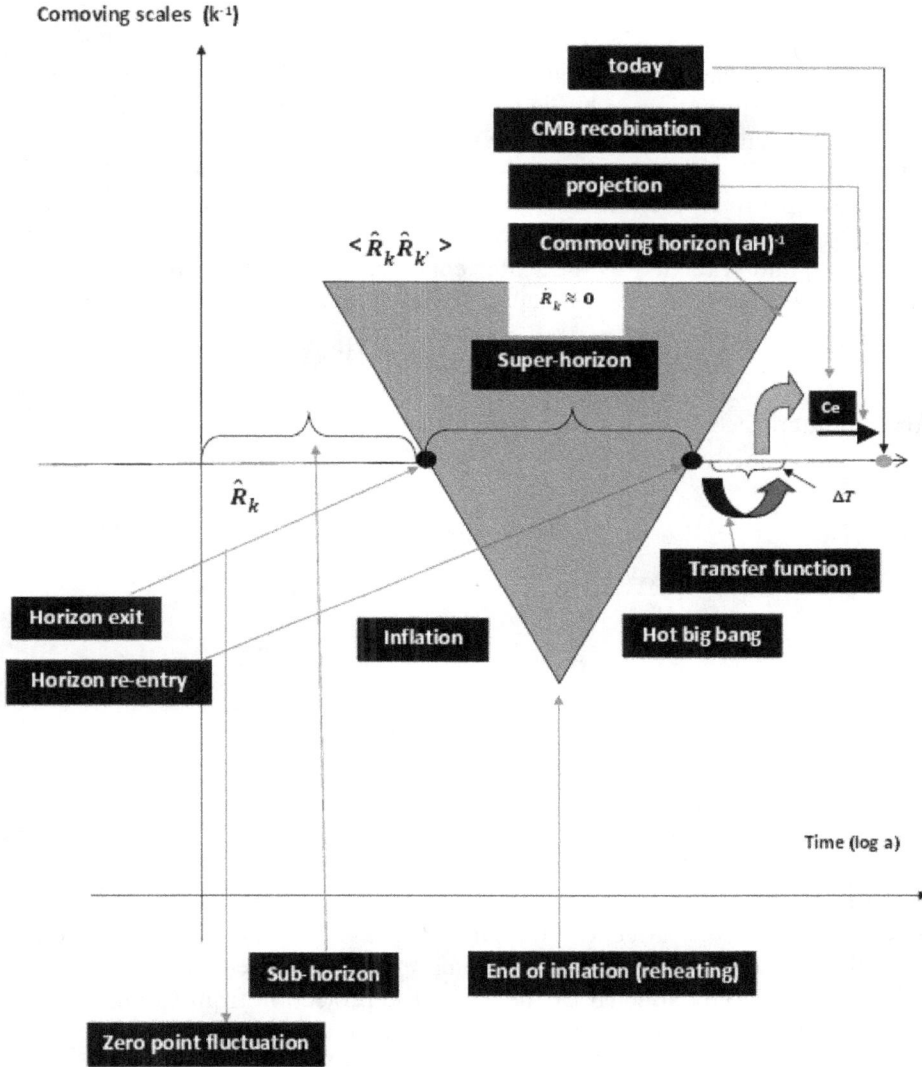

Figure 4.7. Summary of the inflationary scenario.

$$a_{lm} = \int d\Omega\, Y_{lm}^*(\vec{n}) \frac{\Delta T(\vec{n})}{T}. \tag{4.433}$$

The 2-point correlator $\langle a_{l_1 m_1} a_{l_2 m_2} \rangle$ must behave (by rotational invariance) as

$$\langle a_{l_1 m_1}^* a_{l_2 m_2} \rangle = C_l^{TT} \delta_{l_1 l_2} \delta_{m_1 m_2}. \tag{4.434}$$

The rotationally invariant angular power spectrum C_l^{TT} is given by

$$C_l^{TT} = \frac{1}{2l+1} \sum_m \langle a_{l_1 m_1}^* a_{l_2 m_2} \rangle. \tag{4.435}$$

For values of the tensor-to-scalar ratio $r < 0.3$ the CMB temperature fluctuations are dominated by the scalar curvature perturbation $\mathcal{R}$. We have already computed the curvature perturbation at horizon crossing (exit) which then remains constant (freeze at a constant value) until the time of re-entry. From the time of re-entry until the time of CMB recombination the curvature perturbation will evolve in time causing a temperature fluctuation. The temperature fluctuation we observe today as a remnant of last scattering (CMB recombination) is encoded in the multipole moments a_{lm} and is related to the scalar curvature perturbation $\mathcal{R}_k$ at the time of horizon crossing $k = a(t_*)H(t_*)$ through a transfer function $\Delta_{Tl}(k)$ as follows

$$a_{lm} = 4\pi(-i)^l \int \frac{d^3k}{(2\pi)^3}\Delta_{Tl}(k)\mathcal{R}_k Y_{lm}(\vec{k}). \tag{4.436}$$

In the quantum theory $\mathcal{R}_k$ become operators and hence a_{lm} become operators. We compute immediately (with $\mathcal{R}_k^* = \mathcal{R}_{-k}$)

$$
\begin{aligned}
\sum_m <\hat{a}_{lm}^+\hat{a}_{lm}> &= (4\pi)^2\sum_m \int \frac{d^3k}{(2\pi)^3}\Delta_{Tl}(k) Y_{lm}^*(\vec{k}) \int \frac{d^3k'}{(2\pi)^3}\Delta_{Tl}(k') Y_{lm}(\vec{k}') <\hat{\mathcal{R}}_k^+\hat{\mathcal{R}}_{k'}> \\
&= (4\pi)^2 \int \frac{d^3k}{(2\pi)^3}\Delta_{Tl}^2(k)P_{\mathcal{R}}(k)\sum_m Y_{lm}^*(\vec{k}) Y_{lm}(\vec{k}) \\
&= (4\pi)^2 \int \frac{d^3k}{(2\pi)^3}\Delta_{Tl}^2(k)P_{\mathcal{R}}(k)\frac{2l+1}{4\pi}P_l(\hat{k}^2) \\
&= \frac{2}{\pi}(2l+1)\int k^2 dk \Delta_{Tl}^2(k)P_{\mathcal{R}}(k).
\end{aligned}
\tag{4.437}
$$

Hence

$$C_l^{TT} = \frac{2}{\pi}\int k^2 dk \Delta_{Tl}^2(k)P_{\mathcal{R}}(k). \tag{4.438}$$

The term $\Delta_{Tl}^2(k)$ is the anisotropies term.

For large scales, i.e. large k^{-1} we can safely assume that the modes were still outside the horizon at the time of recombination. As a consequence the large-scale CMB spectrum is only affected by the geometric projection from recombination to our current epoch and is not affected by sub-horizon evolution. This is the so-called Sachs–Wolf regime in which the transfer function is a Bessel function, viz

$$\Delta_{Tl}^2(k) = \frac{1}{3}j_l(k(\eta_0 - \eta_{\text{rec}})) + \cdots. \tag{4.439}$$

This term is the monopole contribution to the transfer function. We have neglected a dipole term and the so-called integrated Sachs–Wolfe (ISW) terms.

The Bessel function essentially projects the linear scales with wavenumber k onto angular scales with angular wavenumber l. The angular power spectrum C_l^{TT} on a large scale (corresponding to small l or large angles) is therefore

$$
\begin{aligned}
C_l^{TT} &= \frac{2}{9\pi}\int k^2 dk j_l^2(k(\eta_0 - \eta_{\text{rec}}))P_s(k) \\
&= \frac{4\pi}{9}\int \frac{dk}{k}j_l^2(k(\eta_0 - \eta_{\text{rec}}))\Delta_s^2(k).
\end{aligned}
\tag{4.440}
$$

The Bessel function for large l acts effectively as a delta function since it is peaked around

$$l = k(\eta_0 - \eta_{\text{rec}}). \tag{4.441}$$

We approximate the dimensionless power spectrum $\Delta_s^2(k)$ by the following power law (where n_s is the spectral index evaluated at some reference point k_*)

$$\Delta_s^2(k) = A_s k^{n_s - 1}. \tag{4.442}$$

We obtain then

$$
\begin{aligned}
C_l^{TT} &= \frac{4\pi}{9} A_s \int \frac{dk}{k^{2-n_s}} j_l^2(k(\eta_0 - \eta_{\text{rec}})) \\
&= \frac{4\pi}{9} A_s (\eta_0 - \eta_{\text{rec}})^{1-n_s} \int \frac{dx}{x^{2-n_s}} j_l^2(x) \\
&= 2^{n_s - 4} \frac{4\pi^2}{9} A_s (\eta_0 - \eta_{\text{rec}})^{1-n_s} \frac{\Gamma(l + \frac{n_s}{2} - \frac{1}{2})}{\Gamma(l - \frac{n_s}{2} + \frac{5}{2})} \frac{\Gamma(3 - n_s)}{\Gamma^2(2 - \frac{n_s}{2})}.
\end{aligned}
\tag{4.443}
$$

For a scale-invariant spectrum we have $n_s = 1$. In this case

$$
\begin{aligned}
C_l &\equiv \frac{l(l+1)}{2\pi} C_l^{TT} \\
&= \frac{A_s}{9}.
\end{aligned}
\tag{4.444}
$$

The modified power spectrum C_l is therefore independent of l for small values of l corresponding to the largest scales (largest angles). This is what is observed in the real world. See figure 8.12 of [9]. Thus we conclude that n_s must be indeed very close to 1.

The situation is more involved for intermediate scales where acoustic peaks dominate and for small scales where damping dominates which is an effect due to photon diffusion. The acoustic peaks arise because the early universe was a plasma of photons and baryons forming a single fluid which can oscillate due to the competing forces of radiation pressure and gravitational compression. This struggle between gravity and radiation pressure is what sets up longitudinal acoustic oscillations in the photon–baryon fluid. At recombination the pattern of acoustic oscillations became frozen into the CMB which is what we see today as peaks and troughs in the power spectrum of temperature fluctuations. A proper study of the acoustic peaks seen at intermediate scales and also of the damping seen at small scales is beyond our means at this point.

In conclusion, the predictions of cosmological scalar perturbation theory for the angular power spectrum of CMB temperature anisotropies agrees very well with observations. See figure 4.8.

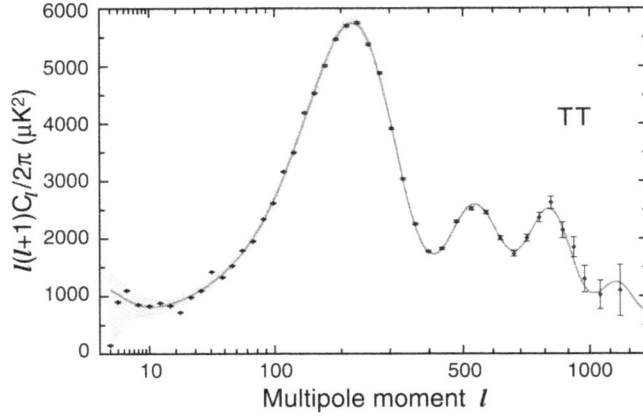

Figure 4.8. The nine-year WMAP TT angular power spectrum. Source: https://lambda.gsfc.nasa.gov/product/map/current/. Credit: NASA/Goddard Space Flight Center/LAMBDA.

4.9 Exercises

Exercise 1: Show that during standard expansion the particle horizon d_{hor} is of the order of the Hubble distance d_H.

Solution 1: The Universe has a finite age and thus photons can only travel a finite distance since the Big Bang singularity. This distance is precisely $d_{\text{hor}}(t)$ which can be rewritten as

$$
\begin{aligned}
d_{\text{hor}}(t) &= a(t) \int_{t_i}^{t} \frac{dt_1}{a(t_1)} \\
&= a(t) \int_{a(t_i)=0}^{a(t)} \frac{1}{a(t_1)H(t_1)} d\ln a(t_1).
\end{aligned}
$$

The number $1/aH$ is precisely the comoving Hubble radius. The distance $d_{\text{hor}}(t_0)$ is effectively the distance to the surface of last scattering which corresponds to the decoupling event.

The first Friedmann equation can be rewritten as $H^2 a^2 = 8\pi G \rho_0 a^{-(1+3w)}/3 - \kappa$. For a flat universe we have

$$
\frac{1}{aH} = \frac{a^{\frac{1}{2}(1+3w)}}{H_0}. \tag{4.445}
$$

It is then clear that the particle horizon is given by

$$
\chi_{\text{hor}} = \frac{2}{H_0(1+3w)} a^{\frac{1}{2}(1+3w)}. \tag{4.446}
$$

For a matter-dominated flat universe we have $w = 0$ and hence $H = H_0 a^{-3/2}$ or equivalently $a = (t/t_0)^{2/3}$. In this case

$$\chi_{\text{hor}} = \frac{2}{H_0} a^{\frac{1}{2}} \Rightarrow d_{\text{hor}} = \frac{2}{H}. \tag{4.447}$$

For a radiation-dominated flat universe we have $w = 1/3$ and hence $H = H_0 a^{-2}$ or equivalently $a = (t/t_0)^{1/2}$. In this case

$$\chi_{\text{hor}} = \frac{1}{H_0} a \Rightarrow d_{\text{hor}} = \frac{1}{H}. \tag{4.448}$$

For a flat universe containing both matter and radiation we should get then

$$d_{\text{hor}} \sim \frac{1}{H}. \tag{4.449}$$

In other words,

$$d_{\text{hor}} \sim d_H. \tag{4.450}$$

Exercise 2: Show that

$$1 + \frac{\bar{\mathcal{P}}_i}{\bar{\rho}_i} < \frac{1}{99} \sim 10^{-2}. \tag{4.451}$$

Solution 2: From the approach that led to equation (4.55) we can get another important estimation. We have

$$\Delta N = \ln \frac{a(t_f)}{a(t_i)} = \int H dt$$

$$= H_i(t_f - t_i) + \frac{H_i^2(t_f - t_i)^2}{2} \frac{\dot{H}_i}{H_i^2} + \cdots$$

We must then have

$$H_i(t_f - t_i) > 66.$$

$$\frac{|\dot{H}_i|}{H_i^2} < \frac{1}{66}.$$

However, from the Friedmann equations $H^2 = 8\pi G \bar{\rho}/3$, $\dot{H} = -4\pi G(\bar{\rho} + \bar{\mathcal{P}})$ we have

$$\frac{\dot{H}}{H^2} = -\frac{3}{2}\left(1 + \frac{\bar{\mathcal{P}}}{\bar{\rho}}\right).$$

We get immediately the estimate

$$1 + \frac{\bar{\mathcal{P}_i}}{\bar{\rho_i}} < \frac{1}{99} \sim 10^{-2}.$$

Exercise 3:
- Derive the Hubble slow-roll conditions

$$\epsilon = -\frac{d \ln H}{dN} \ll 1, \quad \eta = -\frac{\ddot{\phi}}{\dot{\phi} H} \ll 1. \tag{4.452}$$

Explain why the first condition means that the inflaton is moving very slowly, whereas the second condition means that it does so over a wide range (sustained acceleration).
- We define the potential slow-roll parameters ϵ_V and η_V by

$$\epsilon_V = \frac{1}{16\pi G} \frac{(\delta V / \delta \phi)^2}{V^2}. \tag{4.453}$$

$$\eta_V = \frac{1}{8\pi G} \frac{\delta^2 V / \delta \phi^2}{V}. \tag{4.454}$$

Show that the Hubble slow-roll conditions read in terms of the potential slow-roll parameters ϵ_V and η_V as

$$\epsilon_V, \; |\eta_V| \ll 1. \tag{4.455}$$

Exercise 4: Show that for a sufficiently small inflation mass compared to the Planck mass the inflationary phase will last sufficiently long and is followed by a matter-dominated phase.

Solution 4: Alternatively, the Friedmann equation can be immediately solved by the ansatz

$$m\phi = \sqrt{\frac{3}{4\pi G}} H \cos \theta.$$

$$\dot{\phi} = \sqrt{\frac{3}{4\pi G}} H \sin \theta.$$

By taking the time derivative of the first equation and comparing with the second one we get

$$\frac{\dot{H}}{H} \cos \theta - \dot{\theta} \sin \theta = m \sin \theta.$$

By taking the time derivative of the second equation and comparing with the value of $\ddot{\phi}$ obtained from the equation of motion of the inflaton field we get

$$\frac{\dot{H}}{H} \sin\theta + \dot{\theta} \cos\theta = -3H \sin\theta - m\cos\theta.$$

Solving the above two equations for $\dot{H}$ and $\dot{\theta}$ in terms of the original variables H and θ we get

$$\dot{H} = -3H^2 \sin^2\theta, \quad \dot{\theta} = -m - \frac{3}{2}H \sin 2\theta.$$

In terms of α defined by $\theta = -mt + \alpha$ these read

$$\dot{H} = -3H^2 \sin^2(mt - \alpha), \quad \dot{\alpha} = \frac{3}{2}H \sin 2(mt - \alpha).$$

For $mt \gg 1$, i.e. towards the end of inflation, we can neglect α:

$$\dot{H} = -3H^2 \sin^2(mt).$$

The solution is

$$H = \frac{2}{3t}\left(1 - \frac{\sin 2mt}{2mt}\right)^{-1} = \frac{2}{3t}\left(1 + \frac{\sin 2mt}{2mt} + \cdots \right).$$

Now we can check directly that α corresponds to oscillations with decaying amplitude. We can also show that the scalar field oscillates with a frequency $\omega = m$ with slowly decaying amplitude. On the other hand, the scale factor behaves as

$$a = t^{2/3}\left(1 + O\left(\frac{1}{(m^2 t^2)}\right)\right).$$

Thus, if the mass is sufficiently small compared to the Planck mass the inflationary phase will last sufficiently long and is followed by a matter-dominated phase.

Exercise 5: Solve the Poisson's equation

$$\chi''_{\vec{k}} + (c_s^2 \vec{k}^2 - \frac{\theta''}{\theta})\chi_{\vec{k}} = 0 \tag{4.456}$$

for long-wavelengths and determine the Bardeen potential. Assume that the Universe is a mixture of radiation and matter.

Solution 5: In this case we can neglect the spatial derivative in (4.216) and the equation reduces to

$$\chi''_{\vec{k}} - \frac{\theta''}{\theta}\chi_{\vec{k}} = 0.$$

The first solution is obviously $\chi_{\vec{k}} = C_1 \theta$. The second linearly independent solution is

$$\chi_{\vec{k}} = C_2 \theta \int_{\eta_0}^{\eta} \frac{d\eta'}{\theta^2(\eta')}.$$

This can be checked using the Wronskian. The most general solution is a linear combination which is also of the above form (4.9) with a different η_0. It is straightforward to compute

$$\int_{\eta_0}^{\eta} \frac{d\eta'}{\theta^2(\eta')} = \frac{2}{3}\left(\frac{a^2}{\mathcal{H}} - \int a^2 d\eta\right).$$

The gravitational potential is therefore given by

$$\begin{aligned}
\Phi_B &= \frac{\sqrt{\bar{\rho}}}{a\theta} u \\
&= \sqrt{\bar{\rho} + \bar{P}} \exp(i\vec{k}\,\vec{x}) \chi_{\vec{k}} \\
&= C_2 \exp(i\vec{k}\,\vec{x}) \frac{\sqrt{\bar{\rho}}}{a} \int_{\eta_0}^{\eta} \frac{d\eta'}{\theta^2(\eta')} \\
&= \frac{2}{3} C_2 \sqrt{\frac{3}{8\pi G}} \exp(i\vec{k}\,\vec{x})\left(1 - \frac{\mathcal{H}}{a^2}\int a^2 d\eta\right) \\
&= \frac{2}{3} C_2 \sqrt{\frac{3}{8\pi G}} \exp(i\vec{k}\,\vec{x})\left(1 - \frac{\dot{a}}{a^2}\int a\, dt\right) \\
&= \frac{2}{3} C_2 \sqrt{\frac{3}{8\pi G}} \exp(i\vec{k}\,\vec{x}) \frac{d}{dt}\left(\frac{1}{a}\int a\, dt\right).
\end{aligned}$$

Since we are interested in long-wavelengths, i.e. $k \longrightarrow 0$, we can set the plane wave equal 1. The result is then

$$\Phi_B = A\frac{d}{dt}\left(\frac{1}{a}\int a\, dt\right).$$

We assume now that the Universe is a mixture of radiation and matter in the form of, say, cold baryons. The scale factor is then given by

$$a = a_{\mathrm{eq}}\left(\frac{\eta^2}{\eta_*^2} + 2\frac{\eta}{\eta_*}\right).$$

We compute immediately (with $\xi = \eta/\eta_*$)

$$\begin{aligned}
\Phi_B &= \frac{A}{\xi(\xi + 2)}\frac{d}{d\xi}\left(\frac{1}{\xi + 2}\left(\frac{1}{5}\xi^4 + \xi^3 + \frac{4}{3}\xi^2\right) + \frac{A'}{\xi(\xi + 2)}\right) \\
&= \frac{A(\xi + 1)}{(\xi + 2)^3}\left(\frac{3}{5}\xi^2 + 3\xi + \frac{13}{3} + \frac{1}{\xi + 1}\right) + \frac{B(\xi + 1)}{\xi^3(\xi + 2)^3}.
\end{aligned}$$

Exercise 6: Verify equations (4.436) and (4.439).

References

[1] Baumann D 2009 TASI lectures on inflation (arXiv:0907.5424 [hep-th])

[2] Liddle A R 1999 An introduction to cosmological inflation astro-ph/9901124

[3] Weinberg S 2008 *Cosmology* (Oxford: Oxford University Press)

[4] Martin J 2005 *Inflationary Cosmological Perturbations of Quantum-mechanical Origin* (Lecture Notes in Physics vol 669) (Springer)

[5] Bertschinger E 1993 Cosmological dynamics: course 1 ArXiv: astro-ph/9503125

[6] Riotto A 2002 Inflation and the Theory of Cosmological Perturbations *ICTP Lectures*

[7] Mukhanov V 2005 *Physical Foundations of Cosmology* (Cambridge: Cambridge University Press)

[8] Lesgourgues J 2006 Inflationary cosmology *Lecture notes of a course presented in the framework of the'3ieme cycle de physique de Suisse romande'*

[9] Dodelson S 2003 *Modern Cosmology* (Amsterdam: Academic)

[10] Challinor A Part-III Cosmology Course: The Perturbed Universe a.d.challinor@ast.cam.ac.uk

[11] Maldacena J M 2003 Non-Gaussian features of primordial fluctuations in single field inflationary models *J. High Energy Phys.* **2013** JHEP05(2003)013

[12] Ford L H 1987 Gravitational particle creation and inflation *Phys. Rev.* D **35** 2955

[13] Kinney W H, Melchiorri A and Riotto A 2001 New constraints on inflation from the cosmic microwave background *Phys. Rev.* D **63** 023505

[14] Liddle A 2009 *An Introduction to Modern Cosmology* (Weinheim: Wiley-VCH)

[15] Arnowitt R L, Deser S and Misner C W 1959 Dynamical structure and definition of energy in general relativity *Phys. Rev.* **116** 1322

IOP Publishing

Lectures on General Relativity, Cosmology and
Quantum Black Holes (Second Edition)

Badis Ydri

Chapter 5

Quantum field theory on curved backgrounds, vacuum energy and quantum gravity

A standard discussion of the cosmological constant problem is found in the classic paper [1] and a lucid presentation of the current observational and theoretical status of dark energy and its relation to the cosmological constant is found in [2]. In the central analysis of this chapter regarding the calculation of vacuum energy from quantum field theory (QFT) on curved cosmological backgrounds we follow the beautiful books [3, 4]. The topic of quantum field theory on curved backgrounds is discussed for example in [5, 6]. Various calculations of the Casimir force, which is a fundamental measurable quantum effect arising from a non-zero vacuum energy effect, are detailed following the competing/complementary treatments in [7] and [8]. Towards the end of this chapter a complete treatment of the ADM (Arnowitt–Deser–Misner) formulation [9] is given and a brief introduction of the Horava–Lifshitz quantum gravity [10] is presented.

5.1 Dark energy

It is generally accepted now that there is a positive dark energy in the Universe which affects in measurable ways the physics of the expansion. The characteristic feature of dark energy is that it has a negative pressure (tension) smoothly distributed in spacetime so it was proposed that a name like 'smooth tension' is more appropriate to describe it (see reference [11]). The most dramatic consequence of a non-zero value of Ω_Λ is the observation that the Universe appears to be accelerating.

From an observational point of view, astronomical evidence for dark energy comes from various measurements. Here we concentrate, and only briefly, on the the two measurements of CMB anisotropies and type Ia supernovae.

doi:10.1088/978-0-7503-5824-8ch5

- CMB anisotropies: This point has been discussed in great detail from a theoretical point of view in the previous chapter. The main point is as follows. The temperature anisotropies are given by the power spectrum C_l. At intermediate scales (angular scales subtended by H_{CMB}^{-1} where H_{CMB} is the Hubble radius at the time of the formation of the cosmic microwave background (CMB) (decoupling, recombination, last scattering)) we observe peaks in C_l due to acoustic oscillations in the early universe. The first peak is tied directly to the geometry of the Universe. In a negatively curved universe, photon paths diverge leading to a larger apparent angular size compared to flat space, whereas in a positively curved universe photon paths converge leading to a smaller apparent angular size compared to flat space. The spatial curvature as measured by Ω is related to the first peak in the CMB power spectrum by

$$l_{\mathrm{peak}} \sim \frac{220}{\sqrt{\Omega}}. \tag{5.1}$$

The observation indicates that the first peak occurs around $l_{\mathrm{peak}} \sim 200$ which means that the Universe is spatially flat. The Boomerang experiment gives (at the 68% confidence level) the measurement

$$0.85 \leqslant \Omega \leqslant 1.25. \tag{5.2}$$

Since $\Omega = \Omega_M + \Omega_\Lambda$ this is a constraint on the sum of Ω_M and Ω_Λ. The constraints from the CMB in the $\Omega_M - \Omega_\Lambda$ plane using models with different values of Ω_M and Ω_Λ is shown in figure 3 of reference [12]. The best fit is a marginally closed model with

$$\Omega_{\mathrm{CDM}} = 0.26, \quad \Omega_{\mathrm{B}} = 0.05, \quad \Omega_\Lambda = 0.75. \tag{5.3}$$

- **Type Ia supernovae:** This relies on the measurement of the distance modulus $m - M$ of type Ia supernovae where m is the apparent magnitude of the source and M is the absolute magnitude defined by

$$m - M = 5 \log_{10}[(1 + z)d_M(Mpc)] + 25. \tag{5.4}$$

The d_M is the proper distance which is given between any two sources at redshifts z_1 and z_2 by the formula

$$d_M(z_1, z_2) = \frac{1}{H_0\sqrt{|\Omega_{k0}|}} S_k\left(H_0\sqrt{|\Omega_{k0}|} \int_{1/(1+z_1)}^{1/(1+z_2)} \frac{da}{a^2 H(a)} \right). \tag{5.5}$$

Type Ia supernovae are rare events which thought of as standard candles. They are very bright events with almost uniform intrinsic luminosity with absolute brightness comparable to the host galaxies. They result from exploding white dwarfs when they cross the Chandrasekhar limit.

Constraints from type Ia supernovae in the $\Omega_M - \Omega_\Lambda$ plane are consistent with the results obtained from the CMB measurements although the data used is completely independent. In particular, these observations strongly favor a positive cosmological constant.

5.2 The cosmological constant

The cosmological constant was introduced by Einstein in 1917 in order to produce a static Universe. To see this explicitly let us rewrite the Friedmann equations as

$$H^2 = \frac{8\pi G \rho}{3} - \frac{\kappa}{a^2}. \tag{5.6}$$

$$\frac{\ddot{a}}{a} = -\frac{4\pi G}{3}(\rho + 3P). \tag{5.7}$$

The first equation is consistent with a static Universe ($\dot{a} = 0$) if $\kappa > 0$ and $\rho = 3\kappa/(8\pi G a^2)$, whereas the second equation cannot be consistent with a static universe ($\ddot{a} = 0$) containing only ordinary matter and energy which have non negative pressure.

Einstein solved this problem by modifying his equation as follows:

$$R_{\mu\nu} - \frac{1}{2}g_{\mu\nu}R + \Lambda g_{\mu\nu} = 8\pi G T_{\mu\nu}. \tag{5.8}$$

The new free parameter Λ is precisely the cosmological constant. This new equation of motion will entail a modification of the Friedmann equations. To find the modified Friedmann equations we rewrite the modified Einstein equations as

$$R_{\mu\nu} - \frac{1}{2}g_{\mu\nu}R = 8\pi G(T_{\mu\nu} + T^\Lambda_{\mu\nu}). \tag{5.9}$$

$$T^\Lambda_{\mu\nu} = -\rho_\Lambda g_{\mu\nu}, \quad \rho_\Lambda = \frac{\Lambda}{8\pi G}. \tag{5.10}$$

The modified Friedmann equations are then given by (with the substitution $\rho \longrightarrow \rho + \rho_\Lambda$, $P \longrightarrow P - \rho_\Lambda$ in the original Friedmann equations)

$$H^2 = \frac{8\pi G(\rho + \rho_\Lambda)}{3} - \frac{\kappa}{a^2} = \frac{8\pi G\rho}{3} - \frac{\kappa}{a^2} + \frac{\Lambda}{3}. \tag{5.11}$$

$$\frac{\ddot{a}}{a} = -\frac{4\pi G}{3}(\rho - 2\rho_\Lambda + 3P) = -\frac{4\pi G}{3}(\rho + 3P) + \frac{\Lambda}{3}. \tag{5.12}$$

The Einstein static Universe corresponds to $\kappa > 0$ (a 3-sphere S^3) and $\Lambda > 0$ (in the range $\kappa/a^2 \leqslant \Lambda \leqslant 3\kappa/a^2$) with positive mass density and pressure given by

$$\rho = \frac{3\kappa}{8\pi G a^2} - \frac{\Lambda}{8\pi G} > 0, \quad P = \frac{\Lambda}{8\pi G} - \frac{\kappa}{8\pi G a^2} > 0. \tag{5.13}$$

The Universe is in fact expanding and thus this solution is of no physical interest. The cosmological constant is, however, of fundamental importance to cosmology as it might be relevant to dark energy.

It is not difficult to verify that the modified Einstein equations (5.8) can be derived from the action

$$S = \frac{1}{16\pi G} \int d^4x \sqrt{-\det g} \ (R - 2\Lambda) + \int d^4x \sqrt{-\det g} \ \hat{\mathcal{L}}_M. \tag{5.14}$$

Thus the cosmological constant Λ is just a constant term in the Lagrangian density. We call Λ the bare cosmological constant. The effective cosmological constant Λ_{eff} will in general be different from Λ due to possible contribution from matter. Consider for example a scalar field with Lagrangian density

$$\hat{\mathcal{L}}_M = -\frac{1}{2} g^{\mu\nu} \nabla_\mu \phi \nabla_\nu \phi - V(\phi). \tag{5.15}$$

The stress–energy–momentum tensor is calculated to be given by

$$T_{\mu\nu} = \nabla_\mu \phi \nabla_\nu \phi - \frac{1}{2} g_{\mu\nu} g^{\rho\sigma} \nabla_\rho \phi \nabla_\sigma \phi - g_{\mu\nu} V(\phi). \tag{5.16}$$

The configuration ϕ_0 with lowest energy density (the vacuum) is the contribution which minimizes separately the kinetic and potential terms and as a consequence $\partial_\mu \phi_0 = 0$ and $V'(\phi_0) = 0$. The corresponding stress–energy–momentum tensor is therefore $T_{\mu\nu}^{(\phi)} = -g_{\mu\nu} V(\phi_0)$. In other words, the stress–energy–momentum tensor of the vacuum acts precisely like the stress–energy–momentum tensor of a cosmological constant. We write (with $T_{\mu\nu}^{(\phi_0)} \equiv T_{\mu\nu}^{\text{vac}}, V(\phi_0) \equiv \rho_{\text{vac}}$)

$$T_{\mu\nu}^{\text{vac}} = -\rho_{\text{vac}} g_{\mu\nu}. \tag{5.17}$$

The vacuum ϕ_0 is therefore a perfect fluid with pressure given by

$$P_{\text{vac}} = -\rho_{\text{vac}}. \tag{5.18}$$

Thus the vacuum energy acts like a cosmological constant Λ_ϕ given by

$$\Lambda_\phi = 8\pi G \rho_{\text{vac}}. \tag{5.19}$$

In other words, the cosmological constant and the vacuum energy are completely equivalent. We will use the two terms 'cosmological constant' and 'vacuum energy' interchangeably.

The effective cosmological constant Λ_{eff} is therefore given by

$$\Lambda_{\text{eff}} = \Lambda + \Lambda_\phi. \tag{5.20}$$

In other words,

$$\Lambda_{\text{eff}} = \Lambda + 8\pi G \rho_{\text{vac}}. \tag{5.21}$$

This calculation is purely classical.

Quantum mechanics will naturally modify this result. We follow a semi-classical approach in which the gravitational field is treated classically and the scalar field (matter fields in general) are treated quantum mechanically. Thus we need to quantize the scalar field in a background metric $g_{\mu\nu}$ which is here the Robertson–Walker metric. In the quantum vacuum state of the scalar field (assuming that it exists) the expectation value of the stress–energy–momentum tensor $T_{\mu\nu}$ must be, by Lorentz invariance, of the form

$$\langle T_{\mu\nu}\rangle_{\text{vac}} = -\langle \rho\rangle_{\text{vac}} g_{\mu\nu}. \tag{5.22}$$

Einstein's equation in the vacuum state of the scalar field is

$$R_{\mu\nu} - \frac{1}{2}g_{\mu\nu}R + \Lambda g_{\mu\nu} = 8\pi G\langle T_{\mu\nu}\rangle_{\text{vac}}. \tag{5.23}$$

The effective cosmological constant Λ_{eff} must therefore be given by

$$\Lambda_{\text{eff}} = \Lambda + 8\pi G\langle \rho\rangle_{\text{vac}}. \tag{5.24}$$

The energy density of empty space $\langle \rho\rangle_{\text{vac}}$ is the sum of zero-point energies associated with vacuum fluctuations together with other contributions resulting from virtual particles (higher order vacuum fluctuations) and vacuum condensates.

We will assume for simplicity that the bare cosmological constant Λ is zero. Thus the effective cosmological constant is entirely given by vacuum energy, viz

$$\Lambda_{\text{eff}} = 8\pi G\langle \rho\rangle_{\text{vac}}. \tag{5.25}$$

We drop now the subscript 'eff' without fear of confusion. The relation between the density ρ_Λ of the cosmological constant and the density $\langle \rho\rangle_{\text{vac}}$ of the vacuum is then simply

$$\rho_\Lambda = \langle \rho\rangle_{\text{vac}}. \tag{5.26}$$

From the concordance model we know that the favorite estimate for the value of the density parameter of dark energy at this epoch is $\Omega_\Lambda = 0.7$. We recall $G = 6.67 \times 10^{-11}\,\text{m}^3\,\text{kg}^{-1}\,\text{s}^{-2}$ and $H_0 = 70\,\text{km s}^{-1}\,\text{Mpc}^{-1}$ with $\text{Mpc} = 3.09 \times 10^{24}\,\text{cm}$. We compute then the density

$$\rho_\Lambda = \frac{3H_0^2}{8\pi G}\Omega_\Lambda \tag{5.27}$$
$$= 9.19 \times 10^{-27}\Omega_\Lambda \text{kg m}^{-3}.$$

We convert to natural units (1 GeV $= 1.8 \times 10^{-27}$ kg, 1 GeV$^{-1} = 0.197 \times 10^{-15}$ m, 1 GeV$^{-1} = 6.58 \times 10^{-25}$ s) to obtain

$$\rho_\Lambda = 39\Omega_\Lambda(10^{-12}\,\text{GeV})^4. \tag{5.28}$$

To get a theoretical order-of-magnitude estimate of $\langle \rho\rangle_{\text{vac}}$ we use the flat space Hamiltonian operator of a free scalar field given by

$$\hat{H} = \int \frac{d^3p}{(2\pi)^3} \omega(\vec{p}) \left[\hat{a}(\vec{p})^+ \hat{a}(\vec{p}) + \frac{1}{2}(2\pi)^3 \delta^3(0) \right]. \tag{5.29}$$

The vacuum state is defined in this case unambiguously by $\hat{a}(\vec{p})|0\rangle = 0$. We get then in the vacuum state the energy $E_{\text{vac}} = \langle 0|\hat{H}|0\rangle$ where

$$E_{\text{vac}} = \frac{1}{2}(2\pi)^3 \delta^3(0) \int \frac{d^3p}{(2\pi)^3} \omega(\vec{p}). \tag{5.30}$$

If we use box normalization then $(2\pi)^3 \delta^3(\vec{p} - \vec{q})$ will be replaced with $V\delta_{\vec{p},\vec{q}}$ where V is spacetime volume. The vacuum energy density is therefore given by (using also $\omega(\vec{p}) = \sqrt{\vec{p}^2 + m^2}$)

$$\langle \rho \rangle_{\text{vac}} = \frac{1}{2} \int \frac{d^3p}{(2\pi)^3} \sqrt{\vec{p}^2 + m^2}. \tag{5.31}$$

This is clearly divergent. We introduce a cutoff λ and compute

$$\langle \rho \rangle_{\text{vac}} = \frac{1}{4\pi^2} \int_0^\lambda dp p^2 \sqrt{p^2 + m^2}$$
$$= \frac{1}{4\pi^2} \left[\left(\frac{1}{4}\lambda^3 + \frac{m^2}{8}\lambda \right) \sqrt{\lambda^2 + m^2} - \frac{m^4}{8} \ln \left(\frac{\lambda}{m} + \sqrt{1 + \frac{\lambda^2}{m^2}} \right) \right]. \tag{5.32}$$

In the massless limit (the mass is in any case much smaller than the cutoff λ) we obtain the estimate

$$\langle \rho \rangle_{\text{vac}} = \frac{\lambda^4}{16\pi^2}. \tag{5.33}$$

By assuming that quantum field theory calculations are valid up to the Planck scale $M_{\text{pl}} = 1/\sqrt{8\pi G} = 2.42 \times 10^{18}$ GeV then we can take $\lambda = M_{\text{pl}}$ and get the estimate

$$\langle \rho \rangle_{\text{vac}} = 0.22(10^{18} \text{ GeV})^4. \tag{5.34}$$

By taking the ratio of the value (5.28) obtained from cosmological observations and the theoretical value (5.34) we get

$$\left(\frac{\rho_\Lambda}{\langle \rho \rangle_{\text{vac}}} \right)^{1/4} = 3.65 \times \Omega_\Lambda^{1/4} \times 10^{30}. \tag{5.35}$$

For the observed value $\Omega_\Lambda = 0.7$ we see that there is a discrepancy of 30 orders of magnitude between the theoretical and observational mass scales of the vacuum energy which is the famous cosmological constant problem.

Let us note that in flat spacetime we can make the vacuum energy vanish by the usual normal ordering procedure which reflects the fact that only differences in energy have experimental consequences in this case. In curved spacetime this is not, however, possible since general relativity is sensitive to the absolute value of the vacuum energy. In other words, the gravitational effect of vacuum energy will curve spacetime and the above problem of the cosmological constant is certainly genuine.

5.3 Elements of quantum field theory in curved spacetime

Some of the many excellent presentations of QFT on curved backgrounds are [5, 6, 13–15]. The most important case for us in this chapter is QFT on de Sitter spacetime. See for example [16–21].

We start by assuming Friedmann equations with a cosmological constant Λ which are given by (with $H = \dot{a}/a$)

$$H^2 = \frac{8\pi G\rho}{3} - \frac{\kappa}{a^2} + \frac{\Lambda}{3}. \tag{5.36}$$

$$\frac{\ddot{a}}{a} = -\frac{4\pi G}{3}(\rho + 3P) + \frac{\Lambda}{3}. \tag{5.37}$$

We will assume that ρ and P are those of a real scalar field coupled to the metric minimally with action given by

$$S_M = \int d^4x \sqrt{-\det g} \left(-\frac{1}{2}g^{\mu\nu} \nabla_\mu \phi \nabla_\nu \phi - V(\phi)\right). \tag{5.38}$$

If we are interested in an action which is at most quadratic in the scalar field then we must choose $V(\phi) = m^2\phi^2/2$. In curved spacetime there is another term we can add which is quadratic in ϕ, namely $R\phi^2$ where R is the Ricci scalar. The full action should then read (in arbitrary dimension n)

$$S_M = \int d^nx \sqrt{-\det g} \left(-\frac{1}{2}g^{\mu\nu} \nabla_\mu \phi \nabla_\nu \phi - \frac{1}{2}m^2\phi^2 - \frac{1}{2}\xi R\phi^2\right). \tag{5.39}$$

The choice $\xi = (n - 2)/(4(n - 1))$ is called conformal coupling. At this value the action with $m^2 = 0$ is invariant under conformal transformations defined by

$$g_{\mu\nu} \longrightarrow \bar{g}_{\mu\nu} = \Omega^2(x)g_{\mu\nu}(x), \quad \phi \longrightarrow \bar{\phi} = \Omega^{\frac{2-n}{2}}(x)\phi(x). \tag{5.40}$$

The equation of motion derived from this action is (in the following we will keep the metric arbitrary as long as possible)

$$(\nabla_\mu \nabla^\mu - m^2 - \xi R)\phi = 0. \tag{5.41}$$

Let ϕ_1 and ϕ_2 be two solutions of this equation of motion. We define their inner product by

$$(\phi_1, \phi_2) = -i \int_\Sigma \left(\phi_1 \partial_\mu \phi_2^* - \partial_\mu \phi_1 \cdot \phi_2^*\right) d\Sigma n^\mu. \tag{5.42}$$

$d\Sigma$ is the volume element in the space-like hypersurface Σ and n^μ is the time-like unit vector which is normal to this hypersurface. This inner product is independent of the hypersurface Σ. Indeed, let Σ_1 and Σ_2 be two non-intersecting hypersurfaces and let V be the 4-volume bounded by Σ_1, Σ_2 and (if necessary) time-like boundaries on which $\phi_1 = \phi_2 = 0$. We have from one hand

$$i\int_V \nabla^\mu \left(\phi_1 \partial_\mu \phi_2^* - \partial_\mu \phi_1 . \phi_2^*\right) dV = i\oint_{\partial V} \left(\phi_1 \partial_\mu \phi_2^* - \partial_\mu \phi_1 . \phi_2^*\right) d\Sigma^\mu$$

$$= (\phi_1, \phi_2)_{\Sigma_1} - (\phi_1, \phi_2)_{\Sigma_2}. \tag{5.43}$$

From the other hand

$$i\int_V \nabla^\mu \left(\phi_1 \partial_\mu \phi_2^* - \partial_\mu \phi_1 . \phi_2^*\right) dV = i\int_V \left(\phi_1 \nabla^\mu \partial_\mu \phi_2^* - \nabla^\mu \partial_\mu \phi_1 . \phi_2^*\right) dV$$

$$= i\int_V (\phi_1 (m^2 + \xi R)\phi_2^* - (m^2 + \xi R)\phi_1 . \phi_2^*) dV \tag{5.44}$$

$$= 0.$$

Hence

$$(\phi_1, \phi_2)_{\Sigma_1} - (\phi_1, \phi_2)_{\Sigma_2} = 0. \tag{5.45}$$

There is always a complete set of solutions u_i and u_i^* of the equation of motion (6.175) which are orthonormal in the inner product (6.176), i.e. satisfying

$$(u_i, u_j) = \delta_{ij}, \quad (u_i^*, u_j^*) = -\delta_{ij}, \quad (u_i, u_j^*) = 0. \tag{5.46}$$

We can then expand the field as

$$\phi = \sum_i (a_i u_i + a_i^* u_i^*). \tag{5.47}$$

We now canonically quantize this system. We choose a foliation of spacetime into space-like hypersurfaces. Let Σ be a particular hypersurface with unit normal vector n^μ corresponding to a fixed value of the time coordinate $x^0 = t$ and with induced metric h_{ij}. We write the action as $S_M = \int dx^0 L_M$ where $L_M = \int d^{n-1}x \sqrt{-\det g} \ \mathcal{L}_M$. The canonical momentum π is defined by

$$\pi = \frac{\delta L_M}{\delta(\partial_0 \phi)} = -\sqrt{-\det g} \ g^{\mu 0} \partial_\mu \phi$$

$$= -\sqrt{-\det h} \ n^\mu \partial_\mu \phi. \tag{5.48}$$

We promote ϕ and π to hermitian operators $\hat{\phi}$ and $\hat{\pi}$ and then impose the equal time canonical commutation relations

$$[\hat{\phi}(x^0, x^i), \hat{\pi}(x^0, y^i)] = i\delta^{n-1}(x^i - y^i). \tag{5.49}$$

The delta function satisfies the property

$$\int \delta^{n-1}(x^i - y^i) d^{n-1}y = 1. \tag{5.50}$$

The coefficients a_i and a_i^* become annihilation and creation operators $\hat{a}_i$ and $\hat{a}_i^+$ satisfying the commutation relations

$$[\hat{a}_i, \hat{a}_j^+] = \delta_{ij}, \quad [\hat{a}_i, \hat{a}_j] = [\hat{a}_i^+, \hat{a}_j^+] = 0. \tag{5.51}$$

The vacuum state is given by a state $|0\rangle_u$ defined by

$$\hat{a}_i|0_u\rangle = 0. \tag{5.52}$$

The entire Fock basis of the Hilbert space can be constructed from the vacuum state by repeated application of the creation operators $\hat{a}_i^+$.

The solutions u_i, u_i^* are not unique and as a consequence the vacuum state $|0\rangle_u$ is not unique. Let us consider another complete set of solutions v_i and v_i^* of the equation of motion (6.175) which are orthonormal in the inner product (6.176). We can then expand the field as

$$\phi = \sum_i (b_i v_i + b_i^* v_i^*). \tag{5.53}$$

After canonical quantization the coefficients b_i and b_i^* become annihilation and creation operators $\hat{b}_i$ and $\hat{b}_i^+$ satisfying the standard commutation relations with a vacuum state given by $|0\rangle_v$ defined by

$$\hat{b}_i|0_v\rangle = 0. \tag{5.54}$$

We introduce the so-called Bogolubov transformation as the transformation from the set $\{u_i, u_i^*\}$ (which are the set of modes seen by some observer) to the set $\{v_i, v_i^*\}$ (which are the set of modes seen by another observer) as

$$v_i = \sum_j (\alpha_{ij} u_j + \beta_{ij} u_j^*). \tag{5.55}$$

By using orthonormality conditions we find that

$$\alpha_{ij} = (v_i, u_j), \quad \beta_{ij} = -(v_i, u_j^*). \tag{5.56}$$

We can also write

$$u_i = \sum_j (\alpha_{ji}^* v_j + \beta_{ji} v_j^*). \tag{5.57}$$

The Bogolubov coefficients α and β satisfy the normalization conditions

$$\sum_k (\alpha_{ik} \alpha_{jk} - \beta_{ik} \beta_{jk}) = \delta_{ij}, \quad \sum_k (\alpha_{ik} \beta_{jk}^* - \beta_{ik} \alpha_{jk}^*) = 0. \tag{5.58}$$

The Bogolubov coefficients α and β transform also between the creation and annihilation operators $\hat{a}, \hat{a}^+$ and $\hat{b}, \hat{b}^+$. We find

$$\hat{a}_k = \sum_i (\alpha_{ik} \hat{b}_i + \beta_{ik}^* \hat{b}_i^+), \quad \hat{b}_k = \sum_i (\alpha_{ki}^* \hat{a}_i + \beta_{ki}^* \hat{a}_i^+). \tag{5.59}$$

Let N_u be the number operator with respect to the u-observer, viz $N_u = \sum_k \hat{a}_k^+ \hat{a}_k$. Clearly

$$\langle 0_u | N_u | 0_u \rangle = 0. \tag{5.60}$$

We compute

$$\langle 0_v | \hat{a}_k^+ \hat{a}_k | 0_v \rangle = \sum_i \beta_{ik} \beta_{ik}^*. \tag{5.61}$$

Thus

$$\langle 0_v | N_u | 0_v \rangle = tr \, \beta \beta^+. \tag{5.62}$$

In other words, with respect to the v-observer the vacuum state $|0_u\rangle$ is not empty but filled with particles. This opens the door to the possibility of particle creation by a gravitational field.

5.4 Calculation of vacuum energy in curved backgrounds

5.4.1 Quantization in Friedmann–Lemaître–Robertson–Walker (FLRW) universes

We go back to the equation of motion (6.175), viz

$$(\nabla_\mu \nabla^\mu - m^2 - \xi R)\phi = 0. \tag{5.63}$$

The flat FLRW universes are given by

$$ds^2 = -dt^2 + a^2(t)(d\rho^2 + \rho^2 d\Omega^2). \tag{5.64}$$

The conformal time is denoted here by

$$\eta = \int^t \frac{dt_1}{a(t_1)}. \tag{5.65}$$

In terms of η the FLRW universes are manifestly conformally flat, viz

$$ds^2 = a^2(\eta)(-d\eta^2 + d\rho^2 + \rho^2 d\Omega^2). \tag{5.66}$$

The d'Alembertian in FLRW universes is

$$\begin{aligned}
\nabla_\mu \nabla^\mu \phi &= \frac{1}{\sqrt{-\det g}} \partial_\mu (\sqrt{-\det g} \, \partial^\mu \phi) \\
&= \partial_\mu \partial^\mu \phi + \frac{1}{2} g^{\alpha\beta} \partial_\mu g_{\alpha\beta} \partial^\mu \phi \\
&= -\ddot{\phi} + \frac{1}{a^2} \partial_i^2 \phi - 3\frac{\dot{a}}{a}\dot{\phi}.
\end{aligned} \tag{5.67}$$

The Klein–Gordon equation of motion becomes

$$\ddot{\phi} + 3\frac{\dot{a}}{a}\dot{\phi} - \frac{1}{a^2}\partial_i^2 \phi + (m^2 + \xi R)\phi = 0. \tag{5.68}$$

In terms of the conformal time this reads (where $d/d\eta$ is denoted by primes)

$$\phi'' + 2\frac{a'}{a}\phi' - \partial_i^2\phi + a^2(m^2 + \xi R)\phi = 0. \tag{5.69}$$

The positive norm solutions are given by

$$u_k(\eta, x^i) = \frac{e^{i\vec{k}\vec{x}}}{a(\eta)}\chi_k(\eta). \tag{5.70}$$

Indeed, we check that $\phi \equiv u_k(\eta, x^i)$ is a solution of the Klein–Gordon equation of motion provided that χ_k is a solution of the equation of motion (using also $R = 6(\ddot{a}/a + \dot{a}^2/a^2) = 6a''/a^3$)

$$\chi_k'' + \omega_k^2(\eta)\chi_k = 0. \tag{5.71}$$

$$\omega_k^2(\eta) = k^2 + m^2a^2 - (1 - 6\xi)\frac{a''}{a}. \tag{5.72}$$

In the case of conformal coupling $m = 0$ and $\xi = 1/6$ this reduces to a time independent harmonic oscillator. This is similar to flat spacetime and all effects of the curvature are included in the factor $a(\eta)$ in equation (5.70). Thus calculation in a conformally invariant world is very easy.

The condition $(u_k, u_l) = \delta_{kl}$ becomes (with $n^\mu = (1, 0, 0, 0)$, $d\Sigma = \sqrt{-\det h}\ d^3x$ and using box normalization $(2\pi)^3\delta^3(\vec{k} - \vec{p}) \longrightarrow V\delta_{\vec{k},\vec{p}}$ the Wronskian condition

$$iV(\chi_k^*\chi_k' - \chi_k^{*'}\chi_k) = 1. \tag{5.73}$$

The negative norm solutions correspond obviously to u_k^*. Indeed we can check that $(u_k^*, \bar{u}_l) = -\delta_{kl}$ and $(u_k^*, u_l) = 0$.

The modes u_k and $\bar{u}_k$ provide a Fock space representation for field operators. The quantum field operator $\hat{\phi}$ can be expanded in terms of creation and annhiliation operators as

$$\hat{\phi} = \sum_k (\hat{a}_k u_k + \hat{a}_k^+ u_k^*). \tag{5.74}$$

Alternatively, the mode functions satisfy the differential equations (with $\chi_k = v_k^*/\sqrt{2V}$)

$$v_k'' + \omega_k^2(\eta)v_k = 0 \tag{5.75}$$

They must satisfy the normalization condition

$$\frac{1}{2i}(v_k'v_k^* - v_kv_k^{*'}) = 1. \tag{5.76}$$

The scalar field operator is given by $\hat{\phi} = \hat{\chi}/a(\eta)$ where (with $[\bar{a}_k, \bar{a}_{k'}^+] = V\delta_{k,k'}$, etc)

$$\hat{\chi} = \frac{1}{V}\sum_k \frac{1}{\sqrt{2}}\left(\bar{a}_k v_k^* e^{i\vec{k}\vec{x}} + \bar{a}_k^+ v_k e^{-i\vec{k}\vec{x}}\right). \tag{5.77}$$

The stress–energy–momentum tensor in minimal coupling $\xi = 0$ is given by

$$T_{\mu\nu} = \nabla_\mu \phi \nabla_\nu \phi - \frac{1}{2}g_{\mu\nu}g^{\rho\sigma}\nabla_\rho \phi \nabla_\sigma \phi - g_{\mu\nu}V(\phi). \tag{5.78}$$

We compute immediately in the conformal metric $ds^2 = a^2(-d\eta^2 + dx^i dx^i)$ the component

$$\begin{aligned}
T_{00} &= \frac{1}{2}(\partial_\eta\phi)^2 + \frac{1}{2}(\partial_i\phi)^2 + \frac{1}{2}a^2 m^2\phi^2 \\
&= \frac{1}{2a^2}\left[\chi'^2 - 2\frac{a'}{a}\chi\chi' + \frac{a'^2}{a^2}\chi^2\right] + \frac{1}{2a^2}(\partial_i\chi)^2 + \frac{1}{2}m^2\chi^2.
\end{aligned} \tag{5.79}$$

The conjugate momentum (6.182) in our case is $\pi = a^2\partial_\eta\phi$. The Hamiltonian is therefore

$$\begin{aligned}
H &= \int d^{n-1}x \ \pi\partial_0\phi - L_M \\
&= \int d^{n-1}x\sqrt{-\det g}\ \frac{1}{a^2}T_{00} \\
&= -\int d^{n-1}x\sqrt{-\det g}\ T_0^0.
\end{aligned} \tag{5.80}$$

In the quantum theory the stress–energy–momentum tensor in minimal coupling $\xi = 0$ is given by

$$\hat{T}_{00} = \frac{1}{2a^2}\left[\hat{\chi}'^2 - \frac{a'}{a}(\hat{\chi}\hat{\chi}' + \hat{\chi}'\hat{\chi}) + \frac{a'^2}{a^2}\hat{\chi}^2\right] + \frac{1}{2a^2}(\partial_i\hat{\chi})^2 + \frac{1}{2}m^2\hat{\chi}^2. \tag{5.81}$$

We assume the existence of a vacuum state $|0\rangle$ with the properties $a|0\rangle = 0$, $\langle 0|a^+ = 0$ and $\langle 0|0\rangle = 1$. We compute

$$\begin{aligned}
\langle\hat{\chi}'^2\rangle &= \frac{1}{2V^2}\sum_k\sum_p v_k^{*'}v_p' e^{i\vec{k}\vec{x}}e^{-i\vec{p}\vec{x}}\langle 0|\bar{a}_k\bar{a}_p^+|0\rangle \\
&= \frac{1}{2V}\sum_k |v_k'|^2.
\end{aligned} \tag{5.82}$$

$$\begin{aligned}
\langle\hat{\chi}^2\rangle &= \frac{1}{2V^2}\sum_k\sum_p v_k^* v_p e^{i\vec{k}\vec{x}}e^{-i\vec{p}\vec{x}}\langle 0|\bar{a}_k\bar{a}_p^+|0\rangle \\
&= \frac{1}{2V}\sum_k |v_k|^2.
\end{aligned} \tag{5.83}$$

$$\langle (\partial_i \hat{\chi})^2 \rangle = \frac{1}{2V^2} \sum_k \sum_p v_k^* v_p (k_i p_i) e^{i\vec{k}\cdot\vec{x}} e^{-i\vec{p}\cdot\vec{x}} \langle 0 | \bar{a}_k \bar{a}_p^+ | 0 \rangle$$

$$= \frac{1}{2V} \sum_k k^2 |v_k|^2. \tag{5.84}$$

We get then

$$\langle \hat{T}_{00} \rangle = \frac{1}{2a^2} \frac{1}{2V} \sum_k \left[|v_k'|^2 - \frac{a'}{a}(v_k^* v_k' + v_k'^* v_k) + \frac{a'^2}{a^2}|v_k|^2 + k^2|v_k|^2 + a^2 m^2 |v_k|^2 \right]$$

$$= \frac{1}{4a^2} \frac{1}{V} \sum_k \left[|v_k'|^2 + (k^2 + \frac{a''}{a} + a^2 m^2)|v_k|^2 - \partial_\eta \left(\frac{a'}{a}|v_k|^2 \right) \right]. \tag{5.85}$$

The mass density is therefore given by [22]

$$\rho = \frac{1}{a^2} \langle \hat{T}_{00} \rangle = \frac{1}{4a^4} \int \frac{d^3k}{(2\pi)^3} \left[|v_k'|^2 + (k^2 + \frac{a''}{a} + a^2 m^2)|v_k|^2 - \partial_\eta \left(\frac{a'}{a}|v_k|^2 \right) \right]. \tag{5.86}$$

5.4.2 Instantaneous vacuum

Let us do the calculation in a slightly different way [3, 4]. The comoving scalar field $\chi = a\phi$ satisfies the equation of motion

$$\chi'' + m_{\text{eff}}^2 \chi - \partial_i^2 \chi = 0, \quad m_{\text{eff}}^2 = a^2 m^2 - \frac{a''}{a}. \tag{5.87}$$

This can be derived from the action

$$S = \frac{1}{2} \int d\eta d^3x \left[\chi'^2 - (\partial_i \chi)^2 - m_{\text{eff}}^2 \chi^2 \right]. \tag{5.88}$$

We quantize this system now. The conjugate momentum is $\pi = \chi'$. The Hamiltonian is

$$H = \frac{1}{2} \int d^3x \left[\chi'^2 + (\partial_i \chi)^2 + m_{\text{eff}}^2 \chi^2 \right]. \tag{5.89}$$

This is different from the Hamiltonian written down in the previous section. The rest is now the same. For example the field operator can be expanded as (with $[\bar{a}_k, \bar{a}_{k'}^+] = V \delta_{k,k'}$, etc and $v_k' v_k^* - v_k v_k'^* = 2i$)

$$\hat{\chi} = \frac{1}{V} \sum_k \frac{1}{\sqrt{2}} \left(\bar{a}_k v_k^* e^{i\vec{k}\cdot\vec{x}} + \bar{a}_k^+ v_k e^{-i\vec{k}\cdot\vec{x}} \right). \tag{5.90}$$

We compute the Hamiltonian operator (assuming isotropic mode functions, viz $v_k = v_{-k}$)

$$\hat{H} = \frac{1}{4V} \sum_k \left[F_k^* \bar{a}_k \bar{a}_{-k} + F_k \bar{a}_k^+ \bar{a}_{-k}^+ + E_k (\bar{a}_k \bar{a}_k^+ + \bar{a}_k^+ \bar{a}_k) \right]. \tag{5.91}$$

$$F_k = (v_k')^2 + \omega_k^2 v_k^2, \quad E_k = |v_k'|^2 + \omega_k^2 |v_k|^2. \tag{5.92}$$

Let $|0_v\rangle$ be the vacuum state corresponding to the mode functions v_k. Then

$$\langle 0_v | \hat{H} | 0_v \rangle = \frac{1}{4} \sum_k E_k$$
$$= \frac{V}{4} \int \frac{d^3 k}{(2\pi)^3} \left[|v_k'|^2 + \omega_k^2 |v_k|^2 \right]. \tag{5.93}$$

The vacuum energy density is

$$\rho = \frac{1}{4} \int \frac{d^3 k}{(2\pi)^3} \left[|v_k'|^2 + \omega_k^2 |v_k|^2 \right]. \tag{5.94}$$

This clearly depends on the conformal time η. The instantaneous vacuum at a conformal time $\eta = \eta_0$ is the state $|0_{\eta_0}\rangle$ which is the lowest energy eigenstate of the instantaneous Hamiltonian $H(\eta_0)$. Equivalently the instantaneous vacuum at a conformal time $\eta = \eta_0$ is the state in which the vacuum expectation value $\langle 0_v | \hat{H}(\eta_0) | 0_v \rangle$ is minimized with respect to all possible choices of $v_k = v_k(\eta_0)$. The minimization of the energy density ρ corresponds to the minimization of each mode v_k separately. For a given value of $\vec{k}$ we choose $v_k(\eta)$ by imposing at $\eta = \eta_0$ the initial conditions

$$v_k(\eta_0) = q, \quad v_k'(\eta_0) = p. \tag{5.95}$$

The normalization condition $v_k' v_k^* - v_k v_k^{*'} = 2i$ reads therefore

$$q^* p - p^* q = 2i. \tag{5.96}$$

The corresponding energy is $E_k = |p|^2 + \omega_k^2(\eta_0)|q|^2$. By using the symmetry $q \longrightarrow e^{i\lambda} q$ and $p \longrightarrow e^{i\lambda} p$ we can choose q real. If we write $p = p_1 + ip_2$ then the above condition gives immediately $q = 1/p_2$. The energy becomes

$$E_k(\eta_0) = p_1^2 + p_2^2 + \frac{\omega_k^2(\eta_0)}{p_2^2}. \tag{5.97}$$

The minimum of this energy with respect to p_1 is $p_1 = 0$, whereas its minimum with respect to p_2 is $p_2 = \sqrt{\omega_k(\eta_0)}$. The initial conditions become

$$v_k(\eta_0) = \frac{1}{\sqrt{\omega_k(\eta_0)}}, \quad v_k'(\eta_0) = i\omega_k(\eta_0) v_k(\eta_0). \tag{5.98}$$

In Minkowski spacetime we have $a = 1$ and thus $\omega_k = \sqrt{k^2 + m^2}$. We obtain (with $\eta_0 = 0$) the usual result $v_k(\eta) = e^{i\omega_k \eta}/\sqrt{\omega_k}$.

The energy in this minimum reads

$$E_k(\eta_0) = 2\omega_k(\eta_0). \tag{5.99}$$

The vacuum energy density is therefore

$$\rho = \frac{1}{2} \int \frac{d^3k}{(2\pi)^3} \omega_k(\eta_0). \qquad (5.100)$$

This is the usual formula which is clearly divergent so we may proceed in the usual way to perform regularization and renormalization. The problem (which is actually quite severe) is that this energy density is time dependent.

5.4.3 Quantization in de Sitter spacetime and Bunch–Davies vacuum

During inflation and also in the limit $a \longrightarrow \infty$ (the future) it is believed that vacuum dominates and thus spacetime is approximately de Sitter spacetime. Some of the studies concerning QFT on de Sitter spacetime can be found in [16–21]. In this central section we follow again [3, 4].

An interesting solution of the Friedmann equations (5.36) and (5.37) is precisley the maximally symmetric de Sitter space with positive curvature $\kappa > 0$ and positive cosmological constant $\Lambda > 0$ and no matter content $\rho = P = 0$ given by the scale factor

$$a(t) = \frac{\alpha}{R_0} \cosh \frac{t}{\alpha}. \qquad (5.101)$$

$$\alpha = \sqrt{\frac{3}{\Lambda}}, \quad R_0 = \frac{1}{\sqrt{\kappa}}. \qquad (5.102)$$

At large times the Hubble parameter becomes a constant

$$H \simeq \frac{1}{\alpha} = \sqrt{\frac{\Lambda}{3}}. \qquad (5.103)$$

The behavior of the scale factor at large times becomes thus

$$a(t) \simeq a_0 e^{Ht} \quad a_0 = \frac{\alpha}{2R_0}. \qquad (5.104)$$

Thus the scale factor on de Sitter space can be given by $a(t) \simeq a_0 \exp(Ht)$. In this case the curvature is computed to be zero and thus the coordinates t, x, y and z are incomplete in the past. The metric is given explicitly by

$$ds^2 = -dt^2 + a_0^2 e^{2Ht} dx^i dx^i. \qquad (5.105)$$

In this flat patch (upper half) de Sitter space is asymptotically static with respect to conformal time η in the past. This can be seen as follows. First we can compute in closed form that $\eta = -e^{-Ht}/(a_0 H)$ and $a(t) = a(\eta) = -1/(H\eta)$ and thus η is in the interval $[-\infty, 0]$ (and hence the coordinates t, x, y and z are incomplete). We then observe that $H_\eta = a'/a = -1/\eta \longrightarrow 0$ when $\eta \longrightarrow -\infty$ which means that de Sitter is asymptotically static.

The de Sitter space is characterized by the existence of horizons. As usual, null radial geodesics are characterized by $a^2(t)\dot{r}^2 = 1$. The solution is explicitly given by

$$r(t) - r(t_0) = \frac{1}{a_0 H}(e^{-Ht_0} - e^{-Ht}). \tag{5.106}$$

Thus photons emitted at the origin $r(t_0) = 0$ at time t_0 will reach the sphere $r_h = e^{-Ht_0}/(a_0 H)$ at time $t \longrightarrow \infty$ (asymptotically). This sphere is precisely the horizon for the observer at the origin in the sense that signal emitted at the origin cannot reach any point beyond the horizon and similarly any signal emitted at time t_0 at a point $r > r_h$ cannot reach the observer at the origin.

The horizon scale at time t_0 is defined as the proper distance of the horizon from the observer at the origin, viz $a^2(t_0)r_h = 1/H$. This is clearly the same at all times.

The effective frequencies of oscillation in de Sitter space are

$$\omega_k^2(\eta) = k^2 + m^2 a^2 - (1 - 6\xi)\frac{a''}{a}$$

$$= k^2 + \left[\frac{m^2}{H^2} - 2(1 - 6\xi)\right]\frac{1}{\eta^2}. \tag{5.107}$$

These may become imaginary. For example $\omega_0^2(\eta) < 0$ if $m^2 < 2(1 - 6\xi)H^2$. We will take $\xi = 0$ and assume that $m \ll H$.

From the previous section we know that the mode functions must satisfy the differential equations (with $\chi_k = v_k^*/\sqrt{2V}$)

$$v_k'' + \left(k^2 + \left[\frac{m^2}{H^2} - 2\right]\frac{1}{\eta^2}\right)v_k = 0 \tag{5.108}$$

The solution of this equation is given in terms of Bessel functions J_n and Y_n by

$$v_k = \sqrt{k|\eta|}\,[AJ_n(k|\eta|) + BY_n(k|\eta|)], \quad n = \sqrt{\frac{9}{4} - \frac{m^2}{H^2}}. \tag{5.109}$$

The normalization condition (5.76) becomes (with $s = k|\eta|$)

$$ks(A^*B - AB^*)\left(\frac{d}{ds}J_n(s) \cdot Y_n(s) - \frac{d}{ds}Y_n(s) \cdot J_n(s)\right) = 2i. \tag{5.110}$$

We use the result

$$\frac{d}{ds}J_n(s) \cdot Y_n(s) - \frac{d}{ds}Y_n(s) \cdot J_n(s) = -\frac{2}{\pi s}. \tag{5.111}$$

We obtain the constraint

$$AB^* - A^*B = \frac{i\pi}{k}. \tag{5.112}$$

We consider now two limits of interest.

The early time regime $\eta \longrightarrow -\infty$: This corresponds to $\omega_k^2 \longrightarrow k^2$ or equivalently

$$k^2 \gg \left(2 - \frac{m^2}{H^2}\right)\frac{1}{\eta^2}. \tag{5.113}$$

This is a high energy (short distance) limit. The effect of gravity on the modes v_k is therefore negligible and we obtain the Minkowski solutions

$$v_k = \frac{1}{\sqrt{k}}e^{ik\eta}, \quad k|\eta| \gg 1. \tag{5.114}$$

The normalization is chosen in accordance with (5.76).

The late time regime $\eta \longrightarrow 0$: In this limit $\omega_k^2 \longrightarrow (m^2/H^2 - 2)1/\eta^2 < 0$ or equivalently

$$k^2 \ll \left(2 - \frac{m^2}{H^2}\right)\frac{1}{\eta^2}. \tag{5.115}$$

The differential equation becomes

$$v_k'' - \left(2 - \frac{m^2}{H^2}\right)\frac{1}{\eta^2}v_k = 0. \tag{5.116}$$

The solution is immediately given by $v_k = A|\eta|^{n_1} + B|\eta|^{n_2}$ with $n_{1,2} = \pm n + 1/2$. In the limit $\eta \longrightarrow 0$ the dominant solution is obviously associated with the exponent $-n + 1/2$. We have then

$$v_k \sim |\eta|^{\frac{1}{2}-n}, \quad k|\eta| \ll 1. \tag{5.117}$$

Any mode with momentum k is a wave with a comoving wave length $L \sim 1/k$ and a physical wave length $L_p = a(\eta)L$ and hence

$$k|\eta| = \frac{H^{-1}}{L_p}. \tag{5.118}$$

Thus modes with $k|\eta| \gg 1$ corresponds to modes with $L_p \ll H^{-1}$. These are the sub-horizon modes with physical wavelengths much shorter than the horizon scale and which are unaffected by gravity. Similarly, the modes with $k|\eta| \ll 1$ or equivalently $L_p \gg H^{-1}$ are the super-horizon modes with physical wavelengths much larger than the horizon scale. These are the modes which are affected by gravity.

A mode with momentum k which is sub-horizon at early times will become super-horizon at a later time η_k defined by the requirement that $L_p = H^{-1}$ or equivalently $k|\eta_k| = 1$. The time η_k is called the time of horizon crossing of the mode with momentum k.

The behavior $a(\eta) \longrightarrow 0$ when $\eta \longrightarrow -\infty$ allows us to pick a particular vacuum state known as the Bunch–Davies or the Euclidean vacuum. See for example [5]. The

Bunch–Davies vacuum is a de Sitter invariant state and is the initial state used in cosmology.

In the limit $\eta \longrightarrow -\infty$ the frequency approaches the flat space result, i.e. $\omega_k(\eta) \longrightarrow k$ and hence we can choose the vacuum state to be given by the Minkowski vacuum. More precisely, the frequency $\omega_k(\eta)$ is a slowly-varying function for some range of the conformal time η in the limit $\eta \longrightarrow -\infty$. This is called the adiabatic regime of $\omega_k(\eta)$ where it is also assumed that $\omega_k(\eta) > 0$. By applying the Minkowski vacuum prescription in the limit $\eta \longrightarrow -\infty$ we must have

$$v_k = \frac{N}{\sqrt{k}}e^{ik\eta}, \quad \eta \longrightarrow -\infty. \tag{5.119}$$

From the other hand, by using $J_n(s) = \sqrt{2/(\pi s)}\cos \lambda$, $Y_n(s) = \sqrt{2/(\pi s)}\sin \lambda$ with $\lambda = s - n\pi/2 - \pi/4$ we can compute the asymptotic behavior

$$v_k = \sqrt{\frac{2}{\pi}}[A \cos \lambda + B \sin \lambda], \quad \eta \longrightarrow -\infty. \tag{5.120}$$

By choosing $B = -iA$ and employing the normalization condition (5.112) we obtain

$$B = -iA, \quad A = \sqrt{\frac{\pi}{2k}}. \tag{5.121}$$

Thus we have the solution

$$v_k = \frac{1}{\sqrt{k}}e^{i(k\eta + \frac{n\pi}{2} + \frac{\pi}{4})}, \quad \eta \longrightarrow -\infty. \tag{5.122}$$

The Bunch–Davies vacuum corresponds to the choice $N = \exp(i\frac{n\pi}{2} + i\frac{\pi}{4})$. The full solution using this choice becomes

$$v_k = \sqrt{\frac{\pi|\eta|}{2}}[J_n(k|\eta|) - iY_n(k|\eta|)], \quad n = \sqrt{\frac{9}{4} - \frac{m^2}{H^2}}. \tag{5.123}$$

The mass density in FLRW spacetime was already computed in equation (5.86). We have

$$\rho = \frac{1}{4a^4}\int \frac{d^3k}{(2\pi)^3}\left[|v_k'|^2 + (k^2 + \frac{a''}{a} + a^2m^2)|v_k|^2 - \partial_\eta\left(\frac{a'}{a}|v_k|^2\right)\right]. \tag{5.124}$$

For de Sitter space we have $a = -1/(\eta H)$ and thus

$$\rho = \frac{\eta^4 H^4}{4}\int \frac{d^3k}{(2\pi)^3}\left[|v_k'|^2 + (k^2 + \frac{2}{\eta^2} + \frac{m^2}{H^2\eta^2})|v_k|^2 + \partial_\eta\left(\frac{1}{\eta}|v_k|^2\right)\right]. \tag{5.125}$$

For $m = 0$ we have the solutions

$$v_k = \sqrt{\frac{\pi|\eta|}{2}}\left[J_{\frac{3}{2}}(k|\eta|) - iY_{\frac{3}{2}}(k|\eta|)\right]. \tag{5.126}$$

We use the results $(x = k|\eta|)$

$$J_{3/2}(x) = \sqrt{\frac{2}{\pi x}} \left(\frac{\sin x}{x} - \cos x \right), \quad Y_{3/2}(x) = \sqrt{\frac{2}{\pi x}} \left(-\frac{\cos x}{x} - \sin x \right). \quad (5.127)$$

We obtain then

$$v_k = -\frac{i}{k^{\frac{3}{2}}} \frac{e^{ik\eta}}{\eta} - \frac{1}{k^{\frac{1}{2}}} e^{ik\eta}. \quad (5.128)$$

In other words,

$$|v_k|^2 = \frac{1}{k^3} \frac{1}{\eta^2} + \frac{1}{k}, \quad |v_k'|^2 = -\frac{1}{k} \frac{1}{\eta^2} + \frac{1}{k^3} \frac{1}{\eta^4} + k. \quad (5.129)$$

We obtain then (using also a hard cutoff Λ)

$$\rho = \frac{\eta^4 H^4}{4} \int \frac{d^3k}{(2\pi)^3} \left[2k + \frac{1}{k\eta^2} \right]$$

$$= \frac{\eta^4 H^4}{16\pi^2} \left(\Lambda^4 + \frac{\Lambda^2}{\eta^2} \right). \quad (5.130)$$

This goes to zero in the limit $\eta \longrightarrow 0$. However, if we take $\Lambda = \Lambda_0 a$ where Λ_0 is a proper momentum cutoff then the energy density becomes independent of time and we are back to the same problem. We get

$$\rho = \frac{1}{16\pi^2} (\Lambda_0^4 + H^2 \Lambda_0^2). \quad (5.131)$$

We observe that

$$\rho_{\text{de Sitter}} - \rho_{\text{Minkowski}} = \frac{H^2}{\Lambda_0^2} \frac{\Lambda_0^4}{16\pi^2}$$

$$= \frac{H^2}{\Lambda_0^2} \rho_{\text{Minkowski}}. \quad (5.132)$$

We take the value of the Hubble parameter at the current epoch as the value of the Hubble parameter of de Sitter space, viz

$$H = H_0 = \frac{7 \times 6.58}{3.09} 10^{-43} \text{ GeV}. \quad (5.133)$$

We get then

$$\rho_{\text{deSitter}} - \rho_{\text{Minkowski}} = 0.38(10^{-30})^4 . 0.22(10^{18} \text{ GeV})^4$$

$$= 0.084(10^{-12} \text{ GeV})^4. \quad (5.134)$$

5.4.4 QFT on curved background with a cutoff

In [23] a proposal for QFTs on curved backgrounds with a plausible cutoff is put forward.

5.4.5 The conformal limit $\xi \longrightarrow 1/6$

The mode functions χ_k satisfy

$$\chi_k'' + \omega_k^2(\eta)\chi_k = 0, \quad \omega_k^2 = k^2 + m^2 a^2 - (1 - 6\xi)\frac{a''}{a}. \tag{5.135}$$

$$V(\chi_k \chi_k^{*'} - \chi_k^* \chi_k') = i. \tag{5.136}$$

We will consider in this section $m^2 = 0$. We assume now that the Universe is Minkowski in the past $\eta \longrightarrow -\infty$. In other words, in the limit $\eta \longrightarrow -\infty$ the frequency ω_k tends to $\bar{\omega}_k = \sqrt{k^2}$. The corresponding mode function is therefore

$$\chi_k = \chi_k^{(in)} = \frac{1}{\sqrt{2V\bar{\omega}_k}}e^{-i\bar{\omega}_k\eta}. \tag{5.137}$$

We will also assume that the Universe is Minkowski in the future $\eta \longrightarrow +\infty$. The frequency in the limit $\eta \longrightarrow +\infty$ is again given by $\bar{\omega}_k = \sqrt{k^2}$. The corresponding mode function is therefore

$$\chi_k = \chi_k^{(out)} = \frac{\alpha_k}{\sqrt{2V\bar{\omega}_k}}e^{-i\bar{\omega}_k\eta} + \frac{\beta_k}{\sqrt{2V\bar{\omega}_k}}e^{i\bar{\omega}_k\eta}. \tag{5.138}$$

We determine α_k and β_k from solving the equation of motion (5.135) with the initial condition (5.137). We remark that

$$\chi_k^{(out)} = \alpha_k \chi_k^{(in)} + \beta_k \chi_k^{(in)*}. \tag{5.139}$$

We imagine that the out state is the limit $\eta \longrightarrow +\infty$ of some v mode function while the in state is the limit $\eta \longrightarrow -\infty$ of some u mode function. More precisely we are assuming that

$$\begin{aligned} u_i &\longrightarrow \chi_i^{(in)}, \quad \eta \longrightarrow -\infty \\ v_i &\longrightarrow \chi_i^{(out)}, \quad \eta \longrightarrow +\infty. \end{aligned} \tag{5.140}$$

The relation between the u and the v mode functions is given in terms of Bogolubov coefficients by equation (6.189). By comparing with the above relation (5.139) we deduce that

$$\alpha_{ij} = \alpha_i \delta_{ij}, \quad \beta_{ij} = \beta_i \delta_{ij}. \tag{5.141}$$

Let $N_u = \sum_k \hat{a}_k^+ \hat{a}_k$ be the number operator corresponding to the u modes. If $|0_u\rangle$ is the vacuum state corresponding to the u modes then $\langle 0_u|N_u|0_u\rangle=0$. The number of particles created by the gravitational field in the limit $\eta \longrightarrow +\infty$ is precisely

$\langle 0_v | N_u | 0_v \rangle$ where $|0_v\rangle$ is the vacuum state corresponding to the v modes. The number density of created particles is then given by

$$\mathcal{N} = \frac{\langle 0_v | N_u | 0_v \rangle}{V} = \int \frac{d^3 k}{(2\pi)^3} |\beta_k|^2. \tag{5.142}$$

The corresponding energy density is

$$\rho = \int \frac{d^3 k}{(2\pi)^3} \bar{\omega}_k |\beta_k|^2. \tag{5.143}$$

The initial differential equation (5.135) can be rewritten as

$$\chi_k'' + \bar{\omega}_k^2 \chi_k = j_k(\eta), \quad j_k(\eta) = (1 - 6\xi) \frac{a''}{a} \chi_k. \tag{5.144}$$

We can write down immediately the solution as

$$\begin{aligned}
\chi_k &= \chi_k^{(in)} + \frac{1}{\bar{\omega}_k} \int_{-\infty}^{\eta} d\eta' \sin \bar{\omega}_k (\eta - \eta') j_k(\eta') \\
&= \chi_k^{(in)} + \frac{1 - 6\xi}{\bar{\omega}_k} \int_{-\infty}^{\eta} d\eta' \frac{a''(\eta')}{a(\eta')} \sin \bar{\omega}_k (\eta - \eta') \chi_k(\eta').
\end{aligned} \tag{5.145}$$

To lowest order in $1 - 6\xi$ this solution becomes

$$\chi_k = \chi_k^{(in)} + \frac{1 - 6\xi}{\bar{\omega}_k} \int_{-\infty}^{\eta} d\eta' \frac{a''(\eta')}{a(\eta')} \sin \bar{\omega}_k (\eta - \eta') \chi_k^{(in)}(\eta'). \tag{5.146}$$

From this formula we obtain immediately

$$\chi_k^{(out)} = \chi_k^{(in)} + \frac{1 - 6\xi}{\bar{\omega}_k} \int_{-\infty}^{+\infty} d\eta' \frac{a''(\eta')}{a(\eta')} \sin \bar{\omega}_k (\eta - \eta') \chi_k^{(in)}(\eta'). \tag{5.147}$$

By comparing with (5.139) and using (5.137) we get after few more lines (with $a^2 R = 6a''/a$)

$$\alpha_k = 1 + \frac{i}{2\bar{\omega}_k}\left(\frac{1}{6} - \xi\right) \int_{-\infty}^{+\infty} d\eta' a^2(\eta') R(\eta'), \quad \beta_k = -\frac{i}{2\bar{\omega}_k}\left(\frac{1}{6} - \xi\right) \int_{-\infty}^{+\infty} d\eta' a^2(\eta') R(\eta') e^{-2i\bar{\omega}_k \eta'}. \tag{5.148}$$

The number density is given by

$$\begin{aligned}
\mathcal{N} &= \frac{1}{4}\left(\frac{1}{6} - \xi\right)^2 \int_{-\infty}^{+\infty} d\eta_1 \int_{-\infty}^{+\infty} d\eta_2 a^2(\eta_1) R(\eta_1) a^2(\eta_2) R(\eta_2) \int \frac{d^3 k}{(2\pi)^3} \frac{1}{\bar{\omega}_k^2} e^{-2i\bar{\omega}_k(\eta_1 - \eta_2)} \\
&= \frac{1}{4}\left(\frac{1}{6} - \xi\right)^2 \int_{-\infty}^{+\infty} d\eta_1 \int_{-\infty}^{+\infty} d\eta_2 a^2(\eta_1) R(\eta_1) a^2(\eta_2) R(\eta_2) \frac{1}{2\pi} \int_0^{\infty} \frac{dk}{2\pi} e^{-ik(\eta_1 - \eta_2)} \\
&= \frac{1}{4}\left(\frac{1}{6} - \xi\right)^2 \int_{-\infty}^{+\infty} d\eta_1 \int_{-\infty}^{+\infty} d\eta_2 a^2(\eta_1) R(\eta_1) a^2(\eta_2) R(\eta_2) \frac{1}{4\pi} \int_0^{\infty} \frac{dk}{2\pi} e^{-ik(\eta_1 - \eta_2)} \\
&= \frac{1}{16\pi}\left(\frac{1}{6} - \xi\right)^2 \int_{-\infty}^{+\infty} d\eta\, a^4(\eta) R^2(\eta).
\end{aligned} \tag{5.149}$$

The energy density is given by (with the assumption that $a^2(\eta)R(\eta) \longrightarrow 0$ when $\eta \longrightarrow \pm \infty$)

$$
= \frac{1}{4}\left(\frac{1}{6} - \xi\right)^2 \int_{-\infty}^{+\infty} d\eta_1 \int_{-\infty}^{+\infty} d\eta_2 a^2(\eta_1)R(\eta_1)a^2(\eta_2)R(\eta_2) \int \frac{d^3k}{(2\pi)^3} \frac{1}{\bar{\omega}_k} e^{-2i\bar{\omega}_k(\eta_1 - \eta_2)}
$$

$$
= \frac{1}{4}\left(\frac{1}{6} - \xi\right)^2 \int_{-\infty}^{+\infty} d\eta_1 \int_{-\infty}^{+\infty} d\eta_2 a^2(\eta_1)R(\eta_1)a^2(\eta_2)R(\eta_2) \frac{1}{8\pi^2} \int_0^\infty kdk e^{-ik(\eta_1 - \eta_2)}
$$

$$
= \frac{1}{4}\left(\frac{1}{6} - \xi\right)^2 \int_{-\infty}^{+\infty} d\eta_1 \int_{-\infty}^{+\infty} d\eta_2 a^2(\eta_1)R(\eta_1)a^2(\eta_2)R(\eta_2) \frac{1}{8\pi^2} \frac{d^2}{d\eta_1 d\eta_2} \int_0^\infty \frac{dk}{k} e^{-ik(\eta_1 - \eta_2)}
$$

$$
= \frac{1}{4}\left(\frac{1}{6} - \xi\right)^2 \int_{-\infty}^{+\infty} d\eta_1 \int_{-\infty}^{+\infty} d\eta_2 \frac{d}{d\eta_1}(a^2(\eta_1)R(\eta_1))\frac{d}{d\eta_2}(a^2(\eta_2)R(\eta_2))\frac{1}{2\pi} \int_0^\infty \frac{dk}{2\pi} \frac{1}{2k} e^{-ik(\eta_1 - \eta_2)}.
$$

(5.150)

The last factor is precisley one half the Feynamn propagator in $1 + 1$ dimension for $r = 0$ (see equation (4) of [24]). We have then

$$
\rho = \frac{1}{4}\left(\frac{1}{6} - \xi\right)^2 \int_{-\infty}^{+\infty} d\eta_1 \int_{-\infty}^{+\infty} d\eta_2 \frac{d}{d\eta_1}(a^2(\eta_1)R(\eta_1))\frac{d}{d\eta_2}(a^2(\eta_2)R(\eta_2))\frac{1}{2\pi}\frac{-1}{4\pi}\ln|\eta_1 - \eta_2|
$$

$$
= -\frac{1}{32\pi^2}\left(\frac{1}{6} - \xi\right)^2 \int_{-\infty}^{+\infty} d\eta_1 \int_{-\infty}^{+\infty} d\eta_2 \frac{d}{d\eta_1}(a^2(\eta_1)R(\eta_1))\frac{d}{d\eta_2}(a^2(\eta_2)R(\eta_2))\ln|\eta_1 - \eta_2|.
$$

(5.151)

At the end of inflation the Universe transits from a de Sitter spacetime (which is asymptotically static in the infinite past) to a radiation-dominated Robertson–Walker universe (which is asymptotically flat in the infinite future) in a very short time interval. Let us assume that the transition occurs abruptly at a time $\eta_0 < 0$. In de Sitter space ($\eta < \eta_0$) we have $a = -1/(\eta H)$ and $R = 12H^2$. In the radiation-dominated phase ($\eta > \eta_0$) we may assume that $R = 0$. We get immediately

$$
\mathcal{N} = \frac{1}{16\pi}\left(\frac{1}{6} - \xi\right)^2 \int_{-\infty}^{\eta_0} d\eta a^4(\eta)R^2(\eta)
$$

$$
= \frac{H^3}{12\pi}(1 - 6\xi)^2 a^3(\eta_0).
$$

(5.152)

This is the number density of created particles (via gravitational interaction) just after the transition, i.e. during reheating.

To compute the energy density we will assume that the transition from de sitter spacetime to radiation-dominated spacetime is smoother given by the scale factor

$$
a^2(\eta) = f(\eta H). \tag{5.153}
$$

$$
f = \frac{1}{\eta^2 H^2}, \quad \eta < -H^{-1}
$$

$$
= a_0 + a_1 H\eta + a_2 H^2\eta^2 + a_3 H^3\eta^3, \quad -H^{-1} < \eta < (x_0 - 1)H^{-1}
$$

$$
= b_0(H\eta + b_1)^2, \quad \eta > (x_0 - 1)H^{-1}.
$$

(5.154)

In this model the time $\eta = -H^{-1}$ corresponding to $t = 0$ marks the end of the inflationary (de Sitter) phase and the transition to radiation-dominated phase occurs on a time scale given by $\Delta \eta = H^{-1} x_0$. By requiring that f, f' and f'' are continuous at $\eta = -H^{-1}$ and $\eta = (x_0 - 1)H^{-1}$ we can determine the coefficients a_i and b_i uniquely. We compute immediately

$$a^2 R = 3H^2 V, \quad V = f^{-2}\left[f'' f - \frac{1}{2}(f')^2\right].$$ (5.155)

We can then compute in a straightforward manner

$$V = \frac{4}{x^2}, \quad x < -1$$

$$\simeq -\frac{4}{x_0}, \quad -1 < x < x_0 - 1, \quad x_0 \ll 1$$ (5.156)

$$= 0, \quad x > x_0 - 1.$$

The energy density is then given by

$$\rho = -\frac{H^4}{128\pi^2}(1 - 6\xi)^2 \int_{-\infty}^{x_0 - 1} dx_1 \int_{-\infty}^{x_0 - 1} dx_2 \, V'(x_1) V'(x_2) \ln \frac{|x_1 - x_2|}{H}$$

$$= -\frac{H^4}{128\pi^2}(1 - 6\xi)^2 . 16 \ln x_0$$ (5.157)

$$= -\frac{H^4}{8\pi^2}(1 - 6\xi)^2 \ln x_0.$$

In the above model we have chosen the transition time to be $\eta = -H^{-1}$ and thus $a = -1/(\eta H) = +1$ and as a consequence $\Delta \eta = -H\eta \Delta t = \Delta t$. From the other hand, the transition from de Sitter spacetime to radiation-dominated phase occurs on a time scale given by $\Delta \eta = H^{-1} x_0$. From these two facts we obtain $x_0 = H\Delta t$ and hence the energy density becomes

$$\rho = -\frac{H^4}{8\pi^2}(1 - 6\xi)^2 \ln H\Delta t.$$ (5.158)

This is the energy density of the created particles after the end of inflation. The factor $1 - 6\xi$ is small, whereas the factor $\ln H\Delta t$ is large and it is not obvious how they should balance without an extra input.

5.5 Is vacuum energy real?

The Casimir force is a fundamental measurable quantum effect arising from a non-zero vacuum energy effect and its treatment can be found in any book on QFT. See for example [25]. Here, we will mostly follow the detailed careful calculations in [7, 26, 27] and [8, 28].

5.5.1 The Casimir force

We consider two large and perfectly conducting plates of surface area A at a distance L apart with $\sqrt{A} \gg L$ so that we can ignore edge contributions. The plates are in the xy plane at $x = 0$ and $x = L$. In the volume AL the electromagnetic standing waves take the form

$$\psi_n(t, x, y, z) = e^{-i\omega_n t} e^{ik_x x + ik_y y} \sin k_n z. \tag{5.159}$$

They satisfy the Dirichlet boundary conditions

$$\psi_n \mid_{z=0} = \psi_n \mid_{z=L} = 0. \tag{5.160}$$

Thus we must have

$$k_n = \frac{n\pi}{L}, \quad n = 1, 2, \ldots. \tag{5.161}$$

$$\omega_n = \sqrt{k_x^2 + k_y^2 + \frac{n^2 \pi^2}{L^2}}. \tag{5.162}$$

These modes are transverse and thus each value of n is associated with two degrees of freedom. There is also the possibility of

$$k_n = 0. \tag{5.163}$$

In this case there is a corresponding single degree of freedom.

The zero point energy of the electromagnetic field between the plates is

$$
\begin{aligned}
E &= \frac{1}{2} \sum_n \omega_n \\
&= \frac{1}{2} A \int \frac{d^2 k}{(2\pi)^2} \left[k + 2 \sum_{n=1}^{\infty} \left(k^2 + \frac{n^2 \pi^2}{L^2} \right)^{1/2} \right].
\end{aligned}
\tag{5.164}
$$

The zero point energy of the electromagnetic field in the same volume in the absence of the plates is

$$
\begin{aligned}
E_0 &= \frac{1}{2} \sum_n \omega_n \\
&= \frac{1}{2} A \int \frac{d^2 k}{(2\pi)^2} \left[2L \int \frac{dk_n}{2\pi} (k^2 + k_n^2)^{1/2} \right].
\end{aligned}
\tag{5.165}
$$

After the change of variable $k = n\pi/L$ we obtain

$$E_0 = \frac{1}{2} A \int \frac{d^2 k}{(2\pi)^2} \left[2 \int_0^{\infty} dn \left(k^2 + \frac{n^2 \pi^2}{L^2} \right)^{1/2} \right]. \tag{5.166}$$

We have then

$$\mathcal{E} = \frac{E - E_0}{A} = \int \frac{d^2 k}{(2\pi)^2} \left[\frac{1}{2} k + \sum_{n=1}^{\infty} \left(k^2 + \frac{n^2 \pi^2}{L^2} \right)^{1/2} - \int_0^{\infty} dn \left(k^2 + \frac{n^2 \pi^2}{L^2} \right)^{1/2} \right]. \tag{5.167}$$

This is obviously a UV divergent quantity. We regularize this energy density by introducing a cutoff function $f_\Lambda(k)$ which is equal to 1 for $k \ll \Lambda$ and 0 for $k \gg \Lambda$. We have then (with the change of variables $k = \pi x/L$ and $x^2 = t$)

$$
\begin{aligned}
\mathcal{E}_\Lambda = \int \frac{d^2 k}{(2\pi)^2} & \left[\frac{1}{2} f_\Lambda(k) k + \sum_{n=1}^{\infty} f_\Lambda\left(\sqrt{k^2 + \frac{n^2\pi^2}{L^2}} \right)\left(k^2 + \frac{n^2\pi^2}{L^2} \right)^{1/2} \right. \\
& \left. - \int_0^{\infty} dn f_\Lambda\left(\sqrt{k^2 + \frac{n^2\pi^2}{L^2}} \right)\left(k^2 + \frac{n^2\pi^2}{L^2} \right)^{1/2} \right] \\
= \frac{\pi^2}{4L^3} \int dt & \left[\frac{1}{2} f_\Lambda\left(\frac{\pi}{L}\sqrt{t} \right) t^{1/2} + \sum_{n=1}^{\infty} f_\Lambda\left(\frac{\pi}{L}\sqrt{t + n^2} \right)(t + n^2)^{1/2} \right. \\
& \left. - \int_0^{\infty} dn f_\Lambda\left(\frac{\pi}{L}\sqrt{t + n^2} \right)(t + n^2)^{1/2} \right].
\end{aligned}
\tag{5.168}
$$

This is an absolutely convergent quantity and thus we can exchange the sums and the integrals. We obtain

$$
\mathcal{E}_\Lambda = \frac{\pi^2}{4L^3}\left[\frac{1}{2} F(0) + F(1) + F(2) \cdots - \int_0^{\infty} dn F(n) \right].
\tag{5.169}
$$

The function $F(n)$ is defined by

$$
F(n) = \int_0^{\infty} dt f_\Lambda\left(\frac{\pi}{L}\sqrt{t + n^2} \right)(t + n^2)^{1/2}.
\tag{5.170}
$$

Since $f(k) \longrightarrow 0$ when $k \longrightarrow \infty$ we have $F(n) \longrightarrow 0$ when $n \longrightarrow \infty$.
   We use the Euler–MacLaurin formula

$$
\frac{1}{2} F(0) + F(1) + F(2) \cdots - \int_0^{\infty} dn F(n) = -\frac{1}{2!} B_2 F'(0) - \frac{1}{4!} B_4 F'''(0) + \cdots.
\tag{5.171}
$$

The Bernoulli numbers B_i are defined by

$$
\frac{y}{e^y - 1} = \sum_{i=0}^{\infty} B_i \frac{y^i}{i!}.
\tag{5.172}
$$

For example

$$
B_0 = \frac{1}{6}, \quad B_4 = -\frac{1}{30}, \quad \text{etc.}
\tag{5.173}
$$

Thus

$$
\mathcal{E}_\Lambda = \frac{\pi^2}{4L^3}\left[-\frac{1}{12} F'(0) + \frac{1}{720} F'''(0) + \cdots \right].
\tag{5.174}
$$

We can write

$$F(n) = \int_{n^2}^{\infty} dt f_\Lambda\left(\frac{\pi}{L}\sqrt{t}\right)(t)^{1/2}. \tag{5.175}$$

We assume that $f(0) = 1$ while all its derivatives are zero at $n = 0$. Thus

$$F'(n) = -\int_{n^2}^{n^2 + 2n\delta n} dt f_\Lambda\left(\frac{\pi}{L}\sqrt{t}\right)(t)^{1/2} = -2n^2 f_\Lambda\left(\frac{\pi}{L}n\right) \Rightarrow F'(0) = 0. \tag{5.176}$$

$$F''(n) = -4n f_\Lambda\left(\frac{\pi}{L}n\right) - \frac{2\pi}{L}n^2 f'_\Lambda\left(\frac{\pi}{L}n\right) \Rightarrow F''(0) = 0. \tag{5.177}$$

$$F'''(n) = -4f_\Lambda\left(\frac{\pi}{L}n\right) - \frac{8\pi}{L}n f'_\Lambda\left(\frac{\pi}{L}n\right) - \frac{2\pi^2}{L^2}n^2 f''_\Lambda\left(\frac{\pi}{L}n\right) \Rightarrow F'''(0) = -4. \tag{5.178}$$

We can check that all higher derivatives of F are actually 0

$$\mathcal{E}_\Lambda = \frac{\pi^2}{4L^3}\left[-\frac{4}{720}\right] = -\frac{\pi^2}{720L^3}. \tag{5.179}$$

This is the Casimir energy. It corresponds to an attractive force which is the famous Casimir force.

5.5.2 The Dirichlet propagator

We define the propagator by

$$D_F(x, x') = \langle 0|T\hat{\phi}(x)\hat{\phi}(x')|0\rangle. \tag{5.180}$$

It satisfies the inhomogeneous Klein–Gordon equation

$$(\partial_t^2 - \partial_i^2)D_F(x, x') = i\delta^4(x - x'). \tag{5.181}$$

We introduce Fourier transform in the time direction by

$$D_F(\omega, \vec{x}, \vec{x}') = \int dt e^{-i\omega(t-t')}D_F(x, x'), \quad D_F(x, x') = \int \frac{d\omega}{2\pi} e^{i\omega(t-t')}D_F(\omega, \vec{x}, \vec{x}'). \tag{5.182}$$

We have

$$(\partial_i^2 + \omega^2)D_F(\omega, \vec{x}, \vec{x}') = -i\delta^3(\vec{x} - \vec{x}'). \tag{5.183}$$

We expand the reduced Green's function $D_F(\omega, \vec{x}, \vec{x}')$ as

$$D_F(\omega, \vec{x}, \vec{x}') = -i\sum_n \frac{\phi_n(\vec{x})\phi_n^*(\vec{x}')}{\omega^2 - k_n^2}. \tag{5.184}$$

The eigenfunctions $\phi_n(\vec{x})$ satisfy

$$\begin{aligned}
\partial_i^2\phi_n(\vec{x}) &= -k_n^2\phi_n(\vec{x}) \\
\delta^3(\vec{x} - \vec{x}') &= \sum_n \phi_n(\vec{x})\phi_n^*(\vec{x}').
\end{aligned} \tag{5.185}$$

In infinite space we have

$$\phi_i(\vec{x}) \longrightarrow \phi_{\vec{k}}(\vec{x}) = e^{-i\vec{k}\vec{x}}, \quad \sum_i \longrightarrow \int \frac{d^3k}{(2\pi)^3}. \tag{5.186}$$

Thus

$$D_F(\omega, \vec{x}, \vec{x}') = i \int \frac{d^3k}{(2\pi)^3} \frac{e^{-i\vec{k}(\vec{x}-\vec{x}')}}{\vec{k}^2 - \omega^2}. \tag{5.187}$$

We can compute the closed form

$$D_F(\omega, \vec{x}, \vec{x}') = \frac{i}{4\pi} \frac{e^{i\omega|\vec{x}-\vec{x}'|}}{|\vec{x}-\vec{x}'|}. \tag{5.188}$$

Equivalently, we have

$$D_F(x, x') = i \int \frac{d^4k}{(2\pi)^4} \frac{e^{-ik(x-x')}}{k^2}. \tag{5.189}$$

Let us remind ourselves with few more results. We have (with $\omega_k = |\vec{k}|$)

$$D_F(x, x') = \int \frac{d^3k}{(2\pi)^3} \frac{1}{2\omega_k} e^{-ik(x-x')}. \tag{5.190}$$

Recall that $k(x - x') = -k^0(x^0 - x0') + \vec{k}(\vec{x}-\vec{x}')$. After Wick rotation in which $x^0 \longrightarrow -ix_4$ and $k^0 \longrightarrow -ik_4$ we obtain $k(x - x') = k_4(x_4 - x_4') + \vec{k}(\vec{x}-\vec{x}')$. The above integral becomes then

$$D_F(x, x') = \int \frac{d^3k}{(2\pi)^3} \frac{1}{2\omega_k} e^{-i\left(k_4(x_4 - x_4') - \vec{k}(\vec{x}-\vec{x}')\right)}$$
$$= \frac{1}{4\pi^2} \frac{1}{(x - x')^2}. \tag{5.191}$$

We consider now the case of parallel plates separated by a distance L. The plates are in the xy plane. We impose now different boundary conditions on the field by assuming that $\hat{\phi}$ is confined in the z direction between the two plates at $z = 0$ and $z = L$. Thus the field must vanish at these two plates, viz

$$\hat{\phi}\mid_{z=0} = \hat{\phi}\mid_{z=L} = 0. \tag{5.192}$$

As a consequence, the plane wave e^{ik_3z} will be replaced with the standing wave $\sin k_3 z$ where the momentum in the z direction is quantized as

$$k_3 = \frac{n\pi}{L}, \quad n \in Z^+. \tag{5.193}$$

Thus the frequency ω_k becomes

$$\omega_n = \sqrt{k_1^2 + k_2^2 + \left(\frac{n\pi}{L}\right)^2}. \tag{5.194}$$

We will think of the propagator (5.191) as the electrostatic potential (in four dimensions) generated at point y from a unit charge at point x, viz

$$V \equiv D_F(x, x') = \frac{1}{4\pi^2} \frac{1}{(x - x')^2}. \tag{5.195}$$

We will find the propagator between parallel plates starting from this potential using the method of images. It is obvious that this propagator must satisfy

$$D_F(x, x') = 0, \quad z = 0, L \text{ and } z' = 0, L. \tag{5.196}$$

Instead of the two plates at $x = 0$ and $x = L$ we consider image charges (always with respect to the two plates) placed such that the two plates remain grounded. First we place an image charge -1 at $(x, y, -z)$ which makes the potential at the plate $z = 0$ zero. The image of the charge at $(x, y, -z)$ with respect to the plane at $z = L$ is a charge $+1$ at $(x, y, z + 2L)$. This last charge has an image with respect to $z = 0$ equal -1 at $(x, y, -z - 2L)$ which in turn has an image with respect to $z = L$ equal $+1$ at $(x, y, z + 4L)$. This process is to be continued indefinitely. We have then added the following image charges

$$q = +1, \quad (x, y, z + 2nL), \quad n = 0, 1, 2, \ldots \tag{5.197}$$

$$q = -1, \quad (x, y, -z - 2nL), \quad n = 0, 1, 2, \ldots \tag{5.198}$$

The way we did this we are guaranteed that the total potential at $z = 0$ is 0. The contribution of the added image charges to the plate $z = L$ is also zero but this plate is still not balanced properly precisely because of the original charge at (x, y, z).

The image charge of the original charge with respect to the plate at $z = L$ is a charge -1 at $(x, y, 2L - z)$ which has an image with respect to $z = 0$ equal $+1$ at $(x, y, -2L + z)$. This last image has an image with respect to $z = L$ equal -1 at $(x, y, 4L - z)$. This process is to be continued indefinitely with added charges given by

$$q = +1, \quad (x, y, z + 2nL), \quad n = -1, -2, \ldots \tag{5.199}$$

$$q = -1, \quad (x, y, -z - 2nL), \quad n = -1, -2, \ldots \tag{5.200}$$

By the superposition principle the total potential is the sum of the individual potentials. We get immediately

$$V \equiv D_F(x, x') = \frac{1}{4\pi^2} \sum_{n=-\infty}^{+\infty} \left[\frac{1}{(x - x' - 2nLe_3)^2} - \frac{1}{(x - x' - 2(nL + z)e_3)^2} \right]. \tag{5.201}$$

This satisfies the boundary conditions (5.196). By the uniqueness theorem this solution must therefore be the desired propagator. At this point we can undo the Wick rotation and return to Minkowski spacetime.

5.5.3 Another derivation using the energy–momentum tensor

The stress–energy–momentum tensor in flat space with minimal coupling $\xi = 0$ and $m = 0$ is given by

$$T_{\mu\nu} = \partial_\mu\phi\partial_\nu\phi - \frac{1}{2}\eta_{\mu\nu}\partial_\alpha\phi\partial^\alpha\phi. \tag{5.202}$$

The stress–energy–momentum tensor in flat space with conformal coupling $\xi = 1/6$ and $m = 0$ is given by

$$T_{\mu\nu} = \frac{2}{3}\partial_\mu\phi\partial_\nu\phi + \frac{1}{6}\eta_{\mu\nu}\partial_\alpha\phi\partial^\alpha\phi + \frac{1}{3}\phi\partial_\mu\partial_\nu\phi. \tag{5.203}$$

This tensor is traceless, i.e. $T^\mu_\mu = 0$ which reflects the fact that the theory is conformal. This tensor is known as the new improved stress–energy–momentum tensor.

In the quantum theory $T_{\mu\nu}$ becomes an operator $\hat{T}_{\mu\nu}$ and we are interested in the expectation value of $\hat{T}_{\mu\nu}$ in the vacuum state $\langle 0|\hat{T}_{\mu\nu}|0\rangle$. We are of course interested in the energy density which is equal to $\langle 0|\hat{T}_{00}|0\rangle$ in flat spacetime. We compute (using the Klein–Gordon equation $\partial_\mu\partial^\mu\hat{\phi} = 0$)

$$\begin{aligned}
\langle 0|\hat{T}_{00}|0\rangle &= \frac{2}{3}\langle 0|\partial_0\hat{\phi}\partial_0\hat{\phi}|0\rangle - \frac{1}{6}\langle 0|\partial_\alpha\hat{\phi}\partial^\alpha\hat{\phi}|0\rangle + \frac{1}{3}\langle 0|\hat{\phi}\partial_\mu\partial_\nu\hat{\phi}|0\rangle \\
&= \frac{5}{6}\langle 0|\partial_0\hat{\phi}\partial_0\hat{\phi}|0\rangle - \frac{1}{6}\langle 0|\partial_i\hat{\phi}\partial_i\hat{\phi}|0\rangle + \frac{1}{3}\langle 0|\hat{\phi}\partial_0^2\hat{\phi}|0\rangle \\
&= \frac{5}{6}\langle 0|\partial_0\hat{\phi}\partial_0\hat{\phi}|0\rangle - \frac{1}{6}\langle 0|\partial_i\hat{\phi}\partial_i\hat{\phi}|0\rangle + \frac{1}{3}\langle 0|\hat{\phi}\partial_i^2\hat{\phi}|0\rangle \\
&= \frac{5}{6}\langle 0|\partial_0\hat{\phi}\partial_0\hat{\phi}|0\rangle + \frac{1}{6}\langle 0|\partial_i\hat{\phi}\partial_i\hat{\phi}|0\rangle.
\end{aligned} \tag{5.204}$$

We regularize this object by putting the two fields at different points x and y as follows

$$\begin{aligned}
\langle 0|\hat{T}_{00}|0\rangle &= \frac{5}{6}\langle 0|\partial_0\hat{\phi}(x)\partial_0\hat{\phi}(y)|0\rangle + \frac{1}{6}\langle 0|\partial_i\hat{\phi}(x)\partial_i\hat{\phi}(y)|0\rangle \\
&= \left[\frac{5}{6}\partial_0^x\partial_0^y + \frac{1}{6}\partial_i^x\partial_i^y\right]\langle 0|\hat{\phi}(x)\hat{\phi}(y)|0\rangle.
\end{aligned} \tag{5.205}$$

Similarly, we obtain with minimal coupling the result

$$\langle 0|\hat{T}_{00}|0\rangle = \left[\frac{1}{2}\partial_0^x\partial_0^y + \frac{1}{2}\partial_i^x\partial_i^y\right]\langle 0|\hat{\phi}(x)\hat{\phi}(y)|0\rangle. \tag{5.206}$$

In infinite space the scalar field operator has the expansion (with $w_k = |k|$, $[\bar{a}_k, \bar{a}_{k'}^+] = V\delta_{k,k'}$, etc)

$$\hat{\phi} = \int \frac{d^3k}{(2\pi)^3}\frac{1}{\sqrt{2\omega_k}}\left(\bar{a}_k e^{-i\omega_k t + i\vec{k}\vec{x}} + \bar{a}_k^+ e^{i\omega_k t - i\vec{k}\vec{x}}\right). \tag{5.207}$$

In the space between parallel plates the field can then be expanded as

$$\hat{\phi} = \sqrt{\frac{2}{L}} \sum_n \int \frac{d^2k}{(2\pi)^2} \frac{1}{\sqrt{2\omega_n}} \sin\frac{n\pi}{L}z \left(\bar{a}_{k,n} e^{-i\omega_n t + i\vec{k}\vec{x}} + \bar{a}^+_{k,n} e^{i\omega_n t - i\vec{k}\vec{x}} \right). \qquad (5.208)$$

The creation and annihilation operators satisfy the commutation relations $[\bar{a}_{k,n}, \bar{a}^+_{p,m}] = \delta_{nm}(2\pi)^2\delta^2(k - p)$, etc.

We use the result

$$D_F(x - y) = \langle 0|T\hat{\phi}(x)\hat{\phi}(y)|0\rangle$$

$$= \frac{1}{4\pi^2} \sum_{n=-\infty}^{+\infty} \left[\frac{1}{(x - y - 2nLe_3)^2} - \frac{1}{(x - y - 2(nL + x^3)e_3)^2} \right]. \qquad (5.209)$$

We introduce (with $a = -nL, -(nL + x^3)$)

$$\mathcal{D}_a = (x - y + 2ae_3)^2 = -(x^0 - y^0)^2 + (x^1 - y^1)^2 + (x^2 - y^2)^2 + (x^3 - y^3 + 2a)^2. \qquad (5.210)$$

We then compute

$$\partial_0^x \partial_0^y \frac{1}{\mathcal{D}_a} = -\frac{2}{\mathcal{D}_a^2} - 8(x^0 - y^0)^2 \frac{1}{\mathcal{D}_a^3}. \qquad (5.211)$$

$$\partial_i^x \partial_i^y \frac{1}{\mathcal{D}_a} = \frac{2}{\mathcal{D}_a^2} - 8(x^i - y^i)^2 \frac{1}{\mathcal{D}_a^3}, \quad i = 1, 2. \qquad (5.212)$$

$$\partial_3^x \partial_3^y \frac{1}{\mathcal{D}_{-nL}} = \frac{2}{\mathcal{D}_{-nL}^2} - 8(x^3 - y^3 + 2nL)^2 \frac{1}{\mathcal{D}_{-nL}^3}. \qquad (5.213)$$

$$\partial_3^x \partial_3^y \frac{1}{\mathcal{D}_{-(nL+x^3)}} = -\frac{2}{\mathcal{D}_{-(nL+x^3)}^2} + 8(x^3 + y^3 + 2nL)^2 \frac{1}{\mathcal{D}_{-(nL+x^3)}^3}. \qquad (5.214)$$

We can immediately compute

$$\langle 0|\hat{T}_{00}|0\rangle_{\xi=0}^L = \frac{1}{4\pi^2} \sum_{n=-\infty}^{+\infty} \left[\frac{2}{\mathcal{D}_{-nL}^2} - 4(x^3 - y^3 + 2nL)^2\frac{1}{\mathcal{D}_{-nL}^3} - 4(x^3 + y^3 + 2nL)^2\frac{1}{\mathcal{D}_{-(nL+x^3)}^3} \right] \qquad (5.215)$$

$$\longrightarrow -\frac{1}{32\pi^2} \sum_{n=-\infty}^{+\infty} \frac{1}{(nL)^4} - \frac{1}{16\pi^2} \sum_{n=-\infty}^{+\infty} \frac{1}{(nL + x^3)^4}.$$

This is still divergent. The divergence comes from the original charge corresponding to $n = 0$ in the first two terms in the limit $x \longrightarrow y$. All other terms coming from image charges are finite.

The same quantity evaluated in infinite space is

$$\langle 0|\hat{T}_{00}|0\rangle_{\xi=0}^\infty = \int \frac{d^3k}{(2\pi)^3} \frac{\omega_k}{2} e^{-ik(x-y)}. \qquad (5.216)$$

This is divergent and the divergence must be the same divergence as in the case of parallel plates in the limit $L \longrightarrow \infty$, viz

$$\langle 0|\hat{T}_{00}|0\rangle_{\xi=0}^{\infty} = -\frac{1}{32\pi^2}\frac{1}{(nL)^4} \mid_{n=0}. \tag{5.217}$$

Hence the normal ordered vacuum expectation value of the energy–momentum-tensor is given by

$$\langle 0|\hat{T}_{00}|0\rangle_{\xi=0}^{L} - \langle 0|\hat{T}_{00}|0\rangle_{\xi=0}^{\infty} = -\frac{1}{32\pi^2}\sum_{n\neq 0}\frac{1}{(nL)^4} - \frac{1}{16\pi^2}\sum_{n=-\infty}^{+\infty}\frac{1}{(nL + x^3)^4}. \tag{5.218}$$

This is still divergent at the boundaries $x^3 \longrightarrow 0, L$.

In the conformal case we compute in a similar way the vacuum expectation value of the energy–momentum-tensor

$$
\langle 0|\hat{T}_{00}|0\rangle_{\xi=\frac{1}{6}}^{L} = \frac{1}{12\pi^2}\sum_{n=-\infty}^{+\infty}\left[-\frac{2}{\mathcal{D}_{-nL}^2} + \frac{4}{\mathcal{D}_{-(nL+x^3)}^2} - 4(x^3 - y^3 + 2nL)^2\frac{1}{\mathcal{D}_{-nL}^3} \right.
$$
$$
\left. - 4(x^3 + y^3 + 2nL)^2\frac{1}{\mathcal{D}_{-(nL+x^3)}^3} \right] \tag{5.219}
$$
$$
\longrightarrow -\frac{1}{32\pi^2}\sum_{n=-\infty}^{+\infty}\frac{1}{(nL)^4}.
$$

The normal ordered expression is

$$\langle 0|\hat{T}_{00}|0\rangle_{\xi=\frac{1}{6}}^{L} - \langle 0|\hat{T}_{00}|0\rangle_{\xi=\frac{1}{6}}^{\infty} = -\frac{1}{32\pi^2}\sum_{n\neq 0}\frac{1}{(nL)^4}$$
$$= -\frac{1}{16\pi^2 L^4}\sum_{n=1}^{\infty}\frac{1}{n^4} \tag{5.220}$$
$$= -\frac{1}{16\pi^2 L^4}\zeta(4).$$

The zeta function is given by

$$\zeta(4) = \sum_{n=1}^{\infty}\frac{1}{n^4} = \frac{\pi^4}{90}. \tag{5.221}$$

Thus

$$\langle 0|\hat{T}_{00}|0\rangle_{\xi=\frac{1}{6}}^{L} - \langle 0|\hat{T}_{00}|0\rangle_{\xi=\frac{1}{6}}^{\infty} = -\frac{\pi^2}{1440 L^4}. \tag{5.222}$$

This is precisely the vacuum energy density of the conformal scalar field. The electromagnetic field is also a conformal field with two degrees of freedom and thus the corresponding vacuum energy density is

$$\rho_{\text{em}} = -\frac{\pi^2}{720 L^4}. \tag{5.223}$$

This corresponds to the attractive Casimir force. The energy between the two plates (where A is the surface area of the plates) is

$$E_{em} = -\frac{\pi^2}{720L^4}AL. \tag{5.224}$$

The force is defined by

$$F_{em} = -\frac{dE_{em}}{dL}$$
$$= -\frac{\pi^2}{240L^4}A. \tag{5.225}$$

The Casimir force is the force per unit area given by

$$\frac{F_{em}}{A} = -\frac{\pi^2}{240L^4}. \tag{5.226}$$

5.5.4 From renormalizable field theory

In this section we follow [8]. We consider the Lagrangian density (recall the metric is taken to be of signature $-+++\cdots+$ and we will consider mostly $1+2$ dimensions)

$$\mathcal{L} = -\frac{1}{2}\partial_\mu\phi\partial^\mu\phi - \frac{1}{2}m^2\phi^2 - \frac{1}{2}\lambda\phi^2\sigma. \tag{5.227}$$

The static background field σ for parallel plates separated by a distance $2L$ will be chosen to be given by

$$\sigma = \frac{1}{\Delta}\left(\theta(|z| - L + \frac{\Delta}{2}) - \theta(|z| - L - \frac{\Delta}{2})\right). \tag{5.228}$$

Δ is the width of the plates and thus we are naturally interested in the sharp limit $\Delta \longrightarrow 0$. Obviously we have the normalization

$$\Delta \int dz\sigma(z) = 2\int_{-L+\Delta/2}^{0} dz\theta(z) - 2\int_{-L-\Delta/2}^{0} dz\theta(z)$$
$$= 2.\,\Delta. \tag{5.229}$$

We compute the Fourier transform

$$\tilde{\sigma}(q) = \int dz e^{iqz}\sigma(z)$$
$$= \frac{1}{\Delta}\int_{-L-\Delta/2}^{-L+\Delta/2} dz e^{iqz} + \frac{1}{\Delta}\int_{L-\Delta/2}^{L+\Delta/2} dz e^{iqz}$$
$$= \frac{4}{q\Delta}\cos qL \sin\frac{q\Delta}{2}. \tag{5.230}$$

In the limit $\Delta \longrightarrow 0$ we obtain

$$\tilde{\sigma}(q) = 2 \cos qL \rightarrow \sigma(z) = \delta(z - L) + \delta(z + L). \tag{5.231}$$

The 'boundary condition limit' $\phi(\pm L) = 0$ is obtained by letting $\lambda \longrightarrow \infty$. This is the Dirichlet limit.

Before we continue let us give the Casimir force for parallel plates $(\sigma = \delta(z - a) + \delta(z + a))$ in the case of $1 + 1$ dimensions. This is given by

$$F(L, \lambda, m) = -\frac{\lambda^2}{\pi} \int_m^\infty \frac{t^2 dt}{\sqrt{t^2 - m^2}} \frac{e^{-4Lt}}{4t^2 - 4\lambda t + \lambda^2(1 - e^{-4Lt})}. \tag{5.232}$$

It vanishes quadratically in λ when $\lambda \longrightarrow 0$ as it should be since it is a force induced by the coupling of the scalar field ϕ to the background σ. In the boundary condition limit $\lambda \longrightarrow \infty$ we obtain

$$F(L, \infty, m) = -\frac{1}{\pi} \int_m^\infty \frac{t^2 dt}{\sqrt{t^2 - m^2}} \frac{e^{-4Lt}}{1 - e^{-4Lt}}. \tag{5.233}$$

This is independent of the material. Furthermore, it reduces in the massless limit to the usual result, viz (with $a = 2L$)

$$F(L, \infty, 0) = -\frac{\pi}{24a^2}. \tag{5.234}$$

The vacuum polarization energy of the field ϕ in the background σ is the Casimir energy. More precisely the Casimir energy is the vacuum energy in the presence of the boundary minus the vacuum energy without the boundary, viz

$$E[\sigma] = \frac{1}{2} \sum_n \omega_n[\sigma] - \frac{1}{2} \sum_n \omega_n[\sigma = 0]. \tag{5.235}$$

The path integral is given by

$$Z = \int \mathcal{D}\phi e^{i \int d^D x \mathcal{L}}. \tag{5.236}$$

The vacuum energy is given formally by

$$\begin{aligned} W[\sigma] &= \frac{1}{i} \ln Z \\ &= \frac{i}{2} Tr \ln [\partial_\mu \partial^\mu - m^2 - \lambda \sigma] + \text{constant}. \end{aligned} \tag{5.237}$$

Thus

$$W[\sigma] - W[\sigma = 0] = \frac{i}{2} Tr \ln \left[1 - \frac{1}{\partial^\mu \partial_\mu - m^2} \lambda \sigma \right]. \tag{5.238}$$

The diagrammatic expansion of this term is given by the sum of all one-loop Feynman diagrams shown in figure 1 of reference [8]. The two-point function is obtained from W by differentiating with respect to an appropriate source twice, viz

$$G(x, y) = \frac{\delta^2 W[\sigma, J]}{\partial J(x) \partial J(y)}. \tag{5.239}$$

The 2-point function is then what controls the Casimir energy. From the previous section we have for a massless theory the result

$$\begin{aligned} E[\sigma] &= \int d^3x \langle \hat{T}^{00} \rangle_{\xi=0} \\ &= \frac{1}{2} \int d^3x (\partial_x^0 \partial_y^0 - \vec{\nabla}_x^2) D_{F\sigma}(x, y)|_{x=y} \\ &= \int \frac{d\omega}{2\pi} \omega^2 \int d^3x D_{F\sigma}(\omega, \vec{x}, \vec{x}) + \text{constant}. \end{aligned} \tag{5.240}$$

In other words,

$$E[\sigma] - E[\sigma = 0] = \int \frac{d\omega}{2\pi} \omega^2 \int d^3x [D_{F\sigma}(\omega, \vec{x}, \vec{x}) - D_{F0}(\omega, \vec{x}, \vec{x})]. \tag{5.241}$$

As it turns the density of states created by the background is precisely

$$\frac{dN}{d\omega} = \frac{\omega}{\pi} [D_{F\sigma}(\omega, \vec{x}, \vec{x}) - D_{F0}(\omega, \vec{x}, \vec{x})]. \tag{5.242}$$

Using this last equation in the previous one gives precisely (5.235).

Alternatively, we can rewrite the Casimir energy as

$$\begin{aligned} E[\sigma] - E[\sigma = 0] &= \frac{1}{2} \int d^3x (\partial_x^0 \partial_y^0 - \vec{\nabla}_x^2)[D_{F\sigma}(x, y) - D_{F0}(x, y)]|_{x=y} \\ &= \frac{1}{2} \int d^3x (\partial_x^0 \partial_y^0 - \vec{\nabla}_x^2) \frac{1}{\partial_\mu \partial^\mu} \left[\frac{1}{\partial_\mu \partial^\mu} \lambda\sigma + \frac{1}{\partial_\mu \partial^\mu} \lambda\sigma \frac{1}{\partial_\mu \partial^\mu} \lambda\sigma + \cdots \right]|_{x=y} \\ &= -\frac{1}{2} \int d^3x \left[\frac{1}{\partial_\mu \partial^\mu} \lambda\sigma + \frac{1}{\partial_\mu \partial^\mu} \lambda\sigma \frac{1}{\partial_\mu \partial^\mu} \lambda\sigma + \cdots \right]|_{x=y}. \end{aligned} \tag{5.243}$$

This term is again given by the sum of all one-loop Feynman diagrams shown in figure 1 of reference [8]. We observe that

$$E[\sigma] - E[\sigma = 0] = -i\lambda \frac{\partial}{\partial \lambda}[W[\sigma] - W[\sigma = 0]]. \tag{5.244}$$

Both the 1-point function ('tadpole') and the 2-point function (the self-energy) of the sigma field are superficially divergent for $D \leqslant 3$ and thus require renormalization. We introduce a counterterm given by

$$\mathcal{L} = c_1 \sigma + c_2 \sigma^2. \tag{5.245}$$

The coefficients c_1 and c_2 are determined from the renormalization conditions

$$<\sigma> = 0. \tag{5.246}$$

$$<\sigma\sigma> |_{p^2 = -\mu^2} = 0. \tag{5.247}$$

The $<\sigma>$ and $<\sigma\sigma>$ stand for proper vertices and not Green's functions of the field σ.

The total Casimir energy for a smooth background is finite. It can become divergent when the background becomes sharp ($\Delta \longrightarrow 0$) and strong ($\lambda \longrightarrow \infty$). The tadpole is always 0 by the renormalization condition. The 2-point function of the sigma field diverges as we remove Δ and as a consequence the renormalized Casimir energy diverges in the Dirichlet limit. The 3-point function also diverges (logarithmically) in the sharp limit, whereas all higher orders in λ are finite.

Any further study of these issues and a detailed study of the competing perspective of Milton [7, 26, 27] is beyond the scope of these lectures.

5.5.5 Is vacuum energy really real?

The main point of [28] is that experimental confirmation of the Casimir effect does not really establish the reality of zero point fluctuations in QFT. We leave the reader to go through the very sensible argumentation presented in that article.

5.6 The ADM formulation

In this section we will follow the seminal work [9], and the classic book [29].

We consider a fixed spacetime manifold $\mathcal{M}$ of dimension $D + 1$. Let g_{ab} be the 4D metric of the spacetime manifold $\mathcal{M}$. We consider a codimension-one foliation of the spacetime manifold $\mathcal{M}$ given by the spatial hypersurface (Cauchy surfaces) Σ_t of constant time t. Let n^a be the unit normal vector field to the hypersurfaces Σ_t. This induces a 3D metric h_{ab} on each Σ_t given by the formula

$$h_{ab} = g_{ab} + n_a n_b. \tag{5.248}$$

The time flow in this spacetime will be given by a time flow vector field t^a which satisfies $t^a \nabla_a t = 1$. We decompose t^a into its normal and tangential parts with respect to the hypersurface Σ_t. The normal and tangential parts are given by the so-called lapse function N and shift vector N^a, respectively, defined by

$$N = -g_{ab} t^a n^b. \tag{5.249}$$

$$N^a = h_b^a t^b. \tag{5.250}$$

Let us make all this more explicit. Let $t = t(x^\mu)$ be a scalar function on the 4D spacetime manifold $\mathcal{M}$ defined such that constant t gives a family of non-intersecting space-like hypersurfaces Σ_t. Let y^i be the coordinates on the hypersurfaces Σ_t. We introduce a congruence of curves parameterized by t which connect the hypersurfaces Σ_t in such a way that points on each of the hypersurfaces intersected by the same curve are given the same spatial coordinates y^i. We have then

$$x^\mu \longrightarrow y^\mu = (t, y^i). \tag{5.251}$$

The tangent vectors to the hypersurface Σ_t are

$$e_i^\mu = \frac{\partial x^\mu}{\partial y^i}. \tag{5.252}$$

The tangent vector to the congruence of curves is

$$t^\mu = \frac{\partial x^\mu}{\partial t}. \tag{5.253}$$

The vector t^μ satisfies trivially $t^\mu \nabla_\mu t = 1$, i.e. t^μ gives the direction of flow of time. The normal vector to the hypersurface Σ_t is defined by

$$n_\mu = -N\frac{\partial t}{\partial x^\mu}. \tag{5.254}$$

The normalization N is the lapse function. It is given precisely by (8.68), viz $N = -n_\mu t^\mu$. Clearly then N is the normal part of the vector t^μ with respect to the hypersurface Σ_t. Obviously we have $n_\mu e_i^\mu = 0$ and from the normalization $n_\mu n^\mu = -1$ we must also have

$$N^2\frac{\partial t}{\partial x^\mu}\frac{\partial x_\mu}{\partial t} = -1, \quad N = (n^\mu \nabla_\mu t)^{-1}. \tag{5.255}$$

We can decompose t^μ as

$$t^\mu = Nn^\mu + N^i e_i^\mu. \tag{5.256}$$

The three functions N^i define the components of the shift (spatial) vector. We compute immediately that

$$\begin{aligned} dx^\mu &= \frac{\partial x^\mu}{\partial t}dt + \frac{\partial x^\mu}{\partial y^i}dy^i \\ &= t^\mu dt + e_i^\mu dy^i \\ &= (Ndt)n^\mu + (dy^i + N^i dt)e_i^\mu. \end{aligned} \tag{5.257}$$

Also,

$$\begin{aligned} ds^2 &= g_{\mu\nu}dx^\mu dx^\nu \\ &= g_{\mu\nu}\left[N^2 dt^2 n^\mu n^\nu + (dy^i + N^i dt)(dy^j + N^j dt)e_i^\mu e_j^\nu\right] \\ &= -N^2 dt^2 + h_{ij}(dy^i + N^i dt)(dy^j + N^j dt). \end{aligned} \tag{5.258}$$

The 3D metric h_{ij} is the induced metric on the hypersurface Σ_t. It is given explicitly by

$$h_{ij} = g_{\mu\nu}e_i^\mu e_j^\nu. \tag{5.259}$$

From the other hand in the coordinate system y^μ we have

$$\begin{aligned}
ds^2 &= \gamma_{\mu\nu} dy^\mu dy^\nu \\
&= \gamma_{00} dt^2 + 2\gamma_{0j} dt dy^j + \gamma_{ij} dy^i dy^j.
\end{aligned} \tag{5.260}$$

By comparing (8.75) and (5.260) we obtain

$$\gamma_{\mu\nu} = \begin{pmatrix} \gamma_{00} & \gamma_{0j} \\ \gamma_{i0} & \gamma_{ij} \end{pmatrix} = \begin{pmatrix} -N^2 + h_{ij}N^iN^j & h_{ij}N^i \\ h_{ij}N^j & h_{ij} \end{pmatrix} = \begin{pmatrix} -N^2 + N^iN_i & N_j \\ N_i & h_{ij} \end{pmatrix}. \tag{5.261}$$

The condition $\gamma_{\mu\nu}\gamma^{\nu\lambda} = \delta^\lambda_\mu$ reads explicitly

$$\begin{aligned}
(-N^2 + N^iN_i)\gamma^{00} + N_i\gamma^{i0} &= 1 \\
(-N^2 + N^iN_i)\gamma^{0j} + N_i\gamma^{ij} &= 0 \\
N_j\gamma^{00} + h_{ij}\gamma^{i0} &= 0 \\
N_j\gamma^{0k} + h_{ij}\gamma^{ik} &= \delta^k_j.
\end{aligned} \tag{5.262}$$

We define h^{ij} in the usual way, viz $h_{ij}h^{jk} = \delta^k_i$. We get immediately the solution

$$\gamma^{\mu\nu} = \begin{pmatrix} \gamma^{00} & \gamma^{0j} \\ \gamma^{i0} & \gamma^{ij} \end{pmatrix} = \begin{pmatrix} -\dfrac{1}{N^2} & \dfrac{1}{N^2}N^j \\ \dfrac{1}{N^2}N^i & h^{ij} - \dfrac{1}{N^2}N^iN^j \end{pmatrix}. \tag{5.263}$$

We also compute (we work in $1 + 2$ for simplicity)

$$\det\gamma = \det\begin{pmatrix} -N^2 + N^iN_i & N_1 & N_2 \\ N_1 & h_{11} & h_{12} \\ N_2 & h_{21} & h_{22} \end{pmatrix} \tag{5.264}$$

$$= (-N^2 + N^iN_i)\det h - N_1(N_1h_{22} - N_2h_{12}) + N_2(N_1h_{21} - N_2h_{11}).$$

By using $N_i = h_{ij}N^j$ we find

$$\det\gamma = -N^2\det h \tag{5.265}$$

We have then the result

$$\sqrt{-g}\, d^4x = \sqrt{-\gamma}\, d^4y = N\sqrt{h}\, d^4y. \tag{5.266}$$

We conclude that all information about the original 4D metric $g_{\mu\nu}$ is contained in the lapse function N, the shift vector N^i and the 3D metric h_{ij}.

The 3D metric h_{ij} can also be understood in terms of projectors as follows. The projector normal to the hypersurface Σ_t is defined by

$$P^N_{\mu\nu} = -n_\mu n_\nu. \tag{5.267}$$

This satisfies $(P^N)^2 = P^N$ and $P^N n = n$ as it should. The normal component of any vector V^μ with respect to the hypersurface Σ_t is given by $V^\mu n_\mu$. The projector $P^N_{\mu\nu}$ can also be understood as the metric along the normal direction. Indeed we have

$$P^N_{\mu\nu} dx^\mu dx^\nu = -n_\mu n_\nu dx^\mu dx^\nu = -N^2 dt^2. \tag{5.268}$$

The tangent projector is then obviously given by

$$\begin{aligned} P^T_{\mu\nu} &= g_{\mu\nu} - P^N_{\mu\nu} \\ &= g_{\mu\nu} + n_\mu n_\nu. \end{aligned} \tag{5.269}$$

This should be understood as the metric along the tangent directions since

$$P^T_{\mu\nu} dx^\mu dx^\nu = ds^2 + N^2 dt^2 = h_{ij}(dy^i + N^i dt)(dy^j + N^j dt). \tag{5.270}$$

The 3D metric is therefore given by

$$h_{\mu\nu} \equiv P^T_{\mu\nu} = g_{\mu\nu} + n_\mu n_\nu. \tag{5.271}$$

Indeed we have

$$h_{\mu\nu} \frac{\partial x^\mu}{\partial y^\alpha} \frac{\partial x^\nu}{\partial y^\beta} dy^\alpha dy^\beta = h_{ij}(dy^i + N^i dt)(dy^j + N^j dt). \tag{5.272}$$

Or equivalently

$$\begin{aligned} h_{\mu\nu} e_i^\mu e_j^\nu &= h_{ij} \Leftrightarrow g_{\mu\nu} e_i^\mu e_j^\nu = h_{ij} \\ N_i &= h_{\mu\nu} t^\mu e_i^\nu \Leftrightarrow N_i = h_{ij} N^j \\ N_i N^i &\equiv h_{\mu\nu} t^\mu t^\nu. \end{aligned} \tag{5.273}$$

We compute also

$$h^\mu_\nu t^\nu = g^{\mu\alpha} h_{\alpha\nu} t^\nu = N^i e_i^\mu \equiv N^\mu. \tag{5.274}$$

This should be compared with (8.72).

It is a theorem that the 3D metric $h_{\mu\nu}$ will uniquely determine a covariant derivative operator on Σ_t. This will be denoted D_μ and defined in an obvious way by

$$D_\mu X_\nu = h_\mu^\alpha h_\nu^\beta \nabla_\alpha X_\beta. \tag{5.275}$$

In other words, D_μ is the projection of the 4D covariant derivative ∇_μ onto Σ_t.

A central object in the discussion of how the hypersurfaces Σ_t are embedded in the 4D spacetime manifold $\mathcal{M}$ is the extrinsic curvature $K_{\mu\nu}$. This is given essentially by (1) comparing the normal vector n_μ at a point p and the parallel transport of the normal vector n_μ at a nearby point q along a geodesic connecting q to p on the hypersurface Σ_t, and then (2) projecting the result onto the hypersurface Σ_t. The first part is clearly given by the covariant derivative, whereas the projection is done through the 3D metric tensor. Hence the extrinsic curvature must be defined by

$$K_{\mu\nu} = - h_\mu^{\ \alpha} h_\nu^{\ \beta} \ \nabla_\alpha \ n_\beta$$
$$= - h_\mu^{\ \alpha} \ \nabla_\alpha \ n_\nu.$$

(5.276)

In the second line of the above equation we have used $n^\beta \ \nabla_\alpha \ n_\beta = 0$ and $\nabla_\alpha \ g_{\mu\nu} = 0$. We can check that $K_{\mu\nu}$ is symmetric and tangent, viz

$$K_{\mu\nu} = K_{\nu\mu}, \ \ h_\mu^{\ \alpha} K_{\alpha\nu} = K_{\mu\nu}.$$

(5.277)

We recall the definition of the curvature tensor in four dimensions which is given by

$$(\nabla_\alpha \ \nabla_\beta - \nabla_\beta \ \nabla_\alpha) \omega_\mu = R^\nu_{\ \alpha\beta\mu} \omega_\nu.$$

(5.278)

By analogy the curvature tensor of Σ_t can be defined by

$$(D_\alpha D_\beta - D_\beta D_\alpha) \omega_\mu = {}^{(3)} R^\nu_{\ \alpha\beta\mu} \omega_\nu.$$

(5.279)

We compute

$$D_\alpha D_\beta \omega_\mu = D_\alpha (h_\beta^{\ \rho} h_\mu^{\ \nu} \ \nabla_\rho \ \omega_\nu)$$
$$= h_\alpha^{\ \delta} h_\beta^{\ \theta} h_\mu^{\ \gamma} \ \nabla_\delta \ (h_\theta^{\ \rho} h_\gamma^{\ \nu} \ \nabla_\rho \ \omega_\nu)$$
$$= h_\alpha^{\ \delta} h_\beta^{\ \rho} h_\mu^{\ \nu} \ \nabla_\delta \ \nabla_\rho \ \omega_\nu - h_\mu^{\ \nu} K_{\alpha\beta} n^\rho \ \nabla_\rho \ \omega_\nu - h_\beta^{\ \rho} K_{\alpha\mu} n^\nu \ \nabla_\rho \ \omega_\nu.$$

(5.280)

In the last line of the above equation we have used the result

$$h_\alpha^{\ \delta} h_\beta^{\ \theta} \ \nabla_\delta \ h_\theta^{\ \rho} = -K_{\alpha\beta} n^\rho.$$

(5.281)

We also compute

$$h_\beta^{\ \rho} n^\nu \ \nabla_\rho \ \omega_\nu = h_\beta^{\ \rho} \ \nabla_\rho \ (n^\nu \omega_\nu) + K_\beta^{\ \nu} \omega_\nu$$
$$= D_\rho (n^\nu \omega_\nu) + K_\beta^{\ \nu} \omega_\nu$$
$$= K_\beta^{\ \nu} \omega_\nu.$$

(5.282)

Thus

$$D_\alpha D_\beta \omega_\mu = h_\alpha^{\ \delta} h_\beta^{\ \rho} h_\mu^{\ \nu} \ \nabla_\delta \ \nabla_\rho \ \omega_\nu - h_\mu^{\ \nu} K_{\alpha\beta} n^\rho \ \nabla_\rho \ \omega_\nu - K_{\alpha\mu} K_\beta^{\ \nu} \omega_\nu.$$

(5.283)

Similar calculation gives

$$D_\beta D_\alpha \omega_\mu = h_\alpha^{\ \delta} h_\beta^{\ \rho} h_\mu^{\ \nu} \ \nabla_\rho \ \nabla_\delta \ \omega_\nu - h_\mu^{\ \nu} K_{\alpha\beta} n^\rho \ \nabla_\rho \ \omega_\nu - K_{\beta\mu} K_\alpha^{\ \nu} \omega_\nu.$$

(5.284)

Hence we obtain the first Gauss–Codacci relation given by

$${}^{(3)} R^\nu_{\ \alpha\beta\mu} \omega_\nu = h_\alpha^{\ \delta} h_\beta^{\ \rho} h_\mu^{\ \theta} R^\kappa_{\ \delta\rho\theta} \omega_\kappa - K_{\alpha\mu} K_\beta^{\ \nu} \omega_\nu + K_{\beta\mu} K_\alpha^{\ \nu} \omega_\nu.$$

(5.285)

In other words

$${}^{(3)} R^\nu_{\ \alpha\beta\mu} = h_\alpha^{\ \delta} h_\beta^{\ \rho} h_\mu^{\ \theta} R^\kappa_{\ \delta\rho\theta} h_\kappa^{\ \nu} - K_{\alpha\mu} K_\beta^{\ \nu} + K_{\beta\mu} K_\alpha^{\ \nu}.$$

(5.286)

The first term represents the intrinsic part of the 3D curvature obtained by simply projecting out the 4D curvature onto the hypersurface Σ_t, whereas the second term represents the extrinsic part of the 3D curvature which arises from the embedding of Σ_t into the spacetime manifold.

The second Gauss–Codacci relation is given by

$$D_\mu K_\nu^\mu - D_\nu K_\mu^\mu = -h_\nu^\alpha R_{\alpha\kappa} n^\kappa. \tag{5.287}$$

The proof goes as follows. We use $h_\mu^\nu h_\nu^\lambda = h_\mu^\lambda$ and $K_\mu^\lambda = g^{\lambda\nu} K_{\mu\nu} = -h_\mu^\alpha \nabla_\alpha n^\lambda$ to find

$$
\begin{aligned}
D_\mu K_\nu^\mu - D_\nu K_\mu^\mu &= h_\mu^\rho h_\sigma^\mu h_\nu^\lambda \nabla_\rho K_\lambda^\sigma - h_\nu^\rho h_\sigma^\mu h_\mu^\lambda \nabla_\rho K_\lambda^\sigma \\
&= h_\sigma^\rho h_\nu^\lambda \nabla_\rho K_\lambda^\sigma - h_\sigma^\rho h_\nu^\lambda \nabla_\lambda K_\rho^\sigma \\
&= -h_\sigma^\rho h_\nu^\lambda \nabla_\rho (h_\lambda^\alpha \nabla_\alpha n^\sigma) + h_\sigma^\rho h_\nu^\lambda \nabla_\lambda (h_\rho^\alpha \nabla_\alpha n^\sigma) \\
&= -h_\sigma^\rho h_\nu^\lambda \left(\nabla_\rho h_\lambda^\alpha \cdot \nabla_\alpha n^\sigma + h_\lambda^\alpha \nabla_\rho \nabla_\alpha n^\sigma - \nabla_\lambda h_\rho^\alpha \cdot \nabla_\alpha n^\sigma - h_\rho^\alpha \nabla_\lambda \nabla_\alpha n^\sigma \right).
\end{aligned} \tag{5.288}
$$

The first and third terms are zero. Explicitly we have (using $\nabla_\alpha g_{\mu\nu} = 0$ and $n^\mu h_\mu^\nu = 0$)

$$-h_\sigma^\rho h_\nu^\lambda \left(\nabla_\rho h_\lambda^\alpha \cdot \nabla_\alpha n^\sigma - \nabla_\lambda h_\rho^\alpha \cdot \nabla_\alpha n^\sigma \right) = h_\nu^\lambda K_{\sigma\lambda} n^\alpha \nabla_\alpha n^\sigma - h_\sigma^\rho K_{\nu\rho} n^\alpha \nabla_\alpha n^\sigma \tag{5.289}$$
$$= 0.$$

We have then

$$
\begin{aligned}
D_\mu K_\nu^\mu - D_\nu K_\mu^\mu &= -h_\sigma^\rho h_\nu^\lambda \left(h_\lambda^\alpha \nabla_\rho \nabla_\alpha n^\sigma - h_\rho^\alpha \nabla_\lambda \nabla_\alpha n^\sigma \right) \\
&= -h_\sigma^\rho h_\nu^\alpha (\nabla_\rho \nabla_\alpha - \nabla_\alpha \nabla_\rho) n^\sigma \\
&= -h^{\rho\sigma} h_\nu^\alpha R_{\rho\alpha\sigma\kappa} n^\kappa \\
&= h^{\rho\sigma} h_\nu^\alpha R_{\rho\alpha\kappa\sigma} n^\kappa \\
&= g^{\rho\sigma} h_\nu^\alpha R_{\rho\alpha\kappa\sigma} n^\kappa \\
&= h_\nu^\alpha R_{\rho\alpha\kappa}^{\;\;\;\rho} n^\kappa \\
&= -h_\nu^\alpha R_{\alpha\rho\kappa}^{\;\;\;\rho} n^\kappa \\
&= -h_\nu^\alpha R_{\alpha\kappa} n^\kappa.
\end{aligned} \tag{5.290}
$$

The goal now is to compute in terms of 3D quantities the scalar curvature R. We start from

$$
\begin{aligned}
R &= -R g_{\mu\nu} n^\mu n^\nu \\
&= -2(R_{\mu\nu} - G_{\mu\nu}) n^\mu n^\nu \\
&= -2 R_{\mu\nu} n^\mu n^\nu + R_{\mu\nu\alpha\beta} h^{\mu\alpha} h^{\nu\beta}.
\end{aligned} \tag{5.291}
$$

We compute

$$
\begin{aligned}
R_{\mu\nu\alpha\beta}h^{\mu\alpha}h^{\nu\beta} &= h_{\beta\rho}R^{\rho}_{\ \mu\nu\alpha}h^{\mu\alpha}h^{\nu\beta} \\
&= g^{\beta\eta}g^{\kappa\sigma}\left(h^{\mu}_{\kappa}h^{\nu}_{\eta}h^{\alpha}_{\sigma}R^{\rho}_{\ \mu\nu\alpha}h^{\theta}_{\rho}\right)h^{\beta}_{\theta} \\
&= g^{\beta\eta}g^{\kappa\sigma}\left((3)R^{\theta}_{\ \kappa\eta\sigma} + K_{\kappa\sigma}K^{\theta}_{\eta} - K_{\eta\sigma}K^{\theta}_{\kappa}\right)h^{\beta}_{\theta} \\
&= g^{\kappa\sigma}\left((3)R^{\theta}_{\ \kappa\eta\sigma} + K_{\kappa\sigma}K^{\theta}_{\eta} - K_{\eta\sigma}K^{\theta}_{\kappa}\right) \\
&= {}^{(3)}R + K^2 - K_{\mu\nu}K^{\mu\nu}.
\end{aligned}
\tag{5.292}
$$

Next we compute

$$
\begin{aligned}
R_{\mu\nu}n^{\mu}n^{\nu} &= R^{\alpha}_{\ \mu\alpha\nu}n^{\mu}n^{\nu} \\
&= -g^{\alpha\rho}n^{\nu}R_{\nu\rho\alpha\mu}n^{\mu} \\
&= n^{\nu}\ \nabla_{\mu}\ \nabla_{\nu}\ n^{\mu} - n^{\nu}\ \nabla_{\nu}\ \nabla_{\mu}\ n^{\mu} \\
&= \nabla_{\mu}(n^{\nu}\ \nabla_{\nu}\ n^{\mu}) - \nabla_{\nu}(n^{\nu}\ \nabla_{\mu}\ n^{\mu}) - \nabla_{\mu}\ n^{\nu}.\nabla_{\nu}\ n^{\mu} + \nabla_{\nu}\ n^{\nu}.\nabla_{\mu}\ n^{\mu}.
\end{aligned}
\tag{5.293}
$$

The rate of change of the normal vector along the normal direction is expressed by the quantity

$$
a^{\mu} = n^{\nu}\ \nabla_{\nu}\ n^{\mu}.
\tag{5.294}
$$

We have

$$
\begin{aligned}
K = K^{\mu}_{\mu} &= -h^{\alpha}_{\mu}\ \nabla_{\alpha}\ n^{\mu} \\
&= -g^{\alpha}_{\mu}\ \nabla_{\alpha}\ n^{\mu} \\
&= -\nabla_{\mu}\ n^{\mu}.
\end{aligned}
\tag{5.295}
$$

By using now $K_{\mu\nu} = -h^{\alpha}_{\mu}\ \nabla_{\alpha}\ n_{\nu} = -h^{\alpha}_{\nu}\ \nabla_{\alpha}\ n_{\mu}$ and $h^{\nu\beta}K_{\mu\nu} = K^{\beta}_{\mu}$ we can show that

$$
\begin{aligned}
K_{\mu\nu}K^{\mu\nu} &= -K_{\mu\nu}h^{\nu\beta}\ \nabla_{\beta}\ n^{\mu} \\
&= -K^{\beta}_{\mu}\ \nabla_{\beta}\ n^{\mu} \\
&= h^{\rho}_{\mu}\ \nabla_{\rho}\ n^{\beta}\ \nabla_{\beta}\ n^{\mu} \\
&= \nabla_{\mu}\ n^{\beta}.\nabla_{\beta}\ n^{\mu}.
\end{aligned}
\tag{5.296}
$$

We obtain then the result

$$
R_{\mu\nu}n^{\mu}n^{\nu} = \nabla_{\mu}(Kn^{\mu} + a^{\mu}) - K_{\mu\nu}K^{\mu\nu} + K^2.
\tag{5.297}
$$

The end result is given by

$$
R = \mathcal{L}_{\text{ADM}} - 2\ \nabla_{\mu}(Kn^{\mu} + a^{\mu}).
\tag{5.298}
$$

The so-called ADM Lagrangian is given by

$$
\mathcal{L}_{\text{ADM}} = {}^{(3)}R - K^2 + K_{\mu\nu}K^{\mu\nu}.
\tag{5.299}
$$

In other words,

$$\sqrt{-g}\,\mathcal{L}_{\text{ADM}} = \sqrt{h}\,N((3)R - K^2 + K_{\mu\nu}K^{\mu\nu}). \qquad (5.300)$$

The extrinsic curvature $K_{\mu\nu}$ is the covariant analog of the time derivative of the metric as we will now show. First, we recall the definition of the Lie derivative of a tensor T along a vector V. For a function we have obviously $\mathcal{L}_V f = V(f) = V^\mu \partial_\mu f$, whereas for a vector the Lie derivative is defined by $\mathcal{L}_V U^\mu = [V, U]^\mu$. This is essentially the commutator, which is the reason why the commutator is called sometimes the Lie bracket. The Lie derivative of an arbitrary tensor is given by

$$\mathcal{L}_V T^{\mu_1 \dots \mu_k}_{\nu_1 \dots \nu_l} = V^\sigma \nabla_\sigma T^{\mu_1 \dots \mu_k}_{\nu_1 \dots \nu_l} - \nabla_\lambda V^{\mu_1} T^{\lambda \mu_2 \dots \mu_k}_{\nu_1 \dots \nu_l} - \cdots + \nabla_{\nu_1} V^\lambda T^{\mu_1 \dots \mu_k}_{\lambda \nu_2 \dots \nu_l} + \cdots \quad (5.301)$$

A very important example is the Lie derivative of the metric given by

$$\mathcal{L}_V g_{\mu\nu} = \nabla_\mu V_\nu + \nabla_\nu V_\mu. \qquad (5.302)$$

Let us now go back to the extrinsic curvature $K_{\mu\nu}$. We have (using $n^c h_{cb} = 0$, $t^c = Nn^c + N^c$)

$$
\begin{aligned}
K_{ab} &= -h_a^\alpha \nabla_\alpha n_b \\
&= -\frac{1}{2} h_a^\alpha h_b^\beta (\nabla_\alpha n_\beta + \nabla_\beta n_\alpha) \\
&= -\frac{1}{2} h_a^\alpha h_b^\beta \mathcal{L}_n h_{\alpha\beta} \\
&= -\frac{1}{2} h_a^\alpha h_b^\beta (n^c \nabla_c h_{\alpha\beta} + \nabla_\alpha n^c . h_{c\beta} + \nabla_\beta n^c . h_{\alpha c}) \\
&= -\frac{1}{2N} h_a^\alpha h_b^\beta (Nn^c \nabla_c h_{\alpha\beta} + \nabla_\alpha (Nn^c) . h_{c\beta} + \nabla_\beta (Nn^c) . h_{\alpha c}) \\
&= -\frac{1}{2N} h_a^\alpha h_b^\beta (\mathcal{L}_t h_{\alpha\beta} - \mathcal{L}_N h_{\alpha\beta}).
\end{aligned}
\qquad (5.303)
$$

However, we have (using $N^c = h_d^c t^d$)

$$
\begin{aligned}
h_a^\alpha h_b^\beta \mathcal{L}_N h_{\alpha\beta} &= h_a^\alpha h_b^\beta (N^c \nabla_c h_{\alpha\beta} + \nabla_\alpha N^c . h_{c\beta} + \nabla_\beta N^c . h_{c\alpha}) \\
&= t^d D_d h_{ab} + D_a N_b + D_b N_a \\
&= D_a N_b + D_b N_a.
\end{aligned}
\qquad (5.304)
$$

The time derivative of the metric is defined by

$$\dot{h}_{ab} = h_a^\alpha h_b^\beta \mathcal{L}_t h_{\alpha\beta}. \qquad (5.305)$$

Hence

$$K_{ab} = -\frac{1}{2N}(\dot{h}_{ab} - D_a N_b - D_b N_a). \qquad (5.306)$$

5.7 A brief introduction of Horava–Lifshitz quantum gravity

In this section we follow [10, 30, 31] but also [32–39].

We will consider a fixed spacetime manifold $\mathcal{M}$ of dimension $D + 1$ with an extra structure given by a codimension-one foliation $\mathcal{F}$. Each leaf of the foliation is a spatial hypersurface Σ_t of constant time t with local coordinates given by x^i. Obviously, general diffeomorphisms, including Lorentz transformations, do not respect the foliation $\mathcal{F}$. Instead we have invariance under the foliation-preserving diffeomorphism group $\text{Diff}_{\mathcal{F}}(\mathcal{M})$ consisting of space-independent time reparametrizations and time-dependent spatial diffeomorphisms given by

$$t \longrightarrow t'(t), \quad \vec{x} \longrightarrow \vec{x}'(t, \vec{x}). \tag{5.307}$$

The infinitesimal generators are clearly given by

$$\delta t = f(t), \quad \delta x^i = \xi^i(t, \vec{x}). \tag{5.308}$$

The time-dependent spatial diffeomorphisms allow us arbitrary changes of the spatial coordinates x^i on each constant time hypersurfaces Σ_t. The fact that time reparametrization is space-independent means that the foliation of the spacetime manifold $\mathcal{M}$ by the constant time hypersurfaces Σ_t is not a choice of coordinate, as in general relativity, but it is a physical property of spacetime itself.

This property of spacetime is implemented explicitly by positing that spacetime is anisotropic in the sense that time and space do not scale in the same way, viz

$$x^i \longrightarrow b x^i, \quad t \longrightarrow b^z t. \tag{5.309}$$

The exponent z is called the dynamical critical exponent and it measures the degree of anisotropy postulated to exist between space and time. This exponent is a dynamical quantity in the theory which is not determined by the gauge transformations corresponding to the foliation-preserving diffeomorphisms. The above scaling rules (5.309) are not invariant under foliation-preserving diffeomorphisms and they should only be understood as the scaling properties of the theory at the UV free-field fixed point.

5.7.1 Lifshitz scalar field theory

We start by explaining the above point a little further in terms of so-called Lifshitz field theory. Lifshitz scalar field theory describes a tricritical triple point at which three different phases (disorder, uniform (homogeneous) and non-uniform (spatially modulated)) meet. A Lifshitz scalar field is given by the action

$$S = \frac{1}{2} \int dt \int d^D x \left(\dot{\Phi}^2 - \frac{1}{4} (\Delta \Phi)^2 \right). \tag{5.310}$$

This action defines a Gaussian (free) RG fixed point with anisotropic scaling rules (5.309) with $z = 2$. The two terms in the above action must have the same mass dimension and as a consequence we obtain $[t] = [x]^2$. By choosing $\hbar = 1$ the mass dimension of x is P^{-1} where P is some typical momentum and hence the mass dimension of t is P^{-2}. We have then

$$[x] = P^{-1}, \quad [t] = P^{-2}. \tag{5.311}$$

The mass dimension of the scalar field is therefore given by

$$[\Phi] = P^{\frac{D-2}{2}}. \tag{5.312}$$

The values $z = 2$ and $(D - 2)/2$ should be compared with the relativistic values $z = 1$ and $(D - 1)/2$. The lower critical dimension of the Lifshitz scalar at which the two-point function becomes logarithmically divergent is $2 + 1$ instead of the usual $1 + 1$ of the relativistic scalar field.

We can add at the UV free fixed point a relevant perturbation given by

$$W = -\frac{c^2}{2} \int dt \int d^D x \, \partial_i \Phi \partial_i \Phi. \tag{5.313}$$

By using the various mass dimensions at the UV free fixed point the coupling constant c has mass dimension P. The theory will flow in the infrared to the value $z = 1$ since this perturbation dominates the second term of (5.310) at low energies. In other words at large distances Lorentz symmetry emerges accidentally.

This crucial result is also equivalent to the statement that the ground state wave function of the system (5.310) is given essentially by the above relevant perturbation. This can be shown as follows. The Hamiltonian derived from (5.310) is trivially given by

$$H = \frac{1}{2} \int d^D x \left(P^2 + \frac{1}{4} (\Delta \Phi)^2 \right). \tag{5.314}$$

The term $(\Delta \Phi)^2$ appears therefore as the potential. The momentum P can be realized as

$$P = -i \frac{\delta}{\delta \Phi}. \tag{5.315}$$

The Hamiltonian can then be rewritten as

$$H = \frac{1}{2} \int d^D x \, Q^+ Q, \quad Q = iP - \frac{1}{2} \Delta \Phi. \tag{5.316}$$

The ground state wave function is a functional of the scalar field Φ which satisfies $H\Psi_0[\Phi] = 0$ or equivalently

$$Q\Psi_0 = 0 \Rightarrow \left(\frac{\delta}{\delta \Phi} - \frac{1}{2} \Delta \Phi \right) \Psi_0[\Phi] = 0. \tag{5.317}$$

A simple solution is given by

$$\Psi_0[\Phi] = \exp\left(-\frac{1}{4} \int d^D x \, \partial_i \Phi \partial_i \Phi \right). \tag{5.318}$$

The theory given by the action (5.310) satisfy the so-called detailed balance condition in the sense that the potential part can be derived from a variational principle given precisely by the action (5.313), viz

$$\frac{\delta W}{\delta \Phi} = c^2 \Delta \Phi. \tag{5.319}$$

5.7.2 Foliation-preserving diffeomorphisms and kinetic action

We will assume for simplicity that the global topology of spacetime is given by

$$\mathcal{M} = \mathbf{R} \times \Sigma. \tag{5.320}$$

Σ is a compact D-dimensional space with trivial tangent bundle. This is equivalent to the statement that all global topological effects will be ignored and all total derivative and boundary terms are dropped in the action.

The Riemannian structure on the foliation $\mathcal{F}$ is given by the 3D metric g_{ij}, the shift vector N_i and the lapse function N as in the ADM decomposition of general relativity. The lapse function can be either projectable or non-projectable depending on whether or not it depends on time only and thus it is constant on the spatial leafs or it depends on spacetime. As it turns out projectable Horava–Lifshitz gravity contains an extra degree of freedom known as the scalar graviton.

We want here to demonstrate some of the above results. We first write down the metric in the ADM decomposition as

$$ds^2 = -N^2 c^2 dt^2 + g_{ij}(dx^i + N^i dt)(dx^j + N^j dt)$$
$$= (-N^2 + g_{ij} N^i N^j / c^2)(dx^0)^2 + (g_{ij} N^j / c) dx^i dx^0 + (g_{ij} N^i / c) dx^j dx^0 + g_{ij} dx^i dx^j. \tag{5.321}$$

Now we consider the general diffeomorphism transformation

$$x'^0 = x^0 + cf(t, x^i) + O\left(\frac{1}{c}\right)$$
$$x'^i = x^i + \xi^i(t, x^j) + O\left(\frac{1}{c}\right). \tag{5.322}$$

This is an expansion in powers of $1/c$. For simplicity we will also assume that the generators f and ξ^i are small. We compute immediately

$$g_{ij} = \frac{\partial x'^\mu}{\partial x^i} \frac{\partial x'^\nu}{\partial x^j} g'_{\mu\nu}$$
$$= g_i^k g_j^l g'_{kl} + g_i^k \frac{\partial \xi^l}{\partial x^j} g'_{kl} + g_j^l \frac{\partial \xi^k}{\partial x^i} g'_{kl} + c \frac{\partial f}{\partial x^j} g_i^k g'_{k0} + c \frac{\partial f}{\partial x^i} g_j^k g'_{k0} \cdots \tag{5.323}$$

In the limit $c \longrightarrow \infty$ the last two terms diverge and thus one must choose the generator of time reparametrization f such that $f = f(t)$. In this case the above diffeomorphism (5.322) becomes precisely a foliation-preserving diffeomorphism. We obtain in this case

$$g'_{ij} = g_{ij} - g_i^k \frac{\partial \xi^l}{\partial x^j} g_{kl} - g_j^l \frac{\partial \xi^k}{\partial x^i} g_{kl}. \qquad (5.324)$$

Equivalently the gauge transformation of the 3D metric corresponding to a foliation-preserving diffeomorphism is

$$\delta g_{ij} = g'_{ij}(x') - g_{ij}(x)$$

$$= g'_{ij}(x) - g_{ij}(x) + f\frac{\partial g_{ij}}{\partial t} + \xi^k \frac{\partial g_{ij}}{\partial x^k} \qquad (5.325)$$

$$= -\frac{\partial \xi^l}{\partial x^j} g_{il} - \frac{\partial \xi^k}{\partial x^i} g_{kj} + f\frac{\partial g_{ij}}{\partial t} + \xi^k \frac{\partial g_{ij}}{\partial x^k}.$$

Similarly, we compute the gauge transformation of the shift vector corresponding to a foliation-preserving diffeomorphism as follows. We have

$$g_{i0} = \frac{\partial x'^\mu}{\partial x^i} \frac{\partial x'^\nu}{\partial x^0} g'_{\mu\nu}$$

$$= g'_{i0} + \frac{\partial f}{\partial t} g'_{i0} + \frac{1}{c}\frac{\partial \xi^k}{\partial t} g'_{ik} + \frac{\partial \xi^k}{\partial x^i} g'_{k0}. \qquad (5.326)$$

Equivalently we have

$$g'_{ij} N'^j = g_{ij} N^j - \frac{\partial f}{\partial t} N_i - \frac{\partial \xi^k}{\partial t} g_{ik} - \frac{\partial \xi^k}{\partial x^i} N_k. \qquad (5.327)$$

We rewrite this as

$$g_{ij}(N'^j - N^j) = -\frac{\partial f}{\partial t} N_i - \frac{\partial \xi^k}{\partial t} g_{ik} + \frac{\partial \xi^l}{\partial x^j} g_{il} N^j. \qquad (5.328)$$

We have then

$$\delta N_i = g'_{ij}(x') N'^j(x') - g_{ij}(x) N^j(x)$$

$$= -\frac{\partial f}{\partial t} N_i + f\frac{\partial N_i}{\partial t} - \frac{\partial \xi^k}{\partial t} g_{ik} + \xi^k \frac{\partial N_i}{\partial x^k} - \frac{\partial \xi^k}{\partial x^i} N_k. \qquad (5.329)$$

A similar calculation for the lapse function goes as follows. We have

$$g_{00} = \frac{\partial x'^\mu}{\partial x^0} \frac{\partial x'^\nu}{\partial x^0} g'_{\mu\nu}$$

$$= g'_{00} + 2\frac{\partial f}{\partial t} g'_{00} + \frac{2}{c}\frac{\partial \xi^i}{\partial t} g'_{i0}. \qquad (5.330)$$

Explicitly we find from this equation after some calculation (recalling that $g_{ij} N^j = N_i$)

$$N'(x) - N(x) = -\frac{\partial f}{\partial t} N. \qquad (5.331)$$

Thus

$$\delta N = N'(x') - N(x)$$
$$= f\frac{\partial N}{\partial t} + \xi^k\frac{\partial N}{\partial x^k} - \frac{\partial f}{\partial t}N. \tag{5.332}$$

We can use the above gauge transformations to make the choice

$$N = 1, \quad N_i = 0. \tag{5.333}$$

These are called the Gaussian coordinates.

Now we want to write an action principle for this theory. It will be given by the difference of a kinetic term and a potential term as follows

$$S = S_K - S_V. \tag{5.334}$$

The kinetic term is formed from the most general scalar term compatible with foliation-preserving diffeomorphisms which must be quadratic in the time derivative of the 3D metric in order to maintain unitarity. It must be of the canonical form $\int dt d^D x \dot{\Phi}^2$. Explicitly we may write

$$S_K = \frac{1}{2\kappa^2} \int dt d^D x N \sqrt{g} \frac{\partial g_{ij}}{\partial t} G^{ijkl} \frac{\partial g_{kl}}{\partial t}. \tag{5.335}$$

The time derivative of the 3D metric in the above action (5.335) must in fact be replaced by K_{ij} while the metric G^{ijkl} on the space of metrics can be determined from the requirement of invariance under foliation-preserving diffeomorphisms as we will show in the following.

We know from our study of the ADM decomposition of general relativity that the covariant time derivative of the 3D metric is given by the extrinsic curvature, viz

$$K_{ij} = -\frac{1}{2N}(\dot{g}_{ij} - \nabla_i N_j - \nabla_j N_i), \quad \dot{g}_{ij} = g_i^a g_j^b \mathcal{L}_t g_{ab}. \tag{5.336}$$

In this section we have decided to denote the 3D covariant derivative by ∇_i in the same way that we have decided to denote the 3D metric by g_{ij}. We may choose the local coordinates such that the vector field t^a has components $(c, 0, \ldots, 0)$ and as a consequence the diffeomorphism corresponding to time evolution is precisely given by $(x^0, x^1, \ldots, x^D) \longrightarrow (x^0 + \delta x^0, x^1, \ldots, x^D)$ and hence $\dot{g}_{ij} = \partial g_{ij}/\partial t$.

From the ADM decomposition (8.84) we see that the combination $K_{ij}K^{ij} - K^2$ where $K = g^{ij}K_{ij}$ is the only combination which is invariant under 4D diffeomorphisms. Under the 3D (foliation-preserving) diffeomorphisms it is obvious that both terms $K_{ij}K^{ij}$ and K^2 are, by construction, separately invariant. We are led therefore to consider the kinetic action

$$S_K = \frac{1}{2\kappa^2} \int dt d^D x N \sqrt{g}(K_{ij}K^{ij} - \lambda K^2). \tag{5.337}$$

Let us determine the mass dimension of the different objects. Let us set $\hbar = 1$. From the Heisenberg uncertainty principle we know that the mass dimension of x is precisely P^{-1} where P is some typical momentum. In order to reflect the properties of the fixed point we will set a scale Z of dimension $[Z] = [x]^z/[t]$ to be dimensionless, i.e. $[c] = P^{z-1}$. This choice is consistent with the scaling rules (5.309). The mass dimension of t is therefore given by P^{-z}. The volume element is hence of mass dimension

$$[dtd^Dx] = P^{-z-D}. \tag{5.338}$$

Now from the line element (5.321) we see that dx^i and N^idt have the same mass dimension and hence the mass dimension of N^i is P^{z-1}. The mass dimension of the line element ds^2 must be the same as the mass dimension of dx^2, i.e. $[ds] = P^{-1}$ and as a consequence $[g_{ij}] = P^0$. Similarly we can conclude that the mass dimension of N is P^0. In summary we have

$$[g_{ij}] = [N] = P^0, \quad [N^i] = P^{z-1}. \tag{5.339}$$

From the above results we conclude that the mass dimension of the extrinsic curvature is given by

$$[K_{ij}] = P^z. \tag{5.340}$$

We can now derive the mass dimension of the coupling constant κ. We have

$$[S_K] \equiv P^0 = \frac{1}{[\kappa]^2}P^{-z-D}P^{2z} \Rightarrow [\kappa] = P^{\frac{z-D}{2}}. \tag{5.341}$$

Thus in $D = 3$ spatial dimensions we must have $z = 3$ in order for κ to be dimensionless and hence the theory power-counting renormalizable.

The second coupling constant λ is also dimensionless. It only appears because the two terms $K_{ij}K^{ij}$ and K^2 are separately invariant under the 3D (foliation-preserving) diffeomorphisms.

The kinetic action (5.337) can be rewritten in a trivial way as

$$S_K = \frac{1}{2\kappa^2}\int dtd^Dx N\sqrt{g}\,K_{ij}G^{ijkl}K_{kl}. \tag{5.342}$$

The metric on the space of metrics G^{ijkl} is a generalized version of the so-called Wheeler–DeWitt metric given explicitly by

$$G^{ijkl} = \frac{1}{2}(g^{ik}g^{jl} + g^{il}g^{jk}) - \lambda g^{ij}g^{kl}. \tag{5.343}$$

This is the only form consistent with 3D (foliation-preserving) diffeomorphisms. Full spacetime diffeomorphism invariance corresponding to general relativity fixes the value of λ as $\lambda = 1$. The inverse of G is defined by

$$G_{ijmn}G^{mnkl} = \frac{1}{2}(g_i^{\ k}g_j^{\ l} + g_i^{\ l}g_j^{\ k}). \tag{5.344}$$

We find explicitly

$$G_{ijkl} = \frac{1}{2}(g_{ik}g_{jl} - g_{il}g_{jk}) - \frac{\lambda}{D\lambda - 1}g_{ij}g_{kl}. \qquad (5.345)$$

We will always assume for $D = 3$ that $\lambda \neq 1/3$ for obvious reasons. The precise role of λ is still not very clear and we will try to study it more carefully in the following.

5.7.3 Potential action and detailed balance

The total action of Horava–Lifshitz gravity is a difference between the kinetic action constructed above and a potential action, viz

$$S = S_K - S_V. \qquad (5.346)$$

The potential term, in the spirit of effective field theory, must contain all terms consistent with the foliation-preserving diffeomorphisms which are of mass dimensionless or equal than the kinetic action. These potential terms will contain in general spatial derivatives but not time derivatives which are already taken into account in the kinetic action. These potential terms must be obviously scalars under foliation-preserving diffeomorphisms.

The mass dimension of the kinetic term is $[K_{ij}K_{ij}] = P^{2z} = P^6$. Thus the potential action must contain all covariant scalars which are of mass dimensions less than or equal than 6. These terms are built from g_{ij} and N and their spatial derivatives. Because g_{ij} and N are both dimensionless the scalar term of mass dimension n must contain n spatial derivatives since $[x_i] = P^{-1}$. For projectable Horava–Lifshitz gravity the lapse function does not depend on space and hence all terms can only depend on the metric g_{ij} and its spatial derivatives. Obviously terms with an odd number of spatial derivatives are not covariant. There remain terms with mass dimensions 0, 2, 4 and 6.

The term of mass dimension 0 is precisely the cosmological constant while the term of mass dimension 2 is the Ricci scalar, viz.

$$\begin{aligned} &\text{mass dimension} = 0, \quad R^0 \\ &\text{mass dimension} = 2, \quad R. \end{aligned} \qquad (5.347)$$

The terms of mass dimensions 4 and 6 are given by the lists

$$\begin{aligned} &\text{mass dimension} = 4, \quad R^2, R_{ij}R^{ij} \\ &\text{mass dimension} = 6, \quad R^3, RR_i^j R_j^i, R_j^i R_k^j R_i^k, R\,\nabla^2 R, \nabla_i\, R_{jk}\,\nabla^i\, R^{jk}. \end{aligned} \qquad (5.348)$$

The operators of mass dimensions 0, 2 and 4 are relevant (super renormalizable) while the operators of dimension 6 are marginal (renormalizable). The quadratic terms modify the propagator and add interactions while cubic terms in the curvature provide only interaction terms The term $\nabla_i\, R_{jk}\,\nabla^j\, R^{ik}$ is not included in the list because it is given by a linear combination of the above terms up to a total derivative. The potential action of projectable Horava–Lifshitz gravity is then given by

$$S_V = \int dt d^D x \sqrt{g} \, N V[g_{ij}]. \tag{5.349}$$

$$
\begin{aligned}
V[g_{ij}] = {}& g_0 + g_1 R + g_2 R^2 + g_3 R_{ij} R^{ij} + g_4 R^3 \\
& + g_5 R R_{ij} R^{ij} + g_6 R_i^{\,j} R_j^{\,k} R_k^{\,i} + g_7 R \nabla^2 R + g_8 \nabla_i R_{jk} \nabla^i R^{jk}.
\end{aligned}
\tag{5.350}
$$

The lowest order potential coincides with general relativity. In general relativity the projectability condition can always be chosen at least locally as a gauge choice which is not the case for Horava–Lifshitz gravity.

A remark now on non-projectable Horava–Lifshitz gravity is in order. In this case the lapse function depend on time and space which matches the spacetime-dependence of the lapse function in general relativity. Furthermore, it can be shown that $a_i = \partial_i \ln N$ transforms as a vector under the diffeomorphism group $\mathrm{Diff}_{\mathcal{F}}(\mathcal{M})$ and as a consequence more terms such as $a_i a^i$, $\nabla_i a^i$ must be included in the potential action. The lowest order potential in this case is found to be given by

$$V[g_{ij}] = g_0 + g_1 R + \alpha a_i a^i + \beta \nabla_i a^i. \tag{5.351}$$

It is very hard to see whether or not the RG flow of the coupling constants α and β goes to zero in the infrared in order to recover general relativity. In [36] it was shown that the non-vanishing of α and β in the IR leads to the existence of a scalar mode.

Alternatively, we can rewrite the total action as follows. The first part is the Hilbert–Einstein action given by

$$S_{EH} = \frac{1}{2\kappa^2} \int dt \int d^D x N \sqrt{g} \, [K_{ij} K^{ij} - K^2 - 2\kappa^2 g_1 R - 2\kappa^2 g_0]. \tag{5.352}$$

Recall that $[t] = P^{-3}$ and $[x] = P^{-1}$. We scale time as $t' = \zeta^2 t$ where ζ is of mass dimension P. It is clear that $[t'] = P^{-1} = [x]$ and thus in the new system of coordinates (t', x^i) we can choose as usual $c = 1$. We have then

$$S_{EH} = \frac{1}{2(\kappa\zeta)^2} \int dt' \int d^D x N \sqrt{g} \, [K_{ij} K^{ij} - K^2 - 2(\kappa\zeta)^2 g_1 R - 2(\kappa\zeta)^2 g_0]. \tag{5.353}$$

The coupling constant g_1 is of mass dimension P^4. Thus we may choose g_1 or equivalently ζ such that

$$-2(\kappa\zeta)^2 g_1 = 1. \tag{5.354}$$

We can now make the identification

$$\frac{1}{2(\kappa\zeta)^2} = \frac{1}{2} M_{\text{Planck}}^2 = \frac{1}{16\pi G_{\text{Newton}}}, \quad (\kappa\zeta)^2 g_0 = \Lambda. \tag{5.355}$$

Thus the Hilbert–Einstein action is given by

$$S_{EH} = \frac{1}{2} M_{\text{Planck}}^2 \int dt' \int d^D x N \sqrt{g} \, [K_{ij} K^{ij} - K^2 + R - 2\Lambda]. \tag{5.356}$$

To obtain the Horava–Lifshitz action we need to add 8 Lorentz-violating terms given by (with $\xi = 1 - \lambda$ and $g_2 = \hat{g}_2 \zeta^2$, $g_3 = \hat{g}_3 \zeta^2$ since g_2 and g_3 are of mass dimensions P^2)

$$S_{LV} = \frac{1}{2\kappa^2} \int dt \int d^D x N \sqrt{g} \, \xi K^2 + \int dt \int d^D x N \sqrt{g} \Big[-\hat{g}_2 \zeta^2 R^2 - \hat{g}_3 \zeta^2 R_{ij} R^{ij} - g_4 R^3 - g_5 R R_{ij} R^{ij}$$
$$- g_6 R_i^j R_j^k R_k^i - g_7 R \nabla^2 R - g_8 \nabla_i R_{jk} \nabla^i R^{jk} \Big]. \tag{5.357}$$

Equivalently

$$S_{LV} = \frac{1}{2(\kappa\xi)^2} \int dt' \int d^D x N \sqrt{g} \Big[\xi K^2 - 2\hat{g}_2 (\kappa\zeta)^2 R^2 - 2\hat{g}_3 (\kappa\zeta)^2 R_{ij} R^{ij} - 2g_4 \kappa^2 R^3 - 2g_5 \kappa^2 R R_{ij} R^{ij}$$
$$- 2g_6 \kappa^2 R_i^j R_j^k R_k^i - 2g_7 \kappa^2 R \nabla^2 R - 2g_8 \kappa^2 \nabla_i R_{jk} \nabla^i R^{jk} \Big]. \tag{5.358}$$

We may set $\kappa = 1$ for simplicity. These Lorentz-violating terms lead to a scalar mode for the graviton with mass of order $O(\xi)$. Furthermore, these terms are not small since they become comparable to the Einstein–Hilbert action for momenta of the order $M_i = M_{Pl}/g_i^{0.5}$, $i = 2, 3$ and $M_i = M_{Pl}/g_i^{0.25}$, $i = 4, 5, 6, 7$. The Planck scale M_{Pl} is independent of the various Lorentz-violating scales M_i which can be driven arbitrarily high by fine tuning of the dimensionless coupling constants g_i.

We will now impose the condition of detailed balance on the potential action. Thus we require that the potential action is of the special form

$$S_V = \frac{\kappa^2}{8} \int dt d^D x \sqrt{g} \, N E^{ij} G_{ijkl} E^{kl}. \tag{5.359}$$

The tensor E is derived from some Euclidean D-dimensional action W as follows

$$\sqrt{g} E^{ij} = \frac{\delta W}{\delta g_{ij}}. \tag{5.360}$$

It is clear that with the detailed balance condition the potential is a perfect square. As it turns out, detailed balance leads to a cosmological constant of the wrong sign and parity violation. However, it remains true that renormalization with detailed balance condition of the $(D + 1)$-dimensional theory is equivalent to the renormalization of the D-dimensional action W together with the renormalization of the relative couplings between kinetic and scalar terms, which is clearly much simpler than renormalization of a generic theory in $(D + 1)$-dimensions.

For theories which are spatially isotropic we can choose the action W to be precisely the Hilbert–Einstein action in D dimensions. This is a relativistic theory with Euclidean signature given by the action

$$W = \frac{1}{\kappa_W^2} \int d^D x \sqrt{g} (R - 2\Lambda_W). \tag{5.361}$$

A standard calculation gives

$$\delta W = \frac{1}{\kappa_W^2} \int d^D x \sqrt{g} \, \delta g^{ij} \Big(R_{ij} - \frac{1}{2} g_{ij} R + g_{ij} \Lambda_W \Big). \tag{5.362}$$

Equivalently

$$\frac{\delta W}{\delta g^{ij}} = \frac{1}{\kappa_W^2} \sqrt{g} \left(R_{ij} - \frac{1}{2} g_{ij} R + g_{ij} \Lambda_W \right). \tag{5.363}$$

Thus

$$E_{ij} = \frac{1}{\kappa_W^2} \left(R_{ij} - \frac{1}{2} g_{ij} R + g_{ij} \Lambda_W \right). \tag{5.364}$$

The potential action becomes therefore

$$S_V = \frac{\kappa^2}{8\kappa_W^4} \int dt d^D x \sqrt{g} N \left(R^{ij} - \frac{1}{2} g^{ij} R + g^{ij} \Lambda_W \right) G_{ijkl} \left(R^{kl} - \frac{1}{2} g^{kl} R + g^{kl} \Lambda_W \right). \tag{5.365}$$

For very short distances (UV) the curvature is clearly the dominant term in W and thus the potential action S_V is dominated by terms quadratic in the curvature. In this case the mass dimension of the potential action $P^{4-z-D}[\kappa]^2/[\kappa_W]^4$ must be equal to the mass dimension of the kinetic action $P^{z-D}/[\kappa]^2$. This leads to the results

$$[\kappa]^2 = P^{z-D}, \quad \frac{[\kappa]^2}{[\kappa_W]^2} = P^{z-2}. \tag{5.366}$$

We have then anisotropic scaling with $z = 2$ and power counting renormalizability in $1 + 2$ dimensions. In a spacetime with $1 + 3$ dimensions we have $[\kappa]^2 = P^{z-3}$ and $[\kappa_W]^2 = P^{-1}$. The fact that the coupling constant κ_W is dimensionfull means the above theory in $1 + 3$ dimensions can only work as an effective field theory valid for up to energies set by the energy scale $1/[\kappa_W]^2$.

At large distances (IR) the dominant term in W is the cosmological constant Λ_W and thus the potential action is dominated by linear and quadratic terms in Λ_W. This is essentially equivalent to the Einstein–Hilbert gravity theory given by the combination $R - 2\Lambda$ and thus effectively the anisotropic scaling becomes the usual value $z = 1$. In other words in $1 + 3$ dimensions, the above Horava–Lifshitz gravity has a $z = 2$ fixed point in the UV and flows to a $z = 1$ fixed point in the IR.

However, we really need to construct a Horava–Lifshitz gravity with a $z = 3$ fixed point in the UV and flows to a $z = 1$ fixed point in the IR. As explained before, the $z = 3$ anisotropic scaling in $1 + 3$ dimensions is exactly what is needed for power counting renormalizability. The theory must satisfy detailed balance and thus one must look for a tensor E_{ij} which is such that it gives a $z = 3$ scaling. It is easy to convince ourselves that E_{ij} must be third order in spatial derivatives so that the dominant term in the potential action S_V contains six spatial derivatives and hence will balance the two time derivatives in the kinetic action. With such an E_{ij} we will have

$$[\kappa]^2 = P^{z-D}, \quad \frac{[\kappa]^2}{[\kappa_W]^2} = P^{z-3}. \tag{5.367}$$

There is a unique candidate for E_{ij} which is known as the Cotton tensor. This is a tensor which is third order in spatial derivatives given explicitly by

$$C^{ij} = \epsilon^{ikl} \, \nabla_k \left(R_l^{\ j} - \frac{1}{4} R g_l^{\ j} \right). \tag{5.368}$$

We now state some results concerning the Cotton tensor without any proof. This is a symmetric tensor $C^{ij} = C^{ji}$, traceless $g_{ij} C^{ij} = 0$, conserved $\nabla_i \, C^{ij} = 0$ which transforms under Weyl transformations of the metric $g_{ij} \longrightarrow \exp(2\Omega) g_{ij}$ as $C^{ij} \longrightarrow \exp(-5\Omega) C^{ij}$, i.e. it is conformal with weight $-5/2$.

In dimensions $D > 3$ conformal flatness of a Riemannian metric is equivalent to the vanishing of the Weyl tensor defined by

$$\begin{aligned} C_{ijkl} = R_{ijkl} &- \frac{1}{D-2}(g_{ik} R_{jl} - g_{il} R_{jk} - g_{jk} R_{il} + g_{jl} R_{ik}) \\ &+ \frac{1}{(D-1)(D-2)}(g_{ik} g_{jl} - g_{il} g_{jk}) R. \end{aligned} \tag{5.369}$$

We can verify that the Weyl tensor is the completely traceless part of the Riemann tensor. In $D = 3$ the Weyl tensor vanishes identically and conformal flatness becomes equivalent to the vanishing of the Cotton tensor.

The Cotton tensor can be derived from an action principle given precisely by the Chern–Simon gravitational action defined by

$$W = \frac{1}{w^2} \int_\Sigma \omega_3(\Gamma). \tag{5.370}$$

$$\begin{aligned} \omega_3(\Gamma) &= \mathrm{Tr}\left(\Gamma \wedge d\Gamma + \frac{2}{3} \Gamma \wedge \Gamma \wedge \Gamma \right) \\ &= \epsilon^{ijk}\left(\Gamma_{il}^m \partial_j \Gamma_{km}^l + \frac{2}{3} \Gamma_{il}^n \Gamma_{jm}^l \Gamma_{kn}^m \right) d^3x. \end{aligned} \tag{5.371}$$

5.8 Exercises

Exercise 1: Show that the action

$$S_M = \int d^n x \sqrt{-\det g} \left(-\frac{1}{2} g^{\mu\nu} \, \nabla_\mu \, \phi \, \nabla_\nu \, \phi - \frac{1}{2} \xi R \phi^2 \right) \tag{5.372}$$

is invariant under the conformal transformations

$$g_{\mu\nu} \longrightarrow \bar{g}_{\mu\nu} = \Omega^2(x) g_{\mu\nu}(x), \quad \phi \longrightarrow \bar{\phi} = \Omega^{\frac{2-n}{2}}(x)\phi(x), \tag{5.373}$$

provided $\xi = (n-2)/(4(n-1))$.

Exercise 2:

- Show that the canonical momentum π associated with a scalar field ϕ which is minimally coupled to a metric $g_{\mu\nu}$ is given by (with an appropriate foliation of spacetime)

$$\pi = -\sqrt{-\det h} \ n^\mu \partial_\mu \phi. \tag{5.374}$$

- Show that the canonical commutation relations

$$[\hat{\phi}(x^0, x^i), \hat{\pi}(x^0, y^i)] = i\delta^{n-1}(x^i - y^i) \tag{5.375}$$

are equivalent to

$$[\hat{a}_i, \hat{a}_j^+] = \delta_{ij}, \quad [\hat{a}_i, \hat{a}_j] = [\hat{a}_i^+, \hat{a}_j^+] = 0. \tag{5.376}$$

Exercise 3: Show that the Klein–Gordon equation in conformal time takes the form

$$\phi'' + 2\frac{a'}{a}\phi' - \partial_i^2\phi + a^2(m^2 + \xi R)\phi = 0. \tag{5.377}$$

Exercise 4:

- Show that the solution of

$$v_k'' + \left(k^2 + \left[\frac{m^2}{H^2} - 2\right]\frac{1}{\eta^2}\right)v_k = 0 \tag{5.378}$$

is given in terms of Bessel functions by

$$v_k = \sqrt{k|\eta|}\left[AJ_n(k|\eta|) + BY_n(k|\eta|)\right], \quad n = \sqrt{\frac{9}{4} - \frac{m^2}{H^2}}. \tag{5.379}$$

- Show that

$$\frac{d}{ds}J_n(s) \cdot Y_n(s) - \frac{d}{ds}Y_n(s) \cdot J_n(s) = -\frac{2}{\pi s}. \tag{5.380}$$

- Verify that the normalization condition of the above reads

$$AB^* - A^*B = \frac{i\pi}{k}. \tag{5.381}$$

Exercise 5: Verify equations (5.156) and (5.157).

Exercise 6: Show that all higher derivatives of the function

$$F(n) = \int_{n^2}^{\infty} dt f_\Lambda\left(\frac{\pi}{L}\sqrt{t}\right)(t)^{1/2} \tag{5.382}$$

vanish.

Exercise 7: Show the following results

$$D_F(\omega, \vec{x}, \vec{x}') = i \int \frac{d^3k}{(2\pi)^3} \frac{e^{-i\vec{k}\,(\vec{x}-\vec{x}')}}{\vec{k}^2 - \omega^2}$$
$$= \frac{i}{4\pi} \frac{e^{i\omega|\vec{x}-\vec{x}'|}}{|\vec{x}-\vec{x}'|}. \tag{5.383}$$

$$D_F(x, x') = \int \frac{d^3k}{(2\pi)^3} \frac{1}{2\omega_k} e^{-i\left(k_4(x_4 - x_4') - \vec{k}(\vec{x}-\vec{x}')\right)}$$
$$= \frac{1}{4\pi^2} \frac{1}{(x - x')^2}. \tag{5.384}$$

Exercise 8: Show that the stress–energy–momentum tensor in flat space with conformal coupling $\xi = 1/6$ and $m = 0$ is given by

$$T_{\mu\nu} = \frac{2}{3}\partial_\mu\phi\partial_\nu\phi + \frac{1}{6}\eta_{\mu\nu}\partial_\alpha\phi\partial^\alpha\phi + \frac{1}{3}\phi\partial_\mu\partial_\nu\phi. \tag{5.385}$$

Exercise 9: Verify that the density of states created by the background σ in the calculation of the Casimir force from renormalizable quantum field theory is given by

$$\frac{dN}{d\omega} = \frac{\omega}{\pi}[D_{F\sigma}(\omega, \vec{x}, \vec{x}) - D_{F0}(\omega, \vec{x}, \vec{x})]. \tag{5.386}$$

Exercise 10: Verify equation (8.80).

References

[1] Weinberg S 1989 The cosmological constant problem *Rev. Mod. Phys.* **61** 1

[2] Carroll S M 2001 The cosmological constant *Living Rev. Rel.* **4** 1

[3] Mukhanov V and Winitzki S 2007 *Introduction to Quantum Effects in Gravity* (Cambridge: Cambridge University Press)

[4] Mukhanov V 2005 *Physical Foundations of Cosmology* (Cambridge: Cambridge University Press)

[5] Birrell N D and Davies P C W 1982 *Quantum Fields in Curved Space* (Cambridge: Cambridge University Press)

[6] Wald R M 1995 *Quantum Field Theory in Curved Space-Time and Black Hole Thermodynamics* (University of Chicago Press)

[7] Milton K A 2001 *The Casimir Effect: Physical Manifestations of Zero-Point Energy* (Singapore: World Scientific)

[8] Graham N, Jaffe R L, Khemani V, Quandt M, Schroeder O and Weigel H 2004 The Dirichlet Casimir problem *Nucl. Phys.* B **677** 379

[9] Arnowitt R L, Deser S and Misner C W 1959 Dynamical structure and definition of energy in general relativity *Phys. Rev.* **116** 1322

[10] Horava P 2009 Membranes at quantum criticality *J. High Energy Phys.* **0903** 020

[11] Carroll S M 2003 Why is the Universe accelerating? *AIP Conf. Proc.* **743** 16

[12] Melchiorri A[Boomerang Collaboration] *et al* 2000 A measurement of omega from the North American test flight of BOOMERANG *Astrophys. J.* **536** L63

[13] Fulling S A 1989 Aspects of quantum field theory in curved space-time *London Math. Soc. Student Texts* **17** 1

[14] Jacobson T 2000 Introduction to quantum fields in curved space-time and the Hawking effect ArXiv:gr-qc/0308048 https://arxiv.org/abs/gr-qc/0308048

[15] Ford L H 1997 Quantum field theory in curved space-time ArXiv:gr-qc/9707062 https://arxiv.org/abs/gr-qc/9707062

[16] Cortez J, Martin-de Blas D, Mena Marugan G A and Velhinho J 2013 Massless scalar field in de Sitter spacetime: unitary quantum time evolution *Class. Quantum Grav.* **30** 075015

[17] Perez-Nadal G, Roura A and Verdaguer E 2007 Backreaction from weakly and strongly non-conformal fields in de Sitter spacetime *PoS QG* -**PH** 034

[18] Bunch T S and Davies P C W 1978 Quantum field theory in de Sitter space: renormalization by point splitting *Proc. R. Soc. Lond.* A **360** 117

[19] Mottola E 1985 Particle creation in de Sitter space *Phys. Rev.* D **31** 754

[20] Allen B 1985 Vacuum states in de Sitter space *Phys. Rev.* D **32** 3136

[21] Anninos D 2012 De Sitter musings *Int. J. Mod. Phys.* A **27** 1230013

[22] Angus P 2008 Vacuum energy in expanding spacetime and superoscillation induced resonance *Master Thesis* University of Waterloo

[23] Kempf A 2001 Mode generating mechanism in inflation with cutoff *Phys. Rev.* D **63** 083514

[24] Zhang H-H, Feng K-X, Qiu S-W, Zhao A and Li X-S 2010 On analytic formulas of Feynman propagators in position space *Chin. Phys.* C **34** 1576

[25] Siopsis G Quantum Field Theory I (Lecture Notes) siopsis@tennessee.edu

[26] Milton K A 2003 Calculating Casimir energies in renormalizable quantum field theory *Phys. Rev.* D **68** 065020

[27] Milton K A 2011 Local and global Casimir energies: divergences, renormalization, and the coupling to gravity *Casimir Physics* (Lecture Notes in Physics vol 834) (Springer) pp 39–95

[28] Jaffe R L 2005 The Casimir effect and the quantum vacuum *Phys. Rev.* D **72** 021301

[29] Wald R M 1984 *General Relativity* (Chicago, IL: University of Chicago Press)

[30] Horava P 2009 Quantum gravity at a Lifshitz point *Phys. Rev.* D **79** 084008

[31] Horava P 2009 Spectral dimension of the universe in quantum gravity at a Lifshitz point *Phys. Rev. Lett.* **102** 161301

[32] Mukohyama S 2010 Horava–Lifshitz cosmology: a review *Class. Quant. Grav.* **27** 223101

[33] Visser M 2011 Status of Horava gravity: a personal perspective *J. Phys. Conf. Ser.* **314** 012002

[34] Weinfurtner S, Sotiriou T P and Visser M 2010 Projectable Horava–Lifshitz gravity in a nutshell *J. Phys. Conf. Ser.* **222** 012054

[35] Piresa L Hořava-Lifshitz gravity: Hamiltonian formulation and connections with CDT *Master's Thesis* Utrecht University

[36] Bellorin J and Restuccia A 2011 Consistency of the Hamiltonian formulation of the lowest-order effective action of the complete Horava theory *Phys. Rev.* D **84** 104037

[37] Orlando D and Reffert S 2009 On the renormalizability of Horava-Lifshitz-type gravities *Class. Quant. Grav.* **26** 155021

[38] Orlando D and Reffert S 2010 On the perturbative expansion around a Lifshitz point *Phys. Lett.* B **683** 62

[39] Giribet G, Nacir D L and Mazzitelli F D 2010 Counterterms in semiclassical Horava-Lifshitz gravity *J. High Energy Phys.* **1009** 009

IOP Publishing

Lectures on General Relativity, Cosmology and Quantum Black Holes (Second Edition)

Badis Ydri

Chapter 6

Hawking radiation, the information paradox and black hole thermodynamics

6.1 Introduction and summary

String theory provides one of the deepest insights into quantum gravity (QG). Its single most central and profound result is the anti-de Sitter/conformal field theory (AdS/ CFT) correspondence or gauge/gravity duality [1]. See [2, 3] for a pedagogical introduction. As it turns out, this duality allows us to study in novel ways: (i) the physics of gauge theory (QCD, in particular and the existence of Yang–Mills theories in four dimensions), as well as (ii) the physics of black holes (information loss paradox and the reconciliation of general relativity and quantum mechanics). String theory therefore reduces for us to the study of the AdS/CFT correspondence.

Indeed, the fundamental observation which drives the lectures in this chapter is that: 'BFSS matrix model [4] and the AdS/CFT duality [1, 5, 6] relates string theory in certain backgrounds to quantum mechanical systems and quantum field theories' which is a quotation taken from Polchinski [7]. The basic problem which is of paramount interest to QG is Hawking radiation of a black hole and the consequent evaporation of the hole and corresponding information loss [8, 9]. The BFSS (Banks–Fischler–Shenker–Susskind) and the AdS/CFT imply that there is no information loss paradox in the Hawking radiation of a black hole. This is the central question we would like to understand in great detail.

Towards this end, we need to understand first quantum black holes, before we can even touch the AdS/CFT correspondence, which requires in any case a great deal of conformal field theory and string theory as crucial ingredients. Thus, in this last chapter of this book we will only worry about black hole radiation, black hole thermodynamics and the information problem following [7, 10–12].

The main reference, guideline and motivation behind these lectures is the lucid and elegant book by Susskind and Lindesay [10]. The lectures by Jacobson [13] and

doi:10.1088/978-0-7503-5824-8ch6

Harlow [12] played also a major role in many crucial issues throughout. We have also benefited greatly from the books by Mukhanov [14] and Carroll [15]. The reference list at the end of these lectures is very limited and only include articles that were actually consulted by the author in the preparation of this chapter. A far more extensive and exhaustive list of references can be found in Harlow [12] and Jacobson [13].

We summarize the content of this most important chapter of this book as follows.

A systematic derivation of the Hawking radiation is given in three different ways. By employing the fact that the near-horizon geometry of a Schwarzschild black hole is Rindler spacetime and then applying the Unruh effect in Rindler spacetime. Secondly, by considering the eternal black hole geometry and studying the properties of the Kruskal vacuum state with respect to the Schwarzschild observer. Thirdly, by considering a Schwarzschild black hole formed by gravitational collapse and deriving the actual incoming state known as the Unruh vacuum state. Although the actual quantum state of the black hole is pure, the asymptotic Schwarzschild observer registers a thermal mixed state with temperature $T_H = 1/(8\pi GM)$. Indeed, a correlated entangled pure state near the horizon gives rise to a thermal mixed state outside the horizon.

The information loss problem is then discussed in great detail. The black hole starts in a pure state and after its complete evaporation the Hawking radiation is also in a pure state. This is the assumption of unitarity. Thus, the entanglement entropy starts at zero value then it reaches a maximum value at the so-called Page time then drops to zero again. The Page time is the time at which the black hole evaporates around one half of its mass and the information starts to get out with the radiation. Before the Page time only energy gets out with the radiation with little or no information. The behavior of the entanglement entropy with time is called the Page curve and a nice rough derivation of this curve using the so-called Page theorem is outlined.

The last part contains a discussion of the black hole thermodynamics. The thermal entropy is the maximum amount of information contained in the black hole. The entropy is mostly localized near the horizon, but QFT gives a divergent value, instead of the Bekenstein–Hawking value $S = A/4G$. QFT must be replaced by QG near the horizon and this separation of the QFT and QG degrees of freedom can be implemented by the stretched horizon which is a time-like membrane, at a distance of one Planck length $l_P = \sqrt{G\hbar}$ from the actual horizon, and where the temperature gets very large and most of the black hole entropy accumulates.

6.2 Rindler spacetime and general relativity

6.2.1 Rindler spacetime

We start with Minkowski spacetime with metric and interval

$$\eta_{\mu\nu} = (-1, +1, +1, +1), \quad ds^2 = \eta_{\mu\nu} dx^\mu dx^\nu. \tag{6.1}$$

We recall the Planck length

$$l_P = \sqrt{\frac{\hbar G}{c^3}}.$$ (6.2)

Usually we will use the natural units $\hbar = c = 1$.

We will first construct the so-called Rindler spacetime, i.e. a uniformly accelerating (non-inertial) reference frame with respect to (say) Minkowski spacetime. This is characterized by an artificial gravitational field which can be removed (the only known case of its kind) by a coordinates transformation. We will follow the presentation by 't Hooft [16].

Let us consider an elevator in the vicinity of the Earth in free fall. The elevator is assumed to be sufficiently small so that the gravitational field inside can be taken to be uniform. By the equivalence principle all objects inside the elevator will accelerate in the same way. Thus, during the free fall of the elevator the observer inside will not experience any gravitational field at all since they are effectively weightless.

We consider the opposite situation in which an elevator in empty space, where there is no gravitational field, is uniformly accelerated upward. The observer inside will feel pressure from the floor as if they are near the Earth or any other planet. In other words, this observer will be experiencing an artificial uniform gravitational field given precisely by the constant acceleration. The question now is how does this observer inside the elevator see spacetime?

Let ξ^μ be the coordinates system inside the elevator which is uniformly accelerated outward in the x direction in outer space with an acceleration a. The motion of the elevator is given by the functions $x^\mu = x^\mu(\xi)$ where x^μ are the coordinates of Minkowski spacetime. At time $\tau = 0$, as measured by the observer inside the elevator, the two systems coincide. We take the origin to be at the middle floor of the elevator.

During an infinitesimal time $d\tau$ the elevator can be assumed to have a constant velocity $v = ad\tau$. In other words, the motion of the elevator within this time is approximately inertial given by the Lorentz transformation

$$\xi^0 = \gamma\left(x^0 - \frac{v}{c}x^1\right) \Rightarrow d\tau \simeq x^0 - ad\tau x^1$$

$$\xi^1 = \gamma\left(x^1 - \frac{v}{c}x^0\right) \Rightarrow \xi^1 \simeq x^1 - ad\tau x^0$$ (6.3)

$$\xi^2 = x^2$$

$$\xi^3 = x^3.$$

We write this as (by suppressing the transverse directions)

$$\begin{pmatrix} x^0 \\ x^1 \end{pmatrix} - \begin{pmatrix} d\tau \\ 0 \end{pmatrix} = \begin{pmatrix} 1 & ad\tau \\ ad\tau & 1 \end{pmatrix} \begin{pmatrix} 0 \\ \xi^1 \end{pmatrix}.$$ (6.4)

This relates the coordinates $(\vec{\xi}, d\tau)$ as measured by the observer in the elevator to the coordinates $(\vec{x}, t)$ as measured by the Minkowski observer. The above transformation looks like a Poincaré transformation, i.e. a combination of a Lorentz transformation and a translation which is here in time. In many cases Poincaré transformations can be rewritten as Lorentz transformations with respect to a properly chosen reference point as the origin. The reference point here is given by

$$A^\mu = (0, 1/a, 0, 0). \tag{6.5}$$

Indeed,

$$\begin{pmatrix} d\tau \\ 0 \end{pmatrix} = \begin{pmatrix} 0 & ad\tau \\ ad\tau & 0 \end{pmatrix} \begin{pmatrix} 0 \\ 1/a \end{pmatrix}. \tag{6.6}$$

Thus

$$\begin{pmatrix} x^0 \\ x^1 + 1/a \end{pmatrix} = \begin{pmatrix} 1 & ad\tau \\ ad\tau & 1 \end{pmatrix} \begin{pmatrix} 0 \\ \xi^1 + 1/a \end{pmatrix}. \tag{6.7}$$

We rewrite then the Lorentz transformation (6.4) as

$$\begin{pmatrix} x^0 \\ \vec{x} + \vec{A} \end{pmatrix} = (1 + \delta L) \begin{pmatrix} 0 \\ \vec{\xi} + \vec{A} \end{pmatrix}, \quad \delta L = \begin{pmatrix} 0 & ad\tau \\ ad\tau & 0 \end{pmatrix}. \tag{6.8}$$

We repeat this N times. In other words, at time $\tau = Nd\tau$ the Minkowski coordinates $x^\mu = (t, \vec{x})$ are related to the elevator coordinates $\xi^\mu = (\tau, \vec{\xi})$ by

$$\begin{pmatrix} x^0 \\ \vec{x} + \vec{a}/a^2 \end{pmatrix} = L(\tau) \begin{pmatrix} 0 \\ \vec{\xi} + \vec{a}/a^2 \end{pmatrix}, \quad L(\tau) = (1 + \delta L)^N. \tag{6.9}$$

Then we have

$$L(\tau + d\tau) = (1 + \delta L)L(\tau). \tag{6.10}$$

The solution can be put in the form (suppressing again transverse directions)

$$L(\tau) = \begin{pmatrix} A(\tau) & B(\tau) \\ B(\tau) & A(\tau) \end{pmatrix}. \tag{6.11}$$

The initial condition is

$$L(0) = \mathbf{1} \leftrightarrow A(0) = 1, \quad B(0) = 0. \tag{6.12}$$

We have then the differential equation

$$\delta L. \, L(\tau) = L(\tau + d\tau) - L(\tau) = d\tau \begin{pmatrix} \dfrac{dA}{d\tau} & \dfrac{dB}{d\tau} \\ \dfrac{dB}{d\tau} & \dfrac{dA}{d\tau} \end{pmatrix}. \tag{6.13}$$

Equivalently

$$\frac{dA}{d\tau} = aB, \quad \frac{dB}{d\tau} = aA. \tag{6.14}$$

The solution is then

$$A = \cosh a\tau, \quad B = \sinh a\tau. \tag{6.15}$$

Finally, we get the coordinates

$$x^0 = \sinh a\tau \cdot \left(\xi^1 + \frac{1}{a}\right)$$

$$x^1 = \cosh a\tau \cdot \left(\xi^1 + \frac{1}{a}\right) - \frac{1}{a} \tag{6.16}$$

$$x^2 = \xi^2$$

$$x^3 = \xi^3.$$

We compute immediately

$$-(dx^0)^2 + (dx^1)^2 = -a^2\left(\xi^1 + \frac{1}{a}\right)^2 d\tau^2 + (d\xi^1)^2. \tag{6.17}$$

Thus, the metric in Rindler spacetime is given by (with $\xi^0 = \tau$)

$$ds^2 = g_{\mu\nu}d\xi^\mu d\xi^\nu = -a^2(\xi^1 + \frac{1}{a})^2 d\tau^2 + d\vec{\xi}^2. \tag{6.18}$$

This is one of the simplest Riemann spacetimes. More on this spacetime is covered in the following discussion.

6.2.2 Review of general relativity

We consider a Riemannian (curved) manifold $\mathcal{M}$ with a metric $g_{\mu\nu}$. A coordinates transformation is given by

$$x^\mu \longrightarrow x'^\mu = x'^\mu(x). \tag{6.19}$$

The vectors and one-forms on the manifold are quantities which are defined to transform under the above coordinates transformation, respectively, as follows

$$V'^\mu = \frac{\partial x'^\mu}{\partial x^\nu} V^\nu. \tag{6.20}$$

$$V'_\mu = \frac{\partial x^\nu}{\partial x'^\mu} V_\nu. \tag{6.21}$$

The spaces of vectors and one-forms are the tangent and co-tangent bundles.

A tensor is a quantity with multiple indices (covariant and contravariant) transforming in a similar way, i.e. any contravariant index is transforming as (8.3) and any covariant index is transforming as (8.4). For example, the metric $g_{\mu\nu}$ is a second rank symmetric tensor which transforms as

$$g'_{\mu\nu}(x') = \frac{\partial x^\alpha}{\partial x'^\mu} \frac{\partial x^\beta}{\partial x'^\nu} g_{\alpha\beta}(x). \tag{6.22}$$

The interval $ds^2 = g_{\mu\nu} dx^\mu dx^\nu$ is therefore invariant. In fact, all scalar quantities are invariant under coordinate transformations. For example, the volume element $d^4x\sqrt{-\det g}$ is a scalar under coordinate transformation.

The derivative of a tensor does not transform as a tensor. However, the so-called covariant derivative of a tensor will transform as a tensor. The covariant derivatives of vectors and one-forms are given by

$$\nabla_\mu \, V^\nu = \partial_\mu V^\nu + \Gamma^\nu_{\alpha\mu} V^\alpha. \tag{6.23}$$

$$\nabla_\mu \, V_\nu = \partial_\mu V_\nu - \Gamma^\alpha_{\mu\nu} V_\alpha. \tag{6.24}$$

These transform indeed as tensors as one can easily check. Generalization to tensor is obvious. The Christoffel symbols $\Gamma^\alpha_{\mu\nu}$ are given in terms of the metric $g_{\mu\nu}$ by

$$\Gamma^\alpha_{\mu\nu} = \frac{1}{2} g^{\alpha\beta} (\partial_\mu g_{\nu\beta} + \partial_\nu g_{\mu\beta} - \partial_\beta g_{\mu\nu}). \tag{6.25}$$

There exists a unique covariant derivative, and thus a unique choice of Christoffel symbols, for which the metric is covariantly constant, viz

$$\nabla_\mu \, g_{\alpha\beta} = 0. \tag{6.26}$$

The straightest possible lines on the curved manifolds are given by the geodesics. A geodesic is a curve whose tangent vector is parallel transported along itself. It is given explicitly by Newton's second law on the curved manifold

$$\frac{d^2 x^\mu}{d\lambda} + \Gamma^\mu_{\alpha\beta} \frac{dx^\alpha}{d\lambda} \frac{dx^\beta}{d\lambda} = 0. \tag{6.27}$$

The λ is an affine parameter along the curve. The time-like geodesics define the trajectories of freely falling particles in the gravitational field encoded in the curvature of the Riemannian manifold. The Riemann curvature tensor $R^\alpha_{\mu\nu\beta}$ is defined in terms of the covariant derivative by

$$(\nabla_\mu \nabla_\nu - \nabla_\nu \nabla_\mu) t^\alpha = -R^\alpha_{\mu\nu\beta} t^\beta. \tag{6.28}$$

The metric is determined by the Hilbert–Einstein action given by

$$S = \frac{1}{16\pi G} \int d^4x \sqrt{-\det g} \, R, \tag{6.29}$$

where the Ricci scalar R is defined from the Ricci tensor $R_{\mu\nu}$ by

$$R = g^{\mu\nu} R_{\mu\nu}. \tag{6.30}$$

$$R_{\mu\nu} = R^{\alpha}_{\mu\alpha\nu}.$$ (6.31)

The Riemann tensor is given explicitly by

$$R^{\alpha}_{\mu\nu\rho} = \partial_{\nu}\Gamma^{\alpha}_{\mu\rho} - \partial_{\rho}\Gamma^{\alpha}_{\mu\nu} + \Gamma^{\alpha}_{\sigma\nu}\Gamma^{\sigma}_{\mu\rho} - \Gamma^{\alpha}_{\rho\sigma}\Gamma^{\sigma}_{\mu\nu}.$$ (6.32)

Indeed, the Euler–Lagrange equations which follows from the above action are precisely the Einstein equations in vacuum, viz

$$\delta S = \frac{1}{16\pi G}\int d^4x\sqrt{-\det g}\left(R_{\mu\nu} - \frac{1}{2}g_{\mu\nu}R\right)\delta g^{\mu\nu} = 0 \Rightarrow R_{\mu\nu} - \frac{1}{2}g_{\mu\nu}R = 0.$$ (6.33)

If we add matter action S_{matter} we obtain the full Einstein equations of motion, viz

$$R_{\mu\nu} - \frac{1}{2}g_{\mu\nu}R = 8\pi G T_{\mu\nu}.$$ (6.34)

The energy–momentum tensor is defined by the equation

$$T_{\mu\nu} = -\frac{2}{\sqrt{-\det g}}\frac{\delta S_{\mathrm{matter}}}{\delta g^{\mu\nu}}.$$ (6.35)

The cosmological constant is one of the simplest matter actions that one can add to the Hilbert–Einstein action. It is given by

$$S_{\mathrm{cc}} = -\frac{1}{8\pi G}\int d^4x\sqrt{-\det g}\,\Lambda.$$ (6.36)

In this case the energy–momentum tensor and the Einstein equations read

$$T_{\mu\nu} = -\frac{\Lambda}{8\pi G}g_{\mu\nu}.$$ (6.37)

$$R_{\mu\nu} - \frac{1}{2}g_{\mu\nu}R + \Lambda g_{\mu\nu} = 0.$$ (6.38)

6.3 Schwarzschild black holes

6.3.1 Schwarzschild black holes

Without further ado we present our first (eternal) black hole. The Schwarzschild black hole is given by the metric

$$ds^2 = -\left(1 - \frac{2GM}{r}\right)dt^2 + \left(1 - \frac{2GM}{r}\right)^{-1}dr^2 + r^2d\Omega^2.$$ (6.39)

The powerful Birkhoff's theorem states that the Schwarzschild metric is the unique vacuum solution (static or otherwise) to Einstein's equations which is spherically symmetric.

The Schwarzschild radius is given by

$$r_s = 2GM. \tag{6.40}$$

This is the event horizon. We remark that the Schwarzschild metric is apparently singular at $r = 0$ and at $r = r_s$. However, only the singularity at $r = 0$ is a true singularity of the geometry. For example we can check that the scalar quantity $R^{\mu\nu\alpha\beta}R_{\mu\nu\alpha\beta}$ is divergent at $r = 0$ whereas it is perfectly finite at $r = r_s$ since [17]

$$R^{\mu\nu\alpha\beta}R_{\mu\nu\alpha\beta} = \frac{48G^2M^2}{r^6}. \tag{6.41}$$

Indeed, the divergence of the Ricci scalar[1] or any other higher order scalar such as $R^{\mu\nu}R_{\mu\nu}$, $R^{\mu\nu\alpha\beta}R_{\mu\nu\alpha\beta}$, etc at a point is a sufficient condition for that point to be singular. We say that $r = 0$ is an essential singularity. The Schwarzschild radius $r = r_s$ is not a true singularity of the metric and its appearance as such only reflects the fact that the chosen coordinates are behaving badly at $r = r_s$. We say that $r = r_s$ is a coordinate singularity. Indeed, it should appear like any other point if we choose a more appropriate coordinates system.

The Riemann tensor encodes the effect of tidal forces on freely falling objects. Thus, the singularity at $r = 0$ corresponds to infinite tidal forces.

The motion of test particles in (Schwarzschild or otherwise) spacetime is given by the geodesic equation

$$\frac{d^2x^\rho}{d\lambda^2} + \Gamma^\rho{}_{\mu\nu}\frac{dx^\mu}{d\lambda}\frac{dx^\nu}{d\lambda} = 0. \tag{6.42}$$

The Schwarzschild metric is obviously invariant under time translations and space rotations. There will therefore be four corresponding Killing vectors K_μ and four conserved quantities (energy and angular momentum) given by

$$Q = K_\mu \frac{dx^\mu}{d\lambda}. \tag{6.43}$$

The metric is independent of x^0 and ϕ and hence the corresponding Killing vectors are

$$K^\mu = (\partial_{x^0})^\mu = \delta_0^\mu = (1, 0, 0, 0), \quad K_\mu = g_{\mu 0} = \left(-\left(1 - \frac{R_s}{r}\right), 0, 0, 0\right). \tag{6.44}$$

$$R^\mu = (\partial_\phi)^\mu = \delta_\phi^\mu = (0, 0, 0, 1), \quad R_\mu = g_{\mu\phi} = (0, 0, 0, r^2 \sin^2\theta). \tag{6.45}$$

The corresponding conserved quantities are the energy and the magnitude of the angular momentum given by

$$E = -K_\mu \frac{dx^\mu}{d\lambda} = \left(1 - \frac{r_s}{r}\right)\frac{dx^0}{d\lambda}. \tag{6.46}$$

---

[1] Actually, $R = 0$ for the Schwarzschild metric.

$$L = R_\mu \frac{dx^\mu}{d\lambda} = r^2 \sin^2 \theta \frac{d\phi}{d\lambda}. \tag{6.47}$$

There is an extra conserved quantity along the geodesic given by (use the geodesic equation and the fact that the metric is covariantly constant)

$$\varepsilon = -g_{\mu\nu} \frac{dx^\mu}{d\lambda} \frac{dx^\nu}{d\lambda}. \tag{6.48}$$

Clearly,

$$\varepsilon = 1, \quad \text{massive particle.} \tag{6.49}$$

$$\varepsilon = 0, \quad \text{massless particle.} \tag{6.50}$$

This extra conserved quantity leads to the radial equation of motion

$$\frac{1}{2}\left(\frac{dr}{d\lambda}\right)^2 + V(r) = \mathcal{E}, \quad \mathcal{E} = \frac{1}{2}(E^2 - \varepsilon). \tag{6.51}$$

The potential is given by

$$V(r) = -\frac{\varepsilon GM}{r} + \frac{L^2}{2r^2} - \frac{GML^2}{r^3}. \tag{6.52}$$

This is the equation of a particle with unit mass and energy $\mathcal{E}$ in a potential $V(r)$. In this potential only the last term is new compared to Newtonian gravity. Clearly when $r \longrightarrow 0$ this potential will go to $-\infty$, whereas if the last term is absent (the case of Newtonian gravity) the potential will go to $+\infty$ when $r \longrightarrow 0$.

For a radially (vertically) freely object we have $d\phi/d\lambda = 0$ and thus the angular momentum is 0, viz $L = 0$. The radial equation of motion becomes

$$\left(\frac{dr}{d\lambda}\right)^2 - \frac{2GM}{r} = E^2 - 1. \tag{6.53}$$

This is essentially the Newtonian equation of motion. The conserved energy is given by

$$E = \left(1 - \frac{2GM}{r}\right)\frac{dt}{d\lambda}. \tag{6.54}$$

We also consider the situation in which the particle was initially at rest at $r = r_i$, viz

$$\frac{dr}{d\lambda}\Big|_{r=r_i} = 0. \tag{6.55}$$

This means in particular that

$$E^2 - 1 = -\frac{2GM}{r_i}. \tag{6.56}$$

The equation of motion becomes

$$\left(\frac{dr}{d\lambda}\right)^2 = \frac{2GM}{r} - \frac{2GM}{r_i}. \tag{6.57}$$

We can identify the affine parameter λ with the proper time for a massive particle. The proper time required to reach the point $r = r_f$ is

$$\tau = \int_0^\tau d\lambda = -(2GM)^{-\frac{1}{2}} \int_{r_i}^{r_f} dr \sqrt{\frac{rr_i}{r_i - r}}. \tag{6.58}$$

The minus sign is due to the fact that in a free fall $dr/d\lambda < 0$. By performing the change of variables $r = r_i(1 + \cos\alpha)/2$ we find the closed result

$$\tau = \sqrt{\frac{r_i^3}{8GM}}(\alpha_f + \sin\alpha_f). \tag{6.59}$$

This is finite when $r_f \longrightarrow r_s = 2GM$. Thus, a freely falling object will cross the Schwarzschild radius in a finite proper time.

We consider now a distant stationary observer hovering at a fixed radial distance r_∞. His proper time is

$$\tau_\infty = \sqrt{1 - \frac{2GM}{r_\infty}}\, t. \tag{6.60}$$

By using equations (6.53) and (6.54) we can find dr/dt. We get

$$\frac{dr}{dt} = -E^{\frac{1}{2}}\frac{d\lambda}{dt}\left(E - \frac{d\lambda}{dt}\right)^{\frac{1}{2}}$$

$$= -\frac{1}{E}\left(1 - \frac{2GM}{r}\right)\left(E^2 - 1 + \frac{2GM}{r}\right)^{\frac{1}{2}}. \tag{6.61}$$

Near $r = 2GM$ we have

$$\frac{dr}{dt} = -\frac{1}{2GM}(r - r_s). \tag{6.62}$$

The solution is

$$r - r_s = \exp\left(-\frac{t}{2GM}\right). \tag{6.63}$$

Thus when $r \longrightarrow r_s = 2GM$ we have $t \longrightarrow \infty$.

We see that with respect to a stationary distant observer at a fixed radial distance r_∞ the elapsed time τ_∞ goes to infinity as $r \longrightarrow 2GM$. The correct interpretation of this result is to say that the stationary distant observer can never see the particle actually crossing the Schwarzschild radius $r_s = 2GM$ although the particle does cross the Schwarzschild radius in a finite proper time as seen by an observer falling with the particle.

6.3.2 Near horizon coordinates

A proper distance from the horizon can be defined by the formula

$$
\rho = \int_{r_s}^{r} \sqrt{g_{rr}(r')}\, dr' = \int_{r_s}^{r} \frac{dr'}{\sqrt{1 - r_s/r}}
$$
$$
= \sqrt{r(r - r_s)} + r_s \sinh \sqrt{\frac{r}{r_s} - 1}\,.
$$

(6.64)

In terms of ρ the metric becomes

$$
ds^2 = -\left(1 - \frac{r_s}{r(\rho)}\right) dt^2 + d\rho^2 + r^2(\rho) d\Omega^2.
$$

(6.65)

Very near the horizon we write $r = r_s + \delta$ and thus $\rho = 2\sqrt{r_s \delta}$. We get then the metric

$$
ds^2 = -\rho^2 \frac{dt^2}{4r_s^2} + d\rho^2 + r_s^2 d\Omega^2.
$$

(6.66)

The first two terms correspond to two-dimensional Minkowski flat space. Indeed, ρ and $\omega = t/2r_s$ are radial and hyperbolic angle variables for Minkowski spacetime. The Minkowski coordinates X and T are defined by

$$
X = \rho \cosh \frac{t}{2r_s}, \quad T = \rho \sinh \frac{t}{2r_s}.
$$

(6.67)

The metric becomes

$$
ds^2 = -dT^2 + dX^2 + r_s^2 d\Omega^2.
$$

(6.68)

If we are only interested in small angular region of the horizon around $\theta = 0$ we can replace the angular variables by Cartesian coordinates as follows

$$
Y = r_s \theta \cos \phi, \quad Z = r_s \theta \sin \phi.
$$

(6.69)

We have then the metric

$$
\begin{aligned}
ds^2 &= -\rho^2 d\omega^2 + d\rho^2 + dY^2 + dZ^2 \\
&= -dT^2 + dX^2 + dY^2 + dZ^2.
\end{aligned}
$$

(6.70)

By comparing with (6.18), we recognize the first line to be the Rindler metric with the identification $a\tau \leftrightarrow \omega$ and $\xi^1 + 1/a \leftrightarrow \rho$. The time ω is called Rindler time and the time translation $\omega \longrightarrow \omega + c$ corresponds to a Lorentz boost in Minkowski spacetime. This approximation of the black hole near-horizon geometry (valid for $r \simeq r_s$ and small angular region) by a Minkowski spacetime is called the Rindler approximation. It shows explicitly that the event horizon is locally non-singular and in fact it is indistinguishable from flat Minkowski spacetime.

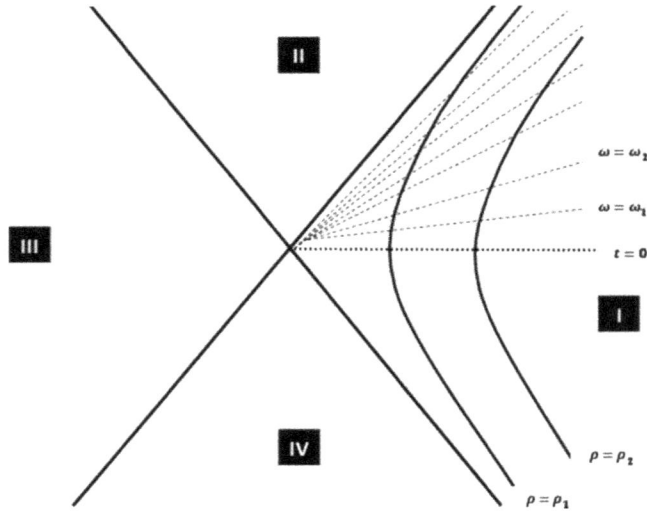

Figure 6.1. Rindler spacetime.

The relation between the Minkowski coordinates $X = \rho \cosh \omega$ and $T = \rho \sinh \omega$ and the Rindler coordinates ρ and ω can also be rewritten as

$$\rho^2 = X^2 - T^2, \quad \frac{T}{X} = \tanh \omega. \tag{6.71}$$

Obviously we must have $X > |T|$. This is called quadrant I or Rindler spacetime. This is the region outside the black hole. The lines of constant ρ are hyperbolae while the lines of constant ω are straight lines through the origin. The horizon lies at the point $\rho = 0$ or $T = X = 0$. The horizon is actually a two-dimensional surface located at $r = r_s$ since $g_{00} = 0$ there and as a consequence this surface has no time extension. See figure 6.1.

6.4 Kruskal–Szekres diagram

6.4.1 Kruskal–Szekres extension and Einstein–Rosen bridge

In this lecture we will follow [15]. The above Schwarzschild geometry can be maximally extended as follows. For a radial null curve, which corresponds to a photon moving radially in Schwarzschild spacetime, the angles θ and ϕ are constants and $ds^2 = 0$, and thus

$$0 = -\left(1 - \frac{2GM}{r}\right)dt^2 + \left(1 - \frac{2GM}{r}\right)^{-1} dr^2. \tag{6.72}$$

In other words,

$$\frac{dt}{dr} = \pm \frac{1}{1 - \frac{2GM}{r}}. \tag{6.73}$$

We integrate the above equation as follows

$$t = \pm \int \frac{dr}{1 - \frac{2GM}{r}}$$

$$= \pm \left(r + 2GM \log \left(\frac{r}{2GM} - 1 \right) \right) + \text{constant}$$

$$= \pm r_* + \text{constant}.$$

(6.74)

We call r_* the tortoise coordinate which makes sense only for $r > 2GM$. The event horizon $r = 2GM$ corresponds to $r_* \longrightarrow \infty$. We compute $dr_* = rdr/(r - 2GM)$ and as a consequence the Schwarzschild metric becomes

$$ds^2 = \left(1 - \frac{2GM}{r} \right)(-dt^2 + dr_*^2) + r^2 d\Omega^2.$$

(6.75)

Next we define $v = t + r_*$ and $u = t - r_*$. Then

$$ds^2 = -\left(1 - \frac{2GM}{r} \right) dv du + r^2 d\Omega^2.$$

(6.76)

For infalling radial null geodesics we have $t = -r_*$ or equivalently $v = \text{constant}$, whereas for outgoing radial null geodesics we have $t = +r_*$ or equivalently $u = \text{constant}$. For every point in spacetime we have two solutions:

- For points outside the event horizon there are two solutions, one infalling and one outgoing.
- For points inside the event horizon there are two solutions which are both infalling.
- For points on the event horizon there are two solutions, one infalling and one trapped.

Next, we will give a maximal extension of the Schwarzschild solution by constructing a coordinate system valid everywhere in Schwarzschild spacetime. We start by noting that the radial coordinate r should be given in terms of u and v by solving the equations

$$\frac{1}{2}(v - u) = r + 2GM \log \left(\frac{r}{2GM} - 1 \right).$$

(6.77)

The event horizon $r = 2GM$ is now either at $v = -\infty$ or $u = +\infty$. The coordinates of the event horizon can be pulled to finite values by defining new coordinates u' and v' as

$$v' = \exp \left(\frac{v}{4GM} \right)$$

$$= \sqrt{\frac{r}{2GM} - 1} \exp \left(\frac{r + t}{4GM} \right).$$

(6.78)

$$u' = -\exp\left(-\frac{u}{4GM}\right)$$

$$= -\sqrt{\frac{r}{2GM} - 1} \, \exp\left(\frac{r - t}{4GM}\right). \tag{6.79}$$

The Schwarzschild metric becomes

$$ds^2 = -\frac{32G^3M^3}{r} \exp\left(-\frac{r}{2GM}\right) dv' du' + r^2 d\Omega^2. \tag{6.80}$$

It is clear that the coordinates u and v are null coordinates and thus u' and v' are also null coordinates. However, we prefer to work with a single time-like coordinate while we prefer the other coordinate to be space-like. We introduce therefore new coordinates T and R defined for $r > 2GM$ by

$$T = \frac{1}{2}(v' + u') = \sqrt{\frac{r}{2GM} - 1} \, \exp\left(\frac{r}{4GM}\right) \sinh\frac{t}{4GM}. \tag{6.81}$$

$$R = \frac{1}{2}(v' - u') = \sqrt{\frac{r}{2GM} - 1} \, \exp\left(\frac{r}{4GM}\right) \cosh\frac{t}{4GM}. \tag{6.82}$$

Clearly, T is time-like while R is space-like. This can be confirmed by computing the metric. This is given by

$$ds^2 = \frac{32G^3M^3}{r} \exp\left(-\frac{r}{2GM}\right)(-dT^2 + dR^2) + r^2 d\Omega^2. \tag{6.83}$$

We see that T is always time-like while R is always space-like since the sign of the components of the metric never get reversed.

We remark that

$$\begin{aligned}
T^2 - R^2 &= v'u' \\
&= -\exp\frac{v - u}{4GM} \\
&= -\exp\frac{r + 2GM \log\left(\frac{r}{2GM} - 1\right)}{2GM} \\
&= \left(1 - \frac{r}{2GM}\right)\exp\frac{r}{2GM}.
\end{aligned} \tag{6.84}$$

The radial coordinate r is determined implicitly in terms of T and R from this equation, i.e. equation (6.84). The coordinates (T, R, θ, ϕ) are called Kruskal–Szekres coordinates. Remarks are now in order:

- The radial null curves in this system of coordinates are given by

$$T = \pm R + \text{constant}. \tag{6.85}$$

All light cones are at ±45 degrees. This 45-degree property means in particular that the radial light cone in the Kruskal–Szekeres diagram has the same form as in special relativity.

- The horizon defined by $r \longrightarrow 2GM$ is seen to appear at $T^2 - R^2 \longrightarrow 0$, i.e. at (6.85) in the new coordinate system. This shows in an elegant way that the event horizon is a null surface.
- The surfaces of constant r are given from (6.84) by $T^2 - R^2 = $ constant which are hyperbolae in the $R - T$ plane.
- For $r > 2GM$ the surfaces of constant t are given by $T/R = \tanh \frac{t}{4GM} = $ constant which are straight lines through the origin. In the limit $t \longrightarrow \pm \infty$ we have $T/R \longrightarrow \pm 1$ which is precisely the horizon $r = 2GM$.

The above solution defines region I of the so-called the Kruskal–Szekres diagram. This solution can be extended to the interior region of the black hole $r < 2GM$ (region II of the Kruskal–Szekres diagram) as follows:

- For $r < 2GM$ we have

$$T = \frac{1}{2}(v' + u') = \sqrt{1 - \frac{r}{2GM}} \, \exp\left(\frac{r}{4GM}\right) \cosh \frac{t}{4GM}. \tag{6.86}$$

$$R = \frac{1}{2}(v' - u') = \sqrt{1 - \frac{r}{2GM}} \, \exp\left(\frac{r}{4GM}\right) \sinh \frac{t}{4GM}. \tag{6.87}$$

The metric and the condition determining r implicitly in terms of T and R do not change form in the (T, R, θ, ϕ) system of coordinates and thus the radial null curves, the horizon as well as the surfaces of constant r are given by the same equation as before.

- For $r < 2GM$ the surfaces of constant t are given by $R/T = \tanh \frac{t}{4GM} = $ constant which are straight lines through the origin.
- It is clear that the allowed range for R and T is (analytic continuation from the region $T^2 - R^2 < 0$ ($r > 2GM$) to the first singularity, which occurs in the region $T^2 - R^2 < 1$ ($r < 2GM$))

$$-\infty \leqslant R \leqslant +\infty, \quad T^2 - R^2 < 1. \tag{6.88}$$

The Kruskal–Szekres diagram gives the maximal extension of the Schwarzschild solution. A Kruskal–Szekres diagram is shown in figure (6.2). Every point in this diagram is actually a 2D sphere since we are suppressing θ and ϕ and drawing only R and T. The Kruskal–Szekres diagram represents the entire Schwarzschild spacetime. It can be divided into four regions:

- Region I: Exterior of black hole with $r > 2GM$ ($R > 0$ and $T^2 - R^2 < 0$). Clearly future-directed time-like (null) worldlines will lead to region II, whereas past-directed time-like (null) worldlines can reach it from region IV. Regions I and III are connected by space-like geodesics.

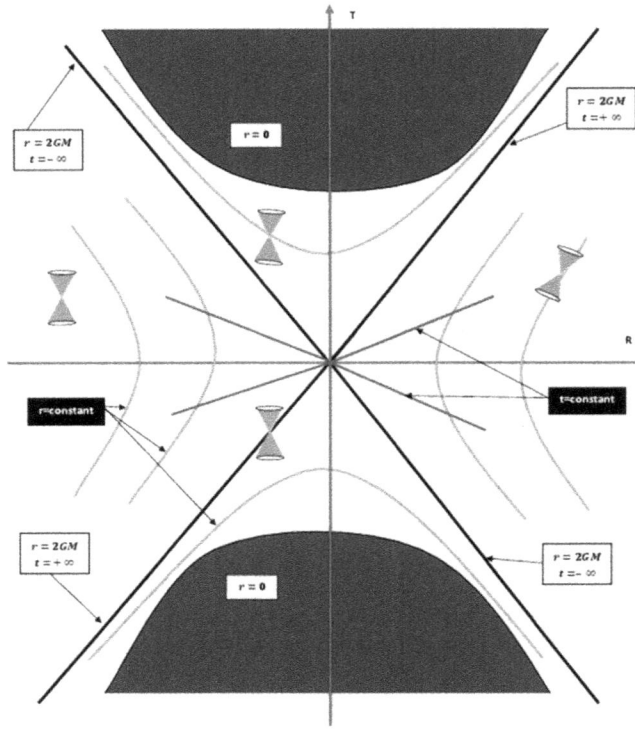

Figure 6.2. Kruskal–Szekres diagram.

- Region II: Inside of black hole with $r < 2GM$ $(T > 0, 0 < T^2 - R^2 < 1)$. Any future-directed path in this region will hit the singularity. In this region r becomes time-like (while t becomes space-like) and thus we cannot stop moving in the direction of decreasing r in the same way that we cannot stop time progression in region I.
- Region III: Parallel exterior region with $r > 2GM$ $(R < 0, T^2 - R^2 < 0)$. This is another asymptotically flat region of spacetime which we cannot access along future or past-directed paths. The Kruskal–Szekres coordinates inside this region are

$$T = -\sqrt{\frac{r}{2GM} - 1} \, \exp\left(\frac{r}{4GM}\right) \sinh \frac{t}{4GM}. \tag{6.89}$$

$$R = -\sqrt{\frac{r}{2GM} - 1} \, \exp\left(\frac{r}{4GM}\right) \cosh \frac{t}{4GM}. \tag{6.90}$$

- Region IV: Inside of white hole with $r < 2GM$ $(T < 0, 0 < T^2 - R^2 < 1)$. The white hole is the time reverse of the black hole. This corresponds to a singularity in the past at which the Universe originated. This is a part of spacetime from which observers can escape to reach us while we cannot go there.

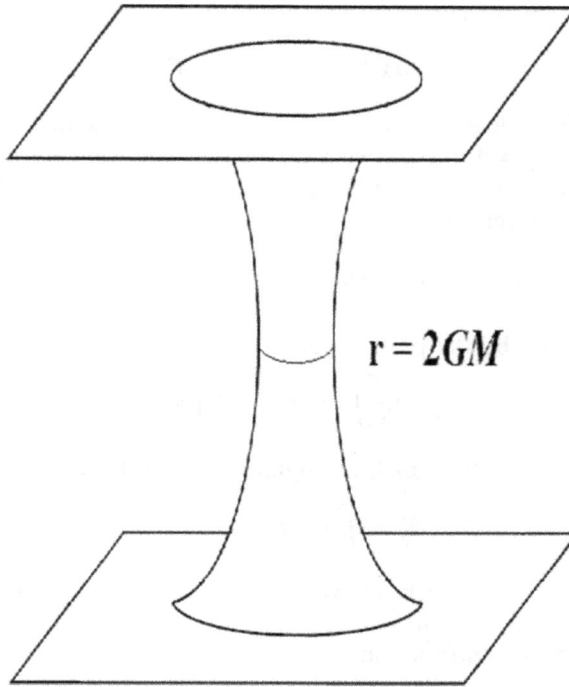

Figure 6.3. The Einstein–Rosen bridge.

The full metric describes therefore two asymptotically flat universes, regions I and III, which are connected by a non-traversable Einstein–Rosen bridge (a wormhole). This is easiest seen at $t = T = 0$ in figure (6.3). However, for constant $T \neq 0$, it is seen that the two asymptotically flat universes disconnect and the wormhole closes up, and thus any time-like observer cannot cross from one region to the other. The singularity $r = 0$ is equivalently given by the hyperboloid $T^2 - R^2 = 1$ which consists of two connected components in regions II (black hole) and IV (white hole) which are called future and past interiors, respectively. The regions I and III are precisely the exterior regions.

6.4.2 Euclidean black hole and thermal field theory

By analytic continuation to Euclidean time $t_E = it$ we obtain

$$ds^2 = \rho^2 \frac{dt_E^2}{4r_s^2} + d\rho^2 + r_s^2 d\Omega^2.$$ (6.91)

The first two terms correspond to 2D flat space, viz

$$X = \rho \cos \frac{t_E}{2r_s}, \quad Y = \rho \sin \frac{t_E}{2r_s}.$$ (6.92)

The metric becomes

$$ds^2 = dX^2 + dY^2 + r_s^2 d\Omega^2. \tag{6.93}$$

In order for the Euclidean metric to be smooth, the Euclidean time t_E must be periodic with period $\beta = 4\pi r_s$ otherwise the metric has a conical singularity at $\rho = 0$.

In quantum mechanics, the transition amplitude between point q at time t and point q' at time t' is given by

$$\langle q', t'|q, t\rangle = \langle q'|\exp(-iH(t' - t))|q\rangle = \sum_n \psi_n(q')\psi_n^*(q)\exp(-iE_n(t' - t)). \tag{6.94}$$

This can also be given by the path integral

$$\langle q', t'|q, t\rangle = \int \mathcal{D}q(t)\exp(iS[q(t)]). \tag{6.95}$$

The action S is given in terms of Lagrangian L by the formula

$$S = \int dt L(q, \dot{q}). \tag{6.96}$$

We perform Wick rotation to Euclidean time $t_E = it$ with $\beta = t_E' - t_E = i(t' - t)$ and we consider closed paths $q' = q(t_E + \beta) = q = q(t_E)$. We get immediately the thermodynamical partition function

$$\begin{aligned} Z = \exp(-\beta F) &= Tr \exp(-\beta H) \\ &= \int dq\langle q|\exp(-\beta H)|q\rangle \\ &= \int dq\langle q, t'|q, t\rangle. \end{aligned} \tag{6.97}$$

The corresponding path integral is (with $iS = -S_E$)

$$Z = \int_{q(t_E + \beta)=q(t_E)} \mathcal{D}q(t)\exp(-S_E[q(t)]). \tag{6.98}$$

The Euclidean action is given in terms of the Lagrangian $L_E = -L$ by the formula

$$S = \int_0^\beta dt_E L_E(q, \dot{q}). \tag{6.99}$$

Thus, a path integral with periodic Euclidean time generates the thermodynamic partition function $Tr \exp(-\beta H)$. This very general and very remarkable result can also be stated by saying that thermal equilibrium is equivalent to summing over all periodic configurations $q(t_E + \beta) = q(t_E)$ in Euclidean time.

The path integral for quantum fields in Euclidean Schwarzschild black hole geometry corresponds to a periodic Euclidean time $t_E \longrightarrow t_E + \beta$ with $\beta = 4\pi r_s$ and thus it describes a gas in equilibrium with the black hole at temperature

$$T_H = \frac{1}{4\pi r_s}. \tag{6.100}$$

The Schwarzschild black hole is thus at equilibrium at the temperature T_H and hence it must emit as many particles as it absorbs.

6.5 Density matrix and entanglement

This section is taken mostly from [10] and [18] but we also found the lecture of [19] on the density matrix very useful.

6.5.1 Density matrix: pure and mixed states

We consider a system consisting of two subsystems A and B. The wave function of this system is written

$$\Psi = \Psi(\alpha, \beta). \tag{6.101}$$

The α and β are two sets of commuting variables relevant for the subsystems A and B separately. If we are only interested in the subsystem A then its complete description is encoded in the density matrix or density operator

$$\rho_A(\alpha, \alpha') = \sum_\beta \Psi^*(\alpha, \beta)\Psi(\alpha', \beta). \tag{6.102}$$

The expectation value of an A-operator a is given by the rule

$$\langle a \rangle = Tr a \rho_A. \tag{6.103}$$

A density matrix ρ satisfies: (I) $Tr\rho = 1$ (sum of probabilities is 1), (II) $\rho = \rho^+$, (III) $\rho_i \geqslant 0$. The eigenvalue ρ_i is the probability that the system A is in the eigenstate $|i\rangle$. The density matrix ρ_A describes therefore a mixed state of the subsystem A, i.e. a statistical ensemble of several quantum states, which arises from the entanglement of the two subsystems A and B, and thus our lack of knowledge of the exact state in which the subsystem A will be found.

This should be contrasted with pure states which are represented by single vectors in Hilbert space. The density matrix associated with a pure state $|i\rangle$ is simply given by $|i\rangle \langle i|$. The complete system formed by A and B is in a pure system although the subsystems A and B are both in mixed states due to entanglement. Another example of a pure state is the case when the density matrix ρ_A has only one non-zero eigenvalue $(\rho_A)_j$ which can only arise from a state of the form

$$\Psi(\alpha, \beta) = \Psi_A(\alpha)\Psi_B(\beta). \tag{6.104}$$

In general, we can write the density matrix corresponding to a mixed state as a convex sum, i.e. a weighted sum with $\sum_i p_i = 1$, of pure state density matrices as follows

$$\rho_A^{\text{mixed}} = \sum_i p_i \rho_i^{\text{pure}} = \sum_i p_i |\psi_i\rangle\langle\psi_i|. \tag{6.105}$$

The states $|\psi_i\rangle$ do not need to be orthogonal. This density matrix satisfies the Liouville–von Neumann equation

$$\frac{\partial \rho}{\partial t} = -\frac{i}{\hbar}[H, \rho]. \tag{6.106}$$

It is better to take an example here. Let us consider a spin 1/2 system. A general pure state of this system is given by

$$|\psi\rangle = \cos\frac{\theta}{2}| + \rangle + \exp(i\phi)\sin\frac{\theta}{2}|-\rangle. \tag{6.107}$$

This state is given by the point on the surface of the unit 2-sphere defined by the vector

$$\overrightarrow{a} = (\sin\theta\cos\phi, \sin\theta\sin\phi, \cos\theta). \tag{6.108}$$

The corresponding density matrix is

$$\rho_{\text{pure}} = |\psi\rangle\langle\psi| = \frac{1}{2}\begin{pmatrix} 1 + \cos\theta & \exp(-i\phi)\sin\theta \\ \exp(i\phi)\sin\theta & 1 - \cos\theta \end{pmatrix} = \frac{1}{2}(1_2 + \overrightarrow{a}.\overrightarrow{\sigma}). \tag{6.109}$$

This is a projector operator, viz

$$\rho_{\text{pure}}^2 = \rho_{\text{pure}}. \tag{6.110}$$

The vector $\overrightarrow{a}$ is called the Bloch vector and the corresponding sphere is called the Bloch sphere. This vector is precisely the expectation value of the spin, viz

$$\overrightarrow{a} = \langle\rho_{\text{pure}}\overrightarrow{\sigma}\rangle. \tag{6.111}$$

Mixed states are given by points inside the Bloch sphere. The corresponding density matrices are given by

$$\rho_{\text{mixed}} = \frac{1}{2}(1_2 + \overrightarrow{a}.\overrightarrow{\sigma}) \neq \rho_{\text{mixed}}^2, \quad \overrightarrow{a}^2 < 1. \tag{6.112}$$

We have then the criterion

$$Tr\rho^2 = 1, \quad \text{pure state.} \tag{6.113}$$

$$0 < Tr\rho^2 = \frac{1 + \overrightarrow{a}^2}{2} < 1, \quad \text{mixed state.} \tag{6.114}$$

The quantity $Tr\rho^2$ is called the purity of the state.

For example, a totally mixed state can have a 50% probability that the electron is in the state $|+\rangle$ and 50% probability that the electron is in the state $|-\rangle$. This corresponds to a completely unpolarized beam, viz $\overrightarrow{a} = 0$. The corresponding density matrix is

$$\rho_{\text{mixed}} = \frac{1}{2}|+\rangle\langle+| + \frac{1}{2}|-\rangle\langle-|$$
$$= \frac{1}{2}1_2. \tag{6.115}$$

This decomposition is not unique. For example, another totally mixed state can have a 50% probability that the electron is in the state $|+\rangle_x$ and 50% probability that the electron is in the state $|-\rangle_x$, viz

$$
\begin{aligned}
\rho_{\text{mixed}} &= \frac{1}{2}\, |+\rangle_x \langle +|_x + \frac{1}{2}\, |-\rangle_x \langle -|_x \\
&= \frac{1}{2}\frac{|+\rangle + |-\rangle}{\sqrt{2}}\frac{\langle +|+\langle -|}{\sqrt{2}} + \frac{1}{2}\frac{|+\rangle - |-\rangle}{\sqrt{2}}\frac{\langle +|-\langle -|}{\sqrt{2}} \\
&= \frac{1}{2}\mathbb{1}_2 .
\end{aligned}
\tag{6.116}
$$

Thus a single density matrix can represent many, infinitely many in fact, different state mixtures.

A partially mixed state, for example, can have a 50% probability that the electron is in the state $|+\rangle$ and 50% probability that the electron is in the state $(|+\rangle + |-\rangle)/\sqrt{2}$, viz

$$
\rho_{\text{mixed}} = \frac{1}{2}|+\rangle\langle +| + \frac{1}{2}\left(\frac{|+\rangle + |-\rangle}{\sqrt{2}}\right)\left(\frac{\langle +|+\langle -|}{\sqrt{2}}\right).
\tag{6.117}
$$

A pure state $|\Phi_c\rangle = (|+\rangle - |-\rangle)/\sqrt{2}$ for example is given by the density matrix

$$
\rho_{\text{pure}} = \frac{|+\rangle - |-\rangle}{\sqrt{2}}\frac{\langle +|-\langle -|}{\sqrt{2}}.
\tag{6.118}
$$

Again this decomposition is not unique. This can be rewritten also as

$$
\rho_{\text{pure}} = \frac{|+\rangle_x - |-\rangle_x}{\sqrt{2}}\frac{\langle +|_x - \langle -|_x}{\sqrt{2}},
\tag{6.119}
$$

since $|\Phi_c\rangle = -(|+\rangle_x - |-\rangle_x)/\sqrt{2}$. Thus the density matrix allows many, possibly infinitely many, different states of the subsystems on the diagonal. This freedom is expected since, by recalling the experiments of Aspect *et al*, which showed that this nonseparable quantum correlation given by the state $|\Phi_c\rangle$ violates Bell's inequalities, we can conclude that: the pure states of the system described by $|\Phi_c\rangle$ are not just unknown but in fact cannot exist before measurement [18].

It is clear from these examples that the relative phases between the basis states in a mixed state are random as opposed to coherent superpositions (pure states). This point is explained in more detail in the following.

A coherent superposition of two states $|\psi_1\rangle$ and $|\psi_2\rangle$ is given by the density matrix

$$
\rho_c = |\alpha|^2 |\psi_1\rangle\langle\psi_1| + |\beta|^2 |\psi_1\rangle\langle\psi_2| + \alpha\beta^* |\psi_1\rangle\langle\psi_2| + \alpha^*\beta |\psi_1\rangle\langle\psi_1|.
\tag{6.120}
$$

However, in the above preceding discussion the mixing is a statistical mixture as opposed to a coherent superposition. A statistical mixture of a state $|\psi_1\rangle$ with a probability $p_1 = |\alpha|^2$ and state $|\psi_1\rangle$ with a probability $p_2 = |\beta|^2$ is given by the density operator

$$\rho_r = p_1 |\psi_1\rangle\langle\psi_1| + p_2 |\psi_1\rangle\langle\psi_2|. \tag{6.121}$$

In other words, it is either $|\psi_1\rangle$ or $|\psi_1\rangle$, whereas in a coherent superposition it is both $|\psi_1\rangle$ and $|\psi_1\rangle$ at the same time. In the first case there is no interference effect (behave as classical probability distribution), while in the second case there is quantum interference. Mixed states are relevant when the exact initial quantum state is not known.

Remark that in the statistical superposition we can change $\alpha \longrightarrow \exp(i\theta)\alpha$ and $\beta \longrightarrow \exp(i\theta')\beta$ without changing the density matrix for θ and θ' arbitrary. In the coherent superposition we must have $\theta = \theta'$.

The probability of obtaining the eigenvalue a_n in the measurement of the observable A is then given by

$$p(a_n) = p_1 |\langle a_n|\psi_1\rangle|^2 + p_2 |\langle a_n|\psi_1\rangle|^2 = Tr\rho|a_n\rangle\langle a_n|. \tag{6.122}$$

In fact, mixed states are incoherent superpositions. The diagonal elements of the density matrix give the probabilities to be in the corresponding states. The off-diagonal elements measure the amount of coherence between the states. The off-diagonal elements are called coherences. Coherence is maximized in a pure state when for every m and n we have

$$\rho_{mn}\rho_{nm} = \rho_{mm}\rho_{nn}. \tag{6.123}$$

A partially mixed state is such that for at least one pair of m and n we have

$$0 < \rho_{mn}\rho_{nm} < \rho_{mm}\rho_{nn}. \tag{6.124}$$

A totally mixed state is such that for at least one pair of m and n we have

$$\rho_{mn} = \rho_{nm} = 0, \quad \rho_{mm}\rho_{nn} \neq 0. \tag{6.125}$$

Coherent superposition means interference, whereas incoherent (mixed) superposition means absence of superposition. Let us take an example. We consider a system described by a coherent superposition of two momentum states k and $-k$ given by the pure state

$$|\psi\rangle = \frac{1}{\sqrt{2}}(|k\rangle + |-k\rangle). \tag{6.126}$$

This quantum coherent superposition corresponds to sending particles through both slits at once. The density matrix is

$$\rho = |\psi\rangle\langle\psi| = \frac{1}{2}|k\rangle\langle k| + \frac{1}{2}|-k\rangle\langle -k| + \frac{1}{2}|k\rangle\langle -k| + \frac{1}{2}|-k\rangle\langle k|. \tag{6.127}$$

The probability of finding the system at x is

$$P(x) = Tr\rho|x\rangle\langle x| = 1 + \cos 2kx. \tag{6.128}$$

These are precisely the fringes (information). If the system is in a mixed (incoherent) state given for example by the density matrix

$$\rho = \frac{1}{2}|k\rangle\langle k| + \frac{1}{2}|-k\rangle\langle -k|. \tag{6.129}$$

This corresponds to the classical case of sending particles at random, i.e. at 50% chance, through either one of the slits (totally mixed state). We get now the probability

$$P(x) = Tr\rho|x\rangle\langle x| = 1. \tag{6.130}$$

So there are no fringes in this case, i.e. the incoherent mixed superposition is characterized by the absence of interference (no information). In a totally mixed state all interference effects are eliminated.

6.5.2 Entanglement, decoherence and von Neumann entropy

We are now in a position to understand better our original definitions (6.101), (6.102), (6.103). The state $\Psi(\alpha, \beta)$ corresponds to a pure state $|\Psi\rangle$, viz $\langle \alpha, \beta|\Psi\rangle = \Psi(\alpha, \beta)$. The corresponding density matrix is $\rho = |\psi\rangle\langle\psi|$. We consider an A-observable $O_A \equiv O_A \otimes 1_B$. The expectation value of O_A is given by

$$\begin{aligned}\langle O_A\rangle &= Tr\rho O_A \otimes 1_B \\ &= \sum_{\alpha,\mu}\rho_A(\alpha, \mu)\langle\mu|O_A|\alpha\rangle \\ &= Tr_A \rho_A O_A.\end{aligned} \tag{6.131}$$

The reduced density matrix ρ_A is precisely given by

$$\rho_A(\alpha, \mu) = \sum_\beta \Psi^*(\alpha, \beta)\Psi(\mu, \beta). \tag{6.132}$$

To finish this important point we consider a system which is initially in a pure state and decoupled from the environment. The initial state of system + environment is then

$$|\psi\rangle^{(s, e)} = \left(\sum_s c_s |s\rangle^{(s)}\right) \otimes |\phi\rangle^{(e)}. \tag{6.133}$$

The coupling between the system and the environment is given by a unitary operator $U^{(s, e)}$, viz

$$|\psi'\rangle^{(s, e)} = U^{(s, e)}|\psi\rangle^{(s, e)}. \tag{6.134}$$

We will assume that the interaction is non-dissipative, i.e. the system does not decay to lower energy states, viz

$$U^{(s, e)}|s\rangle l^{(s)} \otimes |\phi\rangle^{(e)} = |s\rangle^{(s)} \otimes |\phi_s\rangle^{(e)}. \tag{6.135}$$

Also, we assume that the interaction is such that the different system states $|s\rangle$ drive the environment into orthogonal states $|\phi_s(t)\rangle^{(e)}$, viz

$$\langle\phi_s |\phi_{s'}\rangle^{(e)} = \delta_{s,s'}. \tag{6.136}$$

The state of the system + environment becomes

$$|\psi'\rangle^{(s,\,e)} = \sum_s c_s \, |s\rangle^{(s)} \otimes |\phi_s\rangle^{(e)}. \tag{6.137}$$

This is a pure state with a corresponding density matrix $\rho^{(s,\,e)} = |\psi'\rangle\langle\psi'|^{(s,\,e)}$. However, due to entanglement the state of the system is mixed given by tracing over the degrees of freedom of the environment which gives the reduced density matrix

$$\begin{aligned} \rho^{(s)} &= Tr_e \rho^{(s,\,e)} \\ &= \sum_s |c_s|^2 |s\rangle\langle s|^s. \end{aligned} \tag{6.138}$$

The probability of obtaining the system in the state $|s\rangle$ is $|c_s|^2$ which is Born's rule. Hence, entanglement seems to give rise to collapse. The density matrix therefore undergoes the decrease of information $\rho^{(s,\,e)} \longrightarrow \rho^{(s)}$, called also decoherence, through interaction with the environment.

From the above result, entanglement seems also to give rise to decoherence which is actually what is at the origin of the collapse. Indeed, the above state is totally mixed and thus fully decohered since the off-diagonal elements of the density matrix, which are responsible for quantum correlations, are zero. The environment therefore kills the coherence of the state as measured by the off-diagonal elements of the density matrix. The original pure state of the system has evolved into a mixed state because it is an open system, as opposed to being closed, and as such it does not obey the simple form (6.106) of the Liouville–von Neumann equation, but it satisfies instead the so-called master equation which has additional terms, viz

$$\frac{\partial \rho}{\partial t} = -\frac{i}{\hbar}[H,\,\rho] + \cdots. \tag{6.139}$$

The extra terms can be given for example by those found in equation (17) of [18].

We define the von Neumann entropy or the entanglement entropy by the formula

$$S = -Tr\rho \ln \rho = -\sum_i \rho_i \ln \rho_i. \tag{6.140}$$

For a pure state, i.e. when all eigenvalues with the exception of one vanish, we get $S = 0$. For mixed states we have $S > 0$. For example, in the case of a totally incoherent mixed density matrix in which all the eigenvalues are equal to $1/N$ where N is the dimension of the Hilbert space we get the maximum value of the von Neumann entropy given by

$$S = S_{\text{max}} = \ln N. \tag{6.141}$$

In the case where ρ is proportional to a projection operator onto a subspace of dimension n we find

$$S = \ln n. \tag{6.142}$$

In other words, the von Neumann entropy measures the number of important states in the statistical ensemble, i.e. those states which have an appreciable probability. This entropy is also a measure of the degree of entanglement between subsystems A and B and hence its other name entanglement entropy.

The von Neumann entropy is different from the thermodynamic Boltzmann entropy given by the formula

$$S_{\text{thermal}} = -Tr\rho_{\text{MB}} \ln \rho_{\text{MB}}, \tag{6.143}$$

where ρ_{MB} is the usual Maxwell–Boltzmann probability distribution given in terms of the Hamiltonian H and the temperature $T = 1/\beta$ by the formula

$$\rho_{\text{MB}} = \frac{1}{Z} \exp(-\beta H), \quad Z = Tr \exp(-\beta H). \tag{6.144}$$

6.6 Rindler decomposition and Unruh effect

This lecture is based on [10, 12].

6.6.1 Rindler decomposition

We consider quantum field theory (QFT) in Minkowski spacetime. We introduce Rindler decomposition of this spacetime. Quadrant I is Rindler spacetime. Quadrants II and III have no causal relations with quadrant I. Quadrant IV provides initial data for Rindler spacetime. Indeed, signals from region IV must cross the surface $t = -\infty$ ($\omega = -\infty$) in order to reach region I.

We will work near the horizon with the metric (with $\omega = t/4MG$, $T = \rho \sinh \omega$, $Z = \rho \cosh \omega$)

$$\begin{aligned} ds^2 &= \rho^2 d\omega^2 - d\rho^2 - dX^2 - dY^2 \\ &= dT^2 - dZ^2 - dX^2 - dY^2. \end{aligned} \tag{6.145}$$

The light cone is at $X = \pm T$ or $\rho = 0$, $\omega = \pm\infty$. This also corresponds to the event horizon separating between $r < 2GM$ and $r > 2GM$. Remark that $\omega \longrightarrow \infty$ corresponds to $t \longrightarrow \infty$ since an observer falling into the black hole is never seen actually crossing it. Since the Rindler space is only valid near the horizon we have $r \simeq r_s$ or $\delta \simeq 0$ and thus $\rho \simeq 0$. The horizon is actually at $\rho \simeq 0$.

The surface $T = 0$ is divided into two halves. The first half in region I and the second half in region III. The fields in region I ($Z > 0$) act in the Hilbert space $\mathcal{H}_L$ and those in region III ($Z < 0$) act in the Hilbert space $\mathcal{H}_R$. We have then

$$\phi(X, Y, Z) = \phi_L(X, Y, Z), \quad Z > 0. \tag{6.146}$$

$$\phi(X, Y, Z) = \phi_R(X, Y, Z), \quad Z < 0. \tag{6.147}$$

The general wave functional of interest is

$$\Psi = \Psi(\phi_L, \phi_R). \tag{6.148}$$

This is a pure state. But we want to compute the density matrix used by the fiducial observers called FIDOS (static observers at fixed (X, Y, Z) which all measure the time T) in the Rindler quadrant (quadrant I) to describe the system. In other words, we need to compute the reduced density matrix ρ_R which corresponds to the Minkowski vacuum to the FIDOS in Rindler quadrant I.

We have obviously translation invariance along the X and Y axes and thus the reduced density matrix is expected to commute with the momentum operators in the X and Y directions, viz

$$[\rho_R, P_X] = [\rho_R, P_Y] = 0. \tag{6.149}$$

Recall also that a translation in the Rindler time $\omega \longrightarrow \omega + c$ corresponds to a Lorentz boost along the Z direction in Minkowski spacetime. The reduced density matrix ρ_R, since it represents the Minkowski vacuum in quadrant I, must be invariant under Lorentz boosts in quadrant I. In other words, we must have

$$[\rho_R, H_R] = 0. \tag{6.150}$$

H_R is the generator of the Lorentz boosts $\omega \longrightarrow \omega + c$ in quadrant I. This is precisely the Hamiltonian in quadrant I given by

$$H_R = \int_{\rho=0}^{\rho=\infty} \rho d\rho dX dY T^{00}(\rho, X, Y). \tag{6.151}$$

T^{00} is the Hamiltonian density with respect to the Minkowski observer given for example for a scalar field by

$$T^{00}(\rho, X, Y) = \frac{1}{2}\pi^2 + \frac{1}{2}(\nabla\phi)^2 + V(\phi). \tag{6.152}$$

Recall that in the T–X plane the lines of constant ω are straight lines through the origin. The proper time separation between these lines is $\delta\tau = \rho\delta\omega$. This is the origin of the ρ factor multiplying T^{00}. Since $\pi = \dot{\phi}$, the above Hamiltonian corresponds to the action

$$I = \int d^3x dT \left[\frac{1}{2}\dot{\phi}^2 - \frac{1}{2}(\nabla\phi)^2 - V(\phi) \right]. \tag{6.153}$$

After Euclidean rotation $T \longrightarrow iX^0$ we get

$$I_E = \int d^4X \left[\frac{1}{2}(\partial_X\phi)^2 + V(\phi) \right]. \tag{6.154}$$

The original Lorentz invariance is now 4D rotation invariance. In particular, the ω-translation, which is actually a boost in the Z-direction, becomes a rotation in the (Z, X^0) plane.

$\Psi(\phi_L, \phi_R)$ is the ground state of the Minkowski Hamiltonian which can be computed using Euclidean path integrals. We write this as $\Psi(\phi_L, \phi_R) = \langle\phi|\Omega\rangle$. The ground state $|\Omega\rangle$ can be obtained from any other state $|\chi\rangle$ by the action of the Hamiltonian as follows

$$|\Omega\rangle = \frac{1}{\langle\Omega|\chi\rangle} \lim_{T\to\infty} \exp(-TH)|\chi\rangle. \tag{6.155}$$

Thus

$$\langle\phi|\Omega\rangle = \frac{1}{\langle\Omega|\chi\rangle} \lim_{T\to\infty} \langle\phi|\exp(-TH)|\chi\rangle$$

$$\propto \int_{\hat{\phi}(t_E=-\infty)=0}^{\hat{\phi}(t_E=0)=\phi} \mathcal{D}\hat{\phi} \exp(-I_E). \tag{6.156}$$

The boundary condition at $t_E = 0$, viz $\hat{\phi}(t_E = 0) = \phi$, corresponds to the state $|\phi\rangle = |\phi_L\rangle|\phi_R\rangle$, because the states ϕ_L and ϕ_R correspond to $t_E = 0$. The boundary condition at $t_E = -\infty$ is a choice. We could have chosen instead [10]

$$\langle\phi|\Omega\rangle \propto \int_{\hat{\phi}(t_E=0)=\phi}^{\hat{\phi}(t_E=+\infty)=0} \mathcal{D}\hat{\phi} \exp(-I_E). \tag{6.157}$$

Let θ be the angle in the Euclidean plane (Z, X^0) corresponding to the Rindler time ω. We divide the region $T < 0$ into infinitesimal wedges as in figure 6.4.

We integrate from the field ϕ_L at $\theta = 0$ to the field ϕ_R at $\theta = \pi$. The boost operator K_x in the Euclidean plane generates rotations in the (Z, X^0) plane. The restriction of this generator to the right Rindler wedge is given precisely by the Hamiltonian H_R. Thus we can write $\langle\phi|\Omega\rangle$ as a transition matrix element between initial state $|\phi_L\rangle$ and final state $|\phi_R\rangle$. In order to convert $|\phi_L\rangle$ back to a final state we act on it with the CPT operator Θ defined by

$$\Theta^+\phi(X^0, Z, X, Y)\Theta = \Phi^+(-X^0, -Z, X, Y). \tag{6.158}$$

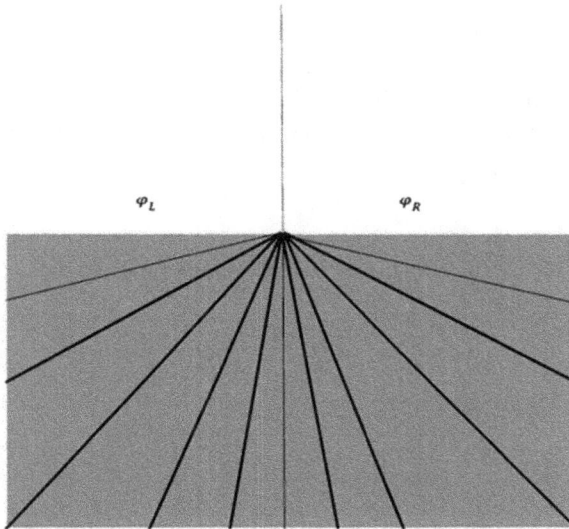

Figure 6.4. Rindler decomposition.

This is an anti-unitary operator which provides a map between the Hilbert spaces $\mathcal{H}_L$ and $\mathcal{H}_R$.

The transfer matrix in an infinitesimal right wedge is $G = \exp(-\delta\theta H_R)$ but we have $n = \pi/\delta\theta$ wedges in total so the total transfer matrix is $G^n = \exp(-\pi H_R)$. In summary we have the result

$$\begin{aligned}\Psi(\phi_L,\ \phi_R) &= \langle\phi_R|\langle\phi_L|\Omega\rangle \\ &\propto \langle\phi_R|\exp(-\pi H_R)\Theta|\phi_L\rangle.\end{aligned} \qquad (6.159)$$

This element is a transition matrix element in the right wedge. Let $|i_R\rangle$ be the eigenstates of H_R with eigenvalues E_i. By inserting a complete set of such eigenstates we get

$$\begin{aligned}\Psi(\phi_L,\ \phi_R) &= \langle\phi_R|\langle\phi_L|\Omega\rangle \\ &\propto \sum_i e^{-\pi E_i}\langle\phi_R|i_R\rangle\langle i_R|\Theta|\phi_L\rangle.\end{aligned} \qquad (6.160)$$

Θ is an anti-unitary operator satisfying $\langle\Theta x|\Theta y\rangle = \langle y|x\rangle$ which should be contrasted with the unitarity property $\langle\Theta x|A\Theta y\rangle = \langle x, y\rangle$. Thus we must have $\langle x|\Theta^+|y\rangle = \langle y|\Theta|x\rangle$. We define the state

$$\Theta^+|i_R\rangle = |i_L^*\rangle. \qquad (6.161)$$

We get then the transition matrix element

$$\begin{aligned}\Psi(\phi_L,\ \phi_R) &= \langle\phi_R|\langle\phi_L|\Omega\rangle \\ &\propto \sum_i e^{-\pi E_i}\langle\phi_R|i_R\rangle\langle\phi_L|i_L^*\rangle.\end{aligned} \qquad (6.162)$$

In other words, we get the ground state

$$|\Omega\rangle = \frac{1}{\sqrt{Z}}\sum_i e^{-\pi E_i}|i_R\rangle|i_L^*\rangle. \qquad (6.163)$$

The entanglement between the left and right wedges is now fully manifest.

We can define immediately the reduced matrix ρ_R by the relation

$$\begin{aligned}\rho_R(\phi_R,\ \phi_R') &= \int \Psi^*(\phi_L,\ \phi_R)\Psi(\phi_L,\ \phi_R')d\phi_L \\ &= \frac{1}{Z}\sum_i e^{-2\pi E_i}\langle i_R|\phi_R\rangle\langle\phi_R'|i_R\rangle \\ &= \frac{1}{Z}\langle\phi_R'|e^{-2\pi H}|\phi_R\rangle.\end{aligned} \qquad (6.164)$$

In the second line we have used the identities

$$\int|\phi_L\rangle\langle\phi_L| = 1, \quad \langle i_L^*|j_L^*\rangle = \delta_{ij}. \qquad (6.165)$$

We get then the reduced density matrix

$$\rho_R = \frac{1}{Z} e^{-2\pi H}. \tag{6.166}$$

Thus the fiducial observers FIDOS see the vacuum as a thermal ensemble with a Maxwell–Boltzmann distribution at a temperature

$$T_R = \frac{1}{2\pi}. \tag{6.167}$$

This is the Unruh effect.

Another derivation is as follows. The state $|\Omega\rangle$ is a pure state. The corresponding density matrix is $|\Omega\rangle\langle\Omega|$. By integrating over the degrees freedom of the left wedge we obtain a mixed state corresponding precisely to the reduced density matrix ρ_R, viz

$$\rho_R = \sum_i \langle i_L^* |\Omega\rangle\langle\Omega| i_L^*\rangle. \tag{6.168}$$

But

$$\langle i_L^* |\Omega\rangle = \frac{1}{\sqrt{Z}} e^{-\pi E_i} |i_R\rangle, \quad \langle\Omega| i_L^*\rangle = \frac{1}{\sqrt{Z}} e^{-\pi E_i} \langle i_R|. \tag{6.169}$$

We get then

$$\rho_R = \frac{1}{Z} \sum_i e^{-2\pi E_i} |i_R\rangle\langle i_R|. \tag{6.170}$$

6.6.2 Unruh temperature

The temperature T_R is dimensionless. We suppose a thermometer at rest with respect to the fiducial observer FIDOS at position ρ, i.e. it has the proper acceleration $a(\rho) = 1/\rho$ (recall that $\rho \leftrightarrow \xi^3 + 1/a$ and $\omega \leftrightarrow a\tau$). The thermometer is also assumed to be in equilibrium with the quantum fields at temperature $T_R = 1/2\pi$. If ε_i are the energy levels of the thermometer at rest then $\rho\varepsilon_i$ are the Rindler energy levels of the thermometer. This is almost obvious from the form of the metric $ds^2 = -\rho^2 d\omega^2 + d\rho^2 + dX^2 + dY^2$. We conclude therefore that the temperature measured by the thermometer is given by

$$T(\rho) = \frac{1}{2\pi\rho} = \frac{a(\rho)}{2\pi}. \tag{6.171}$$

Thus the FIDOS experiences a temperature which increases to infinity as we move towards the horizon at $\rho = 0$. This temperature corresponds to virtual vacuum fluctuations given by particle pairs. Some of these virtual loops are conventional loops created in region I, some of them are of no importance to the FIDOS in region I since they are created in region III, but others are created around the horizon at $\rho = 0$, and thus they are partly in region I partly in region III, and as a consequence cause non trivial entanglement between the degrees of freedom in regions I and III, which leads to a mixed density matrix in region I. Thus, the horizon behaves as a membrane which constantly emits and reabsorbs particles. This membrane is essentially the so-called stretched horizon.

6.7 Quantum field theory in curved spacetime

In this part we will follow briefly [14, 15, 20, 21].

The action of a real scalar field coupled to the metric minimally is given by

$$S_M = \int d^4x \sqrt{-\det g} \left(-\frac{1}{2} g^{\mu\nu} \nabla_\mu \phi \nabla_\nu \phi - V(\phi) \right). \tag{6.172}$$

If we are interested in an action which is at most quadratic in the scalar field then we must choose $V(\phi) = m^2\phi^2/2$. In curved spacetime there is another term we can add which is quadratic in ϕ namely $R\phi^2$ where R is the Ricci scalar. The full action should then read (in arbitrary dimension n)

$$S_M = \int d^nx \sqrt{-\det g} \left(-\frac{1}{2} g^{\mu\nu} \nabla_\mu \phi \nabla_\nu \phi - \frac{1}{2}m^2\phi^2 - \frac{1}{2}\zeta R\phi^2 \right). \tag{6.173}$$

The choice $\zeta = (n - 2)/(4(n - 1))$ is called conformal coupling. At this value the action with $m^2 = 0$ is invariant under conformal transformations defined by

$$g_{\mu\nu} \longrightarrow \bar{g}_{\mu\nu} = \Omega^2(x)g_{\mu\nu}(x), \quad \phi \longrightarrow \bar{\phi} = \Omega^{\frac{2-n}{2}}(x)\phi(x). \tag{6.174}$$

The equation of motion derived from this action are (we will keep in the following the metric arbitrary as long as possible)

$$(\nabla_\mu \nabla^\mu - m^2 - \zeta R)\phi = 0. \tag{6.175}$$

Let ϕ_1 and ϕ_2 be two solutions of this equation of motion. We define their inner product by

$$(\phi_1, \phi_2) = -i \int_\Sigma \left(\phi_1 \partial_\mu \phi_2^* - \partial_\mu \phi_1. \phi_2^* \right) d\Sigma n^\mu. \tag{6.176}$$

$d\Sigma$ is the volume element in the space-like hypersurface Σ and n^μ is the time-like unit vector which is normal to this hypersurface. This inner product is independent of the hypersurface Σ.

Indeed let Σ_1 and Σ_2 be two non-intersecting hypersurfaces and let V be the four-volume bounded by Σ_1, Σ_2 and (if necessary) time-like boundaries on which $\phi_1 = \phi_2 = 0$. We have from one hand

$$i \int_V \nabla^\mu \left(\phi_1 \partial_\mu \phi_2^* - \partial_\mu \phi_1. \phi_2^* \right) dV = i \oint_{\partial V} \left(\phi_1 \partial_\mu \phi_2^* - \partial_\mu \phi_1. \phi_2^* \right) d\Sigma^\mu$$
$$= (\phi_1, \phi_2)_{\Sigma_1} - (\phi_1, \phi_2)_{\Sigma_2}. \tag{6.177}$$

From the other hand

$$i \int_V \nabla^\mu \left(\phi_1 \partial_\mu \phi_2^* - \partial_\mu \phi_1 \cdot \phi_2^* \right) dV = i \int_V \left(\phi_1 \nabla^\mu \partial_\mu \phi_2^* - \nabla^\mu \partial_\mu \phi_1 \cdot \phi_2^* \right) dV$$

$$= i \int_V (\phi_1 (m^2 + \xi R) \phi_2^* - (m^2 + \xi R) \phi_1 \cdot \phi_2^*) dV \quad (6.178)$$

$$= 0.$$

Hence

$$(\phi_1, \phi_2)_{\Sigma_1} - (\phi_1, \phi_2)_{\Sigma_2} = 0. \quad (6.179)$$

There is always a complete set of solutions u_i and u_i^* of the equation of motion (6.175) which are orthonormal in the inner product (6.176), i.e. satisfying

$$(u_i, u_j) = \delta_{ij}, \quad (u_i^*, u_j^*) = -\delta_{ij}, \quad (u_i, u_j^*) = 0. \quad (6.180)$$

We can then expand the field as

$$\phi = \sum_i (a_i u_i + a_i^* u_i^*). \quad (6.181)$$

We now canonically quantize this system. We choose a foliation of spacetime into space-like hypersurfaces. Let Σ be a particular hypersurface with unit normal vector n^μ corresponding to a fixed value of the time coordinate $x^0 = t$ and with induced metric h_{ij}. We write the action as $S_M = \int dx^0 L_M$ where $L_M = \int d^{n-1}x \sqrt{-\det g} \; \mathcal{L}_M$. The canonical momentum π is defined by

$$\pi = \frac{\delta L_M}{\delta(\partial_0 \phi)} = -\sqrt{-\det g} \; g^{\mu 0} \partial_\mu \phi$$

$$= -\sqrt{-\det h} \; n^\mu \partial_\mu \phi. \quad (6.182)$$

We promote ϕ and π to hermitian operators $\hat{\phi}$ and $\hat{\pi}$ and then impose the equal time canonical commutation relations

$$[\hat{\phi}(x^0, x^i), \hat{\pi}(x^0, y^i)] = i\delta^{n-1}(x^i - y^i). \quad (6.183)$$

The delta function satisfies the property

$$\int \delta^{n-1}(x^i - y^i) d^{n-1} y = 1. \quad (6.184)$$

The coefficients a_i and a_i^* become annihilation and creation operators $\hat{a}_i$ and $\hat{a}_i^+$ satisfying the commutation relations

$$[\hat{a}_i, \hat{a}_j^+] = \delta_{ij}, \quad [\hat{a}_i, \hat{a}_j] = [\hat{a}_i^+, \hat{a}_j^+] = 0. \quad (6.185)$$

The vacuum state is given by a state $|0\rangle_u$ defined by

$$\hat{a}_i |0_u\rangle = 0. \quad (6.186)$$

The entire Fock basis of the Hilbert space can be constructed from the vacuum state by repeated application of the creation operators $\hat{a}_i^+$.

The solutions u_i, u_i^* are not unique and as a consequence the vacuum state $|0\rangle_u$ is not unique. Let us consider another complete set of solutions v_i and v_i^* of the equation of motion (6.175) which are orthonormal in the inner product (6.176). We can then expand the field as

$$\phi = \sum_i (b_i v_i + b_i^* v_i^*).\tag{6.187}$$

After canonical quantization the coefficients b_i and b_i^* become annihilation and creation operators $\hat{b}_i$ and $\hat{b}_i^+$ satisfying the standard commutation relations with a vacuum state given by $|0\rangle_v$ defined by

$$\hat{b}_i|0_v\rangle = 0.\tag{6.188}$$

We introduce the so-called Bogolubov transformation as the transformation from the set $\{u_i, u_i^*\}$ (which are the set of modes seen by some observer) to the set $\{v_i, v_i^*\}$ (which are the set of modes seen by another observer) as

$$v_i = \sum_j (\alpha_{ij} u_j + \beta_{ij} u_j^*).\tag{6.189}$$

By using orthonormality conditions we find that

$$\alpha_{ij} = (v_i, u_j), \quad \beta_{ij} = -(v_i, u_j^*).\tag{6.190}$$

We can also write

$$u_i = \sum_j (\alpha_{ji}^* v_j - \beta_{ji} v_j^*).\tag{6.191}$$

The Bogolubov coefficients α and β satisfy the normalization conditions

$$\sum_k (\alpha_{ik} \alpha_{jk} - \beta_{ik} \beta_{jk}) = \delta_{ij}, \quad \sum_k (\alpha_{ik} \beta_{jk}^* - \beta_{ik} \alpha_{jk}^*) = 0.\tag{6.192}$$

The Bogolubov coefficients α and β transform also between the creation and annihilation operators $\hat{a}$, $\hat{a}^+$ and $\hat{b}$, $\hat{b}^+$. We find

$$\hat{a}_k = \sum_i (\alpha_{ik} \hat{b}_i + \beta_{ik}^* \hat{b}_i^+), \quad \hat{b}_k = \sum_i (\alpha_{ki}^* \hat{a}_i - \beta_{ki}^* \hat{a}_i^+).\tag{6.193}$$

Let N_u be the number operator with respect to the u-observer, viz $N_u = \sum_k \hat{a}_k^+ \hat{a}_k$. Clearly

$$\langle 0_u | N_u | 0_u \rangle = 0.\tag{6.194}$$

We compute

$$\langle 0_v | \hat{a}_k^+ \hat{a}_k | 0_v \rangle = \sum_i \beta_{ik} \beta_{ik}^*.\tag{6.195}$$

Thus

$$\langle 0_v | N_u | 0_v \rangle = tr\beta\beta^+. \tag{6.196}$$

In other words with respect to the v-observer the vacuum state $|0_u\rangle$ is not empty but filled with particles. This opens the door to the possibility of particle creation by a gravitational field.

6.8 Hawking radiation

6.8.1 The Unruh effect revisited

In this first part we will follow mostly [15]. We consider 2D spacetime with metric

$$ds^2 = -dt^2 + dx^2. \tag{6.197}$$

We consider a uniformly accelerated, i.e. a Rindler, observer in this spacetime with acceleration α. The trajectory of the Rindler observer is given by the equations (set $\xi^1 = 0$ in (6.16) and discard the constant term $1/a$ in x^1, i.e. the Rindler does not coincide with Minkowski at $\tau = 0$)

$$t = \frac{1}{\alpha} \sinh \alpha\tau, \quad x = \frac{1}{\alpha} \cosh \alpha\tau. \tag{6.198}$$

The trajectory is then a hyperboloid given by

$$x^2 = t^2 + \frac{1}{\alpha^2}. \tag{6.199}$$

Thus the Rindler observer moves from the past null infinity $x = -t$ to the future null infinity $x = +t$ as opposed to the motion of geodesic observers which reaches time-like infinity.

We introduce coordinates in Rindler space (quadrant or wedge I in figure 6.5) by

$$t = \frac{1}{a} \exp(a\xi)\sinh a\eta, \quad x = \frac{1}{a} \exp(a\xi)\cosh a\eta,$$
$$x > |t|, \quad -\infty < \eta, \xi < +\infty. \tag{6.200}$$

The trajectory of the Rindler observer in these coordinates reads

$$\eta = \frac{\alpha}{a}\tau, \quad \xi = \frac{1}{a} \ln \frac{a}{\alpha}. \tag{6.201}$$

In other words,

$$a = \alpha \Rightarrow \eta = \tau, \quad \xi = 0. \tag{6.202}$$

The metric in Rindler space reads

$$ds^2 = \exp(2a\xi)(-d\eta^2 + d\xi^2). \tag{6.203}$$

The metric is independent of η and thus ∂_η is a Killing vector. This is given explicitly by

$$\partial_\eta = a(x\partial_t + t\partial_x). \tag{6.204}$$

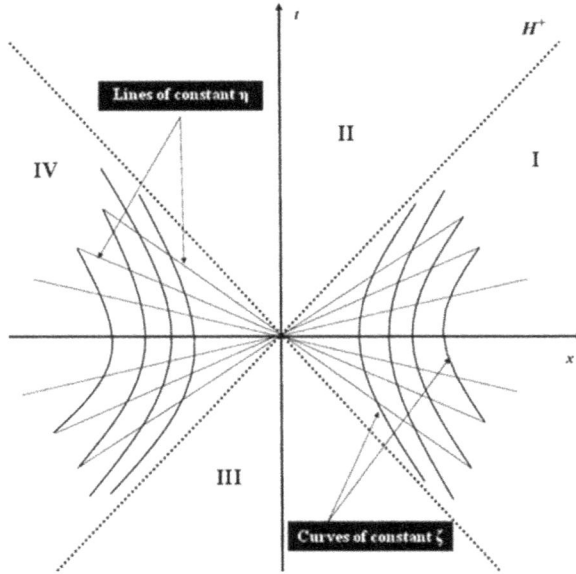

Figure 6.5. Rindler space in two dimensions.

This is then obviously the Killing field associated with a boost in the x-direction. This extends to regions II and III where it is space-like while in region IV it is time-like past-directed[2]. The horizons $x = \pm t$ are actually Killing horizons. Every Killing horizon is associated with an acceleration called the surface gravity κ which is here given exactly by

$$\kappa = a. \tag{6.205}$$

We will also need the coordinates η and ξ in the quadrant IV. They are given by

$$t = -\frac{1}{a}\exp(a\xi)\sinh a\eta, \quad x = -\frac{1}{a}\exp(a\xi)\cosh a\eta, \quad x < |t|. \tag{6.206}$$

The Klein–Gordon equation in Rindler space is (with $m^2 = \zeta = 0$)

$$0 = \nabla_\mu \, \nabla^\mu \, \phi$$

$$= \frac{1}{\sqrt{-\det g}}\partial_\mu(\sqrt{-\det g}\,\partial^\mu\phi) \tag{6.207}$$

$$= e^{-2a\xi}(-\partial_\eta^2 + \partial_\xi^2)\phi.$$

A positive frequency normalized plane wave solution in region I is given by

$$g_k^{(1)} = \frac{1}{\sqrt{4\pi\omega}}\exp(-i\omega\eta + ik\xi), \quad \text{I}$$

$$g_k^{(1)} = 0, \quad \text{IV.} \tag{6.208}$$

---

[2] The labels III and IV are reversed here as compared with the previous discussion.

Indeed,

$$\partial_\eta g_k^{(1)} = -i\omega g_k^{(1)}, \quad \omega = |k|. \tag{6.209}$$

∂_η is a future-director time-like Killing vector in region I. But it is a past-directed time-like Killing vector in region IV. Thus in region IV we should consider the Killing vector $\partial_{-\eta} = -\partial_\eta$ which is future-directed there. A positive frequency normalized plane wave solution in region II is thus given by

$$g_k^{(2)} = 0, \quad \text{I}$$
$$g_k^{(2)} = \frac{1}{\sqrt{4\pi\omega}} \exp(i\omega\eta + ik\xi), \quad \text{IV}. \tag{6.210}$$

Indeed,

$$\partial_{-\eta} g_k^{(2)} = -i\omega g_k^{(2)}, \quad \omega = |k|. \tag{6.211}$$

These two sets of positive frequency modes, together with their negative frequency conjugates, provide a complete set of basis elements for the expansion of any solution of the Klein–Gordon wave equation through spacetime. We denote the associated annihilation operators by $\hat{b}_k^{(1)}$ and $\hat{b}_k^{(2)}$. A general solution of the Klein–Gordon equation takes then the form

$$\phi = \int_k \left(\hat{b}_k^{(1)} g_k^{(1)} + \hat{b}_k^{(2)} g_k^{(2)} + \text{h.c} \right). \tag{6.212}$$

This should be contrasted with the expansion of the same solution in terms of the Minkowski modes $f_k \propto \exp(-i(\omega t - kx))$ with $\omega = |k|$, which we will write as

$$\phi = \int_k \left(\hat{a}_k f_k + \text{h.c} \right). \tag{6.213}$$

The above Rindler modes $g_k^{(1)}$ and $g_k^{(2)}$ are normalized according to the inner product (6.176), viz

$$(\phi_1, \phi_2) = -i \int_\Sigma \left(\phi_1 \partial_\mu \phi_2^* - \partial_\mu \phi_1 . \phi_2^* \right) d\Sigma n^\mu. \tag{6.214}$$

$d\Sigma$ is the volume element in the space-like hypersurface Σ and n^μ is the time-like unit vector which is normal to this hypersurface. Thus $d\Sigma = \sqrt{\det\gamma} \, d^{n-1}x$. In our case, the time-like surface $\eta = 0$ has a unit vector n^μ such as $g_{\mu\nu} n^\mu n^\nu = -1$ and thus $n^0 = \exp(-a\xi)$. Also, we have $\sqrt{\det\gamma} = \exp(a\xi)$ and $x \leftrightarrow \xi$. Hence the inner product becomes

$$(\phi_1, \phi_2) = -i \int \left(\phi_1 \partial_\eta \phi_2^* - \partial_\eta \phi_1 . \phi_2^* \right) d\xi. \tag{6.215}$$

We compute for example

$$(g_{k_1}^{(1)}, g_{k_2}^{(1)}) = -\frac{i}{4\pi\sqrt{\omega_1\omega_2}} \int (i\omega_2 e^{-i\omega_1\eta + ik_1\xi} e^{i\omega_2\eta - ik_2\xi} + i\omega_1 e^{-i\omega_1\eta + ik_1\xi} e^{i\omega_2\eta - ik_2\xi}) d\xi \tag{6.216}$$
$$= \frac{1}{4\pi} . 4\pi\delta(k_1 - k_2).$$

We also show

$$(g_{k_1}^{(2)}, g_{k_2}^{(2)}) = \delta(k_1 - k_2). \tag{6.217}$$

$$(g_{k_1}^{(1)}, g_{k_2}^{(2)}) = 0. \tag{6.218}$$

The Minkowski vacuum $|0_M\rangle$ and the Rindler vacuum $|0_R\rangle$ are defined obviously by

$$\hat{a}_k|0_M\rangle = 0. \tag{6.219}$$

$$\hat{b}_k^{(1)}|0_R\rangle = \hat{b}_k^{(2)}|0_R\rangle = 0. \tag{6.220}$$

However, the Hilbert space is the same. For the Rindler observer the Minkowski vacuum $|0_R\rangle$ is seen as a multi-particle state since she is traveling in Minkowski spacetime with a uniform acceleration, i.e. she is not an inertial observer. The expectation value of the Rindler number operator in the Minkowski vacuum can be calculated using the Bogolubov coefficients as we explained in the previous section.

An alternative method due to Unruh consists in extending the positive frequency modes $g_k^{(1)}$ and $g_k^{(2)}$ to the entire spacetime and thus replacing the corresponding annihilation operators $\hat{b}_k^{(1)}$ and $\hat{b}_k^{(2)}$ by new annihilation operators $\hat{c}_k^{(1)}$ and $\hat{c}_k^{(2)}$ which annihilate the Minkowski vacuum $|0_M\rangle$.

First, the coordinates (t, x) and (η, ξ) are related by

$$-t + x = \frac{1}{a}e^{a(\xi-\eta)} \Rightarrow e^{-a(\eta-\xi)} = a(-t + x), \quad \text{I.} \tag{6.221}$$

$$t - x = \frac{1}{a}e^{a(\xi-\eta)} \Rightarrow e^{-a(\eta-\xi)} = a(t - x), \quad \text{IV.} \tag{6.222}$$

Similarly,

$$e^{a(\eta+\xi)} = a(t + x), \quad \text{I.} \tag{6.223}$$

$$e^{a(\eta+\xi)} = a(-t - x), \quad \text{IV.} \tag{6.224}$$

Thus if we choose $k > 0$ we have in region I ($x > 0$)

$$\sqrt{4\pi\omega}\,g_k^{(1)} = \exp(-i\omega(\eta - \xi)) \\ = e^{i\frac{\omega}{a}}(-t + x)^{i\frac{\omega}{a}}. \tag{6.225}$$

In region IV ($x < 0$) we should instead consider

$$\sqrt{4\pi\omega}\,g_{-k}^{(2)*} = \exp(-i\omega(\eta - \xi)) \\ = e^{i\frac{\omega}{a}}(t - x)^{i\frac{\omega}{a}} \\ = e^{i\frac{\omega}{a}}e^{\frac{\pi\omega}{a}}(-t + x)^{i\frac{\omega}{a}}. \tag{6.226}$$

Thus for all x, i.e. along the surface $t = 0$, we should consider for $k > 0$ the combination

$$\sqrt{4\pi\omega}\left(g_k^{(1)} + e^{-\frac{\pi\omega}{a}}g_{-k}^{(2)*}\right) = e^{i\frac{\omega}{a}}(-t+x)^{i\frac{\omega}{a}}. \tag{6.227}$$

We get the same result for $k < 0$. A normalized analytic extension to the entire spacetime of the positive frequency modes $g_k^{(1)}$ is given by the modes

$$h_k^{(1)} = \frac{1}{\sqrt{2\sinh\dfrac{\pi\omega}{a}}}\left(e^{\frac{\pi\omega}{2a}}g_k^{(1)} + e^{-\frac{\pi\omega}{2a}}g_{-k}^{(2)*}\right). \tag{6.228}$$

Similarly, a normalized analytic extension to the entire spacetime of the positive frequency modes $g_k^{(2)}$ is given by the modes

$$h_k^{(2)} = \frac{1}{\sqrt{2\sinh\dfrac{\pi\omega}{a}}}\left(e^{\frac{\pi\omega}{2a}}g_k^{(2)} + e^{-\frac{\pi\omega}{2a}}g_{-k}^{(1)*}\right). \tag{6.229}$$

The field operator can then be expanded in these modes as

$$\phi = \int_k \left(\hat{c}_k^{(1)}h_k^{(1)} + \hat{c}_k^{(2)}h_k^{(2)} + \text{h.c}\right). \tag{6.230}$$

The relation between the annihilation operators $\hat{b}$ and the annihilation operators $\hat{c}$ is given by the same relation between the modes h and the modes g, viz

$$\hat{b}_k^{(1)} = \frac{1}{\sqrt{2\sinh\dfrac{\pi\omega}{a}}}\left(e^{\frac{\pi\omega}{2a}}\hat{c}_k^{(1)} + e^{-\frac{\pi\omega}{2a}}\hat{c}_{-k}^{(2)+}\right). \tag{6.231}$$

$$\hat{b}_k^{(2)} = \frac{1}{\sqrt{2\sinh\dfrac{\pi\omega}{a}}}\left(e^{\frac{\pi\omega}{2a}}\hat{c}_k^{(2)} + e^{-\frac{\pi\omega}{2a}}\hat{c}_{-k}^{(1)+}\right). \tag{6.232}$$

The modes $h_k^{(1)}$ and $h_k^{(2)}$ are positive frequency modes defined on the entire spacetime and thus they can be expressed entirely in terms of the positive frequency modes of Minkowski spacetime given by the plane waves $f_k \propto \exp(-i(\omega t - kx))$, $\omega = |k|$, where $k > 0$ correspond to right-moving modes and $k < 0$ correspond to left-moving modes. In other words, the modes $h_k^{(1)}$ and $h_k^{(2)}$ share with f_k the same Minkowski vacuum $|0_M\rangle$, viz

$$\hat{c}_k^{(1)}|0_M\rangle = \hat{c}_k^{(2)}|0_M\rangle = 0. \tag{6.233}$$

The Rindler number operator in region I is defined by

$$\hat{N}_R^{(1)}(k) = \hat{b}_k^{(1)+}\hat{b}_k^{(1)}. \tag{6.234}$$

We can now immediately compute the expectation value of the Rindler number operator in region I in the Minkowski vacuum to find

$$\langle 0_M | \hat{N}_R^{(1)}(k) | 0_M \rangle = \langle 0_M | \hat{b}_k^{(1)+} \hat{b}_k^{(1)} | 0_M \rangle$$

$$= \frac{e^{-\frac{\pi\omega}{a}}}{2 \sinh \frac{\pi\omega}{2}} \langle 0_M | \hat{c}_{-k}^{(2)} \hat{c}_{-k}^{(2)+} | 0_M \rangle \qquad (6.235)$$

$$= \frac{1}{e^{\frac{2\pi\omega}{a}} - 1} \delta(0).$$

This is a blackbody Planck spectrum corresponding to the temperature

$$T = \frac{a}{2\pi}. \qquad (6.236)$$

Indeed, this spectrum corresponds to thermal radiation, i.e. to a mixed state, without any correlations. This is the Unruh effect: a uniformly accelerated observer in the Minkowski vacuum observes a thermal spectrum [22].

6.8.2 From quantum scalar field theory in rindler background

We follow in this section the presentation of [10]. We consider Schwarzschild metric in tortoise coordinates, viz

$$ds^2 = F(r_*)(-dt^2 + dr_*^2) + r^2 d\Omega^2$$

$$F(r_*) = 1 - \frac{2GM}{r} \qquad (6.237)$$

$$r_* = r + 2GM \log\left(\frac{r}{2GM} - 1\right).$$

We consider the action of a massless scalar field ϕ in this background given by (with $\psi = r\phi$)

$$I = \int \sqrt{-\det g}\, d^4 x \frac{1}{2} \partial_\mu \phi \partial^\mu \phi$$

$$= \int Fr^2 \sin\theta dt dr_* d\theta d\phi \frac{1}{2}\left(-\frac{1}{F}(\partial_t \phi)^2 + \frac{1}{F}(\partial_{r_*}\phi)^2 + \frac{1}{r^2}(\partial_\theta \phi)^2 + \frac{1}{r^2 \sin^2\theta}(\partial_\phi \phi)^2\right)$$

$$= \int \sin\theta dt dr_* d\theta d\phi \frac{1}{2}\left(-(\partial_t \psi)^2 + (\partial_{r_*}\psi - \partial_{r_*}\ln r. \ \psi)^2 + \frac{F}{r^2}(\partial_\theta \psi)^2 + \frac{F}{r^2 \sin^2\theta}(\partial_\phi \psi)^2\right) \qquad (6.238)$$

$$= \int \sin\theta dt dr_* d\theta d\phi \frac{1}{2}\left(-(\partial_t \psi)^2 + (\partial_{r_*}\psi - \partial_{r_*}\ln r. \ \psi)^2 + \frac{F}{r^2}\psi \mathcal{L}^2 \psi\right),$$

where we have used

$$-\mathcal{L}^2 = \frac{1}{\sin\theta}\frac{\partial}{\partial\theta}(\sin\theta\frac{\partial}{\partial\theta}) + \frac{1}{\sin^2\theta}\frac{\partial^2}{\partial\phi^2}. \qquad (6.239)$$

We expand now in spherical coordinates as

$$\psi = \sum_{lm} \psi_{lm} \, Y_{lm}. \tag{6.240}$$

We get then

$$I = \int dt dr_* \frac{1}{2} \sum_{lm} \psi_{lm}^* \left(\partial_t^2 \psi_{lm} - \partial_{r_*}^2 \psi_{lm} + \left(\partial_{r_*}^2 \ln r + (\partial_{r_*} \ln r)^2 \right) \psi_{lm} + \frac{F}{r^2} l(l+1) \psi_{lm} \right)$$

$$= \int dt dr_* \frac{1}{2} \sum_{lm} \psi_{lm}^* \left(\partial_t^2 \psi_{lm} - \partial_{r_*}^2 \psi_{lm} + V(r_*) \psi_{lm} \right). \tag{6.241}$$

The potential is given by

$$V(r_*) = \partial_{r_*}^2 \ln r + (\partial_{r_*} \ln r)^2 + \frac{F}{r^2} l(l+1)$$

$$= \frac{1}{r} \frac{\partial^2 r}{\partial r_*^2} + \frac{F}{r^2} l(l+1) \tag{6.242}$$

$$= \frac{r - 2GM}{r} \left(\frac{2GM}{r^3} + \frac{l(l+1)}{r^2} \right).$$

The equation of motion reads

$$\partial_t^2 \psi_{lm} = \partial_{r_*}^2 \psi_{lm} - V(r_*) \psi_{lm}. \tag{6.243}$$

The stationary solutions are $\psi_{lm} = \exp(i\nu t) \tilde{\psi}_{lm}$ such that

$$-\tilde{\partial}_{r_*}^2 \tilde{\psi}_{lm} + V(r_*) \tilde{\psi}_{lm} = \nu^2 \tilde{\psi}_{lm}. \tag{6.244}$$

The potential vanishes at the horizon $r = 2GM$ (where the solutions are given by free plane waves) and also vanishes at infinity. Thus it must pass through a maximum given by the condition

$$\frac{dV}{dr} = \frac{1}{r^5} (-2l(l+1). \, r^2 - 6GM(1 - l(l+1)). \, r + 16G^2M^2) = 0. \tag{6.245}$$

We get the solutions

$$r_\pm = 3GM \left(\frac{1}{2} - \frac{1}{2l(l+1)} \pm \frac{1}{2} \sqrt{1 + \frac{7l^2 + 7l + 4}{4l^2(l+1)^2}} \right). \tag{6.246}$$

Obviously, the physical solution is

$$r_{\max} = 3GM \left(\frac{1}{2} - \frac{1}{2l(l+1)} + \frac{1}{2} \sqrt{1 + \frac{7l^2 + 7l + 4}{4l^2(l+1)^2}} \right). \tag{6.247}$$

Thus

$$r_{\max}(l = \infty) = 3GM. \tag{6.248}$$

For very large angular momentum l the maximum of the potential lies at $3GM$. For $r \gg 3GM$ the potential is repulsive, given by a generalization of the centrifugal potential $l(l + 1)/r^2$, whereas for $r < 3GM$ (the region of thermal atmosphere) gravity dominates and the potential becomes attractive. Thus any particle in this region with a zero initial velocity will spiral into the horizon eventually.

The above equation is effectively Schrödinger equation with potential V and energy ν^2. Thus an s-wave ($l = 0$) approaching the barrier $r = 3GM$ from the inside (horizon) with energy satisfying $\omega > V_{\text{max}}$ will be able to escape, whereas if approaching from the outside it will be able to penetrate the barrier and reach the horizon. For energy $\omega < V_{\text{max}}$ the wave needs to tunnel through the barrier.

For higher angular momentum the maximum of the potential is very large, proportional to l^2, and thus it is more difficult to escape or penetrate the barrier.

Near-horizon geometry is given by the metric (with $u = \ln \rho$ the tortoise coordinate in this case)

$$
\begin{aligned}
ds^2 &= -\rho^2 d\omega^2 + d\rho^2 + dY^2 + dZ^2 \\
&= e^{2u}(-d\omega^2 + du^2) + dY^2 + dZ^2.
\end{aligned} \tag{6.249}
$$

The action of a scalar field is given immediately by

$$
I = \int d\omega du dY dZ \frac{1}{2}(-(\partial_\omega \phi)^2 + (\partial_u \phi)^2 + e^{2u}(\partial_Y \phi)^2 + e^{2u}(\partial_Z \phi)^2). \tag{6.250}
$$

We expand the field into transverse plane waves as

$$
\phi = \int \frac{dk_2}{2\pi} \frac{dk_3}{2\pi} e^{i(k_2 Y + k_3 Z)} \psi(k_2, k_3, \omega, u). \tag{6.251}
$$

We get then the action

$$
I = \int d\omega du \frac{1}{2} \psi^* \left(\partial_\omega^2 \psi - \partial_u^2 \psi + e^{2u} \vec{k}^2 \psi \right). \tag{6.252}
$$

The potential is then given by

$$
V = e^{2u} \vec{k}^2. \tag{6.253}
$$

This is proportional to l^2 since $l = |k| r = |k|$. $2MG$ and thus this approximation is not expected to work for small angular momentum. Thus in approximating sums over l and m by integrals over k we should for consistency employ the infrared cutoff $|k| \sim 1/MG$.

The Rindler potential $V = \rho^2 \vec{k}^2$, for $|k| \neq 0$, is confining to the region near the horizon. This is also the situation in the Schwarzschild black hole where the potential confines particles to the region near the horizon. However, in the Schwarzschild black hole the potential becomes repulsive for $r > 3MG$ which is equivalent to $\rho > MG$. Thus the potential barrier for Schwarzschild black hole is the cutoff for $\rho > MG$, as opposed to Rindler space which keeps increasing without bound as ρ^2.

Since (1) a Schwarzschild black hole near the horizon will appear as Rindler, and (2) the Rindler observer will see the Minkowski vacuum as a thermal canonical ensemble with a temperature given by $T = 1/2\pi$, it is expected that an identical thermal effect should be observed near the horizon of the Schwarzschild black hole.

However, there is a crucial difference. In the case of Rindler the thermal atmosphere is fully confined by the potential (6.253) as opposed to the case of the Schwarzschild black hole where the thermal atmosphere is not fully confined by the potential (6.464). This means in particular that particles leak out of the thermal atmosphere in the case of the Schwarzschild black hole and as a consequence the black hole evaporates.

Let us find the temperature as seen by the Schwarzschild observer. The Rindler time ω is related to the Schwarzschild time t by the relation

$$\omega = \frac{t}{4GM}. \tag{6.254}$$

This leads immediately to the fact that frequency as measured by the Schwarzschild observer ν is redshifted compared to the frequency ν_R measured by the Rindler observer given by

$$\nu_R = 4GM. \nu \Rightarrow \nu = \frac{\nu_R}{4GM}. \tag{6.255}$$

Hence the temperature as measured by the Schwarzschild observer is also redshifted as

$$T_R = 4GM. T \Rightarrow T = \frac{T_R}{4GM} = \frac{1}{8\pi GM}. \tag{6.256}$$

This is precisely Hawking temperature.

Next we show how the black hole can radiate particles. The potential (6.464) is not fully confining and it contains only a barrier around $r \simeq 3GM$. The height of the barrier for modes with angular momentum $l = 0$, which corresponds to $r_{\max}(l = 0) = 8GM/3$, is given by

$$V_{\max}(l = 0) = \frac{27}{1024G^2M^2}. \tag{6.257}$$

The energy in the potential (6.464) is $E = \nu^2$. Thus the modes $l = 0$ will escape the potential barrier coming from the horizon if

$$E \geqslant V_{\max}(l = 0) \Rightarrow \nu \geqslant \frac{3\pi\sqrt{3}}{4}T. \tag{6.258}$$

However, these modes since they are in a thermal state with temperature $T = 1/8\pi MG$, their energy is of the order of T, and hence they can quite easily escape the potential barrier. The height of the barrier for modes with higher angular momentum l goes as l^2/G^2M^2, i.e. it is very high compared to the thermal scale set by Hawking radiation, and hence these modes do not escape as easily as the zero modes. This is Hawking radiation.

6.8.3 Summary

In summary, since
- (1) a Schwarzschild black hole near the horizon will appear as Rindler, and
- (2) the Rindler observer will see the Minkowski vacuum as a thermal canonical ensemble with a temperature given by $T = 1/2\pi$ (Unruh effect),

an identical thermal effect is observed near the horizon of the Schwarzschild black hole.

Indeed, to a distant observer the Schwarzschild black hole appears as a body with energy given by its mass M and a temperature T given by Hawking temperature

$$T = \frac{1}{8\pi GM}.$$

However, there is a crucial difference between Rindler space and Schwarzschild black hole. In the case of Rindler the thermal atmosphere (the particles near the horizon) is fully confined by the Rindler potential, as opposed to the case of the Schwarzschild black hole where the thermal atmosphere is not fully confined by the Schwarzschild potential. This means in particular that particles leaks out of the thermal atmosphere in the case of the Schwarzschild black hole and as a consequence the black hole evaporates. The particles which can escape the black hole have zero angular momentum for which the height of the potential barrier at around $r \simeq 3MG$ is of the same order as the thermal scale set by Hawking temperature while particles with larger angular momentum cannot escape because for them the height of the potential barrier is much larger than the thermal scale. See figure 6.6.

The thermodynamical entropy S is related to the energy and the temperature by the formula $dU = TdS$. Thus we obtain for the black hole the entropy

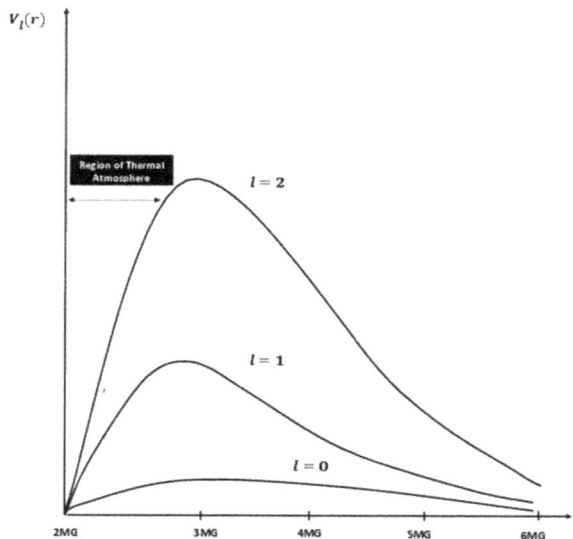

Figure 6.6. Schwarzschild potential.

$$dS = \frac{dM}{T} = 8\pi GMdM \Rightarrow S = 4\pi GM^2. \qquad (6.259)$$

However, the radius of the event horizon of the Schwarzschild black hole is $r_s = 2MG$, and thus the area of the event horizon (which is a sphere) is

$$A = 4\pi(2MG)^2. \qquad (6.260)$$

By dividing the above two equations we get

$$S = \frac{A}{4G}. \qquad (6.261)$$

The entropy of the black hole is proportional to its area. This is the famous Bekenstein–Hawking entropy formula.

6.9 Hawking radiation from quantum field theory in Schwarzschild background

The original derivation of the Hawking radiation is found in [8, 9]. In here we will follow [7, 14] and to a lesser degree [13, 23–25].

6.9.1 Kruskal and Schwarzschild (Boulware) observers and field expansions

Let us start by recalling some formulas. The metric is

$$ds^2 = -\left(1 - \frac{2GM}{r}\right)dt^2 + \frac{dr^2}{1 - \frac{2GM}{r}} + r^2 d\Omega^2. \qquad (6.262)$$

We define the Kruskal ingoing and outgoing null coordinates U and V (scaled versions of our previous u' and v') in region I as

$$U = r_s u' = -\sqrt{r_s(r - r_s)}\, e^{\frac{r-t}{2r_s}}, \quad V = r_s v' = \sqrt{r_s(r - r_s)}\, e^{\frac{r+t}{2r_s}}. \qquad (6.263)$$

They satisfy

$$UV = r_s(r_s - r)e^{\frac{r}{r_s}}, \quad \frac{U}{V} = -e^{-\frac{t}{r_s}}. \qquad (6.264)$$

The metric becomes

$$ds^2 = -\frac{4r_s}{r}e^{-\frac{r}{r_s}}dUdV + r^2 d\Omega^2. \qquad (6.265)$$

This form is valid throughout the spacetime and not only in region I.

We consider now an inertial observer falling through the horizon $r_s = 2GM$. This freely falling observer will cross the horizon in a finite proper time given by (with $r_s = r_i(1 + \cos\alpha_s)/2$)

$$\tau = \sqrt{\frac{r_i^3}{4r_s}}\,(\alpha_s + \sin\alpha_s). \tag{6.266}$$

However, with respect to the Schwarzschild observer the radius r of the freely falling object is related to its time t by the formula (near the horizon)

$$r - r_s = e^{-\frac{t}{r_s}}. \tag{6.267}$$

A distant inertial observer assumed to be hovering at a fixed radial distance r_∞ will observe a proper time τ_∞ related to Schwarzschild time t by the equation

$$\tau_\infty = \sqrt{1 - \frac{r_s}{r_\infty}}\,t. \tag{6.268}$$

Thus, this distant observer will then measure $\tau_\infty \longrightarrow \infty$ as $r \longrightarrow r_s$, i.e. she will never see the falling object actually crossing the horizon. Thus this observer may be interpreted as ending at the horizon.

The discrepancy between the worldviews of the above two inertial observers (the freely falling and the asymptotic fixed observer) is what is at the source of Hawking radiation and all its related paradoxes [7].

We reduce the problem to two dimensions, viz

$$ds^2 = -\left(1 - \frac{2GM}{r}\right)dt^2 + \frac{dr^2}{1 - \frac{2GM}{r}} = -\frac{4r_s}{r}e^{-\frac{r}{r_s}}dUdV. \tag{6.269}$$

The tortoise coordinate (corresponding to a conformally flat metric) is defined by

$$dr = \left(1 - \frac{r_s}{r}\right)dr_* \rightarrow r_* = r - r_s + r_s \ln\left(\frac{r}{r_s} - 1\right). \tag{6.270}$$

We will also work with the ingoing and outgoing null coordinates u and v defined only in quadrant I given by

$$u = t - r_*$$
$$= t - r - r_s \ln\left(\frac{r}{r_s} - 1\right) + r_s$$
$$= -2r_s \ln\left(\frac{-U}{r_s}\right) + r_s. \tag{6.271}$$

$$v = t + r_*$$
$$= t + r + r_s \ln\left(\frac{r}{r_s} - 1\right) - r_s$$
$$= 2r_s \ln\left(\frac{V}{r_s}\right) - r_s. \tag{6.272}$$

The metric in this system is

$$ds^2 = -\left(1 - \frac{2GM}{r}\right)du\,dv = -\frac{r_s}{r}e^{\frac{v-u}{2r_s}}e^{-\frac{r-r_s}{r_s}}du\,dv. \tag{6.273}$$

We will expand the field in modes as usual. The following important points should be taken into consideration.

- For the asymptotic inertial observer the modes will be denoted by the frequencies ω and they are clearly associated with the Schwarzschild time t or equivalently $u = t - r_*$. This what corresponds to the exterior degrees of freedom.

- For the freely falling inertial observer the time is obviously given by the proper time τ. From equation (6.57) (with $\lambda = \tau$) and (6.62) we obtain near the horizon

$$\frac{d\tau}{dt} \sim r - r_s \sim \exp(-t/r_s) \Rightarrow d\tau \sim \exp(-t/r_s)$$
$$dt \Rightarrow \tau \sim -r_s\exp(-t/r_s) + \tau_0. \tag{6.274}$$

We get then near the horizon

$$U \sim \frac{1}{\sqrt{r_s}}\exp(r/2r_s)(\tau - \tau_0), \quad V \sim \sqrt{r_s}\exp(r/2r_s). \tag{6.275}$$

Thus $U \longrightarrow 0$ and $V \longrightarrow$ constant. Also, we conclude that the proper time τ is equivalent to the coordinate U with frequencies denoted by ν. Since U is defined throughout spacetime the frequency ν is what corresponds to the interior degrees of freedom.

- We know already that in the Schwarzschild geometry the solutions of the equation of motion are spherically symmetric which read

$$\psi = \sum_{lm} Y_{lm}\psi_{lm}. \tag{6.276}$$

The ψ_{lm} solves the Schrödinger equation, viz

$$(\partial_t^2 - \partial_{r_*}^2 + V(r_*))\psi_{lm} = 0, \tag{6.277}$$

with a potential function in the tortoise coordinates r_* of the form

$$V(r_*) = \frac{r - r_s}{r}\left(\frac{r_s}{r} + \frac{l(l+1)}{r^2}\right). \tag{6.278}$$

In the limit $r \longrightarrow \infty$ (the asymptotically flat spacetime limit) the tortoise coordinate behaves as $r_* \longrightarrow \infty$ and the potential goes to zero as $V \simeq l(l+1)/r^2$. The particle is therefore free in this limit. Similarly, in the near horizon limit $r \longrightarrow r_s$ the tortoise coordinate behaves as $r_* \longrightarrow -\infty$ and the potential goes to zero again but now as $V \simeq (r - r_s)/r \sim \exp((r_* - r)/r_s)$. The particle is also free in this regime.

Thus near infinity and near the horizon the solutions are plane waves of the form $\exp(ik(t \pm r_*))$ or equivalently $\exp(iku)$ and $\exp(ikv)$.

- The scalar field action is

$$
\begin{aligned}
I &= \frac{1}{2} \int d^2x \sqrt{-\det g}\, g^{\mu\nu} \partial_\mu \phi \partial_\nu \phi \\
&= -\int dU dV\, \partial_U \phi \partial_V \phi \\
&= -\int du dv\, \partial_u \phi \partial_v \phi = \frac{1}{2} \int dt dr_* (-(\partial_t \phi)^2 + (\partial_{r_*} \phi)^2).
\end{aligned}
\tag{6.279}
$$

The equation of motion is

$$
\partial_u \partial_v \phi = \partial_U \partial_V \phi = 0.
\tag{6.280}
$$

The solution is

$$
\begin{aligned}
\phi &= \phi_L(u) + \phi_R(v) \\
&= \phi_L(U) + \phi_R(V).
\end{aligned}
\tag{6.281}
$$

We will only consider the right-moving part.

- We consider a particular foliation of the near horizon geometry. For example, the coordinates u and v in region I are replaced by $\eta = t = (u + v)/2$ and $\xi = r_* = -(u - v)/2$ where η is time. These coordinates near the horizon in region I define the metric of Rindler quadrant with acceleration given formally by $a = 1/2r_s$, viz

$$
ds^2 = \exp(2a\xi)(-d\eta^2 + d\xi^2).
\tag{6.282}
$$

Thus, the Klein–Gordon inner product is precisely given by the formula (6.215), viz

$$
(\phi_1, \phi_2) = -i \int \left(\phi_1 \partial_\eta \phi_2^* - \partial_\eta \phi_1 \cdot \phi_2^* \right) d\xi.
\tag{6.283}
$$

We can check immediately that

$$
(\phi_1, \phi_2) = -(\phi_2^*, \phi_1^*), \quad (\phi_1^*, \phi_2^*) = -(\phi_1, \phi_2)^*.
\tag{6.284}
$$

The positive frequency normalized modes in region I have been already computed. They are given by (6.208)

$$
g_k^{(1)} = \frac{1}{\sqrt{4\pi\Omega}} \exp(-i\Omega\eta + ik\xi), \quad \Omega = |k|.
\tag{6.285}
$$

The right-moving part of this positive frequency mode corresponds to $k > 0$ and it is given explicitly by

$$
g_k^{(1)} = \frac{1}{\sqrt{4\pi\Omega}} \exp(-i\Omega u).
\tag{6.286}
$$

The right-moving part with negative frequency corresponds therefore to $g_k^{(1)*}$. A right-moving field will then be expanded as

$$\phi_R(u) = \int_0^\infty dk \left(\frac{b_k}{\sqrt{4\pi\Omega}} \exp(-i\Omega u) + \frac{b_k^+}{\sqrt{4\pi\Omega}} \exp(i\Omega u) \right). \qquad (6.287)$$

After a change of variable $\omega = k$ and $b_k = b_\omega / \sqrt{2\pi}$ we get

$$\phi_R(u) = \int_0^\infty \frac{d\omega}{2\pi} \left(\frac{b_\omega}{\sqrt{2\omega}} \exp(-i\omega u) + \frac{b_\omega^+}{\sqrt{2\omega}} \exp(i\omega u) \right). \qquad (6.288)$$

Since $g_k^{(1)}$ are normalized such that $(g_k^{(1)}, g_{k'}^{(1)}) = \delta(k - k')$ the annihilation and creation operators b_k and b_k^+ must satisfy $[b_k, b_{k'}^+] = \delta(k - k')$ and thus $[b_\omega, b_{\omega'}^+] = 2\pi\delta(\omega - \omega')$.

- From the above considerations, the field operator in the Schwarzschild tortoise coordinates (t, r_*) is given by the formula

$$\phi(t, r_*) = \int_{-\infty}^{+\infty} \frac{dk}{2\pi} \left(\frac{b_k}{\sqrt{2|k|}} \exp(-i|k|t + ikr_*) + \frac{b_k^+}{\sqrt{2|k|}} \exp(i|k|t - ikr_*) \right). \qquad (6.289)$$

$$[b_k, b_{k'}^+] = 2\pi\delta(k - k'). \qquad (6.290)$$

The frequency is $\omega = |k|$ and t is the proper time at infinity where Schwarzschild becomes Minkowski. The momentum operator is

$$\pi(t, r_*) = \frac{\partial L}{\partial(\partial_t \phi)} = \partial_t \phi(t, r_*)$$

$$= \int_{-\infty}^{+\infty} \frac{dk}{2\pi} \left(\frac{-i|k|b_k}{\sqrt{2|k|}} \exp(-i|k|t + ikr_*) + \frac{i|k|b_k^+}{\sqrt{2|k|}} \exp(i|k|t - ikr_*) \right). \qquad (6.291)$$

We compute immediately

$$[\phi(t, r_*), \pi(t, r_*')] = i \int_{-\infty}^{+\infty} \frac{dk}{2\pi} e^{ik(r_* - r_*')} = i\delta(r_* - r_*'). \qquad (6.292)$$

This confirms our normalization.

The vacuum with respect to the inertial asymptotic tortoise Schwarzschild observer, also called the Boulware vacuum, is given by

$$b_k|0_T\rangle = 0, \quad \forall k. \qquad (6.293)$$

- For an obvious reason, the mode expansion in the Kruskal coordinates (U, V), with proper time given by $T = (U + V)/2$ and space-like coordinate given by $X = -(U - V)/2$, is similar to the above expansion, viz

$$\phi(T, X) = \int_{-\infty}^{+\infty} \frac{dk}{2\pi} \left(\frac{a_k}{\sqrt{2|k|}} \exp(-i|k|T + ikX) + \frac{a_k^+}{\sqrt{2|k|}} \exp(i|k|T - ikX) \right). \qquad (6.294)$$

The frequency here is $\nu = |k|$ and U is equivalent to the proper time τ of an infalling observer. The Kruskal vacuum is defined by

$$a_k |0_K\rangle = 0, \quad \forall\, k. \tag{6.295}$$

- The field decomposes into right-moving field and left-moving field or in the terminology of four dimensions into ingoing and outgoing fields. The right-moving (outgoing) field corresponds to $k > 0$ and the left-moving (ingoing) field corresponds to $k < 0$. We write the field as

$$\phi(T, X) = \int_0^{+\infty} \frac{d\nu}{2\pi}\left(\frac{a_\nu}{\sqrt{2\nu}}\exp(-i\nu U) + \frac{a_{-\nu}}{\sqrt{2\nu}}\exp(-i\nu V) + \text{h.c}\right). \tag{6.296}$$

Similarly,

$$\phi(t, r_*) = \int_0^{+\infty} \frac{d\omega}{2\pi}\left(\frac{b_\omega}{\sqrt{2\omega}}\exp(-i\omega u) + \frac{b_{-\omega}}{\sqrt{2\omega}}\exp(-i\omega v) + \text{h.c}\right). \tag{6.297}$$

6.9.2 Bogolubov coefficients

Let us summarize our main points. We have two observers: the asymptotic Schwarzschild tortoise observer and the freely falling Kruskal observer. The Schwarzschild observer defined for $r > r_s$ is the analog of the accelerating Rindler observer with acceleration given by $a = 1/2r_s$, whereas the Kruskal observer corresponds to the inertial Minkowski observer defined throughout the spacetime manifold.

The asymptotic observer at fixed r ($r > r_s$) expands the right-moving field in terms of the modes v_ω as

$$\phi_R(u) = \int_0^\infty d\omega (v_\omega b_\omega + v_\omega^* b_\omega^\dagger), \quad v_\omega = \frac{1}{\sqrt{4\pi\omega}}\exp(-i\omega u). \tag{6.298}$$

We have the normalization

$$(v_{\omega_1}, v_{\omega_2}) = \delta(\omega_1 - \omega_2), \quad [b_\omega, b_{\omega'}^\dagger] = \delta(\omega - \omega'). \tag{6.299}$$

This observer sees the Schwarzschild tortoise vacuum

$$b_\omega |0_T\rangle = 0. \tag{6.300}$$

The freely falling observer expands the right-moving field in terms of the modes u_ν as

$$\phi_R(U) = \int_0^\infty d\nu (u_\nu a_\nu + u_\nu^* a_\nu^\dagger), \quad u_\nu = \frac{1}{\sqrt{4\pi\nu}}\exp(-i\nu U). \tag{6.301}$$

We have the normalization

$$(u_{\nu_1}, u_{\nu_2}) = \delta(\nu_1 - \nu_2), \quad [a_\nu, a_{\nu'}^\dagger] = \delta(\nu - \nu'). \tag{6.302}$$

This observer sees the Kruskal vacuum

$$a_\nu |0_K\rangle = 0. \tag{6.303}$$

The asymptotic and freely falling objects are related through the Bogolubov transformations

$$v_\omega = \int_0^\infty d\nu (\alpha_{\omega\nu} u_\nu + \beta_{\omega\nu} u_\nu^*), \quad u_\nu = \int_0^\infty d\omega (\alpha_{\omega\nu}^* v_\omega - \beta_{\omega\nu} v_\omega^*). \tag{6.304}$$

$$a_\nu = \int_0^\infty d\omega (\alpha_{\omega\nu} b_\omega + \beta_{\omega\nu}^* b_\omega^\dagger), \quad b_\omega = \int_0^\infty d\nu (\alpha_{\omega\nu}^* a_\nu - \beta_{\omega\nu}^* a_\nu^\dagger). \tag{6.305}$$

The first equation should be corrected by the introduction of the interior degrees of freedom (see next lecture). Nevertheless, the Bogolubov coefficients are

$$\alpha_{\omega\nu} = (v_\omega, u_\nu), \quad \beta_{\omega\nu} = -(v_\omega, u_\nu^*). \tag{6.306}$$

We calculate immediately

$$\alpha_{\omega\nu} = (v_\omega, u_\nu) = -i \int \frac{2}{4\pi\sqrt{\omega\nu}} (i\omega\partial_\eta u) e^{-i\omega u} e^{i\nu U} dr_*$$
$$= -\int_{-\infty}^{+\infty} \frac{du}{2\pi} \sqrt{\frac{\omega}{\nu}} e^{-i\omega u} e^{i\nu U} \Rightarrow \alpha_{\omega\nu}^* = -\sqrt{\frac{\omega}{\nu}} F(\omega, \nu). \tag{6.307}$$

Similarly,

$$\beta_{\omega\nu} = -(v_\omega, u_\nu^*)$$
$$= \int_{-\infty}^{+\infty} \frac{du}{2\pi} \sqrt{\frac{\omega}{\nu}} e^{-i\omega u} e^{-i\nu U} \Rightarrow -\beta_{\omega\nu}^* = -\sqrt{\frac{\omega}{\nu}} F(\omega, -\nu). \tag{6.308}$$

The function F is given by (with $U = U_0 e^{-au}$ where $U_0 = -\sqrt{e}/2a$)

$$F(\omega, \nu) = \int_{-\infty}^{+\infty} \frac{du}{2\pi} e^{i\omega u} e^{-i\nu U} = \int_{-\infty}^{+\infty} \frac{du}{2\pi} e^{i\omega u - i\nu U_0 e^{-au}}. \tag{6.309}$$

This is the Euler gamma function. Indeed, if we make the change of variable $u \longrightarrow z = i\nu U_0 e^{-au}$ we immediately reach the formula

$$F(\omega, \nu) = \frac{1}{2\pi a} \exp\left(\frac{i\omega}{a} \ln i\nu U_0\right) \int_0^{+\infty} e^{-\frac{i\omega}{a}-1} e^{-z} dz$$
$$= \frac{1}{2\pi a} \exp\left(\frac{i\omega}{a} \ln i\nu U_0\right) \Gamma\left(-\frac{i\omega}{a}\right). \tag{6.310}$$

The number of b particles of frequency ω as seen by the Schwarzschild asymptotic fixed observer is given by the expectation value of the number operator $N_\omega = b_\omega^+ b_\omega$. Obviously, the expectation value of this number operator in the tortoise vacuum is zero, viz $\langle 0_T | N_\omega | 0_T \rangle = 0$. However, the actual vacuum state of lowest energy of the quantum scalar field in the presence of a classical black hole is given by the freely falling Kruskal vacuum $|0_K\rangle$. This is because Schwarzschild is the analog of Rindler, whereas Kruskal is the analog of Minkowski. See the nice discussion in [14]. Also, in consideration of the gravitational collapse of a star onto a black hole it has been shown that before the collapse the vacuum state is that of Minkowski and after the collapse the vacuum state becomes that of Kruskal [8, 9]. Thus, the vacuum state is $|0_K\rangle$ and it does actually contain b particles as seen by the asymptotic Schwarzschild observer since

$$\langle 0_K | N_\omega | 0_K \rangle = \langle 0_K | b_\omega^+ b_\omega | 0_K \rangle$$
$$= \int_0^\infty d\nu \, |\beta_{\omega\nu}|^2. \tag{6.311}$$

6.9.3 Hawking radiation and Hawking temperature

The Bogolubov coefficient can be expressed in terms of Euler gamma function as shown above and then integrated over. However, the method outlined in [14] is more illuminating.

We deform the u integral from $-\infty$ to $+\infty$ to the t integral from $-\infty - i\pi/a$ to $+\infty - i\pi/a$ where $u = t + i\pi/a$. See figure 6.7. The integral is not changed because (i) the integrand has no poles which is obvious, (i) the lateral segments are limited in length which is also obvious and (i) the integrand vanishes for $t \longrightarrow \pm \infty - i\alpha$ where $0 < \alpha < \pi/a$. The last point is shown as follows. Firstly,

$$\lim_{t \longrightarrow -\infty - i\alpha} \mathrm{Re}(i\nu U_0 e^{-at}) = \lim_{u \longrightarrow -\infty} \mathrm{Re}(i\nu U_0 e^{ia\alpha} e^{-au})$$
$$= -\lim_{t \longrightarrow -\infty} \mathrm{Re}(\nu U_0 \sin a\alpha \, e^{-au}) \tag{6.312}$$
$$= -\infty.$$

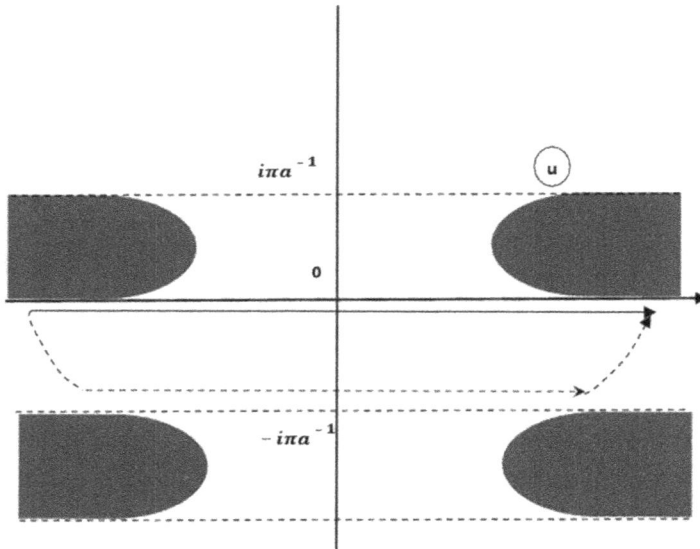

Figure 6.7. The contour of integration.

For the limit $t \longrightarrow +\infty - i\alpha$ the integral diverges and we need to regularize it for example as

$$F(\omega, \nu) = \int_{-\infty}^{+\infty} \frac{du}{2\pi} e^{i\omega u - i\nu U_0 e^{-au}} e^{-bu^2}, \quad b > 0. \qquad (6.313)$$

This integral then for b positive is zero in the limit $t \longrightarrow +\infty - i\alpha$. Since there are no poles inside the closed contour formed by the original contour and the shifted one as in the figure below we conclude immediately that F can be given by the integral

$$F(\omega, \nu) = \int_{-\infty - \frac{i\pi}{a}}^{+\infty - \frac{i\pi}{a}} \frac{dt}{2\pi} e^{i\omega t - i\nu U_0 e^{-at}}$$

$$= \exp\left(\frac{\omega\pi}{a}\right) F(\omega, -\nu). \qquad (6.314)$$

This result should be understood in the sense of distribution. The exhibited contour is the unique possibility allowed to us since we cannot deform the contour to $u = t - i(\pi + 2\pi n)/a$ with $n \neq 0$ because the $\sin a\alpha$ in (6.312) will change sign.

We use now the last formula (6.314) to compute the expectation value of the Schwarzschild asymptotic observer number operator N_ω in the Kruskal (black hole) vacuum $|0\rangle$ as follows. We start from the normalization condition

$$\delta(\omega - \omega') = (v_\omega, v_{\omega'})$$

$$= \int_0^\infty d\nu \left[\alpha_{\omega\nu}(u_\nu, v_{\omega'}) + \beta_{\omega\nu}(u_\nu^*, v_{\omega'})\right]$$

$$= \int_0^\infty d\nu \left[\alpha_{\omega\nu}\alpha_{\omega'\nu}^* - \beta_{\omega\nu}\beta_{\omega'\nu}^*\right]$$

$$= \int_0^\infty d\nu \frac{\sqrt{\omega\omega'}}{\nu} \left[F^*(\omega, \nu)F(\omega', \nu) - F^*(\omega, -\nu)F(\omega', -\nu)\right]$$

$$= \left(e^{\frac{\pi(\omega+\omega')}{a}} - 1\right) \int_0^\infty d\nu \frac{\sqrt{\omega\omega'}}{\nu} F^*(\omega, -\nu)F(\omega', -\nu). \qquad (6.315)$$

We write this equation as

$$\int_0^\infty d\nu \frac{\sqrt{\omega\omega'}}{\nu} F^*(\omega, -\nu)F(\omega', -\nu) = \frac{\delta(\omega - \omega')}{e^{\frac{\pi(\omega+\omega')}{a}} - 1}. \qquad (6.316)$$

For $\omega = \omega'$ we get precisely the desired result

$$\int_0^\infty d\nu \frac{\omega}{\nu} |F(\omega, -\nu)|^2 = \frac{\delta(0)}{e^{\frac{2\pi\omega}{a}} - 1}. \qquad (6.317)$$

In other words,

$$\langle 0_K | N_\omega | 0_K \rangle = \langle 0_K | b_\omega^+ b_\omega | 0_K \rangle$$

$$= \int_0^\infty d\nu |\beta_{\omega\nu}|^2$$

$$= \frac{\delta(0)}{\exp\left(\frac{2\pi\omega}{a}\right) - 1}. \qquad (6.318)$$

The density of b particles in the black hole vacuum state $|0_K\rangle$ is therefore given by[3]

$$n_\omega = \frac{1}{2\pi} \frac{1}{\exp\left(\dfrac{2\pi\omega}{a}\right) - 1}. \tag{6.319}$$

This is a blackbody Planck spectrum with the temperature

$$T_H = \frac{a}{2\pi} = \frac{1}{4\pi r_s} = \frac{1}{8\pi GM}. \tag{6.320}$$

By inserting SI units we obtain

$$T_H = \frac{\hbar c^3}{8\pi GM k_B}. \tag{6.321}$$

This is the famous Hawking temperature. The black hole as seen by a distant observer is radiating energy, thus its mass decreases, and as a consequence its temperature increases, i.e. the black hole becomes hotter, which indicates a negative specific heat.

6.10 The Unruh versuss Boulware vacua: pure to mixed

We will follow here the excellent pedagogical presentation of [13].

The first type of information loss is by falling across the event horizon. The second type which is intimately related concerns Hawking radiation and is equivalent to the evolution of pure states to mixed states which is a process forbidden by quantum mechanics.

6.10.1 The adiabatic principle and trans-Planckian reservoir

We start with the Rindler space (which is the cleanest of the two cases) where we have obtained the Unruh effect by two methods. By computing the density matrix and also the flux formula with respect to the Rindler observer. By using quantum information, we have found that we can put the density matrix into the form

$$\rho_R = \frac{1}{Z} \exp\left(-\frac{2\pi}{a} H_R\right) = \frac{1}{Z} \sum_i \exp\left(-\frac{2\pi}{a} E_i\right) |i_R\rangle \langle i_R|. \tag{6.322}$$

This is a mixed (thermal,random) state obtained by integrating out the left wedge degrees of freedom in the vacuum pure (entangled, correlated) state

$$|\Omega\rangle = \frac{1}{\sqrt{Z}} \sum_i \exp(-\pi E_i) |i_R\rangle |i_L^*\rangle. \tag{6.323}$$

---

[3] In $1 + 3$ dimensions using box normalization we have $(2\pi)^3 \delta^3(0) = V$ where V is the volume of spacetime. In the current $1 + 1$ dimensional case we have $(2\pi)\delta(0) = L$.

On the other hand, by using QFT in curved backgrounds we calculated the number of particles with energy $\omega = |k|$ seen by the Rindler observer in the vacuum Minkowski state $|0_M\rangle \equiv |\Omega\rangle$ to be given by the blackbody spectrum

$$\langle 0_M | \hat{N}_R^{(1)}(k) | 0_M \rangle = \frac{1}{\exp\left(\dfrac{2\pi\omega}{a}\right) - 1} \delta(0). \qquad (6.324)$$

Since the near horizon geometry of the Schwarzschild black hole is Rindler a similar result is expected to hold in the Schwarzschild black hole geometry. Indeed, this is the result of [26] which we will try to derive here following [13].

In discussing the Hawking radiation so far we have omitted several points. First, we have only considered the exterior region. Second, we did not talk about greybody factors and furthermore we have not mentioned at all the underlying adiabatic approximation or the trans-Planckian problem and its so-called nice-slice resolution. All these issues can be remedied somewhat by considering black holes as forming from collapsing shells of matter in some pure quantum state $|\psi\rangle$.

We consider therefore a black hole which had formed during gravitational collapse in a quantum state $|\psi\rangle$. (figure 6.8) The out state corresponds to an outgoing Killing null wave packet P centered around some positive frequency ω with support only at large radii r at late times $t \longrightarrow +\infty$. Recall that ω, r and t relate to the Schwarzschild tortoise coordinates. Obviously, this wave packet is a solution of the Klein–Gordon equation which behaves at infinity as $\exp(-i\omega t)$ and thus near the horizon it can only depend on the outgoing (right-moving) coordinates $u = t - r_*$, viz $P \propto \exp(-i\omega u)$. This wave packet P corresponds to an annihilation operator $a(P)$ given in terms of the field operator ϕ, which solves the Klein–Gordon equation, by the Klein–Gordon inner product

$$a(P) = (\phi, P). \qquad (6.325)$$

We run this wave packet backwards in time towards the black hole. A reflected part R will scatter off the black hole and return to large radii and a transmitted part T with support only immediately outside the event horizon. We write

$$P = R + T. \qquad (6.326)$$

The wave packets R and T have the same positive Killing frequency with respect to the asymptotic Schwarzschild observer as the outgoing wave packet P because the black hole metric is stationary. But with respect to a freely falling observer who intersects the trajectory of the transmitted wave packet T at the event horizon both positive and negative frequency modes will be seen in T. The annihilation operator $a(P)$ decomposes in an almost obvious way as

$$a(P) = a(R) + a(T). \qquad (6.327)$$

Since the reflected wave packet R has only support in the asymptotic flat region very far outside the black hole and since $|\psi\rangle$ contains no positive frequency incoming excitation the annihilation operator $a(R)$ annihilates the state $|\psi\rangle$ exactly

$$a(R)|\psi\rangle = 0. \qquad (6.328)$$

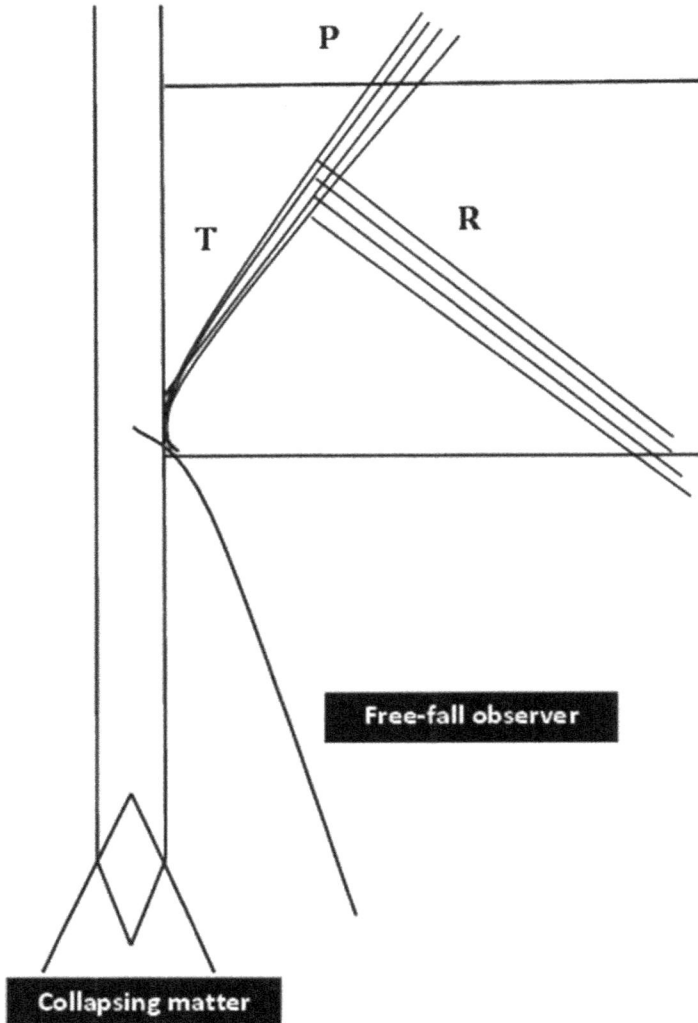

Figure 6.8. The wave packets P, R and T near the horizon.

If the state $|\psi\rangle$ were also annihilated by T it would have been identical with the Boulware vacuum or tortoise vacuum $|0_T\rangle$ introduced in the previous lecture. But T contains positive frequencies as well as negative frequencies with respect to the proper time of the freely falling observer. Thus we decompose it as follows

$$T = T^+ + T^- \Rightarrow a(T) = a(T^+) + a(T^-). \tag{6.329}$$

By using the property of the Klein–Gordon inner product $(\phi_1, \phi_2)^* = -(\phi_1^*, \phi_2^*)$ we derive immediately that $a^\dagger(\bar{T}^-) = -a(T^-)$. Thus

$$a(T) = a(T^+) - a^\dagger(\bar{T}^-). \tag{6.330}$$

We already know that T has only support near the horizon where it behaves as $T \sim \exp(-i\omega u)$. But near the horizon we have $r - r_s = \exp(-t/r_s)$ and $u \sim 2t \sim -2r_s \ln -(\tau - \tau_0)/r_s$. Thus the behavior of T is of the general form $(a = \kappa = 1/2r_s)$

$$T \sim \exp(i\frac{\omega}{a}\ln(-\tau)), \quad \tau < 0.$$
$$T = 0, \quad \tau > 0.$$

(6.331)

Thus near the horizon T consists of rapid oscillations, which means in particular that T^+ and $\bar{T}^-$ are positive high frequency modes. Initially, the black hole state $|\psi\rangle$ does not contain these high energy modes. We say that these modes are in their ground states.

As we evolve backward in time the frequencies blueshift (increase) in the same way that when evolving forward in time they will redshift (decrease). Thus, as we approach the horizon the frequency increases, with respect to the freely falling observer, until it becomes infinitely blueshifted on the horizon. In other words, these modes seem to arise deep in the UV region which is what we call the trans-Planckian reservoir. This could be a problematic issue as discussed in [13] with a proposed resolution which goes under the name of the nice-slice argument given in [27]. Both the potential problem and the proposed resolution are not very essential to us here. Indeed, we are only using the above fact regarding the very large blueshift on the horizon to conclude that the modes T^+ and $\bar{T}^-$ remain high energy modes as we evolve them backward in time. Furthermore, the earlier the infalling observer meets the mode with frequency ω the higher its proper frequency ν will be since the Schwarzschild frequency ω is redshifted with respect to the free fall frequency ν as $\nu = 2r_s\omega$.

Hence, by looking at the black hole after it had formed at times $t \ll r_s$, where the Schwarzschild radius r_s measures the time scale of the collapse process, the high frequency modes with $\omega \gg 1/r_s$ ($\nu \gg 2$) are not excited, which means in particular that the modes T^+ and $\bar{T}^-$ remain unexcited, i.e. they remain in their ground states. We conclude that the black hole state $|\psi\rangle$ does not contain positive high frequency modes throughout, viz

$$a(T^+)|\psi\rangle = a(\bar{T}^-)|\psi\rangle = 0.$$

(6.332)

This is essentially the adiabatic principle. The geometry during the gravitational collapse is obviously time dependent with a time scale given by the Schwarzschild time r_s. Thus, the modes with frequencies $\omega \gg 1/r_s$ see the change of the geometry adiabatically, i.e. very slowly, and hence they remain unexcited.

6.10.2 The Unruh method revisited and greybody factor

If we consider now the expectation value of the number operator $N = a^\dagger(P)a(P)$ in the black hole state $|\psi\rangle$ we find immediately

$$\begin{aligned}
\langle\psi|N|\psi\rangle &= \langle\psi|a^\dagger(P)a(P)|\psi\rangle \\
&= \langle\psi|a^\dagger(T)a(T)|\psi\rangle \\
&= \langle\psi|a(\bar{T}^-)a^\dagger(\bar{T}^-)|\psi\rangle \\
&= \langle\psi|[a(\bar{T}^-), a^\dagger(\bar{T}^-)]|\psi\rangle.
\end{aligned} \tag{6.333}$$

However, we can explicitly expand the field operator in a positive frequency basis $\{f_i\}$ as

$$\phi = \sum_i (a_i f_i + a_i^\dagger f_i^*). \tag{6.334}$$

Also, the positive frequency wave packet $\bar{T}^-$ can be expanded similarly as

$$\bar{T}^- = \sum_i t_i^* f_i. \tag{6.335}$$

The annihilation operator $a(\bar{T}^-)$ is then given explicitly by

$$a(\bar{T}^-) = \sum_i a_i t_i. \tag{6.336}$$

We compute then

$$\begin{aligned}
\langle\psi|[a(\bar{T}^-), a^\dagger(\bar{T}^-)]|\psi\rangle &= \langle\psi\sum_i\sum_j t_i t_j^*[a_i, a_j^\dagger]|\psi\rangle \\
&= \sum_i t_i t_i^* \\
&= (\bar{T}^-, \bar{T}^-) \\
&= -(T^-, T^-).
\end{aligned} \tag{6.337}$$

Thus the expectation value of the number operator becomes

$$\langle\psi|N|\psi\rangle = -(T^-, T^-). \tag{6.338}$$

The transmitted wave packet is given by

$$\begin{aligned}
T(\tau) &= \exp\left(i\frac{\omega}{a}\ln(-\tau)\right), \quad \tau < 0 \\
T &= 0, \quad \tau > 0.
\end{aligned} \tag{6.339}$$

Since $\tau < 0$ this is defined only outside the horizon. Thus this function contains positive and negative frequency modes with respect to the freely falling observer (recall that T is a positive frequency mode with respect to the Schwarzschild observer). This is the analog of $g_k^{(1)}$ in the case of Rindler which was defined as a positive frequency solution only with respect to the Rindler observer in quadrant I but with respect to the Minkowski observer it contains both positive and negative frequencies. As we did in reaching equations (7.186) and (7.188) in the Rindler case, by using the method of [22], we will now extend the solution (6.339) to the region inside the horizon ($\tau > 0$) and obtain in the course the positive frequency and the negative frequency extensions T^+ and T^-.

First, recall that a positive frequency mode can be expanded in terms of $\exp(-i\omega\tau)$, $\omega > 0$. The functions $\exp(-i\omega\tau)$ clearly vanish in the limit $|\tau| \longrightarrow \infty$ in the lower half complex τ plane for $\omega > 0$. Thus the positive frequency extension of T should be obtained by analytic continuation in the lower half complex plane. This extension of T from $\tau < 0$ to $\tau > 0$ is obtained by analytic continuation of $\ln(-\tau)$ from $\tau < 0$ to $\tau > 0$ in the lower half complex plane provided the branch cut of the logarithm is chosen in the upper half complex plane. This continuation of $\ln(-\tau)$ with $\tau < 0$ is given by $\ln\tau + i\pi$ with $\tau > 0$.[4] By replacing in $T(\tau)$ with $\tau < 0$ we get $T(-\tau)\exp(-\pi\omega/a)$ with $\tau > 0$. The wave packet solution inside the horizon is then given by

$$\tilde{T}(\tau) = T(-\tau) = \exp\left(i\frac{\omega}{a}\ln(\tau)\right), \quad \tau > 0.$$

$$\tilde{T} = 0, \quad \tau < 0.$$

(6.340)

The total wave packet

$$T^+ = c_+\left(T + \tilde{T}\exp\left(-\frac{\pi\omega}{a}\right)\right)$$

(6.341)

is clearly analytic in the lower half complex plane and bounded as $|\tau| \longrightarrow \infty$ and as such it can only contain positive frequencies. In other words, T^+ is the desired positive frequency extension of T.

The negative frequency extension of T should be obtained by analytic continuation in the upper half complex plane. This extension of T from $\tau < 0$ to $\tau > 0$ is obtained by analytic continuation of $\ln(-\tau)$ from $\tau < 0$ to $\tau > 0$ in the upper half complex plane provided the branch cut of the logarithm is chosen in the lower half complex plane. This continuation of $\ln(-\tau)$ with $\tau < 0$ is given by $\ln\tau - i\pi$ with $\tau > 0$.[5] By replacing in $T(\tau)$ with $\tau < 0$ we get $T(-\tau)\exp(\pi\omega/a)$ with $\tau > 0$. The total wave packet

---

[4] The function $\ln z$ is multi-valued in the complex plane. To get a single-valued function we introduce a cut line between its two branch points $z = 0$ and $z = \infty$.

For positive frequency modes we will need to extend in the lower half complex plane and choose the branch cut in the upper half complex plane. The function $\ln(-\tau)$ with $\tau < 0$ is analytically continued to $\tau > 0$ by writing $z = -\tau\exp(i\theta)$. Since the branch cut is in the upper half complex plane we can only go from $z = \tau$ to $z = -\tau$ counter-clockwise in the lower half plane, i.e. from $\theta = \pi$ to $\theta = 2\pi$. At $\theta = \pi$ we have $z = \tau < 0$ and $\ln z - i\pi = \ln(-\tau)$, whereas at $\theta = 2\pi$ we have $z = -\tau = \tau' > 0$ and $\ln z - i\pi = \ln z' + i\pi$. Thus the analytic continuation of $\ln(-\tau)$, $\tau < 0$, in the lower half complex plane is given by $\ln\tau + i\pi$, $\tau > 0$, if the branch cut is in the upper half complex plane.

[5] For negative frequency modes we will need to extend in the upper half complex plane and choose the branch cut in the lower half complex plane. The function $\ln(-\tau)$ with $\tau < 0$ is again analytically continued to $\tau > 0$ by writing $z = -\tau\exp(i\theta)$. Since the branch cut now is in the lower half complex plane we can only go from $z = \tau$ to $z = -\tau$ counter-clockwise in the upper half plane, i.e. from $\theta = \pi$ to $\theta = 0$. At $\theta = \pi$ we have $z = \tau < 0$ and $\ln z - i\pi = \ln(-\tau)$ as before, whereas at $\theta = 0$ we have $z = -\tau = \tau' > 0$ and $\ln z - i\pi = \ln z' - i\pi$. Thus the analytic continuation of $\ln(-\tau)$, $\tau < 0$, in the upper half complex plane is given by $\ln\tau - i\pi$, $\tau > 0$, if the branch cut is in the lower half complex plane.

$$T^- = c_- \left(T + \tilde{T} \exp\left(\frac{\pi\omega}{a}\right) \right) \tag{6.342}$$

is clearly analytic in the upper half complex plane and bounded as $|\tau| \longrightarrow \infty$ and as such it can only contain negative frequencies. In other words, T^- is the desired negative frequency extension of T.

The boundary conditions are given by

$$T^+ + T^- = T, \quad \tau < 0 \Rightarrow c_+ + c_- = 1$$

$$T^+ + T^- = 0, \quad \tau > 0 \Rightarrow c_+\exp\left(-\frac{\pi\omega}{a}\right) + c_-\exp\left(\frac{\pi\omega}{a}\right) = 0. \tag{6.343}$$

This gives immediately

$$c_+ = \frac{1}{1 - \exp\left(-\dfrac{2\pi\omega}{a}\right)}, \quad c_- = \frac{1}{1 - \exp\left(\dfrac{2\pi\omega}{a}\right)}. \tag{6.344}$$

By using now the negative frequency extension T^- we can immediately compute the expectation value of the number operator to be given by (using also $(T, T) = -(\tilde{T}, \tilde{T})$ and $(T, \tilde{T}) = 0$)

$$\langle \psi | N | \psi \rangle = \frac{(T, T)}{\exp\left(\dfrac{2\pi\omega}{a}\right) - 1}. \tag{6.345}$$

This is again a blackbody spectrum with the Hawking temperature $T_H = a/2\pi = 1/4\pi r_s$. However, this result is actually reduced by the so-called greybody factor

$$\Gamma = (T, T). \tag{6.346}$$

---

[4] The function $\ln z$ is multi-valued in the complex plane. To get a single-valued function we introduce a cut line between its two branch points $z = 0$ and $z = \infty$.

For positive frequency modes we will need to extend in the lower half complex plane and choose the branch cut in the upper half complex plane. The function $\ln(-\tau)$ with $\tau < 0$ is analytically continued to $\tau > 0$ by writing $z = -\tau \exp(i\theta)$. Since the branch cut is in the upper half complex plane we can only go from $z = \tau$ to $z = -\tau$ counter-clockwise in the lower half plane, i.e. from $\theta = \pi$ to $\theta = 2\pi$. At $\theta = \pi$ we have $z = \tau < 0$ and $\ln z - i\pi = \ln(-\tau)$, whereas at $\theta = 2\pi$ we have $z = -\tau = \tau' > 0$ and $\ln z - i\pi = \ln z' + i\pi$. Thus the analytic continuation of $\ln(-\tau)$, $\tau < 0$, in the lower half complex plane is given by $\ln \tau + i\pi$, $\tau > 0$, if the branch cut is in the upper half complex plane.

[5] For negative frequency modes we will need to extend in the upper half complex plane and choose the branch cut in the lower half complex plane. The function $\ln(-\tau)$ with $\tau < 0$ is again analytically continued to $\tau > 0$ by writing $z = -\tau \exp(i\theta)$. Since the branch cut now is in the lower half complex plane we can only go from $z = \tau$ to $z = -\tau$ counter-clockwise in the upper half plane, i.e. from $\theta = \pi$ to $\theta = 0$. At $\theta = \pi$ we have $z = \tau < 0$ and $\ln z - i\pi = \ln(-\tau)$ as before, whereas at $\theta = 0$ we have $z = -\tau = \tau' > 0$ and $\ln z - i\pi = \ln z' - i\pi$. Thus the analytic continuation of $\ln(-\tau)$, $\tau < 0$, in the upper half complex plane is given by $\ln \tau - i\pi$, $\tau > 0$, if the branch cut is in the lower half complex plane.

This has the normal quantum mechanical interpretation of being the transmission probability, i.e. the probability that the wave packet P when evolved backward in time will become squeezed up against the event horizon.

6.10.3 Unruh vacuum state $|U\rangle$

We will look now at the vacuum conditions $a(T^+)|\psi\rangle = 0$, $a(\bar{T}^-)|\psi\rangle = 0$ more closely. We have (using $(\phi, T) = a(T)$, $-(\phi, \tilde{T}) = a^\dagger(T)$, $(\phi, \tilde{T}) = a(\tilde{T})$, $-(\phi, \tilde{\tilde{T}}) = a^\dagger(\tilde{T})$)

$$a(T^+) = (\phi, T^+) = c_+ a(T) + c_+ e^{-\frac{\pi\omega}{a}} a(\tilde{T}). \tag{6.347}$$

$$a(\bar{T}^-) = (\phi, \bar{T}^-) = -c_- a^\dagger(T) - c_- e^{\frac{\pi\omega}{a}} a^\dagger(\tilde{T}). \tag{6.348}$$

But $\tilde{T}$ is a negative norm solution. Thus, $a(\tilde{T}) = -a^\dagger(\tilde{\tilde{T}})$ and $a^\dagger(\tilde{T}) = -a(\tilde{\tilde{T}})$. The vacuum conditions $a(T^+)|\psi\rangle = 0$, $a(\bar{T}^-)|\psi\rangle = 0$ become

$$\left(a(T) - e^{-\frac{\pi\omega}{a}} a^\dagger(\tilde{\tilde{T}})\right)|\psi\rangle = 0. \tag{6.349}$$

$$\left(-a^\dagger(T) + e^{\frac{\pi\omega}{a}} a(\tilde{\tilde{T}})\right)|\psi\rangle = 0. \tag{6.350}$$

The operator $a(T)$ is the analog of the operator b_ω in (6.305) which is the exterior annihilation operator. The operator $a(\tilde{T})$ is therefore the interior annihilation operator which we will denote by $\tilde{b}_\omega$. The first equation in (6.305) should then be corrected as

$$a_\nu = \int_0^\infty d\omega (\alpha_{\omega\nu} b_\omega + \beta^*_{\omega\nu} b^\dagger_\omega + \tilde{\alpha}_{\omega\nu} \tilde{b}_\omega + \tilde{\beta}^*_{\omega\nu} \tilde{b}^\dagger_\omega). \tag{6.351}$$

The equations (6.349) and (6.350) define the so-called Unruh vacuum $|U\rangle$. As noted before, the Boulware vacuum which we will denote here by $|B\rangle$ should be annihilated by the transmission annihilation operators $a(T)$ and $a(\tilde{\tilde{T}})$, viz

$$a(T)|B\rangle = a(\tilde{\tilde{T}})|B\rangle = 0. \tag{6.352}$$

This state is different from the initial black hole state $|\psi\rangle$. By using the facts $[a(T), a^\dagger(T)] = 1$ and $[a(\tilde{\tilde{T}}), a^\dagger(\tilde{\tilde{T}})] = 1$ (we are assuming that the wave packets T and $\tilde{\tilde{T}}$ are normalized) we can represent the annihilation operators as $a(T) = \partial/\partial a^\dagger(T)$ and $a(\tilde{\tilde{T}}) = \partial/\partial a^\dagger(\tilde{\tilde{T}})$ and as a consequence we can rewrite equations (6.349) and (6.350) in the form

$$a(T)|U\rangle = e^{-\frac{\pi\omega}{a}} a^\dagger(\tilde{\tilde{T}})|U\rangle \Rightarrow \frac{\partial}{\partial a^\dagger(T)}|U\rangle = e^{-\frac{\pi\omega}{a}} a^\dagger(\tilde{\tilde{T}})|U\rangle. \tag{6.353}$$

$$a(\tilde{\tilde{T}})|U\rangle = e^{-\frac{\pi\omega}{a}} a^\dagger(T)|U\rangle \Rightarrow \frac{\partial}{\partial a^\dagger(\tilde{\tilde{T}})}|U\rangle = e^{-\frac{\pi\omega}{a}} a^\dagger(T)|U\rangle. \tag{6.354}$$

A solution is immediately given by the so-called squeezed state

$$|U\rangle = \mathcal{N} \exp\left(e^{-\frac{\pi\omega}{a}} a^{\dagger}(T) a^{\dagger}(\tilde{\tilde{T}})\right) |B\rangle \tag{6.355}$$

Thus the vacuum state of the black hole is the Unruh vacuum $|U\rangle$ and not the Boulware vacuum $|B\rangle$. The Unruh vacuum $|U\rangle$ should be thought of as the in state in the same way that the original black hole state $|\psi\rangle$ should be thought of as the out state.

This squeezed state $|U\rangle$ is a two-mode entangled state. The modes correspond to T (outside horizon) and $\tilde{T}$ (inside horizon). Since the black hole background is invariant under time translations the Hamiltonian must commute with $a^{\dagger}(T) a^{\dagger}(\tilde{\tilde{T}})$. The Killing vector outside the event horizon corresponds to the usual time translation generator and thus $a^{\dagger}(T)$ must raise the Killing energy in the usual way, viz $[H, a^{\dagger}(T)] = \omega a^{\dagger}(T)$ where ω is positive. But inside the black hole the Killing vector reverses signature and it becomes like a momentum and thus its sign can be either positive or negative. We can check that $a^{\dagger}(\tilde{\tilde{T}})$ must in fact lower the energy as $[H, a^{\dagger}(\tilde{\tilde{T}})] = -\omega a^{\dagger}(\tilde{\tilde{T}})$ if we want $[H, a^{\dagger}(T) a^{\dagger}(\tilde{\tilde{T}})] = 0$, which is required by invariance under time translations. This can also be seen from the fact that the interior mode enters through $\tilde{\tilde{T}}$, which has a negative frequency, and not through $\tilde{T}$, which has a positive frequency as the exterior mode T. In conclusion, the total Killing energy of the entangled particle pair T and $\tilde{T}$ is zero.

The Unruh vacuum is an entangled pure state which can also be rewritten, by expanding the exponential, as follows

$$\begin{aligned}
|U\rangle &= \mathcal{N} \sum_{n} \frac{1}{n!} e^{-\frac{n\pi\omega}{a}} (a^{\dagger}(T))^{n} (a^{\dagger}(\tilde{\tilde{T}}))^{n} |B\rangle \\
&\simeq \sum_{n} e^{-\frac{n\pi\omega}{a}} |n_{R}\rangle |n_{L}\rangle.
\end{aligned} \tag{6.356}$$

The states $|n_{R}\rangle$ and $|n_{L}\rangle$ are the level n-excitations of the exterior modes T and the interior modes $\tilde{\tilde{T}}$ given, respectively, by

$$|n_{R}\rangle \simeq \frac{1}{\sqrt{n!}} (a^{\dagger}(T))^{n} |B_{R}\rangle, \quad |n_{L}\rangle \simeq \frac{1}{\sqrt{n!}} (a^{\dagger}(\tilde{\tilde{T}}))^{n} |B_{L}\rangle. \tag{6.357}$$

Hence, this pure state if reduced to the outside of the event horizon we end up with a mixed state given by the density matrix

$$\begin{aligned}
\rho_{R} &= \text{Tr}_{L} |U\rangle\langle U| \\
&= \sum_{n} e^{-\frac{2n\pi\omega}{a}} |n_{R}\rangle\langle n_{R}|.
\end{aligned} \tag{6.358}$$

This is a thermal canonical ensemble. This the most precise statement, in my opinion, of the information loss problem: a correlated entangled pure state near the horizon gives rise to a thermal mixed state outside the horizon.

6.11 The information problem in black hole Hawking radiation

The best presentation of the information problem remains that of Page [11]. This is a very difficult and mysterious topic and we will follow the pedagogical presentation of [12] and the elegant book [10]. We also refer to [7, 28].

6.11.1 Information loss, remnants and unitarity

The transition from a pure state to a mixed state observed in the Hawking radiation and black hole evaporation can be quantified as follows. We start with the Schrödinger equation

$$i\frac{\partial}{\partial t}|\psi\rangle = H|\psi\rangle. \tag{6.359}$$

The integrated form of this equation reads in terms of the unitary scattering matrix

$$|\psi^{\text{final}}\rangle = S|\psi^{\text{initial}}\rangle \Rightarrow \psi_n^{\text{final}} = S_{nm}\psi_m^{\text{initial}}. \tag{6.360}$$

The Schrödinger equation will evolve pure quantum states to pure quantum states. However, black hole radiation takes the pure state (6.356) to the mixed state (6.358). Thus it takes an initial pure state of the form

$$\rho^{\text{initial}} = |\psi^{\text{initial}}\rangle\langle\psi^{\text{initial}}| \tag{6.361}$$

to a final mixed state of the form

$$\rho^{\text{final}} = \sum_i p_i |\psi^{\text{final}}\rangle\langle\psi^{\text{final}}|. \tag{6.362}$$

This can be expressed in terms of the so-called dollar matrix $ as follows

$$\rho_{mm'}^{\text{final}} = \$_{mm',nn'}\rho_{nn'}^{\text{initial}}. \tag{6.363}$$

In the case of the Schrödinger equation we have

$$\$_{mm',nn'} = S_{mn}S_{n'm'}^*, \tag{6.364}$$

whereas in the case of the black hole radiation we have a general dollar matrix which takes pure states to mixed states.

The opinions regrading whether or not black hole radiation corresponds to information loss divides into three possibilities:

- **Information loss:** This is the original stand of Hawking which is based on the conclusion that (6.358) is correct and that the black hole will evaporate completely. In this case, the dollar matrix $ is not given by the Schrödinger equation and there is indeed information loss due to pure states (gravitational collapse and black hole formation) evolving into mixed states (Hawking radiation and black hole evaporation). Since the outgoing Hawking radiation is largely independent of the initial state, i.e. different initial states result in the same final state, black hole evaporation does not conserve information.

If information is really lost then quantum mechanics must be changed in some way. However, there are tight constraints on QG effects arising from the modification of the axioms of field theory [29], and furthermore, any such modification will lead to violation of either locality or energy–momentum conservation [30].

- **Unitarity:** The other possibility is therefore information conservation, i.e. there is a unitary map between the initial state of the collapse to the final state of the outgoing radiation. The black hole will also evaporate completely but (6.358) is only correct in a coarse-grained sense. This means that the final state of the radiation becomes purified and information is carried out with the Hawking radiation in subtle quantum correlations between late and early particles. The final pure state of the radiation is presumably very complicated such that any subsystem will look thermal and as a consequence equation (6.358) is a good approximation [12].

  These pure states are the microstates of the black hole and their counting is given by the exponential of the Bekenstein–Hawking formula.

  The black hole microstates may also be identified with the states of the field (or infalling matter) accumulating on the nice-slice, which is a space-like surface interpolating between a fixed t surface outside the black hole to a fixed r surface inside the black hole, and which gets longer on the inside as the black hole gets older [7].

  This solution, in which unitarity is maintained and information is conserved, if correct, implies, however, a breakdown of the semi-classical description and the machinery of effective field theory.

- **Remnant:** In this case black hole evaporation stops when the decreasing black hole size becomes Planckian. The remaining Planck-sized object is what we call a remnant. This must be characterized by an extremely large entanglement entropy in order for the total state to remain pure. Thus, this is an object with a finite energy but effectively an infinite number of states and thus the connection between Bekenstein–Hawking entropy and number of states is lost.

This situation is the black hole information problem.

6.11.2 Information conservation principle

In this section we will only follow the beautiful presentation of [10].

- **von Neumann entropy:** Information is conserved in classical mechanics (Liouville's theorem)[6] and in quantum mechanics (unitarity of the S-matrix)[7]. As we have already discussed, the von Neumann entropy is the measure of information (or lack of it) which is defined by

---

[6] The volume of the initial phase space region representing the largely unknown state of the system is conserved in time under Hamilton's equations.

[7] In quantum mechanics the initial state of the system if unknown will be represented by a projector on some subspace. The dimension of this subspace, i.e. the rank of the projector, is conserved under the Schrödinger equation.

$$S = -\int dp dq \rho(p, q) \ln \rho(p, q). \tag{6.365}$$

If $\rho = 1/V$, where V the volume of some region in phase space, then $S = \ln V$, i.e. $V = \exp(S)$. In quantum mechanics we use instead the definition

$$S = -\mathrm{Tr}\rho \ln \rho. \tag{6.366}$$

If $\rho = P/\mathrm{Tr}P$, where P is a projector of rank n, then $S = \ln n$. In other words, the number of states is given by the exponential of the entropy, viz

$$n = \exp(S). \tag{6.367}$$

- **Pure states:** We will generally need to separate the system into two subsystems A and B with quantum correlations, i.e. entanglement, between them. The total system is assumed in a pure state $\psi(\alpha, \beta)$. Thus the von Neumann entropy is zero identically, viz

$$S_{A+B} = 0. \tag{6.368}$$

The subsystems considered separately are described by the corresponding density matrices $\rho_A(\alpha)$ and $\rho_B(\beta)$ in which the degrees of freedom of the other system are integrated out. These are generally not pure states.

The density matrix ρ_A is such that: (i) it is Hermitian $\rho_A^\dagger = \rho_A$, (ii) it is positive semi-definite, viz $(\rho_A)_i \geqslant 0$, and (iii) it is normalized, viz $\mathrm{Tr}\rho_A = 1$.

Thus, if just one of the eigenvalues of ρ_A is 1 the rest will vanish identically. In this case the subsystem A is in a pure state which means that the total system pure state factorizes as

$$\psi(\alpha, \beta) = \psi_A(\alpha)\psi_B(\beta). \tag{6.369}$$

The subsystem B is then also in a pure state.
- **Entanglement entropy:** A far more important identity for us here is the equality of the von Neumann entropies of the two subsystems A and B if the total system is described by a pure state, viz

$$S_A = S_B = S_E. \tag{6.370}$$

S_E is precisely the entanglement entropy.

Proof: The density matrix ρ_A is given explicitly by

$$(\rho_A)_{\alpha\alpha'} = \sum_\beta \psi^\star(\alpha, \beta)\psi(\alpha', \beta). \tag{6.371}$$

Let ϕ be an eigenvector of ρ_A with eigenvalue λ, viz

$$(\rho_A)_{\alpha\alpha'}\phi(\alpha') = \sum_\beta \psi^\star(\alpha, \beta)\psi(\alpha', \beta)\phi(\alpha') = \lambda\phi(\alpha). \tag{6.372}$$

We will assume that $\lambda \neq 0$. Similarly, we write explicitly the density matrix ρ_B as

$$(\rho_B)_{\beta\beta'} = \sum_\alpha \psi^\star(\alpha, \beta)\psi(\alpha, \beta'). \tag{6.373}$$

We propose the eigenvector of ρ_B to be of the form

$$\chi(\beta') = \sum_{\alpha'} \psi^\star(\alpha', \beta')\psi^\star(\alpha'). \tag{6.374}$$

Indeed, we compute

$$\begin{aligned}
\sum_{\beta'}(\rho_B)_{\beta\beta'}\chi(\beta') &= \sum_{\beta'}\sum_\alpha\sum_{\alpha'}\psi^\star(\alpha, \beta)\psi(\alpha, \beta')\psi^\star(\alpha', \beta')\phi^\star(\alpha') \\
&= \sum_\alpha\sum_{\alpha'}(\rho_A)_{\alpha'\alpha}\psi^\star(\alpha, \beta)\phi^\star(\alpha') \\
&= \lambda\sum_\alpha\psi^\star(\alpha, \beta)\phi^\star(\alpha) \\
&= \lambda\chi(\beta).
\end{aligned} \tag{6.375}$$

In the above we have also used the result that $(\rho_A)_{\alpha'\alpha}\phi^\star(\alpha') = \lambda\phi^\star(\alpha)$. Thus ρ_A and ρ_B have the same non-zero eigenvalues. Immediately we conclude that

$$S_A = -\sum_i(\rho_A)_i \ln(\rho_A)_i = -\sum_i(\rho_B)_i \ln(\rho_B)_i = S_B. \tag{6.376}$$

Since $S_{A+B} = 0$ and $S_A + S_B = 2S_E$ this shows explicitly that the von Neumann entanglement entropy is not additive. It is a fundamental microscopic fine-grained entropy as opposed to the thermodynamic Boltzmann entropy.

- **Thermal entropy:** The thermodynamic entropy is additive and it can be defined as follows. Let us assume a total system Σ divided into many subsystems σ_i, i.e. a coarse graining. Again we will assume that the total system is in a pure state with vanishing entropy. The subsystems σ_i are supposed to be thermal, i.e. with matrix densities ρ_i given by the Blotzmann distribution

$$\rho_i = \frac{e^{-\beta H_i}}{Z_i}, \tag{6.377}$$

where H_i and Z_i are the Hamiltonian and the partition function of the subsystem σ_i. This is the distribution which maximizes the entropy. The thermodynamic coarse-grained entropy of the total system is then given by the sum of the entropies of the subsystems σ_i, viz

$$S_{\text{therm}} = \sum_i S_i. \tag{6.378}$$

This coarse-grained entropy S_{therm} as opposed to the fine-grained entropy is not conserved. To see this, we assume that initially the pure state of the total system factorizes completely, i.e. the subsystems are in a pure state. Then in this case $S_i = 0$ and hence $S_{\text{therm}} = 0$. After interaction, the pure state of the total system will fail to factorize, i.e. $S_i \neq 0$ and hence $S_{\text{therm}} \neq 0$.

Another important property is the fact that the thermodynamic entropy of a subsystem Σ_1 is always larger than its entanglement entropy, viz

$$S_{\text{therm}}(\Sigma_1) = \sum_i S_i \geqslant S(\Sigma_1) = -\text{Tr}\rho \ln \rho. \tag{6.379}$$

This is almost obvious since from one hand $S_i \geqslant 0$ and thus $S_{\text{therm}} \geqslant 0$, while from the other hand as $\Sigma_1 \longrightarrow \Sigma$ the entanglement entropy approaches zero.

- **Information:** The amount of information in a subsystem Σ_1 is defined as the difference between the coarse-grained entropy (thermodynamic) and the fine-grained entropy (von Neumann), viz

$$I = S_{\text{therm}}(\Sigma_1) - S(\Sigma_1) = \sum_i S_i + \text{Tr}\rho \ln \rho. \tag{6.380}$$

As an example, we take Σ_1 to be the total system Σ. In this case $S(\Sigma) = 0$ and thus the information is given by the thermodynamic entropy. But for a very small subsystem $\Sigma_1 = \sigma_i$ we get $I = 0$ since obviously $S_{\text{therm}} = S$ for such a system. In fact, this is true for all subsystems which are smaller than one half the total system. A nice calculation which attempts to convince us of this result is found in [10].

Let us assume that the total system is composed of two subsystems Σ_1 and $\Sigma - \Sigma_1$. Immediately, we conclude that the von Neumann entropies are equal, viz

$$S(\Sigma - \Sigma_1) = S(\Sigma_1). \tag{6.381}$$

The amounts of information contained in Σ_1 and $\Sigma - \Sigma_1$ are given by

$$I(\Sigma_1) = S_{\text{therm}}(\Sigma_1) - S(\Sigma_1), \quad I(\Sigma - \Sigma_1) = S_{\text{therm}}(\Sigma - \Sigma_1) - S(\Sigma - \Sigma_1). \tag{6.382}$$

If $\Sigma_1 \ll \Sigma/2$ then

$$I(\Sigma_1) = S_{\text{therm}}(\Sigma_1) - S(\Sigma_1) = 0. \tag{6.383}$$

If $\Sigma_1 \gg \Sigma/2$ then $\Sigma - \Sigma_1 \ll \Sigma/2$ and as a consequence $I(\Sigma - \Sigma_1) = S_{\text{therm}}(\Sigma - \Sigma_1) - S(\Sigma - \Sigma_1) = 0$, i.e. there is no information in the smaller subsystem $\Sigma - \Sigma_1$. Also, we will have in this case (with f being the fraction of the total degrees of freedom contained in Σ_1)

$$\begin{aligned}
I(\Sigma_1) &= S_{\text{therm}}(\Sigma_1) - S(\Sigma_1) \\
&= S_{\text{therm}}(\Sigma_1) - S(\Sigma - \Sigma_1) \\
&= S_{\text{therm}}(\Sigma_1) - S_{\text{therm}}(\Sigma - \Sigma_1) \\
&= f S_{\text{therm}}(\Sigma) - (1 - f) S_{\text{therm}}(\Sigma) \\
&= (2f - 1) S_{\text{therm}}(\Sigma).
\end{aligned} \tag{6.384}$$

This result can be clearly continued from $f \simeq 1$ to $f = 1/2$. It vanishes identically for $\Sigma_1 = \Sigma/2$, i.e. $f = 1/2$. Since the amount of information vanishes also for $\Sigma_1 \ll \Sigma$ we conclude, again by continuity, that indeed $I = 0$ for all $\Sigma_1 \leqslant \Sigma/2$.

- **Bomb in a box:** We conclude this section by the illuminating example of [10]. We consider a system Σ_1 consisting of a bomb placed in a box B with reflecting walls and a hole from which electromagnetic radiation can escape. The system $\Sigma - \Sigma_1$ is obviously the environment which will be denoted by A. The bomb will explode and we will watch the system + environment until all thermal radiation inside the box leaks out to the environment. In order to simplify tracking the evolution, we will divide it into four stages:

    - Before the explosion of the bomb, the systems A and B are in their ground (pure) states. The von Neumann fine-grained entropies as well as the Boltzmann thermal coarse-grained entropies all vanish identically and thus the entanglement entropy and the information in the outside radiation also vanish identically, viz

    $$S(A) = S(B) = S_E = 0, \quad S_{\text{therm}}(A) = S_{\text{therm}}(B) = 0 \Rightarrow I(A) = 0. \tag{6.385}$$

    - The bomb explodes and thermal radiation fills the box (no photon has leaked out yet). The thermal entropy inside increases. All others are still zero identically, viz

    $$S(A) = S(B) = S_E = 0, \quad S_{\text{therm}}(A) = 0, \quad S_{\text{therm}}(B) \neq 0 \uparrow \Rightarrow I(A) = 0. \tag{6.386}$$

    The initial information is

    $$I(B) = S_{\text{therm}}(B). \tag{6.387}$$

    - The photons start to leak out. The von Neumann entropies increase and thus the entanglement entropy increases, i.e. entanglement between A and B increases. The thermal entropy of the Box clearly decreases while that of the environment increases. But information in the outside radiation remains negligible.

    $$S(A) = S(B) = S_E \neq 0, \quad S_{\text{therm}}(A) \neq 0 \uparrow, \quad S_{\text{therm}}(B) \neq 0 \downarrow \Rightarrow I(A) = 0. \tag{6.388}$$

    At some point the thermal entropies become equal. This is called the information retention time. This is the time where the entanglement between A and B becomes decreasing and the information starts increasing, i.e. it is the time at which information starts coming out with the radiation. Before the information retention time only energy has come out with the radiation with no or little information. At the information retention time around one half of the radiation inside the box has come out which corresponds to one bit ($\ln 2$) of information encoded in the initial state.

    - When all photons are out, the inside thermal entropy vanishes and since there is no entanglement anymore the von Neumann entropies vanish, viz

    $$S(A) = S(B) = S_E = 0, \quad S_{\text{therm}}(A) \neq 0, \quad S_{\text{therm}}(B) = 0 \Rightarrow I(A) = S_{\text{therm}}(A). \tag{6.389}$$

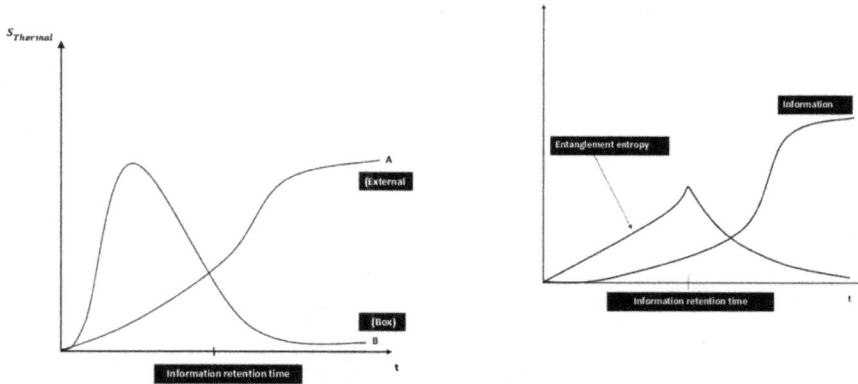

Figure 6.9. The information retention time, the entanglement entropy and the information in the 'bomb in a box' problem.

From the second law of thermodynamics the final value of the outside thermal entropy must be larger than the initial value of the interior thermal entropy, viz $S_{\text{therm}}(A) > S_{\text{therm}}(B)$, i.e. the information in the outgoing radiation is more than the information in the initial state of the box.

– Throughout the process, the entanglement entropy is always smaller than the thermal entropy of A or B. This can be seen as follows. At the beginning most information is in the thermal entropy of B. The information in A is zero, which means that the entanglement entropy is equal to the thermal entropy of A which is less than the thermal entropy of B. At the end we have the reverse situation. Most information is in the thermal entropy of A. The information in B is zero which means that the entanglement entropy is equal to the thermal entropy of B which is less than the thermal entropy of A. In summary, we have

$$S_E \leqslant S_{\text{therm}}(A) \text{ or } S_{\text{therm}}(B). \tag{6.390}$$

Thus information is conserved which means in particular that the final state of the radiation outside the box is pure although it might look thermal at smaller scales. This very clear physical picture is summarized in figure 6.9.

6.11.3 Page curve and Page theorem

We have a quantum system consisting of a black hole and its corresponding Hawking radiation. We split the outgoing Hawking radiation into early and late with corresponding Hilbert spaces $\mathcal{H}_R$ and $\mathcal{H}_{\text{BH}}$, viz

$$\mathcal{H}_{\text{out}} = \mathcal{H}_{\text{R}} \otimes \mathcal{H}_{\text{BH}}. \tag{6.391}$$

The notation $\mathcal{H}_{\text{BH}}$ indicates explicitly that the late Hawking radiation is nothing else but the remaining black hole. The plot of the entanglement entropy S_E of the early radiation as a function of time is called the Page curve [31, 32]. Obviously, $S_E = S(R) = S(BH)$.

The initial state of the black hole is pure. Initially, the thermal entropy of the black hole is non-zero, i.e. $S_{\text{therm}}(BH) = 0$, the entanglement entropy is zero, viz $S_E = 0$, and the information in the Hawking radiation $I(R)$ is zero, i.e. $I(R) = 0$. Hawking radiation starts coming out. The entanglement entropy S_R between the Hawking radiation and the black hole starts increasing, the thermal entropy of the black hole $S_{\text{therm}}(BH)$ decreases while the thermal entropy of the radiation $S_{\text{therm}}(R)$ increases.

At the retention time, also called Page time, the two thermal entropies become identical, viz

$$S_{\text{therm}}(BH) = S_{\text{therm}}(R). \tag{6.392}$$

At this time the entanglement reaches its maximum and starts decreasing, and the information $I(R)$ at the Page time starts increasing, i.e. it starts coming out in the Hawking radiation. The final state is a pure state of the radiation with vanishing entanglement entropy and information at its maximum value.

The expected picture is shown in the second figure of (6.9). However, this is only a sketch of the actual physics by assuming unitarity while the actual calculation of the Page curve remains a major challenge.

Indeed, as reported concisely by Harlow in his lectures [12] he says that 'Andy Strominger has argued that being able to compute the Page curve in some particular theory is what it means to have solved the black hole information problem; even in AdS/CFT or the BFSS model we are far (Harlow stating) from being able to really do this'.

The above picture can, however, be fleshed out a little more by using the elegant Page theorem [33]. This says that for a given bipartite system $\mathcal{H}_{AB} = \mathcal{H}_A \otimes \mathcal{H}_B$ with $|A| = \dim A < |B| = \dim B$ a randomly chosen pure state ρ_{AB} in $\mathcal{H}_{AB}$ is likely to be very close to a maximally entangled state if $|A| \ll |B|$. In other words, if $|A| \ll |B|$, the pure state ρ_{AB} will correspond to a totally mixed state $\rho_A = \text{Tr}_B \rho_{AB}$, i.e. $\rho_A \propto \mathbf{1}_A$.

More precisely, we write this theorem as the inequality

$$\int dU \, ||\rho_A(U) - \frac{\mathbf{1}_A}{|A|}||_1 \leqslant \sqrt{\frac{|A|^2 - 1}{|A||B| + 1}}. \tag{6.393}$$

The norm $|| \cdots ||_1$ is the L_1 operator trace norm defined by $||M||_1 = \text{Tr}\sqrt{M^\dagger M}$. The integration over the Haar measure represents a randomly chosen pure state $|\psi(U)\rangle = U|\psi\rangle$, i.e. $\rho_{AB}(U) = |\psi(U)\rangle\langle\psi(U)|$ and $\rho_A(U) = \text{Tr}_B|\psi(U)\rangle\langle\psi(U)|$. It is clear from the above equation that if $|A| \ll |B|$ then ρ_A is very close to a totally mixed state and as a consequence $|\psi\rangle$ or equivalently ρ_{AB} is a maximally entangled pure state.

Let us compute the behavior of the entanglement entropy. We have (with $\Delta\rho_A = \rho_A - \mathbf{1}_A/|A|$ and $\text{Tr}\Delta\rho_A = 0$)

$$\int dU S_A = - \int dU \text{Tr} \rho_A \ln \rho_A$$

$$= \ln|A| - \frac{1}{2}|A| \int dU \text{Tr} \Delta \rho_A^2 + O(\Delta \rho^3). \tag{6.394}$$

The remaining integral over U can be done exactly using unitary matrix technology (see equation (5.13) of [12]) to find for $|A| \ll |B|$ the result

$$\int dU S_A = \ln|A| - \frac{1}{2} \frac{|A|}{|B|} + \cdots. \tag{6.395}$$

We now apply this theorem to the entanglement entropy of the black hole.

We know that Hawking radiation consists mostly of s-wave quanta, i.e. modes with $l = 0$. These can be described by a $1 + 1$ dimensional free scalar field at a Hawking temperature $T_H = 1/4\pi r_s$. These modes can escape the black hole because the Schwarzschild potential is not fully confining as in the Rindler case. Indeed, the barrier height for s-wave particles is of the same order of magnitude as the Hawking temperature. Thus, each particle which escapes is carrying energy given by Hawking temperature $\nu \sim T_H = 1/(8\pi GM)$.

Further, we will assume that one single quanta will escape (since $l = 0$) per one unit of Rindler time $\omega = t/2r_s$. Thus, $1/2r_s$ quanta per unit Schwarzschild time will escape the barrier. The total energy carried out of the black hole per unit Schwarzschild time is then given by $1/(8\pi GM) \times 1/2r_s \sim 1/G^2M^2$. We write this as

$$\frac{dE_R}{dt} = \frac{C}{G^2M^2}. \tag{6.396}$$

By energy conservation the energy per unit Schwarzschild time lost by the black hole is immediately given by

$$\frac{dM}{dt} = -\frac{C}{G^2M^2} \Rightarrow Cdt = -G^2M^2 dM. \tag{6.397}$$

In the above two equations C is some constant of proportionality.

In order to apply Page's theorem we will first need to assume that the pure state of the Hawking (early) radiation R and the black hole (late radiation) BH is random. Obviously, at early times $|R| \ll |BH|$. The entanglement entropy is then given immediately by the theorem to be given by

$$S_E = S_R \sim \ln|R|. \tag{6.398}$$

On the other hand, the energy carried by the radiation during a small time interval t is obtained by integrating equation (6.396) assuming that the mass M remains constant. This gives

$$E_R = \frac{C}{G^2M^2} t. \tag{6.399}$$

However, from equations (6.449) and (6.450) below, the entropy and energy of the radiation are related by

$$\frac{E_R}{S_R} \sim \frac{1}{r_s}. \tag{6.400}$$

By taking the ratio of the above two results we obtain

$$S_R \sim tT. \tag{6.401}$$

This should be valid only for times such that $S_R \ll S_{\mathrm{BH}} \sim M^2$, i.e. $t \ll M^3$. During these times it is also expected that $S_{\mathrm{BH}} \sim \ln|BH|$. At early times we have then the linear behavior of the entanglement entropy as a function of time, viz

$$S_E \sim tT, \quad t \ll M^3. \tag{6.402}$$

After the Page time t_{Page} defined by

$$\ln|R| \sim \ln|BH|, \tag{6.403}$$

we should apply Page's theorem in the opposite direction since we can assume now that $|BH| \ll |R|$. Thus in this case

$$S_E = S_{BH} \sim \ln|BH|. \tag{6.404}$$

However, by integrating equation (6.396) between t and t_{evap} we obtain

$$C \int_t^{t_{\mathrm{evap}}} dt' = -G^2 \int_M^0 M'^2 dM' \Rightarrow (t_{\mathrm{evap}} - t)^{2/3} \sim M^2. \tag{6.405}$$

However, for the black hole the entropy is proportional to the area which is proportional to its mass squared, thus we obtain immediately

$$S_{BH} \sim (t_{\mathrm{evap}} - t)^{2/3}. \tag{6.406}$$

At late times we have then the behavior of the entanglement entropy as a function of time given by

$$S_E \sim (t_{\mathrm{evap}} - t)^{2/3}, \quad t_{\mathrm{Page}} \leqslant t \leqslant t_{\mathrm{evap}}. \tag{6.407}$$

6.12 Black hole thermodynamics

Again we will follow [12] and the book [10].

6.12.1 Penrose diagrams

The idea of Penrose diagrams relies on the theorem that any two conformally equivalent metrics will have the same null geodesics and thus the same causal structure. Thus, Penrose diagrams represent essentially the causal structure of spacetimes and they involve the so-called conformal compactification. Let us take the example of flat Minkowski spacetime given by the metric

$$ds^2 = -dt^2 + dr^2 + r^2 d\Omega^2. \tag{6.408}$$

The light cone is defined by $dt = \pm dr$. The form of the light cone is therefore preserved if we transform t and r to T and R such that

$$Y^+ = T + R = f(t + r), \quad Y^- = T - R = f(t - r). \tag{6.409}$$

We can map the Minkowski plane $0 \leqslant r \leqslant \infty$ and $-\infty \leqslant t \leqslant +\infty$ to a finite region of the plane with boundaries at finite distance by choosing F to be the function tanh, viz

$$Y^+ = \tanh(t + r), \quad Y^- = \tanh(t - r). \tag{6.410}$$

We have the limiting behaviors

$$Y^+ = +1, \quad Y^- = -1, \quad r \longrightarrow \infty, \quad \forall \, t. \tag{6.411}$$

$$Y^+ = Y^- = \frac{e^t - e^{-t}}{e^t + e^{-t}}, \quad r \longrightarrow 0. \tag{6.412}$$

$$Y^+ = Y^- = 1, \quad t \longrightarrow +\infty. \tag{6.413}$$

$$Y^+ = Y^- = -1, \quad t \longrightarrow -\infty. \tag{6.414}$$

The new coordinates have now the range $|T \pm R| < 1$ and $R \geqslant 0$. The boundary $|T \pm R| = 1$ can be included by dropping the diverging prefactor in the metric when expressed in terms of T and R (by using the above theorem). In other words, spacetime is compactified. The range becomes $|T \pm R| \leqslant 1$ and $R \geqslant 0$. This is a triangle in the TR plane defined by

$$Y^+ = Y^-, \quad Y^+ = +1, \quad Y^- = -1. \tag{6.415}$$

We can discern the following infinities (see figure 6.10):
- The usual future and past time-like infinities at $t = \pm\infty$ denoted by i^- and i^+, respectively. All time-like trajectories begin at i^- and end at i^+.
- The usual space-like infinity at $r = \infty$ denoted by i^0. All space-like trajectories end there. In general relativity, conserved charges such as the energy are written as boundary integrals at this spatial infinity i^0.
- Also, we observe two light-like infinities at $Y^- = -1$ and $Y^+ = +1$ denoted by J^- and J^+, respectively. All light-like trajectories begin at J^- (incoming null rays) and end at J^+ (outgoing null rays). Thus the S-matrix will map incoming states defined on $i^- \cup J^-$ to outgoing states defined on $i^+ \cup J^+$.

Let us consider the more interesting example of Schwarzschild geometry given by the Kruskal–Szekeres metric

$$ds^2 = \frac{32G^3M^3}{r} \exp\left(-\frac{r}{2GM}\right)(-dT^2 + dR) + r^2 d\Omega^2. \tag{6.416}$$

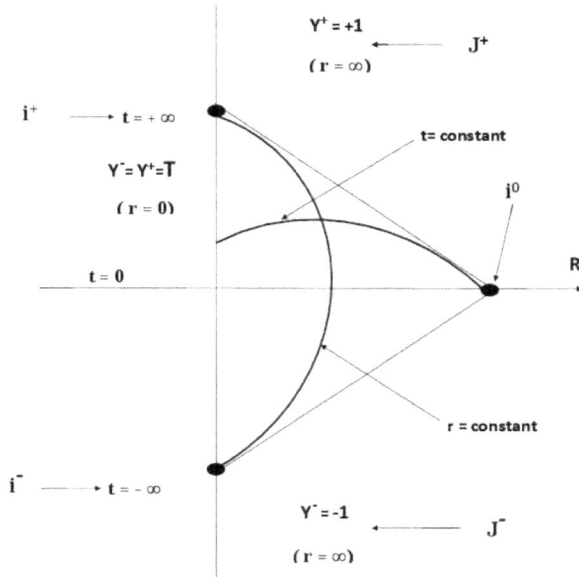

Figure 6.10. Penrose diagram of Minkowski spacetime. The infinities $i^{\mp}$ are at $t = \mp\infty$, the infinity i^0 is at $r = \infty$ while the infinities $\mathcal{J}^{\mp}$ are at $r = \infty$.

The only difference between the Schwarzschild coordinates (T, R, Ω) and the Minkowski coordinates (t, r, Ω) is their range. For the Minkowski coordinates we have $0 \leqslant r \leqslant \infty$ and $-\infty \leqslant t \leqslant +\infty$. For Schwarzschild we have instead (in region I)

$$0 \leqslant R \leqslant \infty, \, -R \leqslant T \leqslant +R. \tag{6.417}$$

The horizon is at $T = \pm R$. Thus, as before, we consider the deformation

$$Y^+ = T' + R' = \tanh(T + R), \quad Y^- = T' - R' = \tanh(T - R). \tag{6.418}$$

We still obtain in the limit $R \longrightarrow +\infty$ the two light-like infinities J^+ ($Y^+ = 1$) and J^- ($Y^- = -1$) and the space-like infinity i^0. We do not now have the boundary $Y^+ = Y^-$ since T does not take the unrestricted values between $-\infty$ and $+\infty$. Since T takes the values between $-R$ and $+R$ we have in the limit $T \longrightarrow R$ the surface

$$Y^+ = \frac{e^{2R} - e^{-2R}}{e^{2R} + e^{-2R}}, \quad Y^- = 0. \tag{6.419}$$

This is the future horizon H^+ which is parallel to J^- and it varies from $Y^+ = 1$ at $R \longrightarrow \infty$ to $Y^+ = 0$ at $R \longrightarrow 0$. The time-like infinity i^+ is at $T = R = +\infty$. Similarly, in the limit $T \longrightarrow -R$ we get the past horizon H^- which is parallel to J^+. The time-like infinity i^- is at $T = R = -\infty$. The Penrose diagram of the full Schwarzschild geometry is shown in figure 6.11.

Let us now consider a real black hole as it forms from the gravitational collapse of a thin spherical shell of massless matter. We start with the Penrose diagram of Minkowski spacetime. The infalling shell is represented by an incoming light-like

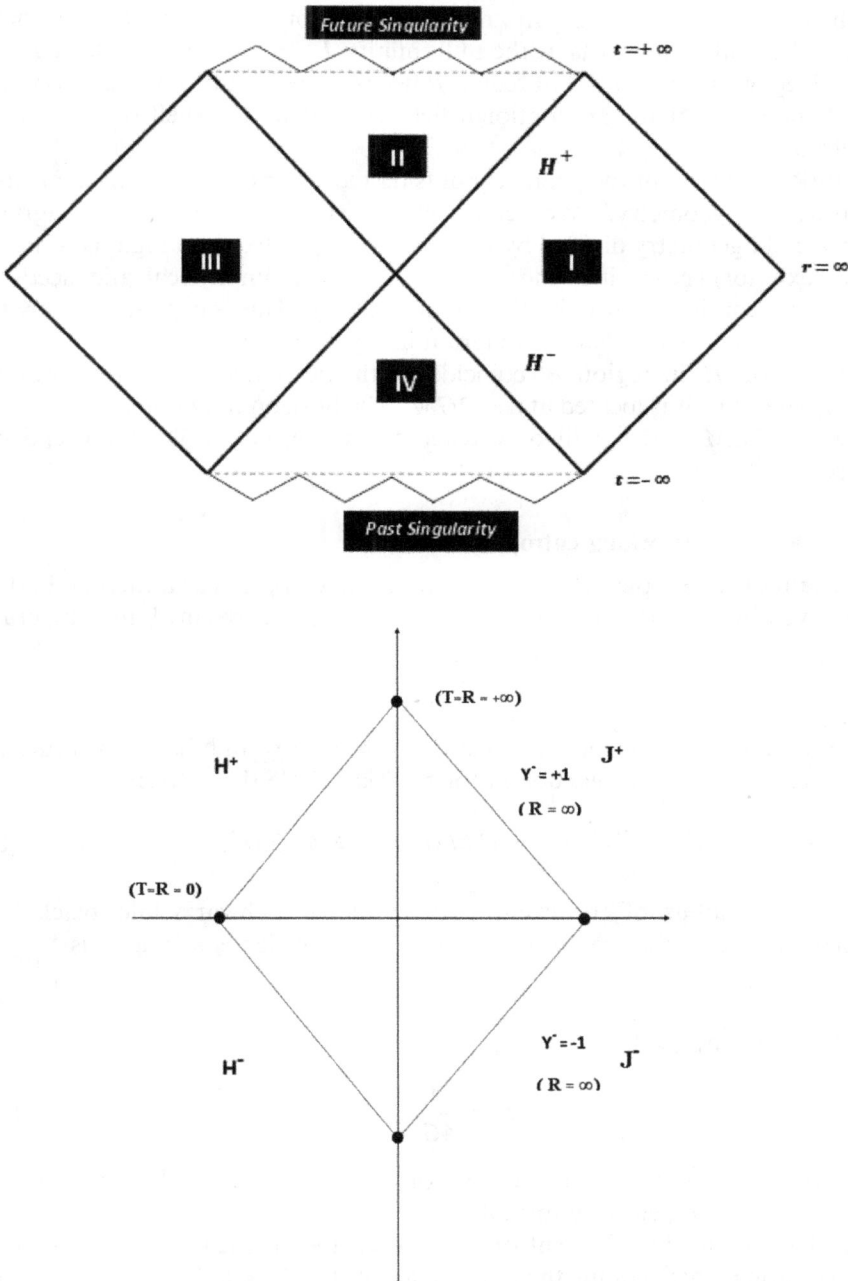

Figure 6.11. Penrose diagram of Schwarzschild metric. The infinities $i^\mp$ are at $t = \mp r = \mp\infty$, the infinity i^0 is at $r = \infty$ while the infinities $\mathcal{J}^\mp$ are at $r = \infty$, $Y^\mp = \mp 1$. The center of the diagram is at $T = R = Y^\pm = 0$ while the horizons $H^\pm$ are at $T = \mp R$.

line which divides the diagram into region A (interior) and region B (exterior). The incoming light-like line starts at the null infinity J^- ($r = \infty$) and ends at $Y^+ = Y^-$ ($r = 0$). Region A is physical but region B needs to be modified in order to take into account the effect of the gravitational field created by the shell on the spacetime geometry.

By Birkoff's theorem the geometry outside the spherical shell is nothing else but Schwarzschild geometry. We consider therefore the Penrose diagram of Schwarzschild geometry divided by the incoming light-like into regions A' (interior) and B' (exterior). Now, it is the region A' which is unphysical and needs to be replaced in a continuous way by the region A above. This is explained nicely in [10] and the end result is the Penrose diagram in figure 6.12.

The horizon H in region B' coincides with the horizon H^+ of Schwarzschild geometry and thus it is located at $r = 2GM$. The horizon in region A is, however, at a value $r < 2GM$ and it will only reach the value $r = 2GM$ at the end of the collapse.

6.12.2 Bekenstein–Hawking entropy formula

To a distant observer, the Schwarzschild black hole appears as a thermal body with energy given by its mass M and a temperature T given by Hawking temperature

$$T = \frac{1}{8\pi GM}. \tag{6.420}$$

The thermodynamical entropy S is related to the energy and the temperature by the formula $dU = TdS$. Thus we obtain for the black hole the entropy

$$dS = \frac{dM}{T} = 8\pi GMdM \Rightarrow S = 4\pi GM^2. \tag{6.421}$$

However, the radius of the event horizon of the Schwarzschild black hole is $r_s = 2MG$, and thus the area of the event horizon (which is a sphere) is

$$A = 4\pi(2MG)^2. \tag{6.422}$$

By dividing the above two equations we get

$$S = \frac{A}{4G}. \tag{6.423}$$

The entropy of the black hole is proportional to its area. This is the famous Bekenstein–Hawking entropy formula.

The Bekenstein–Hawking entropy is a thermodynamical macroscopic coarse-grained entropy which counts the microstates of the black hole. It should satisfy the so-called generalized second law of thermodynamics: 'When common entropy goes down a black hole, the common entropy in the black-hole exterior plus the black-hole entropy never decreases' [34, 35]. But, since the Bekenstein–Hawking entropy is $S = A/4G$ where A is the area of the event horizon, we can see that the area of the event horizon cannot decrease (if there was no radiation) [36].

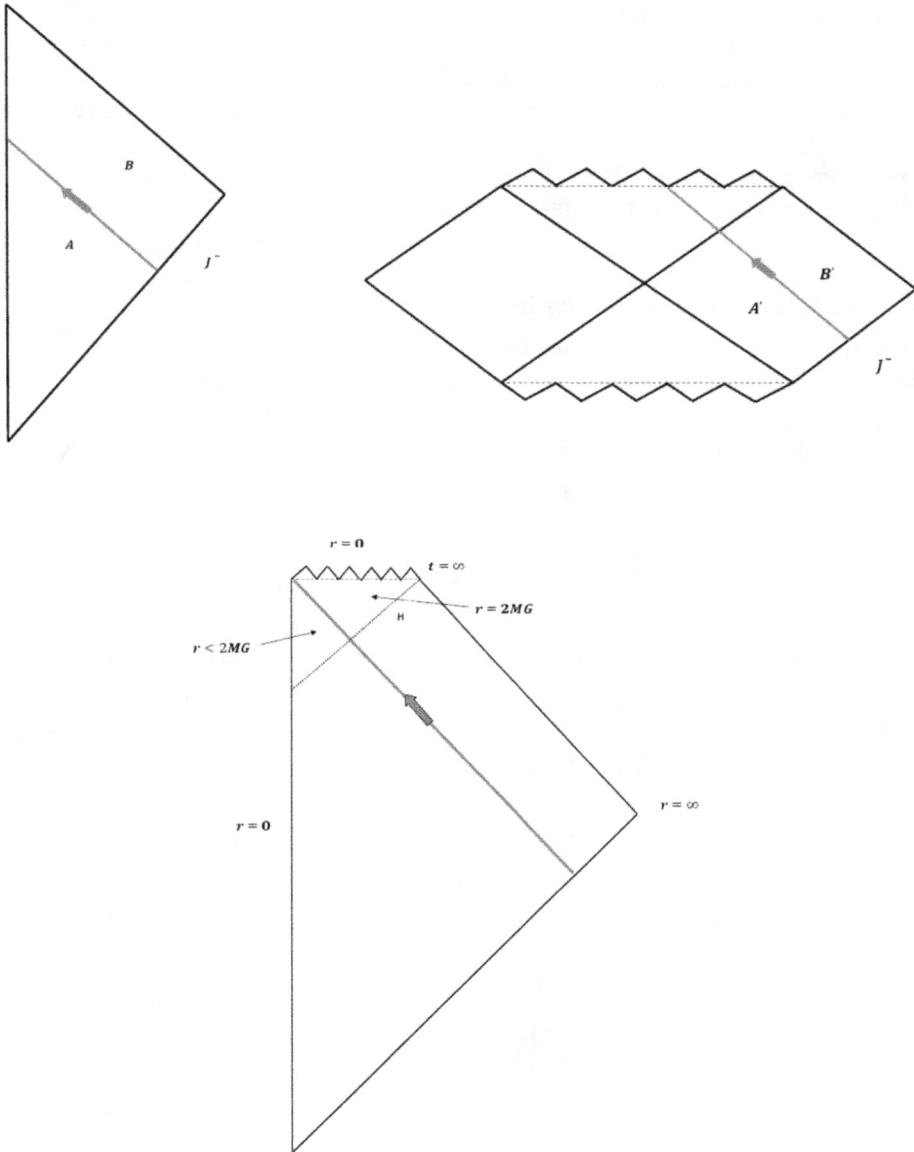

Figure 6.12. Penrose diagram of the formation of a black hole from gravitational collapse. In the last graph the region A is below the red line (infalling shell) and the region B' is above the line. Outside the shell (above the red line) the horizon is at $2GM$ while inside the shell (below the red line) the horizon will start at $r = 0$ and reaches the value $r = 2GM$ at the end of the collapse. Thus, the horizon is a global concept and not a local one, since it forms before the shell reaches the center.

Now, the black hole according to general relativity is only characterized by its temperature and its mass. Thus immediately we conclude that by creating a black hole we loose most of the information about its past, since clearly the initial state cannot be recovered by running the dynamic backward in time starting from the black hole state which is, as we said, characterized only by the mass and the temperature. Thus, the black hole must also be characterized by its microstates which are counted exactly by the exponential of the Bekenstein–Hawking entropy formula.

However, a black hole also evaporates, and thus the above is not sufficient to maintain the principle of information conservation (the first law of Nature in the words of [10]).

6.12.3 Brick wall and stretched horizon

In this final section we will follow the presentation of [10, 12].

We have found that the vacuum of the scalar field in the Schwarzschild geometry is not given by the vacuum state $|B\rangle$, which is annihilated by $a(T)$ and $a(\tilde{\tilde{T}})$, but it is given by a thermal density matrix of the form (with $\beta = 2\pi/a$)

$$
\begin{aligned}
\rho_R &= \underset{\omega,l,m}{\otimes} \rho_R(\omega, l, m) \\
&= \underset{\omega,l,m}{\otimes} \left[(1 - e^{-\beta\omega}) \sum_n e^{-n\beta\omega} |n_R\rangle \langle n_R |_{\omega,l,m} \right].
\end{aligned}
\tag{6.424}
$$

This is diagonal where n is the occupation number and $1 - \exp(-\beta\omega)$ is a normalization constant inserted so that $\mathrm{Tr}\rho_R(\omega, l, m) = 1$ for each mode. The Hamiltonian is given immediately by

$$
\begin{aligned}
H_R &= \underset{\omega,l,m}{\otimes} H_R(\omega, l, m) \\
&= \underset{\omega,l,m}{\otimes} \left[(1 - e^{-\beta\omega}) \sum_n n\omega e^{-n\beta\omega} |n_R\rangle \langle n_R |_{\omega,l,m} \right].
\end{aligned}
\tag{6.425}
$$

Thus the energy is given by

$$
\begin{aligned}
E = \langle H_R \rangle &= \mathrm{Tr}\rho_R H_R \\
&= \sum_{\omega,l,m} \left[(1 - e^{-\beta\omega}) \sum_n n\omega e^{-n\beta\omega} \right] \\
&= \sum_{\omega,l,m} \frac{\omega}{e^{\beta\omega} - 1}.
\end{aligned}
\tag{6.426}
$$

The above state corresponds to the canonical ensemble, i.e. ρ_R can also be rewritten as (see how this was done in the Rindler case)

$$
\rho_R = \underset{\omega,l,m}{\otimes} \frac{e^{-\beta H(\omega, l, m)}}{Z(\omega, l, m)}.
\tag{6.427}
$$

$$
\mathrm{Tr}\rho_R(\omega, l, m) = 1, \quad Z(\omega, l, m) = \mathrm{Tr}e^{-\beta H(\omega, l, m)} = \sum_{n=0} e^{-\beta\omega n} = \frac{1}{1 - e^{-\beta\omega}}.
\tag{6.428}
$$

The entanglement entropy (this is an entanglement entropy because it was obtained by integrating out the interior modes) is then given immediately by

$$S = -\text{Tr}\rho_R \ln \rho_R = -\sum_{\omega,l,m} P_{\omega,l,m} \ln P_{\omega,l,m}. \tag{6.429}$$

We use the identities

$$\frac{\partial}{\partial N}(\rho_i)^N|_{N=1} = \rho_i \ln \rho_i. \tag{6.430}$$

$$\frac{\partial}{\partial N}e^{-N\beta H_i}|_{N=1} = -\beta H_i e^{-\beta H_i}. \tag{6.431}$$

$$\frac{\partial}{\partial N}Z_i^{-N}|_{N=1} = -\frac{\ln Z_i}{Z_i}. \tag{6.432}$$

The entropy then takes the form

$$S_i = \beta E_i + \ln Z_i = \beta E_i - \beta F_i. \tag{6.433}$$

The total entropy, total energy and total free energy are then given simply by

$$S = \sum_{\omega,l,m} S_{\omega,l,m}, \quad E = \sum_{\omega,l,m} E_{\omega,l,m}, \quad F = \sum_{\omega,l,m} F_{\omega,l,m}. \tag{6.434}$$

We have already computed E. The entropy is given on the other hand by

$$S = \sum_{\omega,l,m} \frac{\omega}{e^{\beta\omega} - 1} - \sum_{\omega,l,m} \ln(1 - e^{-\beta\omega}). \tag{6.435}$$

This entropy is clearly an entanglement entropy since it arose from a reduced density matrix.

The expressions for the energy and the entropy are IR divergent due to the infinite volume of space as well as UV divergent due to the presence of the horizon. The $r \longrightarrow \infty$ IR divergent is regulated in the usual way by putting the system in a box while the $r \longrightarrow r_s$ near horizon UV divergence should be regulated by some new unknown physics at the Planck scale near the horizon. Following 't Hooft [37] we will regulate this UV behavior by imposing Dirichelet boundary condition on the scalar field near the horizon, viz

$$\phi = 0 \text{ at } r = r_{\min}. \tag{6.436}$$

In terms of the proper distance ρ this minimum distance from the horizon reads

$$\rho = \sqrt{r_{\min}(r_{\min} - r_s)} + r_s \sinh\sqrt{\frac{r_{\min}}{r_s} - 1} \simeq 2\sqrt{r_s(r_{\min} - r_s)} \Rightarrow r_{\min} = r_s + \frac{\rho^2}{4r_s}. \tag{6.437}$$

This is the so-called brick wall introduced by 't Hooft. In terms of the tortoise coordinate r_* it is situated at

$$r_{*\min} = r_{\min} - r_s + r_s \ln\left(\frac{r_{\min}}{r_s} - 1\right) \simeq 2r_s \ln\frac{\rho}{2r_s}. \tag{6.438}$$

Recall now that every mode ψ_{lm} in the expansion $\psi = \sum_{lm} Y_{lm}\psi_{lm}$ is subjected to the Schrödinger equation

$$(\partial_t^2 - \partial_{r_*}^2 + V(r_*))\psi_{lm} = 0, \tag{6.439}$$

with a potential function in the tortoise coordinates r_* of the form

$$V(r_*) = \frac{r - r_s}{r}\left(\frac{r_s}{r} + \frac{l(l+1)}{r^2}\right). \tag{6.440}$$

We have the behavior

$$V(r_*) = \frac{l(l+1)}{r_*^2}, \quad r_* \longrightarrow \infty, \quad r \longrightarrow \infty. \tag{6.441}$$

$$V(r_*) = \frac{l(l+1) + r_s^2}{r_s^2}\exp\left(\frac{r_* - r_s}{r_s}\right), \quad r_* \longrightarrow -\infty, \quad r \longrightarrow r_s. \tag{6.442}$$

The mode ψ_{lm} comes from the brick wall at r_{*min} until it hits the potential at the turning point r_{*tur} defined by the condition

$$\frac{l(l+1) + r_s^2}{r_s^2}\exp\left(\frac{r_* - r_s}{r_s}\right) = \omega^2. \tag{6.443}$$

Since we are near the horizon, i.e. $r_* \longrightarrow -\infty$, we have $\omega \longrightarrow 0$ unless $l \gg 1$. The modes with small l are also suppressed from entropy consideration, i.e. low degeneracy. Thus, for modes with $l \gg 1$ we obtain the turning point

$$r_{*tur} = 2r_s \ln\frac{r_s\omega}{l}. \tag{6.444}$$

Each mode then moves between the brick wall r_{*min} and its own turning point. These are the zone modes (zero modes with support in the near-horizon region only) which dominates the canonical statistical ensemble. The IR box corresponds to a length

$$L = \Delta r_* = r_{*tur} - r_{*min} = 2r_s \ln\frac{2r_s^2\omega}{\rho l}. \tag{6.445}$$

The quantization of a particle in a box of size L leads immediately to the quantization condition

$$k_n = \frac{n\pi}{L} \Rightarrow \omega_n \simeq \frac{n\pi}{2r_s \ln\dfrac{2r_s^2\omega_n}{\rho l}} \tag{6.446}$$

Obviously, the size of the box shrinks as we increase l until it vanishes when $l = 2r_s^2\omega/\rho$. Since L now depends on the modes, we should make the usual replacement $\sum_{\omega lm}/L \longrightarrow \int d\omega/2\pi$ as follows

$$\sum_{\omega lm} f(\omega) \simeq 2\int_0^\infty \frac{d\omega}{2\pi}f(\omega)\int_0^{2r_s^2\omega/\rho} dl(2l+1)2r_s \ln\frac{2r_s^2\omega}{\rho l}. \tag{6.447}$$

The factor of 2 in front is due to the fact that we only integrate over positive frequencies. We get immediately

$$
\sum_{\omega lm} f(\omega) \simeq 2 \int_0^\infty \frac{d\omega}{2\pi} f(\omega) \left(-\frac{16 r_s^5 \omega^2}{\rho^2} \int_0^1 dx\, x \ln x \right)
$$
$$
\simeq \frac{8 r_s^5}{\rho^2} \int_0^\infty \frac{\omega^2 d\omega}{2\pi} f(\omega). \tag{6.448}
$$

As we can see most contribution comes from large angular momenta $l \sim 2 r_s^2 \omega / \rho$. The energy and the entropy are then given by the estimation (with $\beta = 2\pi/a = 4\pi r_s$)

$$
E \simeq \frac{8 r_s^5}{\rho^2} \int_0^\infty \frac{\omega^3 d\omega}{2\pi} \exp(-\beta\omega)
$$
$$
\simeq \frac{24 r_s^5}{\pi \rho^2 \beta^4} \tag{6.449}
$$
$$
\simeq \frac{24 r_s}{\pi \rho^2 (4\pi)^4}.
$$

$$
S \simeq E + \frac{8 r_s^5}{\rho^2} \int_0^\infty \frac{\omega^2 d\omega}{2\pi} \exp(-\beta\omega)
$$
$$
\simeq E + \frac{8 r_s^5}{\pi \rho^2 \beta^3} \tag{6.450}
$$
$$
\simeq \frac{8 r_s^2}{\pi \rho^2 (4\pi)^3}.
$$

In the above equations we are assuming that Hawking temperature T_H is very small and thus $\beta \longrightarrow \infty$. The energy is proportional to β while the entropy is proportional to β^2. We obtain divergent (as expected) expression in the horizon limit $\rho \longrightarrow 0$. However, if we assume the existence of a stretched horizon away from the mathematical horizon by a distance of the order of the Planck length, then

$$
\rho^2 \leqslant l_P^2 = 8\pi G. \tag{6.451}
$$

We can fix ρ by demanding that the entropy of the field is equal to the full Bekenstein–Hawking entropy, viz

$$
S \equiv \frac{A}{4G} = \frac{8 r_s^2}{\pi \rho^2 (4\pi)^3} \Rightarrow \rho^2 = \frac{G}{8\pi^5}. \tag{6.452}
$$

The energy becomes with this choice

$$
E = \frac{3 r_s}{4G} = \frac{3M}{2}. \tag{6.453}
$$

Thus, indeed, one should take $\rho \leqslant l_P$ in order for the field to carry no more energy and entropy than the black hole itself.

In summary, we have from one hand a divergent entropy in the near-horizon limit $\rho \longrightarrow 0$, while from the other hand the entropy must be, without any doubt, finite equal to the Bekenstein–Hawking value $S = A/4G$. In other words, quantum free field theory gives an overestimation of the entropy. As it turns out, adding interaction will not help but in fact it will make things worse. Indeed, in a $3 + 1$ dimensional interacting scalar field theory the entropy density is always given by a formula of the form

$$S(T) = \gamma(T)T^3, \tag{6.454}$$

where $\gamma(T)$ is the effective number of degrees of freedom at the temperature T and it is a monotonically increasing function of T. Hence, since the proper temperature $T(\rho) = 1/2\pi\rho$ diverges near the horizon we see that QFT gives always a divergent entropy. Furthermore, since the local temperature diverges in the limit $\rho \longrightarrow 0$ the entropy is indeed mostly localized on the horizon.

In the correct quantum theory of gravity it is therefore expected that the number of degrees of freedom decreases drastically as we approach the horizon. In other words, QFT theory should only describe the degrees of freedom at distances much greater than a Planck distance away from the horizon, while at distances less than a Planck distance away from the horizon the degrees of freedom may become sparse or they may even disappear altogether. This separation between QFT degrees of freedom and QG degrees of freedom can be achieved by a stretched horizon, i.e. a physical dynamical membrane, at a distance of one Planck length $l_P = \sqrt{G\hbar}$ from the actual horizon, where the temperature gets very large and most of the black hole entropy accumulates. Thus the stretched horizon is a time-like surface where real dynamics can take place, and where most of the black hole energy and entropy are localized. It is in thermal equilibrium with the thermal atmosphere, and thus it absorbs and then re-emits infalling matter continuously, while evaporation is seen in this case only as a tunneling process.

6.12.4 Conclusion

We consider a black hole formed by gravitational collapse as given by the Penrose diagram (6.12). The Hilbert space $\mathbf{H}_{in}$ of initial states $|\psi_{in}\rangle$ is associated with null rays incoming from $\mathscr{J}^-$ at $r = \infty$, i.e. $\mathbf{H}_{in} = \mathbf{H}_-$. The Hilbert space $\mathbf{H}_{out}$ of final states $|\psi_{out}\rangle$ is clearly a tensor product of the Hilbert space $\mathbf{H}_+$ of the scattered outgoing radiation which escapes to infinity $\mathscr{J}^+$ and the Hilbert space $\mathbf{H}_S$ of the transmitted radiation which falls behind the horizon into the singularity. This is the assumption of locality. Indeed, the outgoing Hawking particle and the lost quantum behind the horizon are maximally entangled, and thus they are space-like separated, and as a consequence localized operators on $\mathscr{J}^+$ and S must commute. We have then

$$H_{in} = H_-, \quad H_{out} = H_+ \otimes H_S. \tag{6.455}$$

From the perspective of observables at $\mathscr{J}^+$ (us), the outgoing Hawking particles can only be described by a reduced density matrix, even though the final state $|\psi_{out}\rangle$ is obtained from the initial state $|\psi_{in}\rangle$ by the action of a unitary S-matrix. This is the assumption of unitarity. This reduced density matrix is completely mixed despite the fact that the final state is a maximally entangled pure state. Eventually, the black hole will evaporate completely and it seems that we will end up only with the mixed state of the radiation. This the information paradox. There are six possibilities here:

1. Information is really lost which is Hawking's original stand.
2. Evaporation stops at a Planck-mass remnant which contains all the information with extremely large entropy.
3. Information is recovered only at the end of the evaporation when the singularity at $r = 0$ becomes a naked singularity. This contradicts the principle of information conservation with respect to the observer at $\mathcal{J}^+$ which states that by the time (Page or retention time) the black hole evaporates around one half of its mass the information must start coming out with the Hawking radiation.
4. Information is not lost during the entire process of formation and evaporation. This is the assumption of unitarity. But how?
5. The horizon is like a brick wall which cannot be penetrated. This contradicts the equivalence principle in an obvious way.
6. The horizon duplicates the information by sending one copy outside the horizon (as required by the principle of information conservation) while sending the other copy inside the horizon (as required by the equivalence principle). This is, however, forbidden by the linearity of quantum mechanics or the so-called quantum xerox principle [10].

6.13 Exercises

Exercise 1: We consider a quantum scalar field in the background geometry of a Schwarzschild black hole reduced to two dimensions.

1. **The asymptotic observer**
   (a) Write down the Schwarzschild metric in the null coordinates u and v and express the foliation t as well as the tortoise coordinate r_* in terms of u and v.
   (b) What are the positive frequency solutions v_ω with respect to the asymptotic static observer where the frequency is denoted by ω? Indicate their normalization.
   (c) Write down the expansion of the scalar field operator in terms of v_ω and v_ω^*. Define the Schwarzschild vacuum $|0_T\rangle$. Indicate the normalization of the creation and annihilation operators in this case which are denoted by b_ω and $b_\omega^\dagger$. Extract the right-moving (outgoing, $k > 0$) and left-moving (ingoing, $k < 0$) parts of the scalar field operator.
2. **The freely falling observer**
   (a) Write down the Schwarzschild metric in the null Kruskal coordinates U and V and express the Kruskal time T as well as the Kruskal space X in terms of U and V.

(b) What are the positive frequency solutions u_ν with respect to the freely falling observer where the frequency is denoted by ν? Indicate their normalization.

(c) Write down the expansion of the scalar field operator in terms of u_ν and u_ν^*. Define the Schwarzschild vacuum $|0_K\rangle$. Indicate the normalization of the creation and annihilation operators in this case, which are denoted by a_ν and $a_\nu^\dagger$. Extract the right-moving (outgoing, $k > 0$) and left-moving (ingoing, $k < 0$) parts of the scalar field operator.

3. Which of the two states $|0_T\rangle$ and $|0_K\rangle$ is the true vacuum of the scalar field and why?

4. Write down the Bogolubov coefficients $\alpha_{\omega\nu}$ and $\beta_{\omega\nu}$ which allow us to expand u in terms of v and vice versa. Expand b_ω in terms of a_ν and $a_\nu^\dagger$.

5. Express the expectation value of the Schwarzschild number operator $N_\omega = b_\omega^\dagger b_\omega$ in the vacuum state $|0_K\rangle$ in terms of the Bogolubov coefficient $\beta_{\omega\nu}$. Express this number in terms of the Euler gamma function

$$F(\omega, \nu) = \int_{-\infty}^{+\infty} \frac{du}{2\pi} e^{i\omega u - i\nu U_0 e^{-au}}. \tag{6.456}$$

6. Prove the following identity

$$F(\omega, \nu) = \exp\left(\frac{\omega\pi}{a}\right) F(\omega, -\nu). \tag{6.457}$$

Hint: Analytically continue the u integral in the lower half complex plane along the contour shown in figure 6.7.

7. Starting from the identity

$$\delta(\omega - \omega') = \int_0^\infty d\nu \left[\alpha_{\omega\nu} \alpha_{\omega'\nu}^* - \beta_{\omega\nu} \beta_{\omega'\nu}^* \right], \tag{6.458}$$

and using the result in the previous question derive the expectation value $\langle 0_K | N_\omega | 0_K \rangle$. What do you conclude? Derive Hawking temperature and insert SI units. What do you observe?

Exercise 2:

1. Write down the transmitted wave packet T which is a positive frequency wave packet defined outside the horizon with respect to Schwarzschild observer.

2. Write down the transmitted wave packet $\tilde{T}$ inside the horizon.

3. Calculate the positive frequency (with respect to the freely falling observer) extension T^+ by extending T to the region inside the horizon via analytical

continuation in the lower half complex plane and choosing the branch cut in the upper half complex plane (explain why?).

4. Calculate the negative frequency (with respect to the freely falling observer) extension T^- by extending T to the region inside the horizon via analytical continuation in the upper half complex plane and choosing the branch cut in the lower half complex plane (explain why).

5. Calculate the expectation value of the number operator $N = a^\dagger(P)a(P)$, where P is the outgoing wave packet, in the black hole initial state $|\psi\rangle$ and determine the greybody factor.

6. Write down the conditions satisfied by the Unruh vacuum $|U\rangle$.

7. Solve these conditions and write down the expression of $|U\rangle$ in terms of the Boulware vacuum $|B\rangle$ as a squeezed state.

8. Show that the Unruh vacuum is an entangled state with a vanishing total Killing energy.

Exercise 3:

- Write down Schwarzschild metric in spherical coordinates r, θ and ϕ.
- Write down Schwarzschild metric in tortoise coordinate given by

$$r_* = r + r_s \ln\left(\frac{r}{r_s} - 1\right), \quad r_s = 2GM. \tag{6.459}$$

- Write down the action for a scalar field

$$I = \int d^4x \sqrt{-\det g}\, \frac{1}{2} \partial_\mu \phi \partial^\mu \phi, \tag{6.460}$$

  in terms of the tortoise coordinate r_* and $\psi = r\phi$. Use the angular momentum operator

$$-\mathcal{L}^2 = \frac{1}{\sin\theta} \frac{\partial}{\partial\theta}\left(\sin\theta \frac{\partial}{\partial\theta}\right) + \frac{1}{\sin^2\theta} \frac{\partial^2}{\partial\phi^2}. \tag{6.461}$$

- We expand now in spherical coordinates as

$$\psi = \sum_{lm} \psi_{lm}\, Y_{lm}. \tag{6.462}$$

  Show that the action I takes the form

$$I = \int dt\, dr_* \frac{1}{2} \sum_{lm} \psi^*_{lm}\left(\partial_t^2 \psi_{lm} - \partial_{r_*}^2 \psi_{lm} + V(r_*)\psi_{lm}\right), \tag{6.463}$$

  where the potential is given by

$$V(r_*) = \frac{r - r_s}{r}\left(\frac{r_s}{r^3} + \frac{l(l+1)}{r^2}\right). \tag{6.464}$$

- Determine the equations of motion. Show that the stationary solution $\psi_{lm} = \exp(i\nu t)\tilde{\psi}_{lm}$ corresponds to a quantum particle with an energy $E = \nu^2$ moving in the potential V.

Exercise 4: In the following there are 30 multiple choice questions and each question has only one correct answer. Determine in each case the correct answer.

1. Near the horizon we have
    - (a) $r - r_s = \exp(-t/r_s), u \sim 2t, t \sim -r_s \ln(-\tau)/r_s$.
    - (b) $r - r_s = \exp(-t/r_s), u \sim 2t, t \sim -r_s \ln(\tau)/r_s$.
    - (c) $r - r_s = \exp(t/r_s), u \sim 2t, t \sim -r_s \ln(-\tau)/r_s$.
    - (d) $r - r_s = \exp(-t/r_s), u \sim t, t \sim -r_s \ln(-\tau)/r_s$.

2. The motion of a scalar field with frequency ω and angular momentum l in the Schwarzschild geometry is equivalent to a scattering problem with energy $E = \omega^2$ and potential
    - (a)
    $$V(r_*) = \frac{r - r_s}{r}\left(\frac{r_s}{r^3} + \frac{l(l + 1)}{r^2}\right).$$

    - (b)
    $$V(r_*) = \frac{l(l + 1)}{r^2}.$$

    - (c)
    $$V(r_*) = \frac{l(l + 1) + r_s^2}{r_s^2}\exp\left(\frac{r_* - r_s}{r_s}\right).$$

3. The null coordinates U and u are related by
    - (a) $U \sim u$.
    - (b) $U \sim \exp(-au)$.
    - (c) $U \sim \exp(au)$.

4. Hawking radiation consists mainly of particle with
    - (a) very large angular momenta $l \gg 1$.
    - (b) zero angular momentum $l = 0$.
    - (c) all possible values of angular momentum.

5. The Schwarzschild geometry near the horizon is Rindler geometry with acceleration
    - (a) $a = 1/r_s$.
    - (b) $a = 1/2r_s$.
    - (c) $a = 2r_s$.

6. Hawking temperature is given by
    - (a) $T_H = 8\pi GM$.
    - (b) $T = 1/2\pi r_s$.

    (c) $T = 1/8\pi GM$.

    (d) $T = a/\pi$.

7. The Bogolubov coefficients $\alpha_{\omega\nu}$ and $\beta_{\omega\nu}$ are given in terms of the function F defined by

    (a)

$$F(\omega, \nu) = \int_{-\infty}^{+\infty} \frac{du}{2\pi} \exp(i\omega U(u) - i\nu u).$$

    (b)

$$F(\omega, \nu) = \int_{-\infty}^{+\infty} \frac{du}{2\pi} \exp(i\omega u - i\nu U(u)).$$

    (c)

$$F(\omega, \nu) = \int_{0}^{+\infty} \frac{du}{2\pi} \exp(i\omega u - i\nu U(u)).$$

8. Which of these statements is correct:

    (a) The Schwarzschild tortoise (Boulware) vacuum $|0_T\rangle$ is defined outside the horizon and the Kruskal vacuum $|0_K\rangle$ is defined inside the horizon.

    (b) The Schwarzschild tortoise vacuum $|0_T\rangle$ is defined inside the horizon and the Kruskal vacuum $|0_K\rangle$ is defined outside.

    (c) The Schwarzschild tortoise vacuum $|0_T\rangle$ is defined outside the horizon and the Kruskal vacuum $|0_K\rangle$ is defined throughout spacetime.

9. The freely falling observer expands the right-moving field as

    (a)

$$\phi_R(u) = \int_0^\infty d\omega (v_\omega(u) b_\omega + \text{h.c.}).$$

    (b)

$$\phi_R(v) = \int_0^\infty d\omega (v_\omega(v) b_{-\omega} + \text{h.c.}).$$

    (c)

$$\phi_R(U) = \int_0^\infty d\nu (u_\nu(U) a_\nu + \text{h.c.}).$$

    (d)

$$\phi_R(V) = \int_0^\infty d\nu (u_\nu(V) a_{-\nu} + \text{h.c.}).$$

10. The positive frequency modes with respect to the Schwarzschild observer are:

    (a)

$$v_\nu = \frac{1}{\sqrt{4\pi\nu}} \exp(-i\nu U).$$

(b)

$$u_\omega = \frac{1}{\sqrt{4\pi\omega}} \exp(-i\omega u).$$

(c)

$$u_\omega = \frac{1}{\sqrt{4\pi\omega}} \exp(-i\omega t).$$

11. Which of these statements is correct:
    (a) The Kruskal observer is a non-inertial observer while the Schwarzschild observer is an inertial observer.
    (b) The Schwarzschild observer is the analog of the Rindler observer while the Kruskal observer is the analog of Minkowski observer.
    (c) The Schwarzschild observer is the analog of the Minkowski observer while the Kruskal observer is the analog of Rindler observer.

12. The tortoise coordinate r_* behaves as
    (a) $r_* \longrightarrow \infty$ when $r \longrightarrow r_s$ and $r_* \longrightarrow -\infty$ when $r \longrightarrow \infty$.
    (b) $r_* \longrightarrow -\infty$ when $r \longrightarrow r_s$ and $r_* \longrightarrow -\infty$ when $r \longrightarrow \infty$.
    (c) $r_* \longrightarrow -\infty$ when $r \longrightarrow r_s$ and $r_* \longrightarrow \infty$ when $r \longrightarrow \infty$.
    (d) $r_* \longrightarrow \infty$ when $r \longrightarrow r_s$ and $r_* \longrightarrow \infty$ when $r \longrightarrow \infty$.

13. A freely falling object through the horizon is seen
    (a) by the Schwarzschild observer to cross the horizon in a finite time.
    (b) by the Kruskal observer to cross the horizon in infinite time.
    (c) by the Schwarzschild observer to cross the horizon in infinite time.

14. Near the horizon the Kruskal coordinates U and V behave as
    (a) $U \longrightarrow 0, V \longrightarrow$ constant.
    (b) $U \longrightarrow$ constant, $V \longrightarrow 0$.
    (c) $U \longrightarrow$ constant, $V \longrightarrow$ constant.

15. The black hole initial quantum state does not contain incoming positive frequency excitation and thus it is annihilated by
    (a) $a(P)$, i.e. $a(P)|\psi\rangle = 0$.
    (b) $a(R)$, i.e. $a(R)|\psi\rangle = 0$.
    (c) $a(T)$, i.e. $a(T)|\psi\rangle = 0$.

16. By comparing Schwarzschild and Rindler geometries:
    (a) The Schwarzschild scattering potential is confining while that of Rindler is not confining.
    (b) The Rindler scattering potential is confining while that of Schwarzschild is not confining.
    (c) They are both confining.
    (d) They are both not confining.

17. We evolve the outgoing wave packet P backward in time. As we approach the horizon the frequency with respect to the freely falling observer
    (a) stays the same.
    (b) decreases (redshift).
    (c) increases (blueshift).

18. Near the horizon the transmitted wave packet T behaves as
    (a)
    $$T(z) \sim \exp\left(i\frac{\omega}{a}\ln(\tau)\right), \quad \tau > 0; \quad T(z) \sim 0, \quad \tau < 0.$$

    (b)
    $$T(z) \sim \exp\left(i\frac{\omega}{a}\ln(-\tau)\right), \quad \tau < 0; \quad T(z) \sim 0, \quad \tau > 0.$$

    (c)
    $$T(z) \sim \exp\left(i\frac{\omega}{a}\ln(-\tau)\right), \quad \tau < 0; \quad T(z) \sim \exp\left(i\frac{\omega}{a}\ln(\tau)\right), \quad \tau > 0.$$

19. The greybody factor is given by
    (a) $\Gamma = (T, T)$.
    (b) $\Gamma = (T^+, T^+)$.
    (c) $\Gamma = (T^-, T^-)$.
20. By comparing with the case of Rindler geometry we can conclude that
    (a) the Kruskal state $|0_K\rangle$ is the true vacuum state.
    (b) the tortoise state $|0_T\rangle$ is the true vacuum state.
    (c) the Kruskal and tortoise states are degenerate.
21. The expectation value of the Schwarzschild number operator $N_\omega = b_\omega^\dagger b_\omega$ in the Kruskal vacuum is given in terms of Bogolubov coefficients by
    (a)
    $$\langle 0_K | N_\omega | 0_K \rangle = \int_0^\infty d\nu \, |F(\omega, \nu)|^2.$$

    (b)
    $$\langle 0_K | N_\omega | 0_K \rangle = \int_0^\infty d\nu \, |\alpha_{\omega\nu}|^2.$$

    (c)
    $$\langle 0_K | N_\omega | 0_K \rangle = \int_0^\infty d\nu \, |\beta_{\omega\nu}|^2.$$

22. Which of these statements is correct:
    (a) a correlated entangled pure state near the horizon gives rise to a correlated entangled state outside the horizon.
    (b) a thermal mixed state near the horizon gives rise to a thermal mixed state outside the horizon.
    (c) a correlated entangled pure state near the horizon gives rise to a thermal mixed state outside the horizon.
    (d) a correlated entangled pure state outside the horizon gives rise to a thermal mixed state near the horizon.

23. Let $\tilde{T}(\tau) = T(-\tau)$ be the solution inside the horizon. The analytic continuation of T in the lower half complex plane with the branch cut in the upper half complex plane gives
    (a) The negative frequency part $\bar{T}^+ = c_+(\bar{T} + \tilde{\bar{T}} \exp(-\frac{\pi\omega}{a}))$.
    (b) The positive frequency part $\bar{T}^- = c_-(\bar{T} + \tilde{\bar{T}} \exp(\frac{\pi\omega}{a}))$.
    (c) The positive frequency part $T^+ = c_+(T + \tilde{T}\exp(-\frac{\pi\omega}{a}))$.
    (d) The negative frequency part $T^- = c_-(T + \tilde{T}\exp(\frac{\pi\omega}{a}))$.

24. The Schwarzschild potential barrier is of the order of
    (a) l.
    (b) l^2.
    (c) l^3.

25. The in state is a squeezed state given by
    (a) The state $|\psi\rangle$ in which the black hole formed satisfying $a(R)|\psi\rangle = 0$.
    (b) The Boulware state $|B\rangle$ satisfying $a(T)|B\rangle = a(\tilde{T})|B\rangle = 0$.
    (c) The Unruh state $|U\rangle$ satisfying $a(T^+)|U\rangle = a(\bar{T}^-)|U\rangle = 0$.

26. The adiabatic principle states that for times much less than r_s (collapse time scale) the modes $\omega \gg 1/r_s$ will see the change in the geometry very slowly and hence remain unexcited. This means that the black hole state $|\psi\rangle$ does not contain positive high frequency modes, viz
    (a) $a(T^+)|\psi\rangle = a(\bar{T}^+)|\psi\rangle = 0$.
    (b) $a(T^+)|\psi\rangle = a(\bar{T}^-)|\psi\rangle = 0$.
    (c) $a(\bar{T}^+)|\psi\rangle = a(\bar{T}^-)|\psi\rangle = 0$.

27. The Unruh state is
    (a) a maximally entangled state with a zero total Killing energy.
    (b) a completely mixed state with a zero total Killing energy.
    (c) an entangled state state with non-zero total Killing energy.
    (d) a mixed state with non-zero total Killing energy.

28. The asymptotic and freely falling objects are related through the Bogolubov transformation
    (a)
    $$a_\nu = \int_0^\infty d\omega(\alpha_{\omega\nu}b_\omega + \beta^*_{\omega\nu}b_\omega^\dagger).$$

    (b)
    $$b_\omega = \int_0^\infty d\nu(\alpha^*_{\omega\nu}a_\nu - \beta^*_{\omega\nu}a_\nu^\dagger).$$

29. The expectation value of the Schwarzschild number operator in the Kruskal vacuum is given by
    (a)
    $$\langle 0_K|N_\omega|0_K\rangle = \frac{1}{\exp\left(-\frac{2\pi\omega}{a}\right) - 1}.$$

(b)

$$\langle 0_K | N_\omega | 0_K \rangle = \frac{1}{\exp\left(\dfrac{2\pi\omega}{a}\right) - 1}.$$

(c)

$$\langle 0_K | N_\omega | 0_K \rangle = \exp\left(-\frac{2\pi\omega}{a}\right).$$

30. The Hawking particles which escape the Schwarzschild potential barrier have energy
    (a) of the same order of magnitude as Hawking temperature T_H.
    (b) of the order of T_H^2.
    (c) much smaller than the thermal scale set by Hawking temperature.

Solution 4:

1. Near the horizon we have
   (a) $r - r_s = \exp(-t/r_s)$, $u \sim 2t$, $t \sim -r_s \ln(-\tau)/r_s$.
2. The motion of a scalar field with frequency ω and angular momentum l in the Schwarzschild geometry is equivalent to a scattering problem with energy $E = \omega^2$ and potential
   (a)

$$V(r_*) = \frac{r - r_s}{r}\left(\frac{r_s}{r^3} + \frac{l(l+1)}{r^2}\right).$$

3. The null coordinates U and u are related by
   (a) $U \sim \exp(-au)$.
4. Hawking radiation consists mainly of particle with
   (a) zero angular momentum $l = 0$.
5. The Schwarzschild geometry near the horizon is Rindler geometry with acceleration
   (a) $a = 1/2r_s$.
6. Hawking temperature is given by
   (a) $T = 1/8\pi GM$.
7. The Bogolubov coefficients $\alpha_{\omega\nu}$ and $\beta_{\omega\nu}$ are given in terms of the function F defined by
   (a)

$$F(\omega, \nu) = \int_{-\infty}^{+\infty} \frac{du}{2\pi} \exp(i\omega u - i\nu U(u)).$$

8. Which of these statements is correct:
   (a) The Schwarzschild tortoise vacuum $|0_T\rangle$ is defined outside the horizon and the Kruskal vacuum $|0_K\rangle$ is defined throughout spacetime.

9. The freely falling observer expands the right-moving field as

   (a)
   $$\phi_R(U) = \int_0^\infty d\nu (u_\nu(U)a_\nu + \text{h.c}).$$

10. The positive frequency modes with respect to the Schwarzschild observer are:

    (a)
    $$u_\omega = \frac{1}{\sqrt{4\pi\omega}} \exp(-i\omega u).$$

11. Which of these statements is correct:
    (a) The Schwarzschild observer is the analog of the Rindler observer while the Kruskal observer is the analog of Minkowski observer.

12. The tortoise coordinate r_* behaves as
    (a) $r_* \longrightarrow -\infty$ when $r \longrightarrow r_s$ and $r_* \longrightarrow \infty$ when $r \longrightarrow \infty$.

13. A freely falling object through the horizon is seen
    (a) by the Schwarzschild observer to cross the horizon in infinite time.

14. Near the horizon the Kruskal coordinates U and V behave as
    (a) $U \longrightarrow 0$, $V \longrightarrow$ constant.

15. The black hole initial quantum state does not contain incoming positive frequency excitation and thus it is annihilated by
    (a) $a(R)$, i.e. $a(R)|\psi\rangle = 0$.

16. By comparing Schwarzschild and Rindler geometries:
    (a) The Rindler scattering potential is confining while that of Schwarzschild is not confining.

17. We evolve the outgoing wave packet P backward in time. As we approach the horizon the frequency with respect to the freely falling observer
    (a) increases (blueshift).

18. Near the horizon the transmitted wave packet T behaves as
    (a)
    $$T(z) \sim \exp(i\frac{\omega}{a}\ln(-\tau)), \quad \tau < 0; \quad T(z) \sim 0, \quad \tau > 0.$$

19. The greybody factor is given by
    (a) $\Gamma = (T, T)$.

20. By comparing with the case of Rindler geometry we can conclude that
    (a) the Kruskal state $|0_K\rangle$ is the true vacuum state.

21. The expectation value of the Schwarzschild number operator $N_\omega = b_\omega^\dagger b_\omega$ in the Kruskal vacuum is given in terms of Bogolubov coefficients by
    (a)
    $$\langle 0_K|N_\omega|0_K\rangle = \int_0^\infty d\nu |\beta_{\omega\nu}|^2.$$

22. Which of these statements is correct:
    (a) a correlated entangled pure state near the horizon gives rise to a thermal mixed state outside the horizon.

23. Let $\tilde{T}(\tau) = T(-\tau)$ be the solution inside the horizon. The analytic continuation of T in the lower half complex plane with the branch cut in the upper half complex plane gives
    (a) The positive frequency part $T^+ = c_+(T + \tilde{T}\exp(-\frac{\pi\omega}{a}))$.

24. The Schwarzschild potential barrier is of the order of
    (a) l^2.

25. The in state is a squeezed state given by
    (a) The Unruh state $|U\rangle$ satisfying $a(T^+)|U\rangle = a(\bar{T}^-)|U\rangle = 0$.

26. The adiabatic principle states that for times much less than r_s (collapse time scale) the modes $\omega \gg 1/r_s$ will see the change in the geometry very slowly and hence remain unexcited. This means that the black hole state $|\psi\rangle$ does not contain positive high frequency modes, viz
    (a) $a(T^+)|\psi\rangle = a(\bar{T}^-)|\psi\rangle = 0$.

27. The Unruh state is
    (a) a maximally entangled state with a zero total Killing energy.

28. The asymptotic and freely falling objects are related through the Bogolubov transformation
    (a)
    $$b_\omega = \int_0^\infty d\nu(\alpha_{\omega\nu}^* a_\nu - \beta_{\omega\nu}^* a_\nu^\dagger).$$

29. The expectation value of the Schwarzschild number operator in the Kruskal vacuum is given by
    (a)
    $$\langle 0_K | N_\omega | 0_K \rangle = \frac{1}{\exp\left(\dfrac{2\pi\omega}{a}\right) - 1}.$$

30. The Hawking particles which escape the Schwarzschild potential barrier have energy
    (a) of the same order of magnitude as Hawking temperature T_H.

References

[1] Maldacena J M 1999 The large N limit of superconformal field theories and supergravity *Int. J. Theor. Phys.* **38**(231) 1113

[2] Natsuume M 2015 *AdS/CFT Duality User Guide* (Lecture Notes in Physics vol 903) (Springer)

[3] Nastase H 2007 Introduction to AdS-CFT arXiv:0712.0689 [hep-th]

[4] Banks T, Fischler W, Shenker S H and Susskind L 1997 M theory as a matrix model: a conjecture *Phys. Rev.* D **55** 5112

[5] Gubser S S, Klebanov I R and Polyakov A M 1998 Gauge theory correlators from noncritical string theory *Phys. Lett.* B **428** 105

[6] Witten E 1998 Anti-de Sitter space and holography *Adv. Theor. Math. Phys.* **2** 253

[7] Polchinski J 1609 The Black Hole information problem arXiv:1609.04036 [hep-th]

[8] Hawking S W 1975 Particle creation by Black Holes *Commun. Math. Phys.* **43** 199 Erratum: [Commun. Math. Phys. **46**, 206 (1976)]

[9] Hawking S W 1976 Breakdown of predictability in gravitational collapse *Phys. Rev.* D **14** 2460

[10] Susskind L and Lindesay J 2005 *An Introduction to Black Holes, Information and the String Theory Revolution: The Holographic Universe* (Hackensack, NJ: World Scientific) p 183

[11] Page D N Black hole information 1993 arXiv:hep-th/9305040 https://arxiv.org/abs/hep-th/9305040

[12] Harlow D 2016 Jerusalem lectures on black holes and quantum information *Rev. Mod. Phys.* **88** 15002

[13] Jacobson T 2003 Introduction to quantum fields in curved space-time and the Hawking effect ArXiv:gr-qc/0308048 https://arxiv.org/abs/gr-qc/0308048

[14] Mukhanov V and Winitzki S 2007 *Introduction to Quantum Effects in Gravity* (Cambridge University Press)

[15] Carroll S M 2004 *Spacetime and Geometry: An Introduction to General Relativity* (San Francisco, CA: Addison-Wesley)

[16] 't Hooft G 2001 *Introduction to General Relativity* (Paramus, NJ: Rinton Press)

[17] Carroll S M 1997 Lecture notes on general relativity ArXiv:gr-qc/9712019 https://arxiv.org/abs/gr-qc/9712019

[18] Zurek W H 1991 Decoherence and the transition from quantum to classical–revisited *Phys. Today* **44** 36–44

[19] Moore M G *Quantum Mechanics I* http://www.pa.msu.edu/mmoore/851.html

[20] Wald R M 1984 *General Relativity* (Chicago, IL: University of Chicago Press)

[21] Birrell N D and Davies P C W *Quantum Fields in Curved Space* (Cambridge University Press)

[22] Unruh W G 1976 Notes on black hole evaporation *Phys. Rev.* D **14** 870

[23] Giddings S B and Nelson W M 1992 Quantum emission from two-dimensional black holes *Phys. Rev.* D **46** 2486

[24] Traschen J H 1999 An introduction to black hole evaporation ArXiv:gr-qc/0010055 https://arxiv.org/abs/gr-qc/0010055

[25] Deeg D 2006 Quantum aspects of black holes *Dissertation* LMU München

[26] Wald R M 1975 On particle creation by black holes *Commun. Math. Phys.* **45** 9

[27] Polchinski J 1995 String theory and black hole complementarity ArXiv:[hep-th/9507094] https://arxiv.org/abs/hep-th/9307168

[28] Mathur S D 2009 The information paradox: a pedagogical introduction *Class. Quant. Grav.* **26** 224001

[29] Ellis J R, Hagelin J S, Nanopoulos D V and Srednicki M 1984 Search for violations of quantum mechanics *Nucl. Phys.* B **241** 381

[30] Banks T, Susskind L and Peskin M E 1984 Difficulties for the evolution of pure states into mixed states *Nucl. Phys.* B **244** 125

[31] Page D N 1993 Information in black hole radiation *Phys. Rev. Lett.* **71** 3743

[32] Page D N 2013 Time dependence of Hawking radiation entropy *J. Cosmol. Astropart. Phys.* **1309** JCAP09(2013)028

[33] Page D N 1993 Average entropy of a subsystem *Phys. Rev. Lett.* **71** 1291

[34] Bekenstein J D 1973 Black holes and entropy *Phys. Rev.* D **7** 2333

[35] Bekenstein J D 1974 Generalized second law of thermodynamics in black hole physics *Phys. Rev.* D **9** 3292

[36] Hawking S W 1971 Gravitational radiation from colliding black holes *Phys. Rev. Lett.* **26** 1344

[37] 't Hooft G 1985 On the quantum structure of a black hole *Nucl. Phys.* B **256** 727

IOP Publishing

Lectures on General Relativity, Cosmology and
Quantum Black Holes (Second Edition)

Badis Ydri

Chapter 7

Quantum black holes and gauge/gravity duality

This chapter contains a detailed overview of the recent work, by Almheiri *et al* and Penington [1, 2], on the **AdS**^{2} black hole information loss paradox and its proposal resolution within the context of the anti-de Sitter/conformal field theory (AdS/CFT) correspondence. The roles of the generalized entanglement entropy, the quantum extremal surfaces, the island conjecture, holography and the replica wormhole in this resolution are discussed. The distinction between the von Neumann entropy and the Bekenstein–Hawking entropy in black hole physics is carefully outlined. A phase transition at the Page time between the trivial quantum extremal surface at the horizon and a non-vanishing quantum extremal surface behind the horizon is shown to lead to the correct Page curve. A simplified version of the information loss problem in the context of the eternal **AdS**^{2} black hole and its resolution along the same lines is also included. The replica trick and its crucial role in computing the entropies of various intervals in **AdS**^{2} is also discussed at some length. However, the use of the replica trick to construct the replica wormholes which dominate the Euclidean path integral, leading to the island rule and the correct quantum extremal surfaces, is only outlined.

7.1 Outline

- The fundamental theory underlying the recent proposed resolution of the black hole information loss paradox is the holographic AdS/CFT correspondence [3–5]. See also [6–8].
- More precisely, we should use, instead of the Bekenstein–Hawking formula [9, 10] for black hole entropy, the fine-grained entropy formula discovered by Ryu and Takayanagi in [11] (see also [12, 13]). This is the correct formula in the case of quantum field theories coupled to gravity.
- The black hole information loss paradox consists of the seemingly non-unitary process of evolving a pure state (black hole) into a mixed state

doi:10.1088/978-0-7503-5824-8ch7

(Hawking radiation) [14–16]. Indeed, Hawking radiation is a thermal black-body radiation. The measurement process in quantum mechanics is the only other known non-unitary process taking a pure state into a mixed state.

- The main result of interest to us here is an argument against the black hole information loss paradox.
- This main result goes as follows: Hawking radiation is computed using the Ryu–Takayanagi formula and shown to be consistent with unitarity in [1, 2]. See also [17–20].
- This result is in fact an evidence for what is termed the 'central dogma' in [20] which can be alternatively stated as follows: From the perspective of an outside observer an evaporating black hole looks like a unitary quantum system with a finite number of degrees of freedom, i.e. a quantum system with a finite-dimensional Hilbert space.
- Evidence for the 'central dogma' comes from the AdS/CFT correspondence [3–5] but also from the Banks–Fischler–Shenker–Susskind (BFSS) matrix quantum mechanics [21].
- Furthermore, it can be shown that the number of degrees of freedom or microstates of the black hole is exactly measured by the Bekenstein–Hawking formula $S = A/4\hbar G_N = A/4l_P^2$ [22].
- The Bekenstein–Hawking entropy is in fact the coarse-grained Boltzmann entropy which is the usual thermodynamic entropy. This thermodynamic entropy is always larger than the fine-grained von Neumann entropy which is the information entropy (density matrix).
- The von Neumann entropy of the outgoing Hawking radiation equals the von Neumann entropy of the black hole because the system (black hole + Hawking radiation) must be described by a pure state according to the 'central dogma'. Thus, the von Neumann entropy of the Hawking radiation must always be smaller than the Bekenstein–Hawking–Boltzmann entropy.
- From an operational point of view the unitarity of the Hawking radiation can be characterized by the so-called Page curve [23, 24] which gives the von Neumann entropy of the outgoing radiation as a function of time. This curve must have a maximum (turnover point) at the so-called Page time, i.e. the von Neumann fine-grained entropy of the radiation must start to decrease at this time so as to not exceed the Boltzmann coarse-grained entropy of the black hole. Thus, Page time is when the von Neumann entropy of radiation equals the Boltzmann entropy of the black hole. See figure 7.1.
- The Hawking curve, in contrast, increases monotonically and thus the von Neumann entropy of the radiation becomes at a certain time larger than the Boltzmann entropy of the black hole. This is the precise statement of the black hole information loss paradox.
- The main results which underlie the resolution of the black hole information loss paradox proposed in [1, 2, 17–20] are as follows:
  (1) The fine-grained von Neumann entropy of an evaporating black hole computed employing the Ryu–Takayanagi formula will follow the Page curve if one uses the so-called quantum extremal surfaces [13].

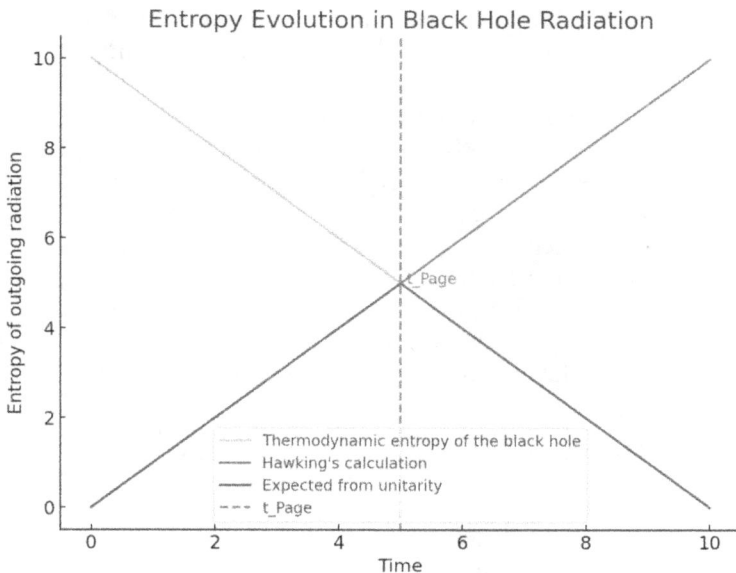

Figure 7.1. Page curve.

(2) The fine-grained von Neumann entropy of Hawking radiation is equal to the entropy of the black hole, and thus it too follows the Page curve, if one uses the so-called 'island conjecture' in conjunction with the Ryu–Takayanagi formula [11, 12]. The boundary of the island is the quantum extremal surface.

(3) The degrees of freedom of the black hole and the Hawking radiation describe the regions of the spacetime given by their respective entanglement wedges where their fine-grained entropies are computed. It is shown that the black hole entanglement wedge describes a portion of the interior, whereas the Hawking radiation entanglement wedge is disconnected as it involves an island behind the horizon.

(4) The island Ryu–Takayanagi formula for the fine-grained von Neumann entropy can be derived from a gravitational path integral using the so-called replica trick where it is found that the replica wormhole (zero entropy) dominates over the Hawking saddle point (large entropy) and thus unitarity indeed holds.

(5) For holographic CFT matter it can be shown that the island is actually connected to the radiation via extra dimensions.

7.2 The JT–CFT$_2$ system: a first look

7.2.1 Eternal AdS^2 black hole

In this first section we describe the Almheiri–Polchinski (AP) model discussed in [25], which is a particular case of the Jackiw–Teitelboim (JT) model of dilaton gravity in two dimensions [26, 27]. This has striking similarities to the Sachdev–Ye–

Kitaev (SYK) model [28, 29] (in the sense that everything is encoded in the boundary theory which is given by a Schwarzian action). Here we will mainly follow [1, 25] but also [30, 31].

The APJT model is a dynamical theory of gravity in two dimensions with no local excitations, i.e. the value of the metric tensor $ds^2 = g_{\mu\nu}dx^\mu dx^\nu$, which is a dynamical variable here (as opposed to pure gravity in two dimensions), and the value of the dilaton field ϕ, which plays a crucial role in the existence of stable black hole configurations in two dimensions, is uniquely fixed in terms of the stress–energy–momentum tensor of the matter field f (through the equations of motion).

We will mostly assume that the matter theory is given by a conformal field theory which can also be holographic. We will also assume that the matter field f interacts with dilaton field Φ only through the metric tensor and not directly. The action reads explicitly

$$S[g, \Phi, f] = S_{JT}[g, \Phi] + D_{CFT}[g, f]$$
$$= \frac{1}{16\pi G} \int d^2x \sqrt{-g}(\Phi^2 R - V(\Phi)) + S_{CFT}[g, f]. \tag{7.1}$$

The AP potential V is given explicitly by

$$V(\Phi) = 2 - 2\Phi^2. \tag{7.2}$$

This potential guarantees that spacetime has a constant negative scalar curvature (integrate out $\phi = \Phi^2$ to obtain the delta function $\delta(R + 2)$), i.e. we must have $R = -2$. In other words, spacetime is precisely anti-de Sitter spacetime $\mathbf{AdS}^2$. In fact, this potential also guarantees that the matter action does not depend directly on the dilaton field but depends on it only indirectly through the metric tensor. The dilaton field itself is the crucial ingredient allowing the existence of black hole configurations on this two-dimensional $\mathbf{AdS}^2$ background.

We check these facts more explicitly as follows. First, we write the metric in the conformal gauge as follows

$$ds^2 = -e^{2\omega(u, v)}dudv. \tag{7.3}$$

The light-cone coordinates u and v will also be denoted as $u = x^+$ and $v = x^-$ where $x^\pm = t \pm z$. The equations of motion (by varying the dilaton field and the metric tensor) will then read

$$4\partial_u\partial_v\omega + e^{2\omega} = 0. \tag{7.4}$$

$$-e^{2\omega}\partial_u(e^{-2\omega}\partial_u\Phi^2) = 8\pi G T_{uu}. \tag{7.5}$$

$$-e^{2\omega}\partial_v(e^{-2\omega}\partial_v\Phi^2) = 8\pi G T_{vv}. \tag{7.6}$$

$$2\partial_u\partial_v\Phi^2 + e^{2\omega}(\Phi^2 - 1) = 16\pi G T_{uv}. \tag{7.7}$$

The stress–energy–momentum tensor of the matter field is of course given by the usual formula, viz

$$T_{ab} = -\frac{2}{\sqrt{-g}}\frac{\delta S_{\mathrm{CFT}}[g, f]}{\delta g^{ab}}. \tag{7.8}$$

For conformal matter, such as a massless free scalar field f given by the Klein–Gordon action $S_{\mathrm{CFT}} = \frac{1}{32\pi G}\int d^2x\sqrt{-g}(\nabla f)^2$, the off-diagonal component T_{uv} of the stress–energy–momentum tensor vanishes classically. We also recall that in this case $T_{uu} = T_{++} = \frac{1}{16\pi G}\partial_+ f\partial_+ f$ and $T_{vv} = T_{--} = \frac{1}{16\pi G}\partial_- f\partial_- f$.

The most general solution of (7.4) is the **AdS**^{2} metric, i.e. $e^{2\omega} = 4/(u - v)^2$. We have explicitly the metric

$$ds^2 = -\frac{4}{(x^+ - x^-)^2}dx^+dx^- = \frac{1}{z^2}(-dt^2 + dz^2), \quad x^\pm = t \pm z. \tag{7.9}$$

The gravitational sector will be treated semi-classically throughout. This means, among many other things, that we will replace the stress–energy–momentum tensor by its expectation value, viz

$$T_{ab} = \langle T_{ab}\rangle. \tag{7.10}$$

For $\langle T_{ab}\rangle = 0$ the most general solution of the equations of motion (7.5), (7.6) and (7.7) is given by the dilaton field

$$\Phi^2 = 1 + \frac{a - \mu x^+ x^-}{x^+ - x^-}. \tag{7.11}$$

This dilaton field represents an eternal black hole with two asymptotic boundaries.

The case $a = 0$ represents pure **AdS**^{2}, or more precisely the Poincaré patch of **AdS**^{2}, whereas the solution with $a > 0$ (and $\mu > 0$) represents an **AdS**^{2} black hole (it is the Rindler wedge of the Poincaré patch).

The most general solution of the equations of motion is a conformal transformation $x = f(y)$ of (7.9) and (7.11), viz

$$ds^2 = -\frac{4f'(y^+)f'(y^-)dy^+dy^-}{(f(y^+) - f(y^-))^2}, \quad \Phi^2 = 1 + \frac{a - f(y^+)f(y^-)}{f(y^+) - f(y^-)}. \tag{7.12}$$

A static form of this black hole configuration is obtained by means of the conformal transformation

$$x = f(y) = \frac{1}{\sqrt{\mu}}\tanh\sqrt{\mu}\,y. \tag{7.13}$$

The black hole solution reads then (we set $a = 1$)

$$ds^2 = -\frac{4\mu dy^+dy^-}{\sinh^2\sqrt{\mu}(y^+ - y^-)}, \quad \Phi^2 = 1 + \sqrt{\mu}\coth\sqrt{\mu}(y^+ - y^-). \tag{7.14}$$

The x coordinates cover the whole geometry of the spacetime manifold, whereas the y coordinates cover only the exterior of the black hole. At high temperature ($\mu \longrightarrow 0$) we can make the identification

$$z = \frac{y^+ - y^-}{2}. \tag{7.15}$$

Thus, the boundary $z = 0$ of **AdS**^{2} is located in the y coordinates at $y^+ - y^- = 0$. On the other hand, the horizon $z \longrightarrow \infty$ of **AdS**^{2} corresponds in the y coordinates either to the future horizon $y^+ \longrightarrow +\infty$ (or equivalently $x^+ \longrightarrow 1/\sqrt{\mu}$) or to the past horizon $y^- \longrightarrow -\infty$ (or equivalently $x^+ \longrightarrow -1/\sqrt{\mu}$).

This black hole configuration after Euclidean rotation becomes periodic in the variable $y^+ - y^-$ with period $\beta_0 = 1/T_0$ given by

$$2\sqrt{\mu} = \frac{2\pi}{\beta_0} \Rightarrow T_0 = \frac{\sqrt{\mu}}{\pi}. \tag{7.16}$$

This is Hawking temperature. This can also be checked in the Schwarzschild coordinates defined by

$$\rho = \sqrt{\mu} \coth \sqrt{\mu}(y^+ - y^-), \quad T = \frac{y^+ + y^-}{2}. \tag{7.17}$$

The metric and the dilaton take then the form

$$ds^2 = -4(\rho^2 - \mu)dT^2 + \frac{d\rho^2}{\rho^2 - \mu}, \quad \Phi^2 = 1 + \rho. \tag{7.18}$$

The Hawking temperature is then given by

$$T_0 = \frac{1}{4\pi}\partial_\rho \sqrt{-\frac{g_{TT}}{g_{\rho\rho}}} \Big|_{\rho=\sqrt{\mu}}$$

$$= \frac{\sqrt{\mu}}{\pi}. \tag{7.19}$$

7.2.2 AdS^2 black hole formed from gravitational collapse

We look now at the equations of motion more carefully. First, we trivially check that $\exp(2\omega) = 4/(x^+ - x^-)^2$ solves the equation of motion (7.4). Next, in terms of the ansatz $\Phi^2 = M/(x^+ - x^-)$ we write the constraints (7.5) and (7.6) in the form

$$\partial_+^2 M = -8\pi G(x^+ - x^-)T_{++}(x^+), \quad \partial_-^2 M = -8\pi G(x^+ - x^-)T_{--}(x^-). \tag{7.20}$$

The final equation of motion (7.7) reads

$$\partial_+\partial_- M + (\partial_+ M - \partial_- M - 2)/(x^+ - x^-) = 8\pi G(x^+ - x^-)T_{+-}. \tag{7.21}$$

The most general solution of (7.20) is

$$M = a + bx^+ + cx^- + dx^+x^- - I^+ + I^-. \tag{7.22}$$

$$I^\pm = 8\pi G \int_{x_0^\pm}^{x^\pm} dx'^\pm (x'^\pm - x^\mp)(x'^\pm - x^\pm)T_{\pm\pm}(x'^\pm). \tag{7.23}$$

For conformal theory we have $T_{+-} = 0$. The solution of equation (7.21) is then given by the requirement

$$b - c = 2. \tag{7.24}$$

The solution is then (with $b = b' + 1$)

$$\Phi^2 = 1 + \frac{a + b'(x^+ + x^-) + dx^+x^-}{x^+ - x^-} - \frac{I^+ - I^-}{x^+ - x^-}. \tag{7.25}$$

For $T_{uv} = \langle T_{uv} \rangle = 0$ we get the solution

$$\Phi^2 = \frac{a + b'(x^+ + x^-) + dx^+x^-}{x^+ - x^-}. \tag{7.26}$$

By an **SL(2,R)** symmetry which acts as $t \longrightarrow t' = (at + b)/(ct + d)$ with $ad - cb = 1$ we can bring this solution to the form

$$\Phi^2 = 1 + \frac{a - \mu x^+x^-}{x^+ - x^-}. \tag{7.27}$$

This is the dilaton profile corresponding to an eternal **AdS**^{2} black hole.

Next, we would like to derive the black hole solution formed from gravitational collapse. We consider then the effect of an infalling matter pulse into the black hole, i.e. the effect of a shockwave of energy E_S traveling on the null curve $x^- = 0$ starting from the boundary $z = 0$ at time $t = 0$. The stress–energy–momentum tensor is given by

$$T_{--} = \langle T_{--} \rangle = E_S\delta(x^-). \tag{7.28}$$

We compute immediately $I^+ = 0$, $I^- = 8\pi GE_S x^+x^-$. The dilaton profile becomes then given by

$$\Phi^2 = \frac{a + b'(x^+ + x^-) + (d + 8\pi GE_S)x^+x^-}{x^+ - x^-}. \tag{7.29}$$

By an **SL(2,R)** symmetry we can bring this solution to the form

$$\Phi^2 = \frac{a - (\mu + 8\pi GE_S)x^+x^-}{x^+ - x^-}. \tag{7.30}$$

We can deduce from this formula the relationship between the mass of the black hole and its temperature. If we start from a pure **AdS**^{2} space we can set $\mu = 0$ and thus we obtain

$$\Phi^2 = \frac{a - 8\pi GE_S x^+x^-}{x^+ - x^-}. \tag{7.31}$$

By comparing (7.31) and (7.27) and using (7.19) we deduce the relationship between the energy and temperature as

$$8\pi GE_S = \mu_S = (\pi T_S)^2. \tag{7.32}$$

Thus, for the eternal **AdS**^{2} black hole we obtain the relationship

$$8\pi G E_0 = \mu_0 = (\pi T_0)^2. \tag{7.33}$$

By considering now the effect of an infalling matter pulse into the black hole (black hole formed by gravitational collapse) we obtain

$$8\pi G E_1 = \mu_1 = (\pi T_1)^2. \tag{7.34}$$

The energy E_1 is simply given by the energy E_0 of the eternal black hole plus the energy E_S of the infalling matter, viz $E_1 = E_0 + E_S$. Thus, the relationship between the the new temperature T_1, the old temperature T_0 and the energy of the pulse E_S is given by

$$(\pi T_1)^2 = (\pi T_0)^2 + 8\pi G E_S. \tag{7.35}$$

7.2.3 Quantum correction and coupling to a heat bath

The next step is to add quantum corrections, due to matter fields, which in the case of a conformal theory are encoded in the conformal anomaly. The Almheiri–Polchinski (AP) model is really characterized by transparent boundary conditions at the boundary as opposed to the reflecting boundary conditions characterizing the usual JT model.

In other words, the (right) boundary of **AdS**^{2} is coupled to a heat bath, at zero temperature, into which Hawking radiation can escape and hence we have a simulated black hole evaporation process with the associated Hawking radiation and the consequent black hole information loss problem.

The matter sector which is independent of the dilaton field (and only interacts with it through the constraints) is treated as a CFT on a fixed **AdS**^{2} background in the coordinates x. The fields are subjected to transparent boundary conditions.

In the external heat bath **M**^{2} we have the same CFT in the coordinates y. The coupling between the **AdS**^{2} space and the heat bath occurs at $t = 0$, i.e. at $x^- = 0$ which results in a shockwave of energy E_S infalling from the boundary into the black hole. We have the following metrics

$$ds^2_{\mathbf{AdS}^2} = -\frac{4dx^+dx^-}{(x^+ - x^-)^2} = -\Omega^{-2}(y)dy^+dy^-, \quad \Omega^{-2}(y) = \frac{4f'(y^+)f'(y^-)}{(f(y^+) - f(y^-))^2} \tag{7.36}$$
$$ds^2_{\mathbf{M}^2} = -dy^+dy^-$$

The boundary conditions of the **AdS**^{2} metric $ds^2_{\mathbf{AdS}^2} = (-dt^2 + dz^2)/z^2$ and the dilaton field $\Phi^2 = 1 + (1 - \mu(t^2 - z^2))/2z$, i.e. their values at the boundary $z = \epsilon$, are given by (where u is the time variable on the boundary)

$$g_{tt}\,|_{\text{boundary}} = \frac{1}{\epsilon^2} = \frac{-t'^2 + z'^2}{z^2}, \quad \Phi^2\,|_{\text{boundary}} = \frac{1}{2\epsilon}. \tag{7.37}$$

The diffeomorphism $x = f(y)$ is chosen such that the boundary is simple at constant value in the y coordinates, viz

$$\epsilon = \frac{y^+ - y^-}{2}. \tag{7.38}$$

Of course, before the coupling between the $\mathbf{AdS}^2$ and the heat bath $\mathbf{M}^2$ is turned on at $t = 0$ this diffeomorphsim is given by (7.103) which corresponds to the static form of the eternal black hole solution.

Approaching the future/past horizon $y^\pm \longrightarrow \pm \infty$ on the boundary $y^+ - y^- = 0$ means that $u = T\,|_{\text{boundary}} = \frac{y^+ + y^-}{2}\,|_{\text{boundary}} \longrightarrow \pm \infty$, which in the x coordinates is equivalent to spending the times $\pm t_\infty = \lim_{u \to \pm \infty} f(u)$.

Before the coupling between $\mathbf{AdS}^2$ and the heat bath is switched on these times $\pm t_\infty$ are precisely the future/past horizon times $\pm t_\infty = f(\pm \infty) = x^\pm = \pm 1/\sqrt{\mu_0} = \pm 1/\pi T_0$.

After the coupling the temperature changes to T_1 and in order for the wormhole to remain not traversable the new event horizon is required to lie outside the original horizon and hence it can be reached in less time, i.e. t_∞ after the coupling must satisfy $t_\infty < 1/\pi T_0$. The idea is that the so-called Averaged Null Energy Condition (ANEC) on the horizon must always be satisfied in order to maintain boundary causality [32, 33].

Let us now introduce the Euclidean time $\tau = it$ and the Euclidean (complex) coordinates x and $\bar{x}$ by

$$x^+ = \bar{x} = t + z, \quad x^- = -x = t - z. \tag{7.39}$$

Thus, $t = (x^+ + x^-)/2 = (\bar{x} - x)/2$ and $z = (x^+ - x^-)/2 = (\bar{x} + x)/2$. The $\mathbf{AdS}^2$ boundary is at $z = 0$, whereas the bulk is $z > 0$. The initial state at $t = 0$ is therefore the Hartle–Hawking state on $\mathbf{AdS}^2$ which is given by the vacuum on the half-line $z > 0$. The heat bath is another half-line $z < 0$ with the same CFT prepared in the same vacuum state. The physical time (time on the boundary and in the heat bath) is u and not t (which is the bulk time). They are related by a diffeomorphism $t = f(u)$ (we choose $0 = f(0)$).

We must therefore go from the coordinates x (defined on $\mathbf{AdS}^2$) to the coordinates y (defined on the heat bath) by the diffeomorphism f, viz $x = f(y)$ and $\bar{x} = f(\bar{y})$. The boundary is simple in the y coordinates located at $y + \bar{y} = 0$ while the physical time is $T = (\bar{y} - y)/2$ (we choose $f(y) = -f(-y)$).

At the initial time $t = u = 0$ we have $y = \bar{y} = f^{-1}(z)$ and $\mathbf{AdS}^2$ corresponds to the right half-line $y > 0$, whereas the heat bath corresponds to the left half-line $y < 0$. In general, $\mathbf{AdS}^2$ corresponds to the right half-plane $y + \bar{y} > 0$, whereas the heat bath corresponds to the left half-plane $y + \bar{y} < 0$.

The initial quantum state of the heat bath is therefore given by the Euclidean path integral on the left lower half-plane (the half-line vacuum). On the other hand, the initial quantum state of $\mathbf{AdS}^2$ is given by a Euclidean path integral on the right lower half-plane with a deformed boundary (a Virasoro descendant of the usual CFT vacuum on the half-line). In conclusion, we have in the y coordinates a combined coupled system evolving in the physical time by the usual Hamiltonian of conformal field theory on the line.

The initial state is then time-reflection symmetric given by a Euclidean path integral over a simply connected space with a single boundary. Therefore, it is a descendent of the half-line vacuum. The Cauchy surface at $t = 0$ can thus be mapped by a means of an appropriate diffeomorphism to a half-line parameterized by a coordinate $w \in [0, \infty]$, i.e. we can map our initial quantum state to the half-line vacuum.

The goal is to derive the diffeomorphism $w = w(x)$ (in the **AdS**^{2} region $x > 0$), the diffeomorphism $w = w(y)$ (in the heat bath **M**^{2} region $y < 0$), the stress–energy–momentum tensor of the conformal matter in both regions and the energy E_S of the initial shockwave due to the turning on of the coupling between **AdS**^{2} and **M**^{2} at time $t = 0$. This will allow us to determine the quantum corrections to the Hawking temperature.

First, at $t = 0$ the stress–energy–momentum tensor of **AdS**^{2} is zero in the physical coordinates y, and also zero in the Poincaré coordinates x since the Weyl anomaly between x and y is zero ($x = x(y)$ is an **SL(2,R)** transformation). Similarly, the heat bath **M**^{2} is at zero tempertaure and thus the corresponding stress–energy–momentum tensor is also zero. We have then

$$\langle T_{xx}(x) \rangle = 0 \ (x > 0), \quad \langle T_{yy}(y) \rangle = 0 \ (y < 0), \quad t = 0. \tag{7.40}$$

At later times we use the transformation law of the stress–energy–momentum tensor under the diffeomorphism w, viz

$$\left(\frac{dw}{dx} \right)^2 \langle T_{ww}(w) \rangle = \langle T_{xx}(x) \rangle - \frac{c}{24\pi} S(w, x). \tag{7.41}$$

$$\left(\frac{dw}{dy} \right)^2 \langle T_{ww}(w) \rangle = \langle T_{yy}(y) \rangle + \frac{c}{24\pi} S(w, y). \tag{7.42}$$

The number c is the central charge of the conformal field theory and S is the so-called Schwarzian which is defined by

$$S(w, x) = \{w, x\} = \frac{w'''(x)}{w'(x)} - \frac{3}{2} \frac{w''^2(x)}{w'^2(x)}$$
$$= \left(\frac{w''}{w'} \right)' - \frac{1}{2} \left(\frac{w''}{w'} \right)^2. \tag{7.43}$$

We take the diffeomorphism w to be an **SL(2,R)** transformation of the relevant coordinates, i.e. a Mobius map which guarantees the vanishing of the stress–energy–momentum tensor in the w coordinates. Hence, we obtain the energy–momentum tensor

$$\langle T_{xx}(x) \rangle = -\frac{c}{24\pi} S(w, x). \tag{7.44}$$

$$\langle T_{yy}(y) \rangle = -\frac{c}{24\pi} S(w, y). \tag{7.45}$$

We go back to the initial time $t = 0$. The diffeomorphism (or conformal transformation) w will map the $\mathbf{AdS}^2$ region $x > 0$ to the interval $[0, w_0]$ while it will map the heat bath $\mathbf{M}^2$ region $y < 0$ to the interval $[w_0, \infty]$ and it is given explicitly by [1]

$$w(x) = \frac{w_0^2}{w_0 + x}, \quad x > 0. \tag{7.46}$$

$$w(y) = w_0 + f^{-1}(-x), \quad x < 0. \tag{7.47}$$

We write this as

$$w(x) = \frac{w_0^2}{w_0 + x}\theta(x) + (w_0 - y)\theta(-x). \tag{7.48}$$

We compute (using $f(0) = 0$)

$$w'(x) = -\frac{w_0^2}{(w_0 + x)^2}\theta(x) - y'(x)\theta(-x). \tag{7.49}$$

Then, we compute (using $y'(x) = 1/f'(y)$ and $f'(0) = 1$ where primes denote derivatives with respect to the appropriate variable)

$$w''(x) = \frac{2w_0^2}{(w_0 + x)^3}\theta(x) - y''(x)\theta(-x). \tag{7.50}$$

Hence, we obtain

$$\frac{w''(x)}{w'(x)} = -\frac{2}{w_0 + x}\theta(x) + \frac{y''(x)}{y'(x)}\theta(-x). \tag{7.51}$$

As a consequence, we have (using $y''/y' = -f''/f'^2$ and hence $\frac{y''}{y'}|_{x=0} = \frac{y''}{y'}|_{y=0}$ $=-f''(0)$)

$$\left(\frac{w''(x)}{w'(x)}\right)' = \frac{2}{(w_0 + x)^2}\theta(x) + \left(\frac{y''(x)}{y'(x)}\right)'\theta(-x) - \frac{2}{w_0}\delta(x) + f''(0)\delta(x). \tag{7.52}$$

We get then the Schwarzian and the energy–momentum tensor which are given explicitly by

$$S(w, x) = \{w, x\} = \{y, x\}\theta(-x) - \left(\frac{2}{w_0} - f''(0)\right)\delta(x). \tag{7.53}$$

$$\langle T_{xx}(x) \rangle = -\frac{c}{24\pi}\{y, x\}\theta(-x) + \frac{c}{24\pi}\left(\frac{2}{w_0} - f''(0)\right)\delta(x). \tag{7.54}$$

Thus, we can make the identification:

$$E_S = \frac{c}{24\pi}\left(\frac{2}{w_0} - f''(0)\right).$$ (7.55)

However, $\mathbf{AdS}^2$ is dual to a conformal quantum mechanics at the boundary and thus it is more natural to map the $\mathbf{AdS}^2$ region to a single point. In other words, we must take the limit $w_0 \longrightarrow 0$ and hence $E_S \longrightarrow \infty$. As it turns out, this is indeed the physically sensible limit in order to avoid acausal correlations [1]. The diffeomorphism w becomes a mapping to the upper-half-plane given by

$$w(x) = \left(\frac{12\pi E_S}{c}\right)^{-1}\frac{1}{x}\theta(x) + f^{-1}(-x)\theta(-x).$$ (7.56)

We also write the result (7.54) in the form

$$\langle T_{x^-x^-}(x^-)\rangle = -\frac{c}{24\pi}\{y^-, x^-\}\theta(x^-) + E_S\delta(x^-).$$ (7.57)

We use this result in (7.25). We compute $I^+ = 0$ as before but now we have

$$I^- = 8\pi G E_S x^+ x^- - \frac{k}{2}\int_0^{x^-} dt(x^+ - t)(x^- - t)\{u, t\}, \quad t = f(u), \quad k = \frac{c.G}{3}.$$ (7.58)

The dilaton field, including quantum corrections, becomes (compare with equation (7.30) and subsequent equations)

$$\Phi^2 = \frac{1 - (\mu_0 + 8\pi G E_S)x^+ x^- + \frac{k}{2}\int_0^{x^-} dt(x^+ - t)(x^- - t)\{u, t\}}{x^+ - x^-}$$

$$= \frac{1 - (8\pi T_1)^2 x^+ x^- + \frac{k}{2}I(x^+, x^-)}{x^+ - x^-}.$$ (7.59)

7.2.4 More on the boundary theory and the Schwarzian

The Schwarzian plays a crucial role in this problem since the underlying dynamics is one-dimensional on the boundary. Indeed, the space $\mathbf{AdS}^2$ is characterized by a boundary and the action should be enhanced by a boundary term, viz

$$S[g, \Phi, f] \longrightarrow S[g, \Phi, f] = S_{JT}[g, \Phi] + D_{CFT}[g, f] + S_b[g, \Phi]$$

$$= \frac{1}{16\pi G}\int_{\mathcal{M}} d^2x\sqrt{-g}(\Phi^2 R - V(\Phi)) + S_{CFT}[g, f] + S_b[g, \Phi].$$ (7.60)

The boundary term is given by

$$S_b[g, \Phi] = \frac{1}{8\pi G}\int_{\partial\mathcal{M}} du\sqrt{-\gamma}\Phi^2 K.$$ (7.61)

Here, K is the scalar extrinsic curvature. More precisely, $K = g^{\mu\nu}K_{\mu\nu} = \gamma^{\mu\nu}K_{\mu\nu}$ where $\gamma_{\mu\nu}$ is the induced metric at the boundary and the extrinsic curvature tensor $K_{\mu\nu}$ describes how the boundary $\partial\mathcal{M}$ is curved with respect to the manifold $\mathcal{M} = \mathbf{AdS}^2$ in which it is embedded.

Since the equation of motion of the dilaton enforces the $\mathbf{AdS}^2$ geometry, the bulk action is zero and we only need to focus on the boundary term.

Let us consider the Euclidean metric $ds^2 = (dt^2 + dz^2)/z^2$. The boundary conditions become then

$$g_{tt}\,|_{\text{boundary}} = \frac{1}{\epsilon^2} = \frac{t'^2 + z'^2}{z^2}, \quad \Phi^2\,|_{\text{boundary}} = \frac{1}{2\epsilon} \Rightarrow \epsilon = \frac{z}{\sqrt{t'^2 + z'^2}}. \tag{7.62}$$

We can solve this equation explicitly in powers of the cutoff ϵ to find

$$z = \epsilon t' + O(\epsilon^2). \tag{7.63}$$

We compute then

$$z' = \epsilon t'' + O(\epsilon^2), \quad z'' = \epsilon t''' + O(\epsilon^2). \tag{7.64}$$

$$\epsilon' = \frac{z'}{\sqrt{t'^2 + z'^2}} - \frac{z(t't'' + z'z'')}{(t'^2 + z'^2)^{3/2}} = 0 + O(\epsilon^2). \tag{7.65}$$

The primes are derivatives with respect to u which is the time parameter on the boundary. The tangent vector at the boundary is $e^\mu = \partial_u x^\mu = (t', z')$, whereas the normal vector is $n^\mu = \epsilon(z', -t')$. By construction these two vectors are orthogonal, i.e. $n_\mu e^\mu = 0$ and furthermore n^μ is normalized, i.e. $n_\mu n^\mu = 1$.

We compute the scalar curvature by the formula

$$K = \nabla_\mu\, n^\mu = \partial_\mu n^\mu + \Gamma^\mu_{\mu\nu}n^\nu$$

$$= \partial_\mu n^\mu + \frac{1}{2}g^{\mu\beta}\partial_\nu g_{\beta\mu}n^\nu$$

$$= -\frac{t'^2 - z'^2}{t'\sqrt{t'^2 + z'^2}} + \frac{2z}{(t'^2 + z'^2)^{3/2}}(z''t' - t''z') + 2\frac{t'^2 - z'^2}{t'\sqrt{t'^2 + z'^2}} \tag{7.66}$$

$$= -\frac{z}{\epsilon t'} + \frac{2}{(t'^2 + z'^2)^{3/2}}(t'(zz'' + t'^2 + z'^2) - zz't'')$$

$$= 1 + 2\epsilon^2\{t, u\}.$$

So, the extrinsic curvature to leading order in ϵ^2 is equal to the Schwarzian. Note, that in going from the second line to the third line we have replaced in both terms the derivatives ∂_t and ∂_z with ∂_u/t' and ∂_u/z', respectively. The boundary term becomes given by

$$S_b[g, \Phi] = (-1)\left(\frac{1}{2}\right)\frac{1}{8\pi G}\int_{\partial\mathcal{M}} du \frac{1}{\epsilon}\frac{1}{2\epsilon}\cdot 2\epsilon^2\{t, u\}$$

$$= -\frac{1}{16\pi G}\int du\{t, u\}. \tag{7.67}$$

The minus sign is due to the Euclidean signature, whereas the factor 1/2 is due to the fact that the boundary of **AdS**^{2} is constituted of two identical disconnected segments.

Thus, we have spontaneous symmetry breaking of conformal symmetry along the boundary down to the **SL(2,R)** Möbius transformations $t \longrightarrow (at + b)/(ct + d)$ with $ad - cb = 1$. Indeed, the Schwarzian is only invariant under these transformations, viz

$$\left\{ \frac{at + b}{ct + d}, u \right\} = \{t, u\}. \tag{7.68}$$

The field $t = t(u)$ acts as the corresponding pseudo Nambu–Goldstone modes associated with this spontaneous breaking [31].

The ADM energy associated with boundary translations $u \longrightarrow u + \delta u$ is immediately given from the above action by the Schwarzian, viz

$$E(u) = -\frac{1}{16\pi G}\{t, u\}. \tag{7.69}$$

This can be obatined by varying the boundary metric and computing the corresponding stress–energy–momentum tensor [25].

We can check that for $u < 0$, where the diffeomorphism $t = f(u) = \frac{1}{\pi T_0} \tanh(\pi T_0 u)$, this ADM energy $E(u)$ is precisely equal to the energy of the eternal black hole $E_0 = \pi T_0^2/8G$.

7.3 More on JT gravity coupled to conformal matter

7.3.1 The eternal AdS^2 black hole

In this section we follow the presentation of [1]. We consider JT gravity coupled to conformal matter given by the action (ϕ being the dilaton field)

$$S = S_0 + S_G + S_M$$
$$S_0 = \frac{1}{16\pi G_N} \int_{\mathcal{M}} d^2x \sqrt{-g}\, \phi_0(R + 2) + \frac{1}{8\pi G_N} \int_{\partial \mathcal{M}} \phi_0 K$$
$$S_G = \frac{1}{16\pi G_N} \int_{\mathcal{M}} d^2x \sqrt{-g}\, \phi(R + 2) + \frac{1}{8\pi G_N} \int_{\partial \mathcal{M}} \phi \mid_b K \tag{7.70}$$
$$S_M = S_{\text{CFT}}[g].$$

This action describes the AP model [25]. As we will show, this theory describes an eternal **AdS**^{2} black hole.

By varying the action with respect to the dilaton field ϕ we obtain the constraint $R + 2 \equiv 0$, i.e. spacetime has a constant negative scalar curvature. In fact, the corresponding spacetime is locally **AdS**^{2} given by the Poincaré metric

$$ds^2 = g_{\mu\nu}dx^\mu dx^\nu = \frac{1}{z^2}(-dt^2 + dz^2) = -\frac{4}{(x^+ - x^-)^2}dx^+dx^-, \quad x^\pm = t \pm z. \tag{7.71}$$

Furthermore, by varying the action with respect to the metric $g_{\mu\nu}$ we obtain equations of motion coupling the dilaton field to the bulk CFT matter (stress–energy–momentum tensor).

The gravitational sector will be treated semi-classically, i.e. we replace the stress–energy–momentum tensor by its expectation value, viz $T_{ab} = \langle T_{ab} \rangle$. For $\langle T_{ab} \rangle = 0$ the dilaton field is found to be given by

$$\phi = 2\bar{\phi}_r \frac{1 - (\pi T_0)^2 x^+ x^-}{x^+ - x^-}. \tag{7.72}$$

This dilaton field ϕ represents an eternal black hole with two asymptotic boundaries and a Hawking temperature T_0. This black hole can be rewritten into a static form by means of the following conformal transformation

$$x^\pm = f(y^\pm) = \frac{1}{\pi T_0} \tanh \pi T_0 y^\pm. \tag{7.73}$$

The coordinates x cover the whole spacetime, whereas y cover the exterior of the black hole. The physical boundary $z = 0$ of **AdS**^{2} is located in the y coordinates at $y^+ - y^- = 0$. The diffeomorphism $x = f(y)$ is in fact such that the boundary is simple at constant value in the y coordinates, viz

$$\frac{y^+ - y^-}{2} = \epsilon. \tag{7.74}$$

The physical boundary proper time u is different from the bulk Poincaré time t near the boundary (boundary particle formulation of JT gravity [30, 31]). The physical time u corresponds to $(y^+ + y^-)/2$. The physical boundary is defined by the boundary conditions (with $\epsilon \longrightarrow 0$)

$$g_{uu}|_b = \frac{1}{z^2}(-t'^2 + z'^2) = -\frac{1}{\epsilon^2}, \quad \phi|_b = \frac{\bar{\phi}_r}{\epsilon}. \tag{7.75}$$

With these boundary conditions the JT action reduces to a boundary term given by the so-called Schwarzian action, viz

$$S_G = \frac{1}{8\pi G_N} \int_{\partial \mathcal{M}} \phi|_b \, K = \frac{\bar{\phi}_r}{8\pi G_N} \int du \{f(u), u\}. \tag{7.76}$$

The diffeomorphism relating the boundary proper time u to the Poincaré time t is precisely given in terms of the function f which converted the black hole into its static form and made the boundary simple located at constant value in the y coordinates. This function f is also what appears in the Schwarzian action.

The Noether charge under physical time translations $u \longrightarrow u + \delta u$ is precisely the ADM energy of the **AdS**^{2} black hole. This is given explicitly by [30, 31]

$$E(u) = -\frac{\bar{\phi}_r}{8\pi G_N} \{f(u), u\} = \frac{\pi \bar{\phi}_r T_0^2}{4 G_N} \equiv E_0. \tag{7.77}$$

7.3.2 The coupling to conformal matter

This **AdS**^{2} black hole is static and does not radiate and evaporate (the black hole is in thermal equilibrium at temperature T_0). In order for Hawking radiation to escape to infinity we couple the right boundary of this **AdS**^{2} black hole to a heat bath B at zero temperature. This heat bath B is given by an identical copy of the bulk conformal field theory. The coupling will produce a transient effect given by an infalling shock of positive energy E_S. After this initial transient effect, which is due to the coupling between the eternal black hole and the heat bath, Hawking evaporation of the black hole begins. The **AdS**^{2} black hole becomes of energy E_1 and temperature T_1 given by

$$E_1 = E_S + E_0, \quad E_1 = \frac{\pi \bar{\phi}_r T_1^2}{4G_N}. \tag{7.78}$$

The initial transient shock is required to satisfy the ANEC on the horizon in order to maintain boundary causality and prevent the formation of a traversable wormhole. In particular, there should be no interaction between the left and right boundaries and the new event horizon must lie outside the original event horizon. For more detail see [32–34].

The stress–energy–momentum tensor determines the dynamics of the dilaton field. The computation of the stress–energy–momentum tensor is greatly simplified by assuming conformally invariant matter (the heat bath is an identical copy of the bulk CFT) and conformally invariant boundary conditions. In fact, the expectation value of the stress–energy–momentum tensor is completely determined by the conformal anomaly c.

As we have already shown, the ingoing stress–energy–momentum tensor is explicitly given by the expectation value [30]

$$\langle T_{x^-x^-}(x^-) \rangle = E_S \delta(x^-) - \frac{c}{24\pi}\{y^-, x^-\}\theta(x^-). \tag{7.79}$$

The energy of the black hole $E(u)$ can be obtained by varying the action $S_G + S_{CFT}[g]$ with respect to the boundary time u giving the change in energy $\partial_u E(u)$ as the difference between the ingoing and incoming fluxes, viz

$$\partial_u E(u) = f'^2(u)(T_{x^-x^-}(x^-) - T_{x^+x^+}(x^+)). \tag{7.80}$$

Explicitly, the energy of the black hole is found to be given by

$$E(u) = E_0\theta(-u) + E_1\theta(u)e^{-ku}, \quad k = \frac{G_N c}{3\bar{\phi}_r} \ll 1. \tag{7.81}$$

For $u < 0$ the diffeomorphism f is given explicitly by

$$t = f(u) = \frac{1}{\pi T_0}\tanh \pi T_0 u, \quad u < 0 \Rightarrow \{u, t\} = 2(\pi T_0)^2. \tag{7.82}$$

The horizon $z \longrightarrow \infty$ of **AdS**^{2} corresponds in the y coordinates either to the future horizon $y^+ \longrightarrow +\infty$ or to the past horizon $y^- \longrightarrow -\infty$. In the x coordinates this horizon corresponds to

$$x^{\pm} = \pm \frac{1}{\pi T_0}. \tag{7.83}$$

This is precisely the Poincaré time t_{∞} at which the boundary particle reaches the horizon $u \longrightarrow \pm \infty$, viz

$$t_{\infty} = \lim_{u \longrightarrow \pm \infty} f(u) = \pm \frac{1}{\pi T_0}. \tag{7.84}$$

For $u > 0$ this diffeomorphism f is given explicitly by a complicated expression in terms of Bessel functions. This expression is crucial at very late times $u \sim k^{-1} \ln k$ when the black hole is almost extremal and the semi-classical treatment is no longer trusted. The Poincaré time t_{∞} at which the boundary particle reaches the horizon $u \longrightarrow + \infty$ is given now by

$$t_{\infty} = \lim_{u \longrightarrow +\infty} f(u) = \frac{1}{\pi T_1} + \frac{k}{4\pi^2 T_1^2} + O(k^2). \tag{7.85}$$

The causality requirement $t_{\infty} < 1/\pi T_0$ (the new event horizon is required to lie outside the original horizon and hence it can be reached in less time) gives the lower bound $E_S > c T_0 / 24$.

At early times $u \longrightarrow 0$ the black hole starts giving up Hawking radiation through the right boundary. At these times the diffeomorphism f is given by

$$f(u) = \frac{1}{\pi T_1} \tanh \pi T_1 u, \quad u < 0 \Rightarrow \{u, t\} = \frac{2(\pi T_1)^2}{(1 - (\pi T_1 t)^2)^2}. \tag{7.86}$$

At late times $u \sim k^{-1}$ the black hole is still far from being extremal despite the continued Hawking radiation through the right boundary. At these times the diffeomorphism f is given through the equation

$$\{u, t\} = \frac{1}{2(t_{\infty} - t)^2}(1 + O(k^2 e^{-ku})). \tag{7.87}$$

The double pole is shifted to t_{∞} which is crucial for the behavior of the quantum extremal surfaces.

The **AdS**^{2} black hole solution after the shock involves a different dilaton field configuration which is given explicitly by [25]

$$\phi = 2\bar{\phi}_r \frac{1 - (\pi T_1)^2 x^+ x^- + \frac{1}{2} k I(x^+, x^-)}{x^+ - x^-}, \quad I(x^+, x^-) = \int_0^{x^-} dt (x^+ - t)(x^- - t)\{u, t\}. \tag{7.88}$$

7.3.3 The Hartle–Hawking state

Starting in this section, we revert to the notation of [1].

The matter sector is given by a two-dimensional conformal field theory which couples to the dilaton field only through the constraints. Thus, this CFT$_2$ will be treated as a quantum field theory on a fixed **AdS**^{2} background.

The boundary conditions at the boundary of **AdS**^{2} are also chosen to be conformally invariant, e.g. reflecting boundary conditions.

The initial state is prepared by a path integral in Euclidean signature (Wick rotation to imaginary time: $t \longrightarrow \tau = it$). The Euclidean metric can be given by

$$ds^2 = \frac{4dxd\bar{x}}{(x + \bar{x})^2}, \quad -x \equiv x^- = -z - i\tau, \quad \bar{x} \equiv x^+ = z - i\tau. \tag{7.89}$$

The initial state on **AdS**^{2} (after a conformal scale transformation to the metric $dxd\bar{x}$) is therefore given by the vacuum on the half-line $z > 0$ with conformally invariant boundary conditions at $z = 0$. This is the Hartle–Hawking state on the **AdS**^{2} black hole at $t = 0$.

The construction of the initial state of the total system formed by the **AdS**^{2} black hole and the heat bath is much more involved.

First, recall that the heat bath contains an identical copy of the bulk CFT$_2$ with identical boundary conditions. The time parameter in the bath is precisely the physical time u at the boundary, whereas the time parameter in the bulk is the Poincaré time $t = f(u)$. We can use the diffeomorphism f to define the coordinates $x = f(y)$ and $\bar{x} = f(\bar{y})$ in which the metric is given by

$$ds^2 = \Omega^{-2}(y, \bar{y})dyd\bar{y}, \quad \Omega^{-2} = \frac{4f'(y)f'(\bar{y})}{(f(y) + f(\bar{y}))^2}. \tag{7.90}$$

Thus, by a Weyl transformation Ω_y we can go to the flat metric $dyd\bar{y}$.

Recall also that in the y coordinates the **AdS**^{2} boundary is at $(y + \bar{y})/2 = 0$, whereas $(y - \bar{y})/2$ corresponds to the physical time u. The left half-line $y + \bar{y} < 0$ corresponds to the bath while the right half-line $y + \bar{y} > 0$ corresponds to the bulk or **AdS**^{2} black hole.

Obviously, the evaporation process of the black hole corresponds to $u > 0$. We extend f to $u < 0$ by imposing $f(-u) = -f(u)$. Thus, we obtain a time-reversal invariant Hamiltonian of the total system formed by the **AdS**^{2} black hole and the heat bath. This choice makes also the dynamics highly non-trivial, i.e. it makes the Hamiltonian of the total system genuinely coupled.

The initial state on **AdS**^{2} is given by the vacuum on the right half-line $y + \bar{y} > 0$ with conformally invariant boundary conditions at $y + \bar{y} = 0$. In general, the state on **AdS**^{2} is given by the vacuum on the right half-cylinder Re $y > 0$ (since the matter sector, after Wick rotation, is in fact given by a thermal field theory which is periodic with period T^{-1}).

The initial state in the heat bath is given by the vacuum on the left half-line $y + \bar{y} < 0$ with conformally invariant boundary conditions at $y + \bar{y} = 0$. In general, the state in the heat bath is given by the vacuum on the left half-plane Re $y < 0$.

The total Hamiltonian of the coupled system formed by the **AdS**^{2} black hole and the heat bath is then given by the standard CFT$_2$ Hamiltonian on the line with the times identified in such a way that the state in the right half-line is a descendant of the half-line vacuum with some complicated boundary.

This state is then mapped to the half-space vacuum by means of the diffeo-morphism $x \longrightarrow w = \omega(x)$, $\bar{x} \longrightarrow \bar{w} = \bar{\omega}(\bar{x})$, i.e. this state is defined in the half-space $w + \bar{w} > 0$. The metric in the coordinate w, $\bar{w}$ is defined by

$$ds^2 = \Omega^{-2}(w, \bar{w})dwd\bar{w}, \quad \Omega = \frac{x + \bar{x}}{2}\sqrt{w'(x)\bar{w}'(\bar{x})}. \tag{7.91}$$

At $t = 0$, the diffeomorphism or conformal transformation $w(x)$ to the coordinates w is determined by means of conformal invariance to be a Möbius map of $x = f(y)$. The **AdS^2** region $x > 0$ is mapped to $w \in [0, w_0]$ and the bath region $y < 0$ is mapped to $[w_0, \infty]$ under the following transformation to the upper-half-plane:

$$w(x) = \frac{w_0^2}{w_0 + x}, \quad x > 0. \tag{7.92}$$

$$w(x) = w_0 + f^{-1}(-x), \quad x < 0. \tag{7.93}$$

By using the conformal anomaly we can show that the scale w_0 is related to the energy E_S of the shock by the relation

$$E_S = \frac{c}{24\pi}\left(\frac{2}{w_0} - f''(0)\right). \tag{7.94}$$

In order to avoid acausal influences we must take $E_S \longrightarrow \infty$ or $w_0 \longrightarrow 0$. Indeed, **AdS^2** is dual to a conformal quantum mechanics at the boundary and thus it is more natural to map the **AdS^2** region to a single point. In this limit the above conformal transformation becomes

$$w(x) = \left(\frac{c}{12\pi}E_S\right)^2\frac{1}{x}, \quad x > 0. \tag{7.95}$$

$$w(x) = f^{-1}(-x), \quad x < 0. \tag{7.96}$$

In summary, the **AdS^2** black hole is described by the Poincaré coordinates x. We first work in the conformally flat coordinates y (in the right half-line $y > 0$) and then transform back to x at the end of the calculation. The initial state here is a Virasoro descendant of the half-line vacuum (Euclidean path integral with a deformed boundary). In fact, the **AdS^2** state on the half-line $y > 0$ is given by an Euclidean path integral on a half-cylinder (since we are dealing with a thermal field theory with a Hawking temperature T_1, i.e. the Euclidean time is periodic with period T_1^{-1}).

The bath is described by the coordinates y (in the left half-line $y < 0$). The initial state here is the Hartle–Hawking state which is given by the Euclidean path integral on the left half-line $y < 0$.

The state of the combined system is then described by the standard CFT Hamiltonian on a line (with a complicated shape).

The state of the combined system is described, in the coordinates w, by a Euclidean path integral on the half-line. In other words, the state in the coordinate

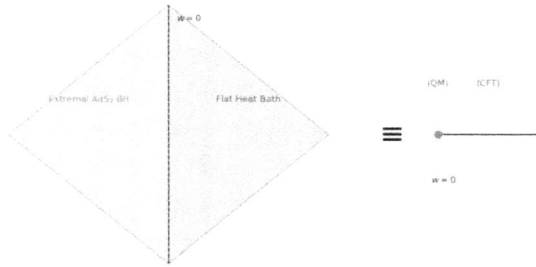

Figure 7.2. The whole **AdS²** theory maps to the point $\sigma = 0$ where the boundary is located and where the dual quantum mechanics lives (holographic correspondence).

w is the half-line vacuum (in the region $w + \bar{w} > 0$). Everything reduces to calculation in the half-space vacuum in the metric $dwd\bar{w}$ followed by a Weyl transformation to the physical metric $\Omega_w^{-2}dwd\bar{w}$.

The diffeomorphism $w = w(x)$ is derived from the fact that the stress–energy tensor vanishes in the half-line vacuum. The stress–energy tensor vanishes also in the coordinates x (**AdS²**) and the coordinates y (heat bath). The stress–energy tensor picks up an anomaly under the diffeomorphisms $w = w(x)$ and $w = w(y)$. This allows us to conclude that w is Möbius map of x and y.

The **AdS²** is mapped to the region $[0, w_0]$ while the heat bath is mapped to the region $[w_0, \infty]$. In order to avoid acausal action we take $w_0 \longrightarrow 0$. Hence, the whole **AdS²** theory maps to the point $w = 0$ where the boundary is located and where the dual quantum mechanics lives (holographic correspondence). See figure 7.2.

7.3.4 An evaporating AdS² black hole

The dilaton configuration including the perturbation ($E_S = E_1 - E_0 \neq 0$ and $T_1 \neq T_0$) and the backreaction ($I \neq 0$) to the future of the shock $x^- > 0$ is given by

$$\phi = 2\phi_r \frac{1 - (\pi T_1)^2 x^+ x^- + \dfrac{k}{2}I(x^+, x^-)}{x^+ - x^-}. \tag{7.97}$$

The backreaction integral I is given in terms of the Schwarzian $\{u, t\}$, where $u = f^{-1}(t)$, by the relation

$$I(x^+, x^-) = \int_0^\infty dt(x^+ - t)(x^- - t)\{u, t\}. \tag{7.98}$$

The expansion parameter k is given in terms of the central charge c and Newton's constant G_N by

$$k = \frac{c}{12}\frac{4G_N}{\bar{\phi}_r}. \tag{7.99}$$

Recall that the diffeomorphism $t = f(u)$ (which relates the boundary proper time u to the Poincaré time $t = (x^+ + x^-)/2$) is given for $u < 0$ (bath) by the simple expression

$$t = f(u) = \frac{1}{\pi T_0} \tanh(\pi T_0 u). \tag{7.100}$$

In the region $u > 0$ this diffeomorphism is more complicated (it is required to solve the condition $\{t, u\} = 2(\pi T_1)^2 \exp(-ku)$). For the unperturbed black hole ($E_S = 0$ and $k = 0$) we get the same reparameterization. But for the perturbed black hole with backreaction the reparameterization $f(u)$ is a complicated expression given in terms of Bessel functions [1].

The new future horizon in the x-coordinates is located at

$$x^+ = t_\infty = \lim_{u \to \infty} f(u). \tag{7.101}$$

This is the required time for a boundary particle to reach the horizon. Indeed, the physical time u, which is given by $u = \frac{1}{2}(y^+ + y^-)|_{\text{boundary}}$, approaches ∞ when we approach the future horizon $y^+ \longrightarrow +\infty$.

Recall also that the boundary in the y-coordinates is located at the surface $\frac{1}{2}(y^+ - y^-) = \epsilon \longrightarrow 0$.

By using the expression of $f(u)$ in the region $u > 0$ and expanding in k we obtain the horizon

$$x^+ = t_\infty = \frac{1}{\pi T_1} + \frac{k}{4(\pi T_1)^2} + \cdots \tag{7.102}$$

In order to maintain causality we must have $t_\infty < 1/\pi T_0$ which is the old horizon.

The relevant Bessel functions found are $K_{0,1}(z)$ and $I_{0,1}(z)$ where the variable z is either $z = 2\pi T_1/k$ or $z = 2\pi T_1 \exp(-ku/2)/k$. This means that the regime of very late times $u \gg 1$ corresponds to $e^{-ku/2}/k \ll 1$ or equivalently $u \gg -\log k/k$. At these times the **AdS**^{2} black hole becomes extremal and the semi-classical description becomes unreliable.

We are more interested in late times $u \gg 1$ corresponding to $u \gg 1/k$ where the **AdS**^{2} black hole has evaporated a sufficient fraction of its mass but remains non-extremal. In this regime of interest the diffeomorphism $t = f(u)$ is given by the Schwarzian equation

$$\log \frac{t_\infty - f(u)}{2t_\infty} = -\frac{4\pi T_1}{k}\left(1 - e^{-\frac{ku}{2}}\right) \Rightarrow \frac{f'(u)}{t_\infty - t} = 2\pi T_1 e^{-\frac{ku}{2}}. \tag{7.103}$$

This expression is actually obtained at small k keeping ku fixed.

At early times the diffeomorphism $t = f(u)$ is simply given by

$$t = f(u) = \frac{1}{\pi T_1} \tanh(\pi T_1 u). \tag{7.104}$$

7.4 Entropy and quantum extremal surface in JT gravity

7.4.1 Entropies, Ryu–Takayanagi formula and the island conjecture

In this section we follow the beautiful review [20].

The most fundamental result concerning classical black holes is the Bekenstein–Hawking entropy formula [9, 10]. This states that the entropy S_{B-H} of a classical black hole is proportional to the area A of the event horizon, viz

$$S_{B-H} = \frac{A}{4\hbar G_N}. \tag{7.105}$$

By including the entropy $S_{outside}$ of matter and gravitons in the outside region of the black hole we obtain the Bekenstein–Hawking total entropy

$$S_{gen} = \frac{A}{4\hbar G_N} + S_{outside}. \tag{7.106}$$

For classical black holes the area always increases in time and hence we obtain the second law of thermodynamics. In fact, we obtain the second law of thermodynamics even if we include the outside entropy, viz

$$\Delta S_{gen} \geqslant 0. \tag{7.107}$$

However, it was shown by Hawking that a quantum black hole evaporates at a temperature T proportional to the surface gravity κ of the black hole [14, 15]. More precisely, we have

$$T = \frac{\hbar \kappa}{2\pi}. \tag{7.108}$$

In other words, surface gravity and the area of the horizon are conjugate variables in general relativity in the same way that temperature and entropy are conjugate variables in thermodynamics.

Thus, an evaporating quantum black hole should be described instead by the quantum von Neumann–Landau entropy defined in terms of the density matrix ρ by the standard formula

$$S_{vN} = -\text{Tr}\rho ln\rho. \tag{7.109}$$

This vanishes for a pure state, i.e. for $\rho = |\psi\rangle\langle\psi|$ we have $S_{vN} = 0$. Thus, S_{vN} measures the degree of mixing of the quantum state, i.e. our ignorance about the quantum state of the system. This entropy is precisely Shannon's entropy of information.

The Bekenstein–Hawking entropy should be thought of as the coarse-grained Gibbs thermodynamic entropy (measures Boltzmann's number of microscopic states of the black hole). In contrast von Neumann entropy is thought of as the fine-grained entropy of the black hole (measures Bell's quantum entanglement characterizing the quantum state of the black hole). The coarse-grained entropy is obtained from the fine-grained entropy by a maximization procedure over all possible choices of density matrices. Thus, we must have

$$S_{vN} \leqslant S_{\text{B--H}}. \tag{7.110}$$

The generalized entropy should also be thought of as a coarse-grained thermodynamic entropy.

In contrast to the coarse-grained Bekenstein–Hawking entropy which can only increase in time, if there is no black hole evaporation, the fine-grained von Neumann entropy can both increase or decrease in time after the start of the evaporation process of the black hole.

An evaporating black hole can thus be viewed, by an outside observer located at infinity, as a unitary quantum system with a total number of degrees of freedom given by the area term $A/4\hbar G_N$ of the Bekenstein–Hawking entropy. This area term is precisely the logarithm of the dimension of the Hilbert space of quantum states of the black hole.

The most important discovery regarding von Neumann entropy in the past 20 years, which was originally motivated by the gauge/gravity duality and quantum entanglement, is the result that the fine-grained von Neumann entropy of quantum systems coupled to gravitational theories can be computed using the so-called Ryu–Takayanagi formula [11, 12], which is a quantum generalization of the Bekenstein–Hawking formula where the horizon is replaced with the so-called quantum extremal surfaces.

In this formulation a codimension-2 surface X of area $A(X)$ is used to compute the generalized entropy given by the formula

$$S_{\text{gen}}(X) = \frac{A(X)}{4\hbar G_N} + S_{\text{semi--cl}}(\Sigma_X). \tag{7.111}$$

Here, Σ_X is the region bounded by a cutoff surface (an arbitrarily chosen surface demarcating the black hole system and separating it from its Hawking radiation) and the codimension-two surface X. The entropy $S_{\text{semi--cl}}(\Sigma_X)$ is then the von Neumann entropy of the quantum fields of matter and gravitons, which are propagating on the classical geometry of the region Σ_X (semi-classical approximation). This fine-grained quantum entropy is therefore computed using the density matrix ρ_{Σ_X} of the region Σ_X, viz

$$S_{\text{semi--cl}}(\Sigma_X) = S_{vN}(\rho_{\Sigma_X}). \tag{7.112}$$

The codimension-2 surface X is a surface, which has two dimensions less than the embedding spacetime and it is chosen in such a way that the generalized entropy is minimized in the spatial direction but maximized in the temporal direction. In other words, this surface is in fact an extremal surface, i.e. the generalized entropy takes an extremal value on this so-called 'quantum extremal surface'. This extremal value of the generalized entropy is precisely the fine-grained von Neumann entropy of the black hole system, viz

$$S_{b-h} = \min_X \left\{ \text{ext}_X [S_{\text{gen}}[X]] \right\}$$
$$= \min_X \left\{ \text{ext}_X \left[\frac{A(X)}{4\hbar G_N} + S_{\text{semi--cl}}(\Sigma_X) \right] \right\}. \tag{7.113}$$

The quantum extremal surface can be shown to lie behind the event horizon. There are two different surfaces which dominate, respectively, the early and late times of the evaporation process. At early times the quantum extremal surface is found to be the vanishing surface, i.e. a trivial surface of zero size. At late times the quantum extremal surface is found to be a non-vanishing surface which lies just behind the event horizon. See figure 7.3.

The so-called 'entanglement wedge' of this fine-grained entropy is the region bounded between the cutoff surface and the quantum extremal surface, i.e. it includes only a portion of the interior of the black hole. The degrees of freedom of the black hole in this region describe the geometry up to the extremal surface.

In more detail, at very early times there is only the vanishing quantum extremal surface. Thus, at these very early times, the entropy of the area term is zero. Furthermore, since no Hawking modes has enough time to escape the black hole region at these very early times, the semi-classical von Neumann entropy of the matter enclosed by the cutoff and the vanishing surfaces is zero because there is no quantum entanglement.

At early times when some Hawking radiation has the chance to escape the black hole region the semi-classical fine-grained von Neumann entropy of the matter modes enclosed by the cutoff and the vanishing surfaces becomes non-zero due to the quantum entanglement between these inner modes and the outer modes of the Hawking radiation. This entropy increases as the black hole evaporates and thus as more Hawking modes escape or more matter modes accumulate. This increasing semi-classical entropy dominates the generalized entropy as the entropy of the area term always vanishes for the vanishing surface. In other words, the generalized entropy of the vanishing quantum extremal surface is increasing in time.

However, at some early time another quantum extremal surface appears. This surface is non-vanishing and time-dependent and lies close to the event horizon. This surface can be found as follows. At time t on the cutoff surface (which determines how much Hawking radiation has escaped) we must go backward an amount of time of the order of the so-called srcambling time $r_s \ln S_{B-H}$ and then shoot a light ray towards the event horizon. It is near the intersection point behind the event horizon where the non-vanishing surface is found. The entropy of the area term is now non-zero given precisely by the entropy of the black hole which is decreasing in time as the area of the black hole is constantly shrinking due to the evaporation process. The semi-classical von Neumann entropy is clearly negligible compared to the Bekenstein–Hawking entropy. In other words, the generalized entropy of the non-vanishing quantum extremal surface is decreasing in time.

In summary, the generalized entropy of the vanishing surface is dominated by the area term and is increasing while the generalized entropy of the non-vanishing surface is dominated by the semi-classical entropy and is decreasing. Thus, the vanishing surface gives the minimum at early times, whereas the non-vanishing surface gives the minimum at late times and as a consequence the generalized entropy goes through a maximum at some time called the Page time. In other words, the generalized entropy follows the so-called Page curve where a phase transition at

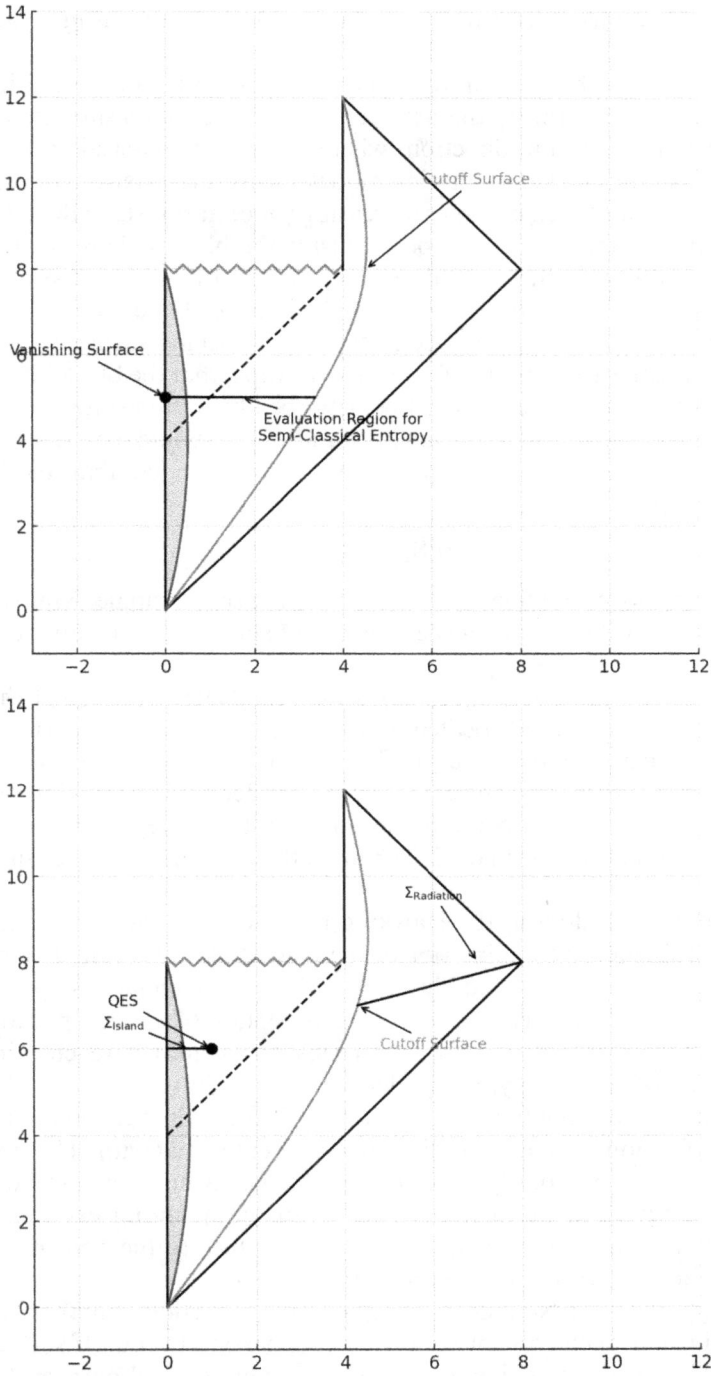

Figure 7.3. The vanishing surface (before Page time) and quantum extremal surface or QES (after Page time).

the Page time occurs between the vanishing and non-vanishing surfaces. See figure 7.1.

This behavior can be seen more explicitly as follows. For a generic surface X (which starts on the horizon) the area term is always decreasing as we move the surface inward along a null direction, whereas the semi-classical entropy starts by decreasing (since the included interior modes purify the exterior modes inside the black hole region) and then becomes increasing (since at this stage the interior modes are entangled with the Hawking modes outside the black hole region). It is in this regime where the area term is decreasing while the semi-classical entropy is increasing and thus their derivatives can be balanced, i.e. the switch or phase transition from the vanishing surface to the non-vanishing surface occurs.

By employing the assumption of unitarity we know that the black hole system and the Hawking radiation system should be described by a pure state. This means that the fine-grained entropy of the Hawking radiation should be equal to the fine-grained entropy of the black hole which must always be less than the Bekenstein–Hawking entropy, viz

$$S_{\rm rad} = {\rm S}_{\rm b-h} \leqslant S_{B-H} \qquad (7.114)$$

As we have discussed, the fine-grained entropy region (entanglement wedge) of the black hole system is bounded between the cutoff surface and the quantum extremal surface X.

Naively, the fine-grained entropy region (entanglement wedge) of the Hawking radiation system should be located beyond the cutoff surface where Hawking radiation has escaped and where gravity can be neglected and spacetime is flat. However, and as it turns out, the entanglement wedge of the Hawking radiation is a disconnected surface which contains, in addition to the region beyond the cutoff surface, the region inside the black hole behind the quantum extremal surface X. See figure 7.4.

The degrees of freedom in the Hawking radiation are obviously entangled with the degrees of freedom in the interior of the black hole. Thus, the fine-grained von Neumann entropy of the emitted Hawking radiation which has escaped beyond the cutoff surface increases steadily in the early stages of the evaporation. As we accumulate more radiation the entropy keeps rising until it reaches in value the Bekenstein–Hawking entropy which defines the maximum number of degrees of freedom contained originally in the black hole, i.e. it defines the maximum number of degrees of freedom which the radiation can be entangled with. The time at which the fine-grained entropy of the radiation ceases increasing and starts decreasing is precisely the Page time and it is the time when the quantum extremal surface X changes or jumps from the trivial or vanishing surface to the non-trivial quantum extremal surface which lies just behind the horizon.

The fine-grained von Neumann entropy of the radiation contains no area term and the semi-classical entropy of the region $\Sigma_{\rm rad}$ beyond the cutoff surface is always increasing. The area can be increased and the semi-classical entropy $S_{\rm semi-cl}(\Sigma_{\rm rad})$ can be decreased if we modify the region where we compute the entropy, i.e. the

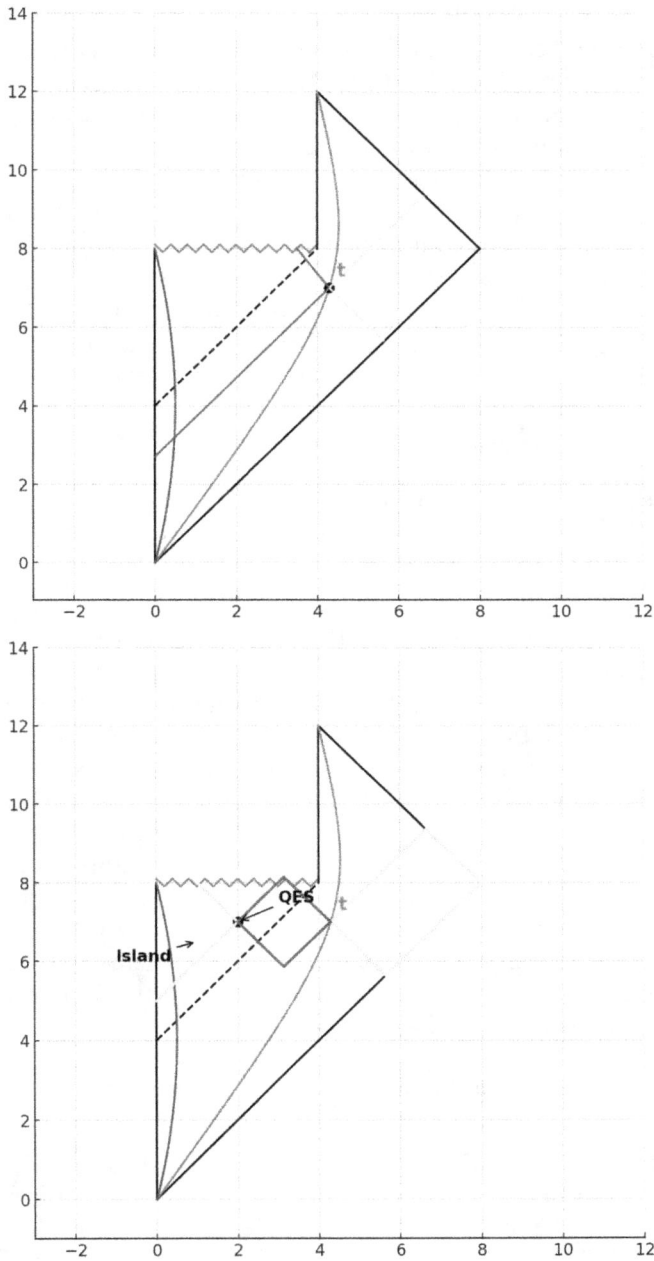

Figure 7.4. Entanglement wedges before (first diagram) and after (second diagram) the Page time. The radiation wedge is drawn in yellow, whereas the black hole wedge is drawn in purple. The island is counted in the radiation wedge after the Page time.

entanglement wedge of the radiation in such a way that it becomes a disconnected region (the area term becomes thus increasing) which contains entangled matter in far away parts (the semi-classical entropy becomes increasing).

The only obvious solution to achieve this outcome is to modify the entanglement wedge of the radiation in such a way that it includes a portion of the interior of the black hole. More precisely, the region Σ_{rad} is replaced with the disconnected region $\Sigma_X' = \Sigma_{\text{rad}} + \Sigma_{\text{island}}$ where Σ_{island} represents the portion of the black hole interior which must be included in the entanglement wedge of the radiation. This region Σ_{island} is also known as an 'island' and it is centered around the origin. It appears a time $r_s \log S_{B-H}$ (scrambling time) after the formation of the black hole. The boundary of the island, i.e. the boundary of the disconnected region Σ_X' is precisely given by the quantum extremal surface X and thus the area term is given as before by $A(X)/4\hbar G_N$.

The fine-grained von Neumann entropy of the radiation is then given explicitly by

$$S_{\text{rad}} = \min_X \left\{ \text{ext}_X \left[S_{\text{gen}}[X] \right] \right\}$$
$$= \min_X \left\{ \text{ext}_X \left[\frac{A(X)}{4\hbar G_N} + S_{\text{semi-cl}}(\Sigma_{\text{rad}} + \Sigma_{\text{island}}) \right] \right\}. \tag{7.115}$$

This is the von Neumann entropy of the exact quantum state of the radiation computed using only the state of the radiation in the semi-classical approximation. This island formula can be derived from the gravitational path integral. Also, it is not difficult to show that the von Neumann entropies of the Hawing radiation and of the black hole system are identically equal. Indeed, we have $S_{\text{rad}} = S_{\text{b-h}}$ since the area term is the same in both cases and by using the properties of quantum entanglement we have $S_{\text{semi-cl}}(\Sigma_{\text{rad}} + \Sigma_{\text{island}}) = S_{\text{semi-cl}}(\Sigma_X)$.

We can check that the von Neumann entropy of the radiation as defined in terms of the island by the above formula will decrease at late times. In fact, this entropy follows exactly the Page curve. This von Neumann entropy is the minimum of the vanishing and non-vanishing islands contributions. The increasing segment for early times corresponds to the no-island contribution, whereas the decreasing segment for late times corresponds to the with-island contribution.

7.4.2 Entanglement entropy of an interval in CFT$_2$ and AdS^2

In this section we follow [1].

The computation of the fine-grained von Neumann entropy in **AdS**^{2} is equivalent to the use of conformal field theory techniques to calculate entropies in curved spacetime settings. The basic technique is the replica trick [35–38]. The von Neumann entropy S is obtained as the limit $n \longrightarrow 1$ of the so-called Renyi entropy S_n, viz

$$S = -\rho \log \rho = S_n, \quad n \longrightarrow 1$$
$$S_n = \frac{1}{1-n} Tr \log \rho^n. \tag{7.116}$$

To find the von Neumann entropy S, we use the replica trick. The idea is to first compute the Renyi entropy S_n for integer n, and then analytically continue the expression to non-integer n, eventually taking the limit as $n \longrightarrow 1$. We can apply L'Hopital's rule to the limit $n \longrightarrow 1$ as follows

$$S = \frac{f(n)}{1 - n} = \frac{f'(n)}{-1} = -\frac{d}{dn} \log Tr\rho^n = -\frac{Tr\rho \log \rho}{Tr\rho} = -Tr\rho \log \rho, \quad Tr\rho = 1. \quad (7.117)$$

The Renyi entropy can also be given by

$$S_n = \frac{1}{1 - n} Tr\rho^n = -\frac{\partial}{\partial n} Tr\rho^n, \quad n \longrightarrow 1. \quad (7.118)$$

The most important thing in the replica trick is the fact that the calculation of the Renyi entropy involves considering n copies (called replicas) of the system and then cyclically gluing them together along the interval whose entropy is being calculated. This procedure creates a new manifold $\tilde{\mathcal{M}}_n$ with a specific structure where the field configurations on the different replicas are connected. The replicated geometry will be discussed further when we discuss the replica wormhole picture.

The Renyi entropy is given in terms of the effective action $I_n = -\ln Z_n$ by the relation

$$S_n = \partial_n\left(\frac{I_n}{n}\right) = (1 - n\frac{\partial}{\partial n})\ln Z_n = -\frac{1}{n - 1} \ln Z_n, \quad n \longrightarrow 1. \quad (7.119)$$

Here, Z_n is the partition function on the replicated n geometry $\tilde{\mathcal{M}}_n$.

The replicated geometry $\tilde{\mathcal{M}}_n$ is symmetric under cyclic permutation. This replica symmetry reduces the replicated geometry $\tilde{\mathcal{M}}_n$ to a single manifold given by the orbifold $\mathcal{M}_n = \tilde{\mathcal{M}}_n/Z_n$. In other words, correlation functions are computed on this orbifold.

We have also to insert twist operators in the replicated geometry at the endpoints $(x_1, \bar{x}_1)$ and $(x_2, \bar{x}_2)$ of the interval of interest. These twist fields are primary operators in the orbifold theory $\mathcal{M}_n = \tilde{\mathcal{M}}_n/Z_n$ with scaling dimensions

$$\Delta_n = \frac{c}{12} \frac{n^2 - 1}{n}. \quad (7.120)$$

The partition function and the entropy of an interval on the half-plane are then given by

$$Z_n = \langle \sigma(x_1, \bar{x}_1)\tilde{\sigma}(x_2, \bar{x}_2)\rangle_{\mathcal{M}_n}, \quad S_n = -\frac{1}{n - 1} \ln\langle \sigma(x_1, \bar{x}_1)\sigma(x_2, \bar{x}_2)\rangle_{\mathcal{M}_n}. \quad (7.121)$$

The replicated geometry is singular at the endpoints. The entropies are renormalized by normalizing the operator product expansion (OPE) in such a way that the identity operator appears with coefficient 1, i.e. $\sigma(x_1, \bar{x}_1)\tilde{\sigma}(x_2, \bar{x}_2) \sim |x_1 - x_2|^{-2\Delta_n}$ when $x_1 - x_2 \longrightarrow 0$. This normalization gives the standard result $S = \frac{c}{3} \log l$ for the entropy of an interval of length l in two-dimensional conformal field theory [39, 40].

As we have shown, the half-plane and $\mathbf{AdS}^2$ are essentially related by a Weyl transformation. The partition function and the entropy of an interval in $\mathbf{AdS}^2$ are then obtained by using the transformation law under Weyl transformations $g \longrightarrow \Omega^{-2}g$ given by

$$\langle \sigma(x_1, \bar{x}_1)\sigma(x_2, \bar{x}_2)\rangle_{\Omega^{-2}g} = \Omega(x_1, \bar{x}_1)^{\Delta_n}\Omega(x_2, \bar{x}_2)^{\Delta_n}\langle \sigma(x_1, \bar{x}_1)\sigma(x_2, \bar{x}_2)\rangle_g. \tag{7.122}$$

$$S_{\Omega^{-2}g} = S_g - \frac{c}{6}\sum_{\text{endpoints}} \log \Omega. \tag{7.123}$$

We are mostly interested in an interval for which both endpoints are in the bulk and they are spacelike separated. The entropy of such an interval on the half-plane is given by [1]

$$S = \frac{c}{6}\log((w_1 - w_2)(\bar{w}_1 - \bar{w}_2)\eta) + \log G(\eta). \tag{7.124}$$

Here, η is the cross-ratio and $G(\eta)$ is defined in terms of the function $G_n(\eta)$ by the relation $\log G(\eta) = -\partial_n \log G_n(\eta)$, $n \longrightarrow 1$. The function $G_n(\eta)$, which depends on the theory and the boundary conditions, is such that $G(1) = 1$. More importantly, the function $G_n(\eta)$ determines the two-point function of twist operators. Explicitly, we have

$$\eta = \frac{(w_1 + \bar{w}_1)(w_2 + \bar{w}_2)}{(w_1 + \bar{w}_2)(w_2 + \bar{w}_1)}. \tag{7.125}$$

$$\langle \sigma(w_1, \bar{w}_1)\sigma(w_2, \bar{w}_2)\rangle = \frac{G_n(\eta)}{(w_1 - w_2)^{\Delta_n}(\bar{w}_1 - \bar{w}_2)^{\Delta_n}\eta^{\Delta_n}}. \tag{7.126}$$

We can now compute the entropy of an interval in $\mathbf{AdS}^2$ by using the transformation law of the entropy under Weyl transformations $g \longrightarrow \Omega^{-2}g$.

We have two cases to consider here. We can have an interval with one endpoint to the future of the shock $x_1^{\pm} > 0$ and the other endpoint to the past of the shock $x_2^+ > 0$, $x_2^- < 0$. Recall that $x = f(y)$. The entropy in this case is given by [1]

$$S = \frac{c}{6}\log\left[\frac{48\pi E_S}{c}\frac{-y_1^- x_1^+ x_2^-(x_2^- - x_1^+)\sqrt{f'(y_1^-)}}{x_2^+(x_1^+ - x_1^-)(x_1^+ - x_2^-)}\right] + \log G(\eta). \tag{7.127}$$

In the second case, we can have an interval where both endpoints are to the future of the shock $x_i^{\pm} > 0$. The entropy in this case is given by [1]

$$S = \frac{c}{6}\log\left[\frac{4(y_1^- - y_2^-)(x_2^+ - x_1^+)\sqrt{f'(y_1^-)f'(y_2^-)}}{(x_1^+ - x_1^-)(x_2^+ - x_2^-)}\right]. \tag{7.128}$$

Finally, we take the limit in which one of the endpoints is located on the cutoff surface (regularized boundary) $z = \epsilon f'(u)$ at the physical time $u = f^{-1}(t)$. We obtain the two equations [1]

$$S = \frac{c}{6} \log\left[\frac{24\pi E_S}{\epsilon c} \frac{-utx^-(x^+ - t)}{x^+(t - x^-)\sqrt{f'(u)}}\right] + \log G\left(\frac{t(x^+ - x^-)}{x^+(t - x^-)}\right), \quad x^- < 0 < t < x^+. \quad (7.129)$$

$$S = \frac{c}{6} \log\left[\frac{2(u - y^-)(x^+ - t)}{\epsilon(x^+ - x^-)} \sqrt{\frac{f'(y^-)}{f'(u)}}\right], \quad 0 < x^- < t < x^+. \quad (7.130)$$

7.4.3 The late time quantum extremal surfaces

First, we are interested in the behavior at late times to the future of the shock $x^+ > t > x^- > 0$. The central quantity is the generalized entropy defined by

$$S_{\text{gen}} = \frac{A}{4G_N} + S_{\text{bulk}} = \frac{\phi + \phi_0}{4G_N} + S_{\text{CFT}}. \quad (7.131)$$

In two dimensions, the area of a point is given by the coefficient of the Ricci scalar in the action.

The bulk term S_{bulk} is the entanglement entropy S_{CFT} of conformal matter in the region between the cutoff surface (near the boundary of $\mathbf{AdS}^2$) and the quantum extremal surface determined by the coordinates (x^+, x^-). This matter entropy S_{CFT} is computed using the formula (7.130).

The area term A is the surface area of the quantum extremal surface which is given by the dilaton field.

The entanglement region corresponding to the correct value of the entropy, i.e. the value which is consistent with unitarity (Page curve) and thus resolving the information loss problem, is called the entanglement wedge of the black hole. This entanglement region is found by extremizing the generalized entropy over (x^+, x^-). In other words, the quantum extremal surfaces are given by the conditions

$$\partial_\pm S_{\text{gen}} = 0 \Rightarrow \partial_\pm \phi = -4G_N \partial_\pm S_{\text{CFT}}. \quad (7.132)$$

In the late time regime the generalized entropy is dominated by the area term yet the derivative of the bulk term plays an essential role in shifting the extremal surface from the location of the trivial vanishing surface which dominates at early times.

- **The variations $\partial_\pm\phi$:** We start by computing the derivatives of the dilaton field $\partial_\pm\phi$.
  - We will assume that the sought after quantum extremal surface is near the horizon $x^+ = t_\infty$.
  - We will start on the boundary $y^+ - y^- = 0$ or equivalently $x^+ - x^- = 0$ where the dilaton field is infinite and then move inward.
  - We expand then the dilaton as follows
    $$\phi\mid_{x^+} = \phi\mid_{x^+ = x^-} + (x^+ - x^-)\partial_+\phi + \cdots \quad (7.133)$$
  - The **variation of the dilaton field** (7.97) **in the x^+ direction** is given by
    $$(x^+ - x^-)^2\partial_+\phi = \bar{\phi}_r\left(-2 + 2(\pi T_i x^-)^2 - k\int_0^{x^-} dt(x^- - t)^2\{u, t\}\right) \equiv F(x^-). \quad (7.134)$$

The right-hand side is a function F of only x^-. In other words, we have

$$\phi \mid_{x^+} = \phi \mid_{x^+ = x^-} + \frac{F(x^-)}{x^+ - x^-} + \cdots \tag{7.135}$$

This shows that the function $F(x^-)$ is always negative for $x^- < t < x^+ < t_\infty$. Clearly, for $x^- = x^+ = t_\infty$ this function must necessarily vanish. We obtain then the integral

$$2(\pi T_1 t_\infty)^2 - 2 = k \int_0^{t_\infty} dt(t_\infty - t)^2 \{u, t\} \equiv k I_\infty. \tag{7.136}$$

– At late times $u \gg 1/k$ the diffeomorphism $t = f(u)$ is given by equation (7.103) from which we can derive the crucial double pole at $t = t_\infty$ in the Schwarzian at late times. Indeed, we compute

$$\{u, t\} = \frac{1}{2(t_\infty - t)^2}. \tag{7.137}$$

– We use this formula to compute the following derivative

$$\partial_- \int_0^{x^-} dt(x^- - t)^2 \{u, t\} = \frac{t_\infty - x^-}{t_\infty} - 1 - \log\left(\frac{t_\infty - x^-}{t_\infty}\right). \tag{7.138}$$

– Then, by integrating both sides of this equation and using the integral (7.136), we get

$$\int_0^{x^-} dt(x^- - t)^2 \{u, t\} = -\frac{(t_\infty - x^-)^2}{2t_\infty} + (t_\infty - x^-)\log\left(\frac{t_\infty - x^-}{t_\infty}\right)$$
$$+ \frac{2}{k}((\pi T_1 t_\infty)^2 - 1). \tag{7.139}$$

– Next, by substituting this result back in the derivative $\partial_+ \phi$ and use $x^+ \simeq t_\infty$ we get the desired result

$$\Rightarrow \partial_+ \phi = \frac{\bar{\phi}_r}{t_\infty - x^-}\left[-2(\pi T_1)^2(t_\infty + x^-) - k \log\frac{t_\infty - x^-}{t_\infty} + \frac{k}{2t_\infty}(t_\infty - x^-)\right]$$
$$= \frac{\bar{\phi}_r}{t_\infty - x^-}\left[-(2\pi T_1)^2 t_\infty - k \log\frac{t_\infty - x^-}{t_\infty} + (2\pi^2 T_1^2 + \frac{k}{2t_\infty})(t_\infty - x^-)\right] \tag{7.140}$$
$$= \frac{\bar{\phi}_r}{t_\infty - x^-}\left[-4\pi T_1 e^{-\frac{k}{2}y^-} - k(1 + \log 2) + (2\pi^2 T_1^2 + \frac{k}{2t_\infty})(t_\infty - x^-)\right].$$

In the last line we have used $t_\infty = 1/(\pi T_1) + k/(2\pi T_1)^2 \ldots$ and also used equation (7.103) in the form

$$\log \frac{t_\infty - x^-}{2t_\infty} = -\frac{4\pi T_1}{k}\left(1 - e^{-\frac{ky^-}{2}}\right). \tag{7.141}$$

- In the above variation we can neglect the second term compared to the first one since k is small with fixed ky^-. And also neglect the third term since $t_\infty - x^-$ is sufficiently small. We get the final variation

$$\Rightarrow \partial_+\phi = -4\pi T_1 \bar{\phi}_r \frac{e^{-\frac{k}{2}y^-}}{t_\infty - x^-}. \tag{7.142}$$

- We consider now the **variation of the dilaton field** (7.97) **in the** x^- **direction** given by

$$(x^+ - x^-)^2\partial_-\phi = \bar{\phi}_r\left(2 - 2(\pi T_1 x^+)^2 + k \int_0^{x^-} dt(x^+ - t)^2\{u, t\}\right). \tag{7.143}$$

- The integral is now computed by expanding in powers of $t_\infty - x^+$ (since $x^+ \simeq t_\infty$) and neglecting quadratic powers in this variable as follows

$$\int_0^{x^-} dt(x^+ - t)^2\{u, t\} - I_\infty = \int_0^{x^-} dt(x^+ - t)^2\{u, t\} - \int_0^{t_\infty} dt(t_\infty - t)^2\{u, t\}$$

$$= \int_0^{x^-} dt[(x^+ - t_\infty)^2 + (t_\infty - t)^2 + 2(x^+ - t_\infty)(t_\infty - t)]\{u, t\}$$

$$- \int_0^{x^-} dt(t_\infty - t)^2\{u, t\} - \int_{x^-}^{t_\infty} dt(t_\infty - t)^2\{u, t\}$$

$$= \frac{1}{2}\frac{(t_\infty - x^+)^2}{t_\infty - x^-} + (t_\infty - x^+)\log\frac{t_\infty - x^-}{t_\infty} - \frac{(t_\infty - x^+)^2}{2t_\infty} \tag{7.144}$$

$$- \frac{1}{2}(t_\infty - x^-)$$

$$= (t_\infty - x^+)\log\frac{t_\infty - x^-}{t_\infty} - \frac{1}{2}(t_\infty - x^-).$$

- The derivative $\partial_-\phi$ becomes

$$\partial_-\phi = \frac{\bar{\phi}_r}{(t_\infty - x^-)^2}\left(4\pi T_1(t_\infty - x^+) + k(t_\infty - x^+)\log\frac{t_\infty - x^-}{t_\infty} - \frac{k}{2}(t_\infty - x^-)\right)$$

$$= \frac{\bar{\phi}_r}{(t_\infty - x^-)^2}\left(4\pi T_1(t_\infty - x^+)e^{-\frac{ky^-}{2}} + k(t_\infty - x^+)\log 2 - \frac{k}{2}(t_\infty - x^-)\right) \tag{7.145}$$

$$= \frac{\bar{\phi}_r}{(t_\infty - x^-)^2}\left(4\pi T_1(t_\infty - x^+)e^{-\frac{ky^-}{2}} - \frac{k}{2}(t_\infty - x^-)\right).$$

In the second line we can neglect the second term compared to the first one since k is small with fixed ky^-. The first term is more important since $t_\infty - x^-$ is sufficiently small.

– From this equation we can determine the apparent horizon to be given by

$$\partial_-\phi = 0 \Rightarrow x^+ = t_\infty - \frac{1}{3}(t_\infty - t). \qquad (7.146)$$

In other words, the new horizon is shifted outside the old horizon $x^+ = t_\infty$.

• **The variations** $\partial_\pm S_{\text{CFT}}$ and $\partial_\pm S_{\text{gen}}$: Now we compute the derivatives of the bulk entropy $\partial_\pm S_{\text{CFT}}$ and the derivatives of the generalized entropy $\partial_\pm S_{\text{gen}}$. Next, we extremize by setting $\partial_\pm S_{\text{gen}} = 0$ or equivalently (7.132) and then using equations (7.142) and (7.145).

– First, the bulk entropy of an interval to the future of the shock between the point $t = f(u)$ on the boundary and the point (x^+, x^-) in the bulk is given explicitly by

$$S_{\text{CFT}} = \frac{c}{6} \log \left[\frac{2(u - y^-)(x^+ - t)}{\epsilon(x^+ - x^-)} \sqrt{\frac{f'(y^-)}{f'(u)}} \right]. \qquad (7.147)$$

– The **derivative in the x^+ direction of the bulk entropy** is dominated by the light-cone singularity at $x^+ - t \longrightarrow 0$. We compute then

$$\partial_+ S_{\text{CFT}} = \frac{c}{6}\partial_+ \log \left[\frac{(x^+ - t)}{(x^+ - x^-)} \right] = \frac{c}{6}\left[\frac{1}{x^+ - t} + \frac{1}{x^+ - x^-} \right]$$

$$\partial_+\phi = -4G_N\partial_+ S_{\text{CFT}} = -2\bar{\phi}_r k \left[\frac{1}{x^+ - t} - \frac{1}{t_\infty - x^-} \right] \qquad (7.148)$$

$$2\pi T_1 \frac{e^{-\frac{k}{2}y^-}}{t_\infty - x^-} + \frac{k}{t_\infty - x^-} = \frac{k}{x^+ - t}.$$

– We can neglect the second term compared to the first one since k is small with fixed ky^-. We get then the result

$$2\pi T_1 \frac{e^{-\frac{k}{2}y^-}}{t_\infty - x^-} = \frac{k}{x^+ - t}. \qquad (7.149)$$

– The **derivative in the x^- direction of the bulk entropy** will involve the second derivative of the diffeomorphism $f(y^-)$. By taking the derivative with respect of x^- twice of equation (7.141) we obtain

$$\frac{1}{t_\infty - x^-} = 2\pi T_1 \frac{1}{f'(y^-)}e^{-\frac{k}{2}y^-}$$

$$\partial_- \log f'(y^-) = -\frac{1}{t_\infty - x^-} - \frac{k}{2}\frac{1}{f'(y)}. \qquad (7.150)$$

- We can now compute the derivative in the x^- direction of the bulk entropy and then solve the extremum condition $\partial_- S_{\text{gen}} = 0$ by using equation (7.145) as follows

$$
\begin{aligned}
\partial_- S_{\text{SFT}} &= \frac{c}{6} \partial_- \log \left[\frac{u - y^-}{x^+ - x^-} \sqrt{f'(y^-)} \right] \\
&= \frac{c}{6} \left[-\frac{1}{f'(y^-)} \frac{1}{u - y^-} + \frac{1}{x^+ - x^-} + \frac{1}{2} \partial_- \log f'(y^-) \right] \\
\partial_- \phi &= -4 G_N \partial_- S_{\text{CFT}} = -2 \bar{\phi}_r k \left[-\frac{1}{f'(y^-)} \frac{1}{u - y^-} + \frac{1}{2} \frac{1}{t_\infty - x^-} - \frac{k}{4} \frac{1}{f'(y^-)} \right] \\
4\pi T_1 \frac{(t_\infty - x^+) e^{-\frac{k}{2} y^-}}{(t_\infty - x^-)^2} &+ \frac{k}{2(t_\infty - x^-)} = \frac{2k}{f'(y^-)} \frac{1}{u - y^-} + \frac{k^2}{2} \frac{1}{f'(y^-)}.
\end{aligned}
$$

(7.151)

- Both terms in the right-hand side are neglected since k is small with fixed ky^-. We get then the result

$$
4\pi T_1 \frac{(t_\infty - x^+) e^{-\frac{k}{2} y^-}}{(t_\infty - x^-)^2} + \frac{k}{2(t_\infty - x^-)} = 0.
\tag{7.152}
$$

- The two conditions (7.149) and (7.152) lead immediately to the location of the quantum extremal surface which is given by

$$
x^+ = t_\infty + \frac{1}{3}(t_\infty - t).
\tag{7.153}
$$

In other words, the quantum extremal surface is located inside the old horizon $x^+ = t_\infty$.

- This surface can be rewritten in terms of the proper time $u = f^{-1}(t)$ and the corresponding coordinate $y^- = f^{-1}(x^-)$ as follows. First, the two condition (7.149) and (7.152) lead to the equation

$$
t_\infty - x^- = \frac{8\pi T_1}{3k}(t_\infty - t) e^{-\frac{ky^-}{2}} \Rightarrow \log \frac{t_\infty - x^-}{t_\infty - t} = \log \left(\frac{8\pi T_1}{3k} e^{-\frac{ky^-}{2}} \right).
\tag{7.154}
$$

But from equations (7.103) and (7.141) we have

$$
\frac{e^{\frac{ky^-}{2}}}{2\pi T_1} \log \frac{t_\infty - x^-}{t_\infty - t} = \frac{2}{k} \left(1 - \exp \left(\frac{k}{2}(y^- - u) \right) \right).
\tag{7.155}
$$

Hence, we obtain

$$
u = y^- + \frac{e^{\frac{ky^-}{2}}}{2\pi T_1} \log \left(\frac{8\pi T_1}{3k} e^{-\frac{ky^-}{2}} \right).
\tag{7.156}
$$

- The generalized entropy of this quantum extremal surface is dominated by the area term. Indeed, the value of the dilaton field at the quantum extremal surface is equal to the thermodynamic entropy at the corresponding coordinate y^-, viz

$$A = \phi_0 + \phi = \phi_0 + \bar\phi_r 2\pi T_1 e^{-\frac{ky^-}{2}} \Rightarrow S_{\text{gen}} - S_0 = \frac{\bar\phi_r}{4G_N} 2\pi T_1 e^{-\frac{ky^-}{2}}. \qquad (7.157)$$

This entropy is a linearly decreasing function of time which starts at the Page time (which is to be defined shortly) from the entropy of the perturbed black hole at temperature T_1 and then goes to zero at infinite time.

- This late time behavior should be contrasted with the early time behavior which, as we will see in the next section, gives a linearly increasing function of time which starts at time zero from the entropy of the unperturbed black hole at temperature T_0 then reaches after the Page time the entropy corresponding to temperature T_1.

- By using (7.156) and (7.157) we obtain the delay time

$$t_{\text{HP}} = u - y^- = \frac{\beta}{2\pi} \log \frac{16}{c} (S_{\text{gen}} - S_0), \quad \beta = \frac{1}{T_1(y^-)} = \frac{1}{T_1 e^{-\frac{ky^-}{2}}}. \qquad (7.158)$$

- This delay is precisely the scrambling time. The quantum extremal surface at the boundary proper time u corresponds actually to the intersection with the horizon of the ingoing null geodesic which was emitted from the boundary at the earlier time $u - t_{\text{HP}} = y^-$. This quantum extremal surface, as we have seen, lies immediately behind the horizon. The scrambling time is also related to the Hayden–Preskill effect as it demarcates the region in spacetime beyond which the system has no access. Thus, information which falls behind the horizon before this time, although it is certainly lost to the system, can still be recovered from the Hawking radiation.

7.4.4 The early time quantum extremal surfaces

- Now, we are interested in the behavior at early times to the past of the shock $x^+ > t > 0 > x^-$.
- The central quantity is still given by the generalized entropy defined by

$$S_{\text{gen}} = \frac{A}{4G_N} + S_{\text{bulk}} = \frac{\phi + \phi_0}{4G_N} + S_{\text{CFT}}. \qquad (7.159)$$

- At early times the area term (dilaton field) corresponds to the original black hole with temperature T_0 with no gravitational backreaction.
- However, the bulk term S_{bulk}, which is the entanglement entropy S_{CFT} of conformal matter in the region between the cutoff surface and the quantum extremal surface determined by the coordinates (x^+, x^-), is now computed using the first formula (7.129).

- This entanglement region is found by extremizing the generalized entropy over (x^+, x^-). In other words, the quantum extremal surfaces are given by the conditions

$$\partial_\pm S_{\text{gen}} = 0 \Rightarrow \partial_\pm \phi = -4G_N \partial_\pm S_{\text{CFT}}. \tag{7.160}$$

- The solution is given by a quantum extremal surface close to the horizon, viz [1]

$$x^\pm \mp \frac{1}{\pi T_0} = \frac{k}{(\pi T_0)^2} f_\pm(\eta), \quad \eta = \frac{2\pi T_0 t}{1 + \pi T_0 t}. \tag{7.161}$$

Here, we will not write down the functions $f_\pm(\eta)$ explicitly.

Thus, the quantum extremal surface is close to the classical bifurcation surface of the original black hole horizon. It moves out towards the boundary in a spacelike, but nearly-null, direction.

- Recall that $t_\infty = \frac{1}{\pi T_1} + O(k)$ and from causality we must have $t_\infty \leqslant \hat{t}_\infty$ where $\hat{t}_\infty = \frac{1}{\pi T_0}$. The Poincaré time $t = f(u)$ is in the range $t \leqslant t_\infty \leqslant \hat{t}_\infty$ where it reaches the value t_∞ when the physical/boundary time u goes to ∞.

- Early times correspond to u of order 1, i.e. ku small and hence equation (7.103) takes the form

$$\frac{1}{2} - \frac{t}{2t_\infty} = \exp(-2\pi T_1 u). \tag{7.162}$$

- The regime of this approximation is defined by the range of η given by

$$1 - \eta = \frac{\hat{t}_\infty - t}{\hat{t}_\infty + t} \geqslant \frac{\hat{t}_\infty - t_\infty}{\hat{t}_\infty + t_\infty} = \frac{T_1 - T_0}{T_1 + T_0} - O(k). \tag{7.163}$$

In fact, since $T_1 - T_0 \gg k$, we have simply

$$1 - \eta \geqslant \frac{T_1 - T_0}{T_1 + T_0}. \tag{7.164}$$

- Alternatively, η must be in the range

$$1 - \eta = \frac{\hat{t}_\infty - t}{\hat{t}_\infty + t} \geqslant \frac{1}{2} - \frac{t}{2\hat{t}_\infty} \geqslant \frac{1}{2} - \frac{t}{2t_\infty} = \exp(-2\pi T_1 u)$$

$$u \geqslant -\frac{1}{2\pi T_1} \log(1 - \eta). \tag{7.165}$$

- For very early times we have $\eta \longrightarrow 0$ and $u \longrightarrow 0$ or equivalently

$$u = -\frac{1}{2\pi T_1} \log(1 - \eta) \leqslant \frac{1}{2\pi T_1} \log \frac{T_1 + T_0}{T_1 - T_0}. \tag{7.166}$$

This corresponds to the regime of transient state of the black hole.

- The regime of the steady state of the black hole at early times corresponds to the regime when η reaches its maximum value 1 when the physical time u is in the range

$$u \geqslant \frac{1}{2\pi T_1} \log \frac{T_1 + T_0}{T_1 - T_0} \simeq \frac{\beta_1}{2\pi} \log \frac{4E_1}{E_s} = t_{\text{HP}}. \qquad (7.167)$$

In this equation, we have used $\pi^2 T^2 = \frac{4\pi G_N}{\bar{\phi}_r} E$ and $(\pi T_1)^2 = (\pi T_0)^2 + \frac{4\pi G_N}{\bar{\phi}_r} E_S$. We have also made the approximation $T_1 + T_0 \simeq 2T_1$.

The time t_{HP} is the scrambling time which marks here the conclusion of the transient phase and the start of the Page curve behavior of the 'young' black hole.

- Indeed, the corresponding entropy to the quantum external surface (7.161) is computed from (7.129) and is given by

$$S_{\text{gen}} = \cdots + \frac{c}{6} \log(u(1 - \eta)\sinh(\pi T_1 u)) + \cdots \qquad (7.168)$$

The contribution from the area term is negligible at early times.

- Thus, at very early times we have a logarithmic increase $\log u$ due to the switching on of the coupling with the bath, i.e. to the addition of the entropy of the shock to the equilibrium entropy. Then, this is followed by a linear decrease $\log(1 - \eta) = -2\pi T_1 u$ of the entropy due to the initial Hawking modes which did not have enough time to escape the black hole region and reach the bath. This process continues until η reaches its maximum value 1 at the scrambling time t_{HP} at which point the entropy reaches a minimum and the transient phase concludes.

- After this transient phase, the entropy starts to increase linearly as $\frac{c}{6} \log \sinh \pi T_1 u \sim \frac{c \pi T_1}{6} u$ due to the escape of Hawking modes into the bath. This is the true early time behavior of the entropy for a 'young' black hole (although this is 'young' in the sense that we have a recent coupling of an 'old' black hole to the bath and not to a recent formation of a black hole from gravitational collapse). This steady increase of the entropy due to the build up of entanglement of the Hawking radiation is the first leg of the unitary Page curve.

7.5 The 'island' conjecture

In most of this section we follow [19]. First, we start with a digression.

7.5.1 The Randall–Sundrum model: a digression

We present here the Randall–Sundrum (RS) model [41, 42] which is a more successful extension of the Arkani-Hamed–Dimopoulos–Dvali (ADD) model [43, 44]. These modes are inspired by string theory and they aim to solve among other things the hierarchy problem, i.e. why gravity is so weak compared to the other forces.

The most important ingredients of these models are (1) extra dimensions and (2) Dp-branes.

We consider one extra dimension labeled by z given by the orbifold $\mathbf{S}^1/\mathbf{Z}_2$ and NOT by the circle $\mathbf{S}^1$. In other words, we impose the two compactification requirements:

- Periodicity $y \longrightarrow y + 2y_c$, $y_c = \pi r_c$. The range of the extra dimension is $[0, 2\pi r_c]$.
- Orbifold symmetry $y \longrightarrow -y$. In other words, we identify $(x, y) = (x, -y)$ which allows us to take the range $[0, +\pi r_c]$.

Now, we place two four-dimensional 3-branes located at the two fixed points of the orbifold $y = y_c$ (the IR-brane where the standard model and dark matter live) and $y = 0$ (the UV-brane where quantum gravity lives). We will also use $\varphi = y/r_c$. The five-dimensional spacetime (the bulk) is then bounded between these two branes. See figure 7.5.

The action is given by

$$
\begin{aligned}
S &= S_{\text{bulk}} + S_{\text{IR}} + S_{\text{UV}} \\
S_{\text{bulk}} &= \frac{1}{2}M_5^3 \int d^4x \int_{-\pi}^{+\pi} d\varphi \sqrt{-G}(R - 2\Lambda_{\text{bulk}}) \\
S_{\text{IR}} &= \int d^4x \int_{-\pi}^{+\pi} d\varphi \sqrt{-g_{\text{IR}}}(-V_{\text{IR}} + \mathcal{L}_{\text{IR}})\delta(\varphi - \pi) \\
S_{\text{UV}} &= \int d^4x \int_{-\pi}^{+\pi} d\varphi \sqrt{-g_{\text{UV}}}(-V_{\text{UV}} + \mathcal{L}_{\text{UV}})\delta(\varphi).
\end{aligned}
\tag{7.169}
$$

The dark matter fields on the IR-brane are denoted by ϕ and they can only interact gravitationlly with the fields of the standard model. They are both contained in the Lagrangian density $\mathcal{L}_{\text{IR}}$. This is the basic idea.

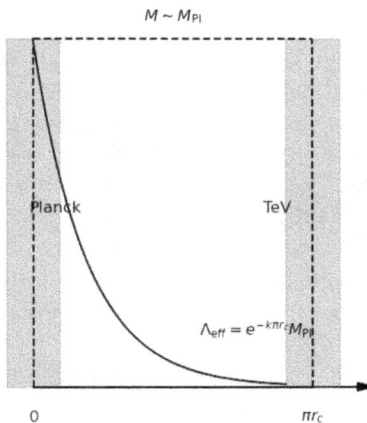

Figure 7.5. The RS model.

The Einstein equations of motion derived from the above action can be solved by a a four-dimensional Poincaré invariant background given by

$$ds^2 = \exp(-2\sigma(\varphi))g_{\mu\nu}dx^\mu dx^\nu + dy^2. \tag{7.170}$$

The warp factor $\exp(-2\sigma)$ is given in terms of the warping parameter σ which is such that

$$\sigma = |y|\sqrt{\frac{-\Lambda_{\text{bulk}}}{6}}. \tag{7.171}$$

The vacuum energies on the branes must be related by

$$V_{\text{UV}} = -V_{\text{IR}} = 6M_5^3 k^2, \quad k^2 = \frac{-\Lambda_{\text{bulk}}}{6}. \tag{7.172}$$

The background solution is a five-dimensional anti-de Sitter spacetime **AdS**^{5} with radius $R = 1/k$.

We substitute the solution in the action S_{bulk} and by assuming that the fields do not depend on the extra dimension y we obtain the effective action

$$S_{\text{bulk}} = \int d^4x \int_{-y_c}^{+y_c} dy \frac{1}{2} M_5^3 e^{-4k|y|} \sqrt{-g^{(4)}} e^{2k|y|} R^{(4)} = \frac{M_5^3}{2k}(1 - e^{-2k|y_c|}) \int d^4x \sqrt{-g^{(4)}} R^{(4)}. \tag{7.173}$$

Thus, the four-dimensional Planck mass is given by

$$M_{\text{Pl}}^2 = \frac{M_5^3}{k}(1 - e^{-2k|y_c|}). \tag{7.174}$$

The same considerations on the IR-brane gives the result that the effective gravitational coupling is suppressed by the warp factor, viz

$$\Lambda_{\text{eff}} = \frac{1}{\sqrt{G}} = M_{\text{Pl}}e^{-k|y_c|} = M_{\text{Pl}}e^{-\pi kr_c}. \tag{7.175}$$

In fact all VEVs (vacuum expectation values) are suppressed, or more precisely redshifted, on the IR-brane. This is the core idea behind brane world scenarios or brane cosmology.

The Planck mass M_{Pl} is obviously at the Planck scale which is of order 10^{19} GeV. The hierarchy problem can then be solved by choosing M_5, $k = -\Lambda_{\text{bulk}}/6$ and y_c appropriately. But in this case (as opposed to the ADD model) the size of the extra dimensions has no real impact on the ratio M_{Pl}/M_5.

However, we can choose $\Lambda_{\text{eff}} \ll M_{\text{Pl}}$ even for moderate choices of kr_c. For example, for $kr_c \sim 10$ the RS model can solve the hierarchy problem, i.e. Λ_{eff} is at the TeV scale with the choice $M_5 \sim M_{\text{Pl}} \sim k$.

These models provide also the possibility that dark matter can be explained by the massive Kaluza–Klein gravitons obtained by compactification of the extra dimensions. For example, see [45–47].

The gravitational content of the Randall–Sundrum model is obtained by expanding the five-dimensional metric as follows

$$G_{MN} \longrightarrow G_{MN} + \kappa h_{MN}, \quad \kappa = 2/M_5^{3/2}. \tag{7.176}$$

The field content of the theory is given by

- (1) A spin two-tensor field (graviton).
- (2) A spin one-vector field (can be made to vanish).
- (3) A spin zero scalar field (radion).

These fields are five-dimensional fields. The graviton and the radion can also be made to decouple. Explicitly, we write

$$G_{\mu\nu} = e^{-2k|y|-2\hat{u}}(\eta_{\mu\nu} + \kappa\hat{h}_{\mu\nu}), \quad G_{55} = -(1 + 2\hat{u})^2, \quad G_{\mu 5} = G_{5\mu} = 0. \tag{7.177}$$

We will not discuss the radion field which measures the width of the extra dimension.

By integrating the extra dimension we get four-dimensional fields. This is the Kaluza–Klein reduction which is given for the metric by the field expansion

$$\hat{h}_{\mu\nu}(x, y) = \sum_{n=0}^{\infty} \frac{1}{\sqrt{r_c}} h_{\mu\nu}^{(n)}(x)\psi_n(\varphi). \tag{7.178}$$

The massless single 5D-graviton is transformed into a tower of massive 4D-gravitons (Kaluza–Klein modes). These are the particles of the dark matter. The mode $n = 0$ is precisely the massless graviton of the theory of general relativity. The wave functions ψ_n are determined in terms of the mass m_n of the n-th graviton by the equation

$$\frac{d}{d\varphi}\left(e^{-4kr_c|\varphi|}\frac{d\psi_n}{d\varphi}\right) = -r_c^2 m_n^2 e^{-2kr_c|\varphi|}\psi_n. \tag{7.179}$$

The equation of motion of the nth massive graviton is precisely the Pauli–Fierz equation of massive gravity given by

$$(\eta_{\mu\nu}\partial^\mu\partial^\nu + m_n^2)h_{\mu\nu}^{(n)}(x) = 0. \tag{7.180}$$

The masses of the KK-graviton modes are given by

$$m_n = kx_n e^{-\pi kr_c}, \quad J_1(x_n) = 0. \tag{7.181}$$

In other words, the masses m_n (or more precisely x_n) are the roots of the Bessel function J_1.

This shows how dark matter candidates can be KK-gravitons in extra dimensions which is another very interesting idea worth pursuing.

7.5.2 Holographic conformal matter

Again, we consider two-dimensional JT gravity coupled to a matter bulk theory given by a CFT_2 theory. The JT gravity theory is also coupled to another copy of the same CFT_2 representing the heat bath. The fields in the JT theory are the dilaton field ϕ and the metric $g_{ij}^{(2)}$ while the fields in the CFT_2 theory are the matter fields χ and the metric $g_{ij}^{(2)}$. The action is given by

$$S[g_{ij}^{(2)}, \phi, \chi] = S_{JT}[g_{ij}^{(2)}, \phi] + S_{CFT\,2}[g_{ij}^{(2)}, \chi].$$ (7.182)

We write the two-dimensional metric explicitly as

$$ds^2 = -\exp(2\rho(x))dx^- dx^+.$$ (7.183)

As we have seen, we can introduce new coordinates $w = w(x)$ in which the stress–energy–momentum tensor vanishes locally. The stress–energy–momentum tensor in the coordinates x is determined by the conformal anomaly and it is given in terms of the diffeomorphism $w = w(x)$ by the relations

$$T_{x^+ x^+} = -\frac{c}{24\pi}\{w^+, x^+\}, \quad T_{x^- x^-} = -\frac{c}{24\pi}\{w^-, x^-\}.$$ (7.184)

These equations actually determine the diffeomorphism $w = w(x)$. By an additional Weyl transformation we can bring the above metric into the flat space form, viz $ds^2 = -dw^- dw^+$. In fact, the stress–energy–momentum tensor vanishes identically in this flat metric. We have then the vacuum solution on flat space given by

$$ds^2 = -dw^- dw^+, \quad T_{w^+ w^+} = T_{w^- w^-} = 0.$$ (7.185)

Now we will assume that the above bulk CFT_2 theory has a three-dimensional holographic gravity dual. In other words, the CFT_2 theory lives on the boundary of some three-dimensional bulk theory. The three-dimensional metric is given by

$$g_{ij}^{(3)}\big|_{\text{boundary}} = \frac{1}{\epsilon^2}g_{ij}^{(2)}.$$ (7.186)

The three-dimensional bulk geometry is **AdS**^{3}. Indeed, the second copy of the CFT_2, representing the heat bath, will live on the flat fixed boundary of a pure **AdS**^{3}.

However, with respect to the first copy of the CFT_2 representing matter fields, there are two differences with the usual AdS/CFT correspondence. First, the corresponding three-dimensional bulk theory does not have a fixed boundary metric since $g_{ij}^{(2)}$ is a dynamical field. Second, there is another scalar field (the dilaton ϕ) propagating on this two-dimensional boundary in addition to the matter scalar fields χ. The boundary itself in this case is therefore a dynamical space since we are integrating over the metric $g_{ij}^{(2)}$ (and the dilaton field ϕ which determines the metric). This dynamical space is analogous to the Planck brane of the Randall–Sundrum model [41, 42]. Hence, the three-dimensional bulk is locally **AdS**^{3} with a dynamical boundary where the two-dimensional theory with action (7.182) lives. The bulk metric near the Planck brane is given explicitly by the **AdS**^{3} metric

$$ds^2 = \frac{-dw^+ dw^- + dz_w^2}{z_w^2}.$$ (7.187)

The Planck brane is located at

$$z_w = \epsilon e^{-\rho(x)}\sqrt{\frac{dw^+}{dx^+}\frac{dw^-}{dx^-}}.$$ (7.188)

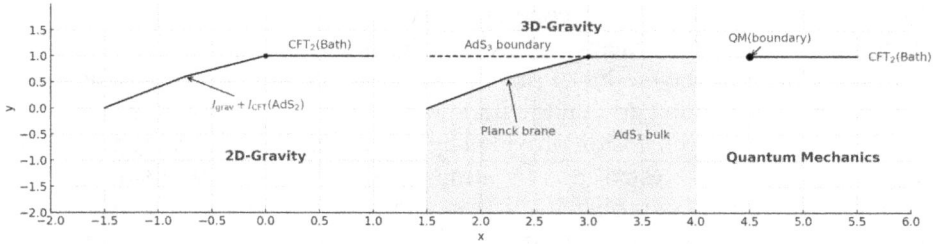

Figure 7.6. Three equivalent description of **AdS**^{2} black hole.

The evaporation of a two-dimensional **AdS**^{2} black hole (into a heat bath) can then be described by three different but equivalent systems (where $\sigma_y = (y^+ - y^-)/2$):

- **2D-gravity:** A two-dimensional JT/CFT model of gravity–matter interaction living in $\sigma_y < 0$ coupled to a two-dimensional CFT living in $\sigma_y > 0$.
- **3D-gravity:** A three-dimensional theory of gravity in **AdS**^{3} spacetime with a dynamical boundary (Planck brane) in the region $\sigma_y < 0$ and a fixed boundary in the region $\sigma_y > 0$.
- **Quantum mechanics:** The dAFF/Yang–Mills conformal/matrix quantum mechanics (the holographic dual of **AdS**^{2} or noncommutative **AdS**$^2_\theta$ space) living at the point $\sigma_y = 0$ coupled to a two-dimensional conformal field theory living in $\sigma_y > 0$.

See figure 7.6.

7.5.3 The island and entanglement wedges at late times

We have already shown how to obtain the correct Page curve of an evaporating black hole by computing the quantum extremal surfaces which extremize the generalized entropy of the black hole by following the prescription of [13].

By assuming that the state of the black hole and the Hawking radiation is pure we can immediately conclude that the Hawking radiation is also characterized by the correct Page curve. However, this does not solve the information loss paradox.

A real solution of the information loss paradox amounts to the direct calculation of the Page curve of the Hawking radiation by computing its quantum extremal surfaces and showing that they coincide with the quantum extremal surfaces of the black hole. This direct calculation will involve in a crucial way the so-called 'quantum extremal islands' which are located deep inside the black hole interior behind the event horizon.

The necessity of these islands in the correct calculation of the von Neumann entropy of the radiation has already being discussed but their construction will be made explicit now by means of holography. Indeed, by assuming that the bulk matter is holographic we can escalate the problem from two-dimensional gravity to three-dimensional gravity where the physical meaning of these islands is much more transparent.

In fact, these islands provide a concrete realization of the ER=EPR proposal [48]. In other words, the entanglement between the interior modes of the black hole within the island and the modes of the Hawking radiation is precisely equivalent to the geometric connection between the island and the asymptotic region of the black hole through the extra dimension provided by the higher dimensional gravity theory.

We are therefore interested in computing the quantum extremal surfaces of the Hawking radiation holographically, i.e. in the three-dimensional gravity theory. This calculation is equivalent to leading order to the Ryu–Takayanagi formula for extremizing areas.

Let us start by outlining the holographic calculation, i.e. the calculation from the perspective of the three-dimensional gravity of the generalized entropy in one interval corresponding to the quantum extremal surfaces of the black hole system. First, recall that from the perspective of the two-dimensional gravity, the generalized entropy of an interval I_y (the accessible region) extending from the boundary to the point y in the bulk (which is the boundary of the accessible region in this case) is given by

$$S[\mathbf{Bl-Ho}] \equiv S_{\text{gen}}(y) = \frac{\phi(y)}{4G_N^{(2)}} + S_{\text{2dbulk}}[I_y]. \tag{7.189}$$

The bulk entropy $S_{\text{2dbulk}}[I_y]$ is the von Neumann entanglement or fine-grained entropy of the bulk conformal matter fields χ, dilaton field ϕ and the bulk two-dimensional metric $g_{ij}^{(2)}$. This entropy is dominated by the matter fields χ in the limit of a large number of degrees of freedom of the CFT2.

As we have seen, the generalized entropy $S_{\text{gen}}(y)$ is minimized in the spatial direction but maximized in the temporal direction. In other words, the obtained surface is in fact an extremal surface found by extremizing the generalized entropy over the point y.

The Ryu–Takayanagi formula instructs us to extend the point y to the one-dimensional minimal area surface Σ_y in the bulk which bounds the accessible region in $\mathbf{AdS}^3$. Here, in the context of three-dimensional gravity, Σ_y is an interval. The generalized entropy $S_{\text{gen}}(y)$ is then given by the Ryu–Takayanagi formula as follows

$$S_{\text{gen}}(y) = \frac{\phi(y)}{4G_N^{(2)}} + \frac{\text{Area}(\Sigma_y)}{3G_N^{(3)}}. \tag{7.190}$$

At late times we consider the spatial slice Σ_{late} which is an interval between the point $0 < \sigma_y \equiv \sigma_0 \ll 1$ (in the heat bath) and the point $y^+ \equiv y_e^+$ (at the quantum extremal surface). As we have shown, the quantum extremal surface lies immediately behind the horizon. The quantum extremal surface at time u corresponds to the intersection with the horizon of the future directed light ray which was emitted from the boundary at the earlier time $y^- = u - t_{\text{HP}}$ where t_{HP} is the scrambling time.

The spatial slice Σ_{late} corresponds in the three-dimensional geometry to the bulk region bounded by the minimal surface area Σ_y and Σ_{late}.

The entanglement wedge of the black hole at late times is therefore given in the two-dimensional picture by the causal domain of the spacelike slice connecting the two points (u, σ_0) and (y_e^+, y_e^-) while in the three-dimensional picture it is given by the bulk region bounded by the minimal surface area Σ_y and the spatial slice Σ_{late}. See figure 7.7.

For Hawking radiation, at late times, we consider the disconnected spatial slice Σ'_{late} which corresponds to the union of two disconnected intervals. The first interval, in the heat bath, starts at the point $0 < \sigma_y \equiv \sigma_0 \ll 1$ and goes to infinity and is the complement of the interval considered in the black hole case. The second interval, which corresponds to the island, is found deep inside the black hole beyond the point $y^+ \equiv y_e^+$ (where the quantum extremal surface is located).

Clearly, the region in the three-dimensional bulk corresponding to the radiation is complement to the region corresponding to the black hole. Thus, from the three-dimensional perspective the above two intervals correspond to a simply connected bulk region. Indeed, the minimization of the entanglement entropy according to the Ryu–Takayanagi formula will show that the accessible spatial slice Σ'_{late} of the radiation will connect the exterior region (Hawking radiation) to the interior region (island).

Hence, the entanglement wedge of the radiation contains the interior region that was not covered by the entanglement wedge of the black hole and thus the quantum extremal surface is found to be the same.

We use now the fact that the interval $I_y \equiv [y_e^+, \sigma_0]$ (black hole) is complementary to the interval $I'_y \equiv [0, y_e^+] \cup [\sigma_0, \infty]$ (radiation+island). In other words, the corresponding von Neumann entropies are the same since these two intervals correspond to the entanglement wedges Σ_{late} and Σ'_{late} of the black hole and the radiation + island, respectively.

From the three-dimensional perspective it is much easier to see that the entanglement entropy of the radiation + island is equal to the entanglement entropy of the black hole as they correspond to two regions which are complementary. These two regions (the black hole and the radiation + island regions) are thus found in a pure state.

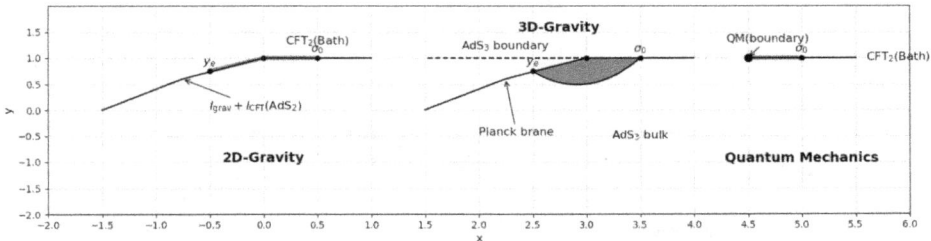

Figure 7.7. Entanglement wedges in the three theories for the black hole.

The entanglement entropy of the radiations is then computed as follows:

$$S_{2\text{dbulk}}[I_y] = S_{2\text{dbulk}}[I'_y] \Rightarrow$$

$$S_{\text{gen}}(y) = \frac{\phi(y)}{4G_N^{(2)}} + S_{2\text{dbulk}}[I_y] = \frac{\phi(y)}{4G_N^{(2)}} + S_{2\text{dbulk}}[I'_y] \Rightarrow$$

$$S[\mathbf{RAD}] \equiv S_{\text{gen}}(y) = \frac{\phi(y)}{4G_N^{(2)}} + S_{2\text{dbulk}}[I'_y] \Rightarrow \tag{7.191}$$

$$S[\mathbf{RAD}] = \frac{\text{Area}(\partial\mathcal{I})}{4G_N^{(2)}} + S[\mathbf{rad} \bigcup \mathcal{I}].$$

In general, the von Neumann entropy of the Hawking radiation involves an island $\mathcal{I}$ behind the horizon and is given by

$$S[\mathbf{RAD}] = \min\left\{\text{ext}\left[\frac{\text{Area}\partial\mathcal{I}}{4G_N} + S[\mathbf{rad} \bigcup \mathcal{I}]\right]\right\}. \tag{7.192}$$

The extremization is done over the choices of islands $\mathcal{I}$ followed by a minimization over all these extrema. We insist, following [19], that **RAD** represents the quantum state of Hawking radiation (which is not known), whereas **rad** represents the state of the radiation in the semi-classical description. This rule gives us then the entropy of the exact quantum state (not the quantum state itself) by using only semi-classical physics.

The entanglement entropy of the radiations at late times is then given explicitly by the result (7.157), i.e. it is dominated by the area term of the quantum extremal surface and hence it is given by the Bekenstein–Hawking entropy S_{B-H} at the corresponding temperature. Indeed, equation (7.157) can be put in the form

$$S[\mathbf{RAD}] = S[\mathbf{Bl-Ho}] = S_{\text{gen}} - S_0 = S_{B-H}(T_l(y^-)) = \frac{A(y^-)}{4G_N} \tag{7.193}$$

$$A(y^-) \equiv \phi(y^-) = \bar{\phi}_r 2\pi T_l(y^-), \quad T_l(y^-) = T_i e^{-\frac{ky^-}{2}}.$$

The scrambling time t_{HP} is given in terms of this entropy $S_{\text{gen}} - S_0 \equiv S[\mathbf{RAD}] = S[\mathbf{Bl-Ho}]$ by the relation

$$t_{\text{HP}} = u - y^- = \frac{\beta}{2\pi} \log \frac{16}{c}(S_{\text{gen}} - S_0), \quad \beta = \frac{1}{T_l(y^-)}. \tag{7.194}$$

See figure 7.8.

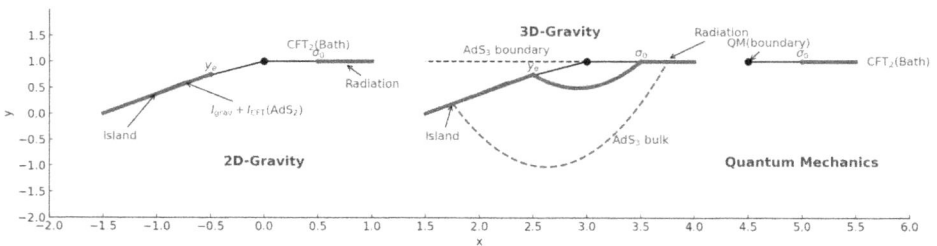

Figure 7.8. Entanglement wedges in the three theories for the radiation + island.

7.5.4 The entanglement wedges at early times and Page curve

- Thus, the entropy of the black hole at late times corresponds to a non-trivial quantum extremal surface and is dominated by the area term. This entropy is found to be a linearly decreasing function of time which starts at the Page time from the Bekenstein–Hawking entropy $S_{B-H} \mid_{T=T_1}$ associated with the perturbed temperature T_1 and then approaches zero at infinite time. We have then

$$S[\mathbf{Bl-Ho}]_{\text{late times}} = S_{\text{gen}} - S_0 = S_{B-H}(T_1(y^-)). \tag{7.195}$$

- As we have seen, at early times, the quantum extremal surface of the black hole is found to be the vanishing surface.
- Thus, the entanglement entropy of the black hole is dominated at early times by the bulk term (computed for the black hole region) and is a linearly increasing function of time starting at zero time from the Bekenstein–Hawking entropy $S_{B-H} \mid_{T=T_0}$ associated with the unperturbed temperature T_0. Explicitly, we have

$$S[\mathbf{Bl-Ho}]_{\text{early times}} = S_{\text{gen}} - S_0 = \frac{\bar{\phi}_r}{4G_N} 2\pi T_0 + k\frac{\bar{\phi}_r}{4G_N} 2\pi T_1 u. \tag{7.196}$$

- The correct Page curve is thus reproduced for the black hole.
- The entanglement entropy of the radiation is controlled by the island which is intimately connected to the quantum extremal surface.
- A non-vanishing island appears in the spacetime after a scrambling time. The bulk entropy involves therefore both the radiation and the island and is always small since the island degrees of freedom purify the Hawking modes in the radiation region. In other words, the generalized entropy of the radiation is controlled by the area term (boundary of the island) and thus starts large from the Bekenstein–Hawking entropy of the black hole and decreases to zero as the area decreases with time.
- However, the non-vanishing quantum extremal surface corresponds to the global minimum of the entropy only at late times.
- Hence, the entanglement entropy of the radiation at late times is dominated by the area term and is a decreasing function of time which starts from the Bekenstein–Hawking entropy $S_{B-H} \mid_{T=T_1}$ associated with the perturbed temperature T_1 and decreases to zero as the area decreases with time. We have then the purity condition

$$S[\mathbf{RAD}]_{\text{late times}} = S[\mathbf{Bl-Ho}]_{\text{late times}} = S_{B-H}(T_1(y^-)). \tag{7.197}$$

- There is no islands at very early times since the quantum extremal surface is vanishing. Hence, the entanglement entropy of the radiation at these very early times is solely given by the bulk term (computed for the radiation region) which is an increasing function of time.

- In fact, if we simply compute the entropy of the radiation without the island we will always find an increasing function of time (representing the continued accumulation in the radiation region of the Hawking modes which are entangled with their interior partners).
- However, the vanishing quantum extremal surface corresponds to the global minimum of the entropy only at early times.
- Hence, the entanglement entropy of the radiation at early times is dominated by the bulk term and is an increasing function of time which starts from 0 and increases to $2S_{B-H}(T_1)$ as time increases to infinity. Indeed, we have

$$
\begin{aligned}
S[\mathbf{RAD}]_{\text{early times}} &= S[\mathbf{Bl-Ho}]_{\text{early times}} - \frac{\bar{\phi}_r}{4G_N} 2\pi T_0 \\
&= S_{\text{gen}} - S_0 - \frac{\bar{\phi}_r}{4G_N} 2\pi T_0 \\
&= k \frac{\bar{\phi}_r}{4G_N} 2\pi T_1 y^- \\
&= 2S_{B-H}(T_1)(1 - e^{-\frac{ky^-}{2}}) \\
&= 2(S_{B-H}(T_1) - S_{B-H}(T_1(y^-))).
\end{aligned}
\tag{7.198}
$$

- Purity requires that we have

$$
S[\mathbf{RAD}]_{\text{early times}} + \frac{\bar{\phi}_r}{4G_N} 2\pi T_0 = S[\mathbf{Bl-Ho}]_{\text{early times}}.
\tag{7.199}
$$

- We reproduce therefore the same Page curve for the Hawking radiation.
- Page time is the time at which the late time behavior equals the early time behavior. At this time a phase transition occurs from the vanishing quantum extremal surface (no island) to the non-vanishing quantum extremal surface (island). The Page time is thus determined as follows:

$$
\begin{aligned}
S_{B-H}(T_1(y^-)) &= 2S_{B-H}(T_1)(1 - e^{-\frac{ky^-}{2}}) + \frac{\bar{\phi}_r}{4G_N} 2\pi T_0 \Rightarrow \\
e^{-\frac{ky^-}{2}} &= 2(1 - e^{-\frac{ky^-}{2}}) + \frac{S_{B-H}(T_0)}{S_{B-H}(T_1)} \Rightarrow \\
y^- &= \frac{2 \ln \frac{3}{2}}{k}, \quad \frac{S_{B-H}(T_0)}{S_{B-H}(T_1)} \ll 1.
\end{aligned}
\tag{7.200}
$$

- The entanglement wedges of the black hole and the radiation before and after the Page time are fundamentally distinguished from each other by the presence of the island in the black hole interior which becomes, at late times after the Page time, causally connected to the Hawking radiation at infinity.

7.5.5 The AdS^2 eternal black hole: the two-intervals entropy of radiation

We consider here an eternal **AdS**^{2} black hole, in JT gravity with matter given by a two-dimensional CFT with central charge c, in thermal equilibrium. This involves two entangled black holes, interacting with two copies of a heat bath given by the same CFT, in a thermofield double state represented by the Hartle–Hawking state.

This system is much simpler than the case of an evaporating **AdS**^{2} black hole since there is no gravitational backreaction.

This system is also dual to two copies of the same conformal quantum mechanics residing at the two boundaries. See figure 7.9.

We work in the coordinates $y^{\pm}$ in which the right black hole has the metric and dilaton profiles given by

$$ds^2 = -\frac{4\pi^2}{\beta^2}\frac{dy^+dy^-}{\sinh^2\frac{\pi}{\beta}(y^+ - y^-)}, \quad \phi = \phi_0 - \frac{2\pi\phi_r}{\beta}\frac{1}{\tanh\frac{\pi}{\beta}(y^+ - y^-)}. \quad (7.201)$$

Here, β is the inverse temperature.

This system provides a new version of the information paradox for a black hole in contact with a heat bath in the Hartle–Hawking state. See [49] and [17, 50]. In this version of the problem, the entropy of the combined system of black hole and bath initially grows but eventually stabilizes (at twice the Bekenstein–Hawking value of the entropy of one of the boundaries) due to the presence of islands. Indeed, the

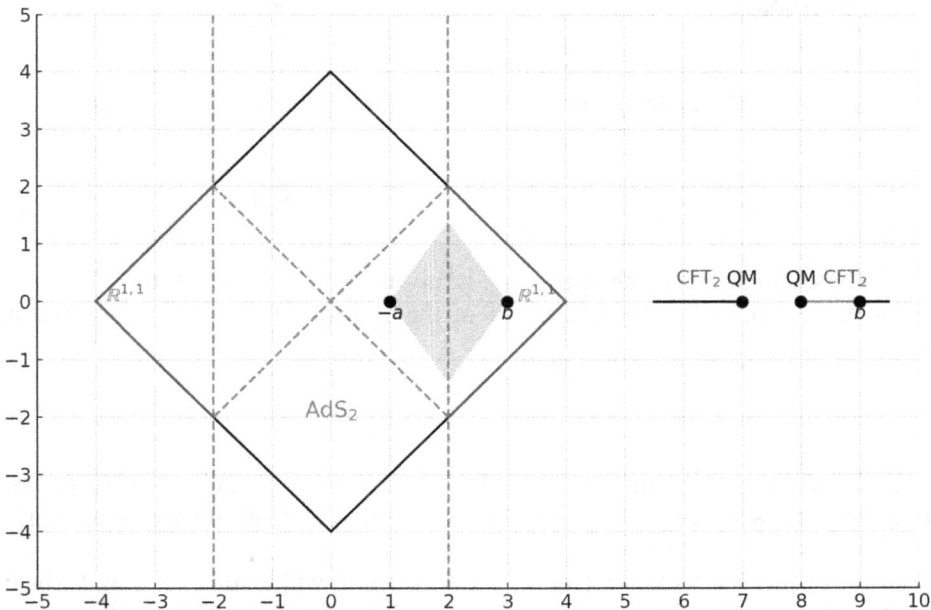

Figure 7.9. A simplified version of the information paradox: An eternal **AdS**^{2} black hole in thermal equilibrium.

energy absorbed by the black hole from the bath is re-emitted by the black hole into the bath without any correlation, i.e. as thermal radiation and thus the von Neumann entropy increases with time indefinitely until it exceeds twice the Bekenstein–Hawking entropy (corresponding to the two boundaries). This non-unitary behavior is impossible which is the paradox and is only cured by the presence of the island which modifies the quantum extremal surface dominating the generalized entropy at late times.

However, in this case the islands can extend outside the black hole horizon which may conflict with causality if not for the so-called quantum focusing conjecture which ensures that the entanglement wedge remains within the causal wedge [51].

The entropy of the black hole is computed in a single interval $[0, b]_R$ in the quantum mechanical description. Clearly, the point b_R is in the right conformal quantum mechanics. This interval corresponds in the gravitational description to the region $[-a, b]_R$. The generalized entropy is a function of a and is given by [17, 50]

$$S_{\text{gen}}(a) = \frac{2\pi\phi_r}{\beta} \frac{1}{\tanh\frac{2\pi a}{\beta}} + \frac{c}{6}\log\frac{\sinh^2\frac{\pi(a+b)}{\beta}}{\sinh\frac{2\pi a}{\beta}}. \qquad (7.202)$$

The location a of the quantum extremal surface is determined from the extremum value, viz

$$\partial_a S_{\text{gen}}(a) = 0 \Rightarrow \frac{\sinh\frac{\pi(a-b)}{\beta}}{\sinh\frac{\pi(a+b)}{\beta}} = \frac{12\pi\phi_r}{c\beta}\frac{1}{\sinh\frac{2\pi a}{\beta}}. \qquad (7.203)$$

This location a of the quantum extremal surface, in the limit $\phi_r/c\beta \gg 1$, is such that we have the entropy

$$S_{\text{BH}} = S_0 + \frac{c}{12}\exp\left(\frac{2\pi}{\beta}(a-b)\right), \quad \frac{\phi_r}{c\beta} \gg 1. \qquad (7.204)$$

The Bekenstein–Hawking entropy S_{BH} is given by the value of the generalized entropy S_{gen} evaluated at the location a of the quantum extremal surface and it is given by

$$S_{\text{BH}} = S_0 + \frac{2\pi\phi_r}{\beta}. \qquad (7.205)$$

Thus, $a - b$ is the scrambling time in this context. Clearly, the value of a corresponds to a point outside the horizon, i.e. the quantum extremal surface (island) is outside the horizon.

The entropy of the radiation (collected on the right) will involve an island which is contained in the entanglement wedge of the region complementary to $[0, b]_R$. This complementary region is the two-intervals given by the union $[-\infty, 0]_L \cup [b, +\infty]_R$.

This two-intervals corresponds, in the gravitational description, to the region $[-\infty, \infty]_L \cup [-\infty, -a]_R \cup [b, +\infty]_R$. In other words, the island corresponds to the right bulk region $[-\infty, -a]_R$.

By construction, the entropy of the complementary region $[-\infty, 0]_L \cup [b, +\infty]_R$ is equal to the entropy of the single interval $[0, b]_R$.

A more realistic two-intervals for the calculation of the entropy of radiation starts with the union $[-\infty, -b]_L \cup [b, +\infty]$ in the quantum mechanical description. This two-intervals corresponds, in the gravitational region, to the region $R = [-\infty, -b]_L \cup [b, +\infty]$ which is also the entanglement wedge in this case. See figure 7.10.

This region $R = [-\infty, -b]_L \cup [b, +\infty]$ lies on the $t = 0$ slice in the bath region. This is the case which will give an information paradox for the eternal **AdS**^{2} black hole. The two intervals $[-\infty, -b]_L$ and $[b, +\infty]_R$ can be made to lie on the same non-zero time slice t by means of the isometries under time translations in the two black holes.

The entropy in this case is the entropy of two intervals in the thermofield double state which is found to be given by [17, 50]

$$S = \frac{c}{3} \log \left(\frac{\pi}{\beta} \cosh \frac{2\pi t}{\beta} \right) = \frac{2\pi c}{3\beta} t + \cdots, \quad t \gg \beta. \qquad (7.206)$$

This is the no-island result which provides a new version of the information loss paradox. See figure 7.11.

However, the complement interval, to the above considered two-intervals, is $[-b, 0]_L \cup [0, b]_R$ with an entanglement wedge given by the region $[-b_L, b_R]$. Thus, this no-island result corresponds really to a single interval $[-b_L, b_R]$.

We need to include the island $I = [a, -\infty] \cup [-\infty, -a]$ in the calculation which will turn the problem into a genuine two-intervals calculation. The single interval and the two-intervals cases are shown in figure 7.11. The value of a is obtained from extremizing the generalized entropy for the two-intervals case given by the island rule

$$S_{\text{gen}} = \frac{\text{Area}(\partial I)}{4G_N} + S_{\text{matter}}(R \cup I). \qquad (7.207)$$

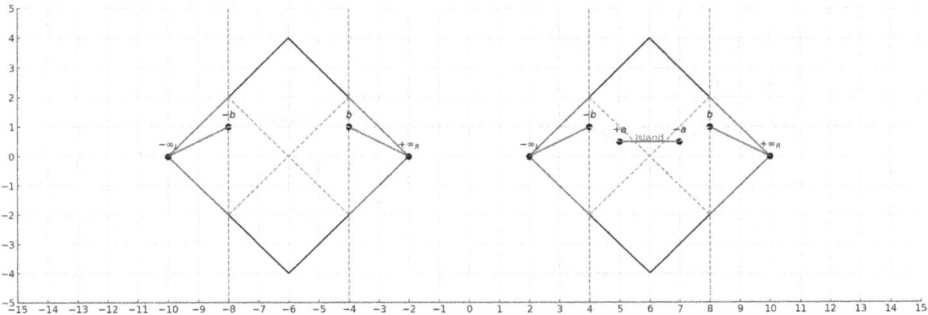

Figure 7.10. The one-interval (no island) and the two-intervals (island) cases for an eternal black hole in thermal equilibrium.

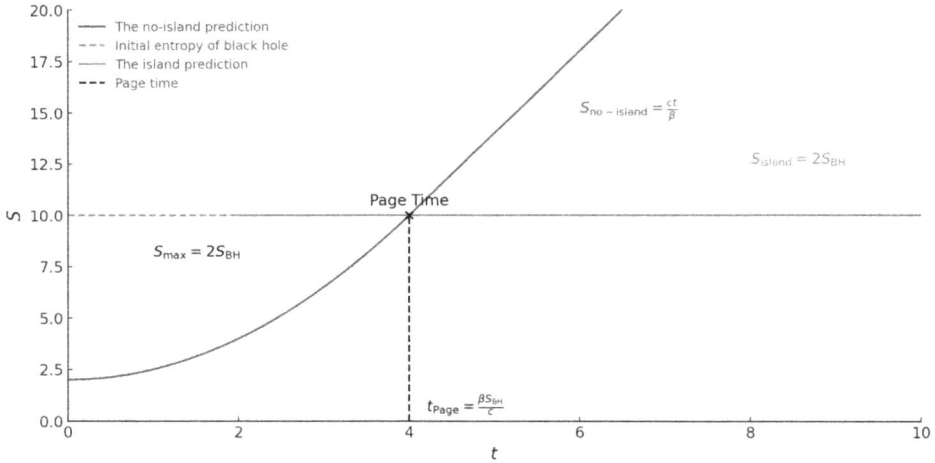

Figure 7.11. The Page curve of an eternal black hole in thermal equilibrium.

Luckily, the extremization of this entropy, which depends on the cross-ratio of the four endpoints $-b_L$, a_L, b_R, $-a_R$ of the two intervals $[-b, a]_L$ and $[-a, b]_R$, is not required since we already know the answer. In fact, it is expected that the quantum extremal surfaces coming from the two-intervals calculation should agree with the single integral calculation. In other words, we expect that the location a of the quantum extremal surface to be given by the same formula found in the single interval case, i.e. by (7.203).

Indeed, in a replicated n geometry (discussed in the next section), we can perform an operator product expansion (OPE) when the two endpoints $-a$ and b are close to each other to find that the generalized entropy is dominated by the identity operator in the OPE.

This gives the result that the generalized entropy for the two-intervals case equals twice the generalized entropy of the single interval case, given by equation (7.202), evaluated at the quantum extremal surface. This means in particular that the location a of the quantum extremal surface is indeed given by the formula (7.203). This means also that at late times $t \gg \beta$, the generalized entropy of the radiation, identified with the generalized entropy of the two-intervals, is constant in time given by twice the Bekenstein–Hawking value S_{BH} in (7.205), viz

$$S_{RAD} = 2S_{BH} = 2(S_0 + \frac{2\pi\phi_r}{\beta}). \tag{7.208}$$

See figure 7.11.

7.5.6 The replica wormhole picture

- The replica trick involves computing the entropy by considering n copies of the system, each in the same quantum state, and then cyclically gluing these copies together along the region of interest, i.e. along the interval whose entropy is being calculated, creating thus a new manifold with a specific structure where the field configurations on the different copies (replicas) are connected.

- In fact, the quantum extremal surface prescription can be derived from a gravitational theory by means of the replica trick [35–38].
- We are interested here in the construction outlined in [17] which shows that the off-shell replicated action in the limit $n \longrightarrow 1$ is precisely the generalized fine-grained von Neumann entropy. See also [18].
- First, we note that the von Neumann entanglement entropy S can be computed in terms of the so-called Renyi entropy S_n by means of the relations

$$S = - Tr\rho \log \rho = S_n, \quad n \longrightarrow 1$$

$$S_n = \frac{1}{1-n} \log(Tr\rho^n) = -\frac{\partial}{\partial n}(Tr\rho^n). \qquad (7.209)$$

Here, ρ is the density matrix of the radiation or the black hole. Thus, we need to compute $Tr\rho^n$ by pasting together n copies (replicas) of the original system, which are glued cyclically along the intervals whose entropy is being computed, and then extend the resulting function to non-integer values of n through analytic continuation [52, 53].

- This can be connected to the action as follows

$$Tr\rho^n = \frac{Z_n}{Z_1^n} \Rightarrow S = (1 - n\frac{\partial}{\partial n})\ln Z_n, \quad n \longrightarrow 1. \qquad (7.210)$$

Clearly, Z_n is the partition function of the n-replicated system.

- The replica trick involves manifolds $\tilde{\mathcal{M}}_n$ with fixed geometry in the non-gravitational region while in the gravitational region any geometry which satisfies the boundary conditions is allowed.
- Thus, the partition function of the n-replicated system is given by $Z_n = Z_n(\tilde{\mathcal{M}}_n)$. This corresponds to an effective action for the geometry which combines gravitational and quantum field contributions. The gravitational part is in fact evaluated at a saddle point resulting in a classical metric with quantum matter contributions.
- By imposing an extra replica symmetry the manifold $\tilde{\mathcal{M}}_n$ reduces to the single manifold $\mathcal{M}_n = \tilde{\mathcal{M}}_n/Z_n$ which has conical singularities and twist fields. Furthermore, if the saddle point respects the replica symmetry, the geometry and fields are periodic under the cyclic permutation of the replicas.
- However, the single manifold $\mathcal{M}_n = \tilde{\mathcal{M}}_n/Z_n$ involves, as we have just mentioned, conical singularities and twist fields.
- The conical singularities are enforced, in the gravitational region, by means of codimension-two cosmic branes with tension $4G_N T_n = 1 - 1/n$. In two dimensions these branes are just points while in four dimensions they are strings. We must also have twist operators, inserted at the positions of these branes, for the n copies of the matter theory. The positions of these branes (which are the fixed points of the Z_n action) are determined using Einstein's equations.

- There are additional twist fields inserted at the endpoints of intervals in the non-gravitational region. These twist fields, which encode the entanglement structure of the matter fields, create branch cuts where the matter quantum fields are required to match across different replicas in order to enforce the cyclic identification of the replica trick.
- The original problem on the replicated manifold $\tilde{\mathcal{M}}_n$ is rewritten as a problem on the single manifold $\mathcal{M}_n$ where n copies of the matter field theory are living.
- The gravitational part of the action is essentially changed by the addition of the cosmic branes action, viz

$$\frac{1}{n} S_{\text{grav}}[\tilde{\mathcal{M}}_n] = S_{\text{grav}}[\mathcal{M}_n] + T_n \int_{\Sigma_{d-2}} \sqrt{g}. \qquad (7.211)$$

- The n copies of the quantum field theory living on $\mathcal{M}_n$ are connected through the branch cuts or twist operators. These cuts enforce boundary conditions such that the field on the i-th replica is connected to the field on the $(i + 1)$-th replica across the cut which means that the matter degrees of freedom on different replicas are not independent but are correlated through these boundary conditions.
- The partition function on the replicated manifold Z_n is not simply the product of the partition functions of n independent copies of the system but includes additional terms that account for the correlations induced by the twist operators. For instance, the partition function can be written as $Z_n = \langle \mathcal{T}_1 \dots \mathcal{T}_k \rangle$, i.e. it is the expectation value of the product of twist operators reflecting the correlations between the replicas.
- The total effective action for the system is a sum of the above gravitational action and the action of the quantum fields on the replicated manifold $\tilde{\mathcal{M}}_n$ which is equal to the action of n copies of the quantum fields on the manifold $\mathcal{M}_n$, viz

$$\begin{aligned} \frac{1}{n} S_n^{\text{tot}} &= \frac{1}{n} S_{\text{grav}}[\tilde{\mathcal{M}}_n] + \frac{1}{n} S_{\text{matt}}[\tilde{\mathcal{M}}_n] \\ &= S_{\text{grav}}[\mathcal{M}_n] + T_n \int_{\Sigma_{d-2}} \sqrt{g} + \frac{1}{n} \sum_{i=1}^{n} S_{\text{matt}}^i[\mathcal{M}_n]. \end{aligned} \qquad (7.212)$$

- For $n \simeq 1$, the action is then expanded perturbatively, starting from the original solution $\mathcal{M}_1 = \tilde{\mathcal{M}}_1$, as follows

$$\frac{S_n^{\text{tot}}}{n} = S_1^{\text{tot}} + \delta\left(\frac{S_n^{\text{tot}}}{n}\right), \quad n \longrightarrow 1. \qquad (7.213)$$

The perturbative correction includes contributions from the tension of the cosmic branes and from the twist fields inserted at the positions of these branes both of which are evaluated on $\mathcal{M}_1$ since these two effects are already of order $n - 1$. We have then

$$\frac{S_n^{\text{tot}}}{n} = S_1^{\text{tot}} + (n - 1)S_{\text{gener}}(w_i). \tag{7.214}$$

$S_{\text{gener}}(w_i)$ is the generalized von Neumann entropy evaluated at the positions of the cosmic branes given by

$$S_{\text{gener}} = \frac{\text{Area}}{4G_N} + S_{\text{matter}}. \tag{7.215}$$

This combines the area term (for holographic theories) and quantum corrections from matter fields. The extremization of the generalized entropy functional (coming from the extremization of the action functional) determines the correct entropy.

- The cosmic branes (conical singularities) and twist fields together ensure that the correct boundary conditions and singularities are imposed on the replicated geometry allowing therefore for the computation of the Renyi entropy S_n and, in the limit $n \longrightarrow \infty$, of the von Neumann entropy. This procedure leads to the quantum extremal surface prescription, showing that the off-shell action evaluated near $n \simeq 1$ yields the generalized entropy, thus connecting the replica trick with holographic entropy calculations.

- In the case when the interior of the black hole is fully connected to the n replicas we get the so-called 'replica wormholes', which leads to the island rule, whereas in the case when the interior is fully disconnected from the replicas we get the usual result of Hawking.

- The 'replica wormholes' are new gravitational saddles of the gravitational path integral that appear when calculating the entropy of Hawking radiation in a gravitational setting [17, 18]. These wormholes connect different replicas leading to non-trivial contributions to the path integral.

- The 'replica wormholes' are the physical principle underlying the island rule which in turn leads to a unitary Page curve. Hence, certain regions (islands) inside the black hole contribute to the entanglement entropy of the radiation which implies that parts of the black hole interior are encoded in the radiation resolving many aspects of the black hole information paradox.

- In [17], the replica trick is applied to the calculation of the entanglement entropy for the eternal $\mathbf{AdS}^2$ black hole. The entanglement entropy was calculated for both the single interval case (black hole) and the two-intervals case (Hawking radiation) where 'replica wormholes' are constructed explicitly and the correct quantum extremal surfaces are obtained.

- In particular, when replica wormholes are included in the two-intervals case, the saddle point analysis shows that islands, which refers to those regions of spacetime that are disconnected from the asymptotic boundary, become relevant, i.e. they contribute to the entanglement entropy of the radiation. These islands appear as regions inside the black hole horizon that are part of the entanglement wedge of the radiation. The generalized entropy in their presence becomes

$$S_{\text{gen}} = \frac{\text{Area}(\partial I)}{4G_N} + S_{\text{matter}}(R \bigcup I). \qquad (7.216)$$

In this equation ∂I is the boundary of the island I and R is the radiation region. Thus, the inclusion of islands means that information about the interior of the black hole can be encoded in the radiation observed at infinity.

- This shows explicitly how to obtain a unitary Page curve from a gravitational path integral by including 'replica wormholes' in the saddle point calculation. This also shows the validity of the notion of entanglement wedge reconstruction in AdS/CFT correspondence which means that the entire bulk geometry, including regions inside the event horizon, can be reconstructed from the boundary state, implying that the information about the interior is accessible from the radiation.

7.5.7 Replica wormhole versus Hawking saddle

- As an explicit example, we consider here a replicated geometry with $n = 2$ which is relevant for the calculation of the purity of the state of the black hole. This schematic, but very illuminating, discussion is taken from [20].
- An evaporating black hole can be viewed as having two future regions, the future region of the outside universe, and the future of the interior where the singularity lies.
- A Euclidean black hole (for example the Schwarzschild metric after Wick rotation) has the geometry of a cigar where the tip is the horizon $r = r_s$ and far away, for $r \gg r_s$, the Euclidean time is a circle of circumference $\beta = 4\pi r_s$ which is the inverse temperature of the partition function calculating the black hole Euclidean path integral. The periodicity in time is required in order to avoid a conical singularity at $r = r_s$. Thus, a Euclidean black hole has no interior, as shown in figure 7.12.
- The initial state of the black hole is a pure state $|\Psi(t_0)\rangle$. The final state $|\Psi(t)\rangle = U(t, t_0)|\Psi(t_0)\rangle$ is obtained by evolving $|\Psi(t_0)\rangle$ using the gravitational path integral. The corresponding density matrix is given by $\rho(t) = |\Psi(t)\rangle\langle\Psi(t)|$ with matrix elements $\rho_{ij}(t) = \langle i|\rho(t)|j\rangle = \langle i|\Psi(t)\rangle\langle\Psi(t)|j\rangle$ computed by means of the same gravitational path integral. See figure 7.13.

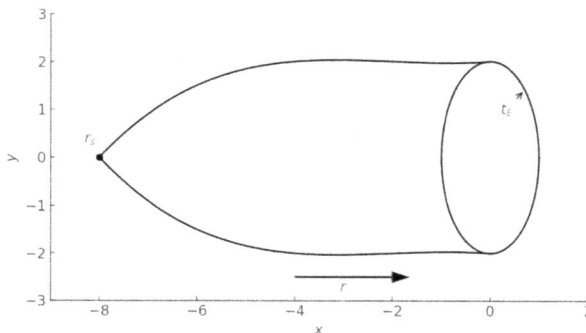

Figure 7.12. Euclidean black hole: The cigar configuration.

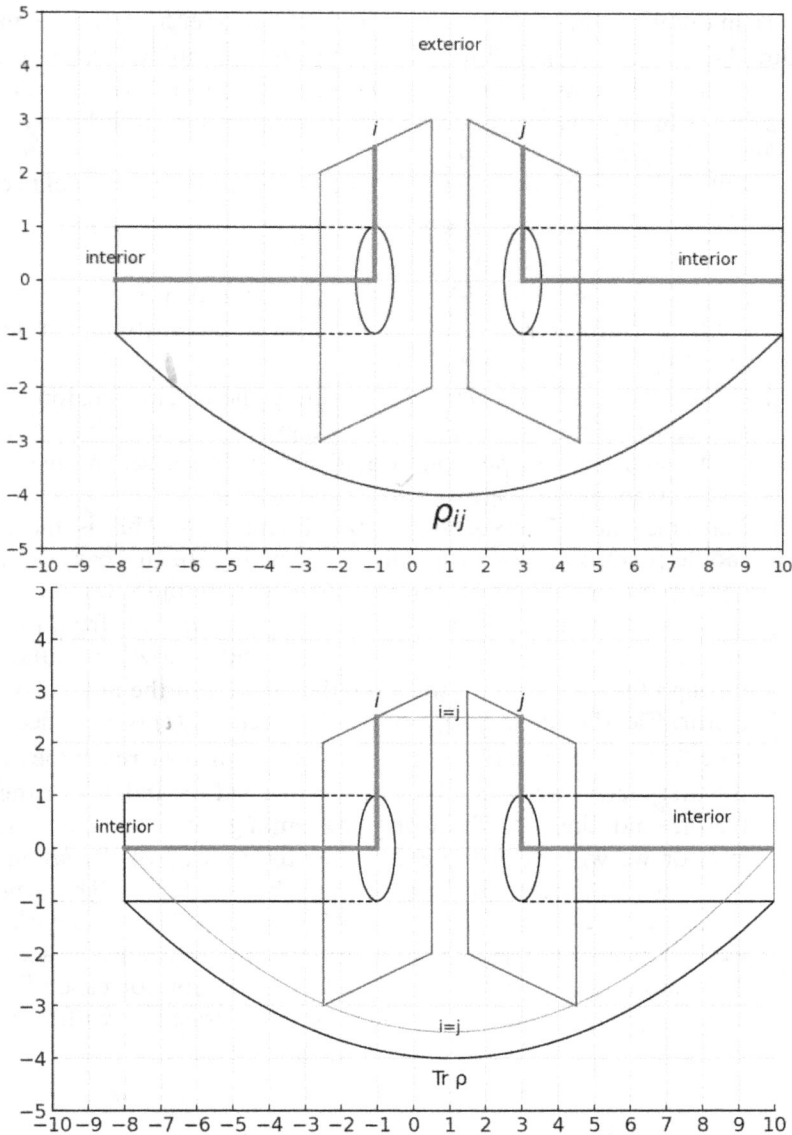

Figure 7.13. The matrix element ρ_{ij} and the trace $Tr\rho$ in the gravitational path integral. The red line represents entanglement.

Similarly, we compute $Tr\rho = \sum_i \rho_{ii}(t)$ by connecting the exterior regions, i.e. by identifying the final and initial states $|i\rangle$ and $|j\rangle$ and then summing over them, which amounts to gluing the interiors together as required by entanglement. See figure 7.13.

Here, $|i\rangle$ are taken to be states of radiation since we are assuming that the black hole has evaporated completely and we are interested in the entanglement entropy of radiation.

- Thus, in order to compute ρ_{ij} we start at the exterior either from the index i or from the index j then follow the entanglement until we reach either the interior corresponding to the ket $|\Psi(t)\rangle$ or the interior corresponding to the bras $\langle\Psi(t)|$. In $Tr\rho$ the two indices i and j are identified and the two interiors are naturally glued together.

- We want really to compute the purity of the black hole state defined by

$$Tr\rho^2(t) = \sum_{i,j}\rho_{ij}(t)\rho_{ji}(t). \tag{7.217}$$

We have two possibilities. Either ρ is a pure state density matrix for which $Tr\rho^2 = (Tr\rho)^2$ (zero entropy) or ρ is a mixed state density matrix for which $Tr\rho^2 \ll (Tr\rho)^2$ (large entropy).

- As before, we compute $Tr\rho^2(t)$ by connecting the exterior regions and then gluing together the interior regions. But in a gravitational path integral we are required to sum over all possible topologies. In this case, the two topologically distinct possibilities are:

    1. **Hawking saddle (mixed state):** See figure 7.14. This is the standard saddle point obtained originally by Hawking. In this case the interiors corresponding to each of the two factors $\rho_{ij}(t)$ and $\rho_{ij}(t)$ are separately glued together, i.e. we have two interiors. We can start from the exterior at the index i of the first factor $\rho_{ij}(t)$ and follow the entanglement through the interior until we reach the exterior at the index j of the first factor. Clearly, the index i/j of the first factor $\rho_{ij}(t)$ is identified with the index i/j of the second factor $\rho_{ji}(t)$. Hence, when we reach the index j of the first factor, we would have really reached the index j of the second factor, and then by following the entanglement through the other interior we will reach the exterior at the index i of the second factor which is identified with the index i of the first factor. This whole path completes one closed loop for which obviously $Tr\rho^2 \neq (Tr\rho)^2$.

    2. **Replica wormhole (pure state):** See figure 7.15. This is actually the dominating saddle point. In this case the interior corresponding to the ket $|\Psi(t)\rangle$ in the first factor $\rho_{ij}(t)$ is glued together with the interior

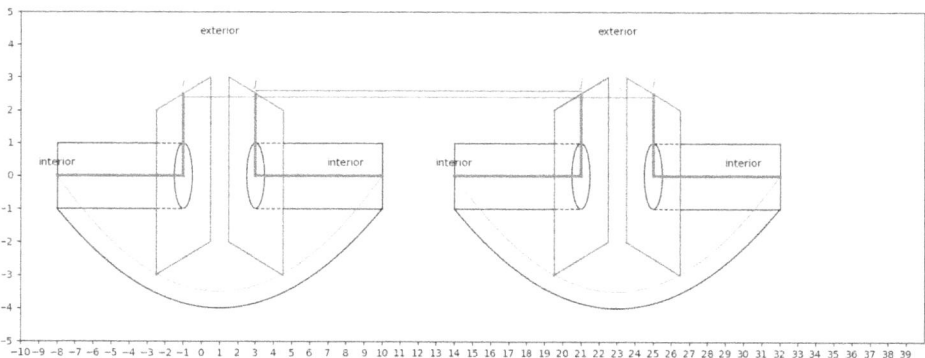

Figure 7.14. The Hawking saddle point (first configuration).

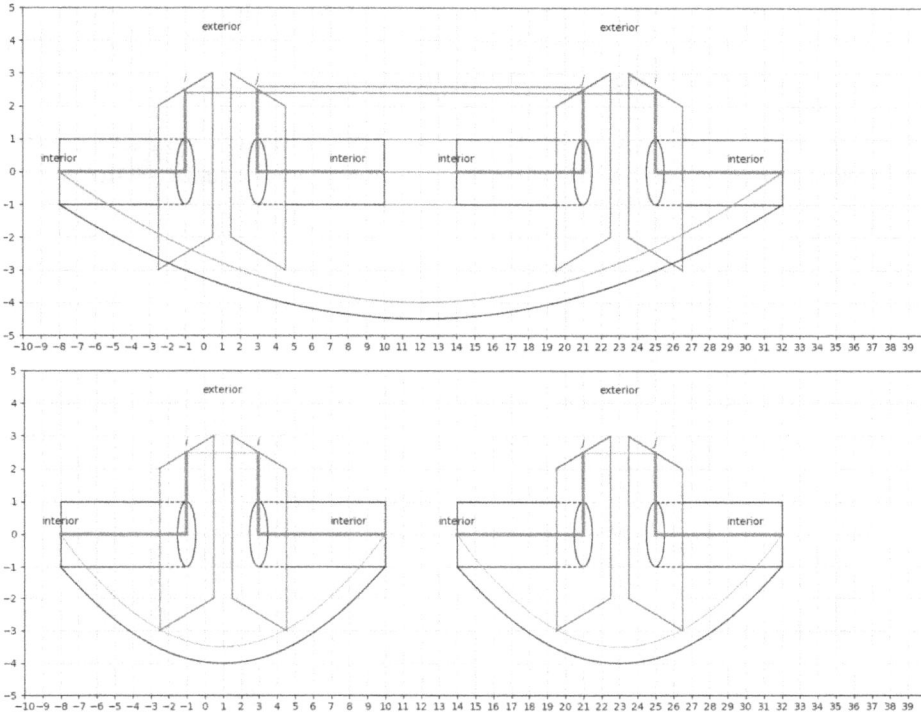

Figure 7.15. The replica wormhole saddle point (second configuration) which is also given by the third configuration.

corresponding to the bras $\langle \Psi(t)|$ in the second factor $\rho_{ji}(t)$. Similarly, the interior corresponding to the bras $\langle \Psi(t)|$ in the first factor $\rho_{ij}(t)$ is glued together with the interior corresponding to the ket $|\Psi(t)\rangle$ in the second factor $\rho_{ji}(t)$. Thus, in this case we have a single interior. Again, the index i/j of the first factor $\rho_{ij}(t)$ is identified with the index i/j of the second factor $\rho_{ji}(t)$.

We can then start from the exterior at the index i of the first factor $\rho_{ij}(t)$ and follow the entanglement through the glued interior until we reach the exterior at the index i of the second factor $\rho_{ji}(t)$. Since the indices i in the two factors are identified this path by itself corresponds to a closed loop. Similarly, we can start from the exterior at the index j of the first factor $\rho_{ij}(t)$ and follow the entanglement through the glued interior until we reach the exterior at the index j of the second factor $\rho_{ji}(t)$. Since the indices j in the two factors are identified, this path corresponds also to a closed loop. Then, we have obviously $Tr\rho^2 = (Tr\rho)^2$ in this case.

- The Hawking saddle is thus characterized by a very large entropy (mixed state), whereas the replica wormhole is characterized by zero entropy (pure state). Hence, the replica wormhole dominates over the Hawking saddle in the path integral.

- In summary, in the case of the Hawking saddle, we have the identifications $i/j \longrightarrow i/j$ and the entanglement relations $i/j \longrightarrow j/i$. But, in the case of the replica wormhole, we have the identifications $i/j \longrightarrow i/j$ as well as the entanglement relations $i/j \longrightarrow i/j$.
- Thus, in the Euclidean formulation, the Hawking saddle will correspond to two disconnected copies of the cigar geometry while the replica wormhole will correspond to two Euclidean black holes joined through the interior. See figure 7.16.

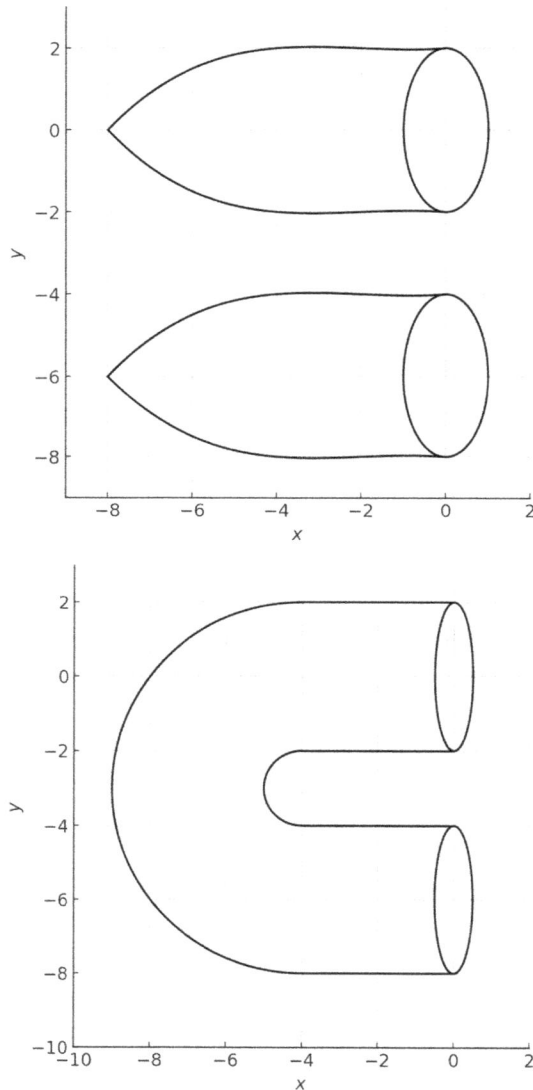

Figure 7.16. The Hawking saddle point (cigar configurations) and the replica wormhole saddle point (two connected cigars).

7.6 Summary

JT gravity, quantum extremal surfaces, and the island conjecture provide a robust framework to resolve the black hole information paradox. The distinction between von Neumann entropy and Bekenstein–Hawking entropy highlights the differences in fine-grained and coarse-grained descriptions of entropy, crucial for understanding information preservation in black hole dynamics. The replica trick is an essential tool in computing entropy, facilitating the understanding of complex entanglement structures in quantum gravity.

References

[1] Almheiri A, Engelhardt N, Marolf D and Maxfield H 2019 The entropy of bulk quantum fields and the entanglement wedge of an evaporating black hole *J. High Energy Phys.* **12** 063

[2] Penington G 2020 Entanglement wedge reconstruction and the information paradox *J. High Energy Phys.* **09** 002

[3] Maldacena J M 1999 The large N limit of superconformal field theories and supergravity *Int. J. Theor. Phys.* **38** 1113

[4] Witten E 1998 Anti-de Sitter space and holography *Adv. Theor. Math. Phys.* **2** 253–91

[5] Gubser S S, Klebanov I R and Polyakov A M 1998 Gauge theory correlators from noncritical string theory *Phys. Lett.* B **428** 105–14

[6] Itzhaki N, Maldacena J M, Sonnenschein J and Yankielowicz S 1998 Supergravity and the large N limit of theories with sixteen supercharges *Phys. Rev.* D **58** 046004

[7] Polchinski J 1995 Dirichlet branes and Ramond-Ramond charges *Phys. Rev. Lett.* **75** 4724

[8] Witten E 1996 Bound states of strings and p-branes *Nucl. Phys.* B **460** 335

[9] Bekenstein J D 1972 Black holes and the second law *Lett. Nuovo Cim.* **4** 737–40

[10] Bekenstein J D 1973 Black holes and entropy *Phys. Rev.* D **7** 2333–46

[11] Ryu S and Takayanagi T 2006 Holographic derivation of entanglement entropy from AdS/CFT *Phys. Rev. Lett.* **96** 181602

[12] Hubeny V E, Rangamani M and Takayanagi T 2007 A covariant holographic entanglement entropy proposal *J. High Energy Phys.* **07** 062

[13] Engelhardt N and Wall A C 2015 Quantum extremal surfaces: holographic entanglement entropy beyond the classical regime *J. High Energy Phys.* **01** 073

[14] Hawking S W 1974 Black hole explosions *Nature* **248** 30–1

[15] Hawking S W 1975 Particle creation by black holes *Commun. Math. Phys.* **43** 199 [erratum: Commun. Math. Phys. **46**, 206 (1976)]

[16] Hawking S W 1976 Breakdown of predictability in gravitational collapse *Phys. Rev.* D **14** 2460

[17] Almheiri A, Hartman T, Maldacena J, Shaghoulian E and Tajdini A 2020 Replica wormholes and the entropy of Hawking radiation *J. High Energy Phys.* **05** 013

[18] Penington G, Shenker S H, Stanford D and Yang Z 2022 Replica wormholes and the black hole interior *J. High Energy Phys.* **03** 205

[19] Almheiri A, Mahajan R, Maldacena J and Zhao Y 2020 The page curve of Hawking radiation from semiclassical geometry *J. High Energy Phys.* **03** 149

[20] Almheiri A, Hartman T, Maldacena J, Shaghoulian E and Tajdini A 2021 The entropy of Hawking radiation *Rev. Mod. Phys.* **93** 035002

[21] Banks T, Fischler W, Shenker S H and Susskind L 1997 M theory as a matrix model: a conjecture *Phys. Rev.* D **55** 5112–28

[22] Strominger A and Vafa C 1996 Microscopic origin of the Bekenstein-Hawking entropy *Phys. Lett.* B **379** 99–104

[23] Page D N 1993 Information in black hole radiation *Phys. Rev. Lett.* **71** 3743–6

[24] Page D N 2013 Time dependence of Hawking radiation entropy *JCAP* **09** 028

[25] Almheiri A and Polchinski J 2015 Models of AdS$_2$ backreaction and holography *J. High Energy Phys.* **11** 014

[26] Jackiw R 1985 Lower dimensional gravity *Nucl. Phys.* B **252** 343–56

[27] Teitelboim C 1983 Gravitation and Hamiltonian structure in two space-time dimensions *Phys. Lett.* B **126** 41–5

[28] Kitaev A A simple model of quantum holography KITP strings seminar and Entanglement 2015 program (Feb. 12, April 7, and May 27, 2015) http://online.kitp.ucsb.edu/online/entangled15/

[29] Sachdev S and Ye J 1993 Gapless spin-fluid ground state in a random quantum Heisenberg magnet *Phys. Rev. Lett.* **70** 3339–42

[30] Engelsöy J, Mertens T G and Verlinde H 2016 An investigation of AdS$_2$ backreaction and holography *J. High Energy Phys.* **07** 139

[31] Maldacena J, Stanford D and Yang Z 2016 Conformal symmetry and its breaking in two-dimensional nearly anti-de-Sitter space *PTEP* **2016** 12C104

[32] Maldacena J and Qi X L 2018 Eternal traversable wormhole ArXiv:1804.00491 https://doi.org/10.48550/arXiv.1804.00491

[33] Galloway G J and Graf M 2019 Rigidity of asymptotically $AdS_2 \times S^2$ spacetimes *Adv. Theor. Math. Phys.* **23** 403–35

[34] Faulkner T, Leigh R G, Parrikar O and Wang H 2016 Modular Hamiltonians for deformed half-spaces and the averaged null energy condition *J. High Energy Phys.* **09** 038

[35] Faulkner T, Lewkowycz A and Maldacena J 2013 Quantum corrections to holographic entanglement entropy *J. High Energy Phys.* **11** 074

[36] Lewkowycz A and Maldacena J 2013 Generalized gravitational entropy *J. High Energy Phys.* **08** 090

[37] Dong X, Lewkowycz A and Rangamani M 2016 Deriving covariant holographic entanglement *J. High Energy Phys.* **11** 028

[38] Dong X and Lewkowycz A 2018 Entropy, extremality, euclidean variations, and the equations of motion *J. High Energy Phys.* **01** 081

[39] Holzhey C, Larsen F and Wilczek F 1994 Geometric and renormalized entropy in conformal field theory *Nucl. Phys.* B **424** 443–67

[40] Calabrese P and Cardy J L 2004 Entanglement entropy and quantum field theory *J. Stat. Mech.* **0406** P06002

[41] Randall L and Sundrum R 1999 An alternative to compactification *Phys. Rev. Lett.* **83** 4690–3

[42] Randall L and Sundrum R 1999 A large mass hierarchy from a small extra dimension *Phys. Rev. Lett.* **83** 3370–3

[43] Arkani-Hamed N, Dimopoulos S and Dvali G R 1998 The hierarchy problem and new dimensions at a millimeter *Phys. Lett.* B **429** 263–72

[44] Arkani-Hamed N, Dimopoulos S and Dvali G R 1999 Phenomenology, astrophysics and cosmology of theories with submillimeter dimensions and TeV scale quantum gravity *Phys. Rev.* D **59** 086004

[45] Folgado M G, Donini A and Rius N 2019 Gravity-mediated scalar dark matter in warped extra-dimensions *J. High Energy Phys.* **2020** 161 [erratum: J. High Energy Phys. **2**, 129 (2022)]

[46] de Giorgi A and Vogl S 2021 Unitarity in KK-graviton production: a case study in warped extra-dimensions *J. High Energy Phys.* **4** 143

[47] Cai H, Cacciapaglia G and Lee S J 2022 Massive gravitons as feebly interacting dark matter candidates *Phys. Rev. Lett.* **128** 081806 [erratum: Phys. Rev. Lett. **132**, 169901 (2024)] [arXiv:2107.14548 [hep-ph]]

[48] Maldacena J and Susskind L 2013 Cool horizons for entangled black holes *Fortsch. Phys.* **61** 781–811

[49] Mathur S D 2014 What is the dual of two entangled CFTs? arXiv:1402.6378 https://doi.org/10.48550/arXiv.1402.6378

[50] Almheiri A, Mahajan R and Maldacena J 2019 Islands outside the horizon ArXiv:1910.11077 https://doi.org/10.48550/arXiv.1910.11077

[51] Bousso R, Fisher Z, Leichenauer S and Wall A C 2016 Quantum focusing conjecture *Phys. Rev.* D **93** 064044

[52] Callan C G Jr and Wilczek F 1994 On geometric entropy *Phys. Lett.* B **333** 55–61

[53] Casini H and Huerta M 2009 Entanglement entropy in free quantum field theory *J. Phys.* A **42** 504007

IOP Publishing

Lectures on General Relativity, Cosmology and
Quantum Black Holes (Second Edition)

Badis Ydri

Chapter 8

Loop quantum gravity and Banks–Fischler–Shenker–Susskind approaches to quantum gravity

8.1 Loop quantum gravity

We present a brief initiation to the loop quantum gravity (LQG) approach to canonical quantization of quantum gravity and to the Banks–Fischler–Shenker–Susskind (BFSS) model which provides the quantum mechanical description of M-theory. In the final section we put forward a thought-provoking idea in which we posited that causal dynamical triangulation (CDT) provides a non-perturbative definition of LQG, whereas multitrace matrix models are more fundamental than BFSS-like matrix models. We then observe that both the CDT and multitrace matrix model enjoy the same phase structure.

8.1.1 Outline

General relativity can be formulated either using the metric tensor $g_{\mu\nu}$ (and implicitly the affine Levi-Civita connection defined in terms of Christoffel symbols $\Gamma^{\rho}_{\mu\nu}$) or in terms of the vielbein field or tetrad e^m_μ and the spin connection ω^{mn}_μ. The two formulations are equivalent but contrary to widespread belief, the formulation based on the spin connection is more fundamental for two reasons. First, fundamental matter fields in Nature are described by chiral fermions (leptons and quarks) which when propagating in a curved spacetime feel the metric only through the vielbein field. The second reason is the fact that the most successful canonical quantization of general relativity today, i.e. LQG uses as canonical conjugate variables the (densitized) vielbein field and the (self-dual part of the) spin connection as the canonical momentum and the configuration variable, respectively. This is in contrast to the

doi:10.1088/978-0-7503-5824-8ch8
8-1

use of the three-dimensional metric h_{ij} (or equivalently the vielbein field or triads e_i^m) and the extrinsic curvature K_{ij} in the classic Arnowitt–Deser–Misner (ADM) formulation. It will be seen that general relativity can then be reformulated as a complex $SU(2)$ gauge theory of the self-dual spin connections and as a consequence general relativity becomes a dynamical theory for three-dimensional connections and not for three-dimensional geometries and this makes it embeddable into Yang–Mills gauge theory. Loop representation and spin network states are then briefly discussed.

8.1.2 Review of general relativity

We start with a recollection of some facts from general relativity.

The equivalence principle states that spacetime must be a manifold, i.e, locally it must look like flat Minkowski spacetime. Thus, curvature can be canceled by freely falling observers in the gravitational field associated with the metric and as a consequence the spacetime manifold will be seen as flat by these observers.

Indeed, by choosing the so-called normal coordinates near any point of the spacetime manifold it is seen that the trajectories of freely falling objects looks locally like straight lines. Effectively, the manifold is approximated there by its tangent vector space. The difference between the metric $g_{\mu\nu}$ and the Minkowski metric $\eta_{\mu\nu}$ vanish to first order while the difference at second order is characterized by the so-called Riemann curvature tensor $R_{\mu\alpha\nu\beta}$. Explicitly, we have

$$g_{\mu\nu} = \eta_{\mu\nu} - \frac{1}{3}R_{\mu\alpha\nu\beta}x^\alpha x^\beta + \cdots \tag{8.1}$$

We consider a Riemannian (curved) manifold $\mathcal{M}$ with a metric $g_{\mu\nu}$. A coordinates transformation is given by

$$x^\mu \longrightarrow x'^\mu = x'^\mu(x). \tag{8.2}$$

The vectors and one-forms on the manifold are quantities which are defined to transform under the above coordinates transformation, respectively, as follows

$$V'^\mu = \frac{\partial x'^\mu}{\partial x^\nu}V^\nu. \tag{8.3}$$

$$V'_\mu = \frac{\partial x^\nu}{\partial x'^\mu}V_\nu. \tag{8.4}$$

The spaces of vectors and one-forms are the tangent and co-tangent bundles.

A tensor is a quantity with multiple indices (covariant and contravariant) transforming in a similar way, i.e. any contravariant index is transforming as (8.3) and any covariant index is transforming as (8.4). For example, the metric $g_{\mu\nu}$ is a second rank symmetric tensor which transforms as

$$g'_{\mu\nu}(x') = \frac{\partial x^\alpha}{\partial x'^\mu}\frac{\partial x^\beta}{\partial x'^\nu}g_{\alpha\beta}(x). \tag{8.5}$$

The interval $ds^2 = g_{\mu\nu}dx^\mu dx^\nu$ is therefore invariant. In fact, all scalar quantities are invariant under coordinate transformations. For example, the volume element $d^4x\sqrt{-\det g}$ is a scalar under coordinate transformation.

The derivative of a tensor does not transform as a tensor. However, the so-called covariant derivative of a tensor will transform as a tensor. The covariant derivatives of vectors and one-forms are given by

$$\nabla_\mu V^\nu = \partial_\mu V^\nu + \Gamma^\nu_{\alpha\mu} V^\alpha. \tag{8.6}$$

$$\nabla_\mu V_\nu = \partial_\mu V_\nu - \Gamma^\alpha_{\mu\nu} V_\alpha. \tag{8.7}$$

These transform indeed as tensors as one can easily check. Generalization to tensors is obvious. The Christoffel symbols $\Gamma^\alpha_{\mu\nu}$ are given in terms of the metric $g_{\mu\nu}$ by

$$\Gamma^\alpha_{\mu\nu} = \frac{1}{2}g^{\alpha\beta}(\partial_\mu g_{\nu\beta} + \partial_\nu g_{\mu\beta} - \partial_\beta g_{\mu\nu}). \tag{8.8}$$

There exists a unique covariant derivative, and thus a unique choice of Christoffel symbols, for which the metric is covariantly constant, viz

$$\nabla_\mu g_{\alpha\beta} = 0. \tag{8.9}$$

The straightest possible lines on the curved manifolds are given by the geodesics. A geodesic is a curve whose tangent vector is parallel transported along itself. It is given explicitly by the Newton's second law on the curved manifold

$$\frac{d^2x^\mu}{d\lambda} + \Gamma^\mu_{\alpha\beta}\frac{dx^\alpha}{d\lambda}\frac{dx^\beta}{d\lambda} = 0. \tag{8.10}$$

The λ is an affine parameter along the curve. The time-like geodesics define the trajectories of freely falling particles in the gravitational field encoded in the curvature of the Riemannian manifold.

The Riemann curvature tensor $R^\alpha_{\mu\nu\beta}$ is defined in terms of the covariant derivative by

$$(\nabla_\mu \nabla_\nu - \nabla_\nu \nabla_\mu)t^\alpha = -R^\alpha_{\mu\nu\rho}t^\rho. \tag{8.11}$$

It is given explicitly by

$$R^\alpha_{\mu\nu\rho} = \partial_\nu\Gamma^\alpha_{\mu\rho} - \partial_\rho\Gamma^\alpha_{\mu\nu} + \Gamma^\alpha_{\sigma\nu}\Gamma^\sigma_{\mu\rho} - \Gamma^\alpha_{\rho\sigma}\Gamma^\sigma_{\mu\nu}. \tag{8.12}$$

We define the Ricci tensor $R_{\mu\nu}$ and the Ricci scalar R by the equations

$$R = g^{\mu\nu}R_{\mu\nu}. \tag{8.13}$$

$$R_{\mu\nu} = R^\alpha_{\mu\alpha\nu}. \tag{8.14}$$

The Einstein's equations for general relativity reads (with $T_{\mu\nu}$ being the energy–momentum tensor and G is the Newton's constant)

$$R_{\mu\nu} - \frac{1}{2}g_{\mu\nu}R = 8\pi G T_{\mu\nu}. \tag{8.15}$$

8.1.3 The Hilbert–Einstein action

The dynamical variable is obviously the metric $g_{\mu\nu}$. The goal is to construct an action principle from which the Einstein's equations follow as the Euler–Lagrange equations of motion for the metric. This action principle will read as

$$S = \int d^n x \, \mathcal{L}(g). \tag{8.16}$$

The first problem with this way of writing is that both $d^n x$ and $\mathcal{L}$ are tensor densities rather than tensors. We digress briefly to explain this important different.

Let us recall the familiar Levi-Civita symbol in n dimensions defined by

$$\tilde{\epsilon}_{\mu_1 \ldots \mu_n} = +1 \text{ even permutation}$$
$$= -1 \text{ odd permutation} \tag{8.17}$$
$$= 0 \text{ otherwise.}$$

This is a symbol and not a tensor since it does not change under coordinate transformations. The determinant of a matrix M can be given by the formula

$$\tilde{\epsilon}_{\nu_1 \ldots \nu_n} \det M = \tilde{\epsilon}_{\mu_1 \ldots \mu_n} M^{\mu_1}{}_{\nu_1} \ldots M^{\mu_n}{}_{\nu_n}. \tag{8.18}$$

By choosing $M^\mu{}_\nu = \partial x^\mu / \partial y^\nu$ we get the transformation law

$$\tilde{\epsilon}_{\nu_1 \ldots \nu_n} = \det \frac{\partial y}{\partial x} \tilde{\epsilon}_{\mu_1 \ldots \mu_n} \frac{\partial x^{\mu_1}}{\partial y^{\nu_1}} \ldots \frac{\partial x^{\mu_n}}{\partial y^{\nu_n}}. \tag{8.19}$$

In other words $\tilde{\epsilon}_{\mu_1 \ldots \mu_n}$ is not a tensor because of the determinant appearing in this equation. This is an example of a tensor density. Another example of a tensor density is $\det g$. Indeed from the tensor transformation law of the metric $g'_{\alpha\beta} = g_{\mu\nu}(\partial x^\mu / \partial y^\alpha)(\partial x^\nu / \partial y^\beta)$ we can show in a straightforward way that

$$\det g' = (\det \frac{\partial y}{\partial x})^{-2} \det g. \tag{8.20}$$

The actual Levi-Civita tensor can then be defined by

$$\epsilon_{\mu_1 \ldots \mu_n} = \sqrt{\det g} \, \tilde{\epsilon}_{\mu_1 \ldots \mu_n}. \tag{8.21}$$

Next, under a coordinate transformation $x \longrightarrow y$ the volume element transforms as

$$d^n x \longrightarrow d^n y = \det \frac{\partial y}{\partial x} d^n x. \tag{8.22}$$

In other words, the volume element transforms as a tensor density and not as a tensor. We verify this important point in our language as follows. We write

$$d^n x = dx^0 \wedge dx^1 \wedge \cdots \wedge dx^{n-1}$$
$$= \frac{1}{n!} \tilde{\epsilon}_{\mu_1 \ldots \mu_n} dx^{\mu_1} \wedge \cdots \wedge dx^{\mu_n}. \tag{8.23}$$

Recall that a differential p-form is a $(0, p)$ tensor which is completely antisymmetric. For example scalars are 0-forms and dual co-tangent vectors are 1-forms The Levi-Civita tensor $\epsilon_{\mu_1 \ldots \mu_n}$ is a 4-form. The differentials dx^μ appearing in the second line of equation (8.23) are 1-forms and hence under a coordinate transformation $x \longrightarrow y$ we have $dx^{\mu} \longrightarrow dy^{\mu} = dx^{\nu} \partial y^{\mu}/\partial x^{\nu}$. By using this transformation law we can immediately show that dx^n transforms to $d^n y$ exactly as in equation (8.22).

It is not difficult to see now that an invariant volume element can be given by the n-form defined by the equation

$$dV = \sqrt{\det g} \, d^n x. \tag{8.24}$$

We can show that

$$
\begin{aligned}
dV &= \frac{1}{n!} \sqrt{\det g} \, \tilde{\epsilon}_{\mu_1 \cdots \mu_n} dx^{\mu_1} \wedge \cdots \wedge dx^{\mu_n} \\
&= \frac{1}{n!} \epsilon_{\mu_1 \cdots \mu_n} dx^{\mu_1} \wedge \cdots \wedge dx^{\mu_n} \\
&= \epsilon_{\mu_1 \cdots \mu_n} dx^{\mu_1} \otimes \cdots \otimes dx^{\mu_n} \\
&= \epsilon.
\end{aligned}
\tag{8.25}
$$

In other words, the invariant volume element is precisely the Levi-Civita tensor. In the case of Lorentzian signature we replace $\det g$ with $-\det g$.

We go back now to equation (8.16) and rewrite it as

$$
\begin{aligned}
S &= \int d^n x \, \mathcal{L}(g) \\
&= \int d^n x \sqrt{-\det g} \, \hat{\mathcal{L}}(g).
\end{aligned}
\tag{8.26}
$$

Clearly, $\mathcal{L} = \sqrt{-\det g} \, \hat{\mathcal{L}}$. Since the invariant volume element $d^n x \sqrt{-\det g}$ is a scalar the function $\hat{\mathcal{L}}$ must also be a scalar and as such can be identified with the Lagrangian density.

We use the result that the only independent scalar quantity which is constructed from the metric and which is at most second order in its derivatives is the Ricci scalar R. In other words, the simplest choice for the Lagrangian density $\hat{\mathcal{L}}$ is

$$\hat{\mathcal{L}}(g) = R. \tag{8.27}$$

The corresponding action is called the Hilbert–Einstein action. We compute

$$\delta S = \int d^n x \delta \sqrt{-\det g} \, g^{\mu\nu} R_{\mu\nu} + \int d^n x \sqrt{-\det g} \, \delta g^{\mu\nu} R_{\mu\nu} + \int d^n x \sqrt{-\det g} \, g^{\mu\nu} \delta R_{\mu\nu}. \tag{8.28}$$

We have

$$
\begin{aligned}
\delta R_{\mu\nu} &= \delta R_{\mu\rho\nu}{}^{\rho} \\
&= \partial_{\rho} \delta \Gamma^{\rho}{}_{\mu\nu} - \partial_{\mu} \delta \Gamma^{\rho}{}_{\rho\nu} + \delta(\Gamma^{\lambda}{}_{\mu\nu} \Gamma^{\rho}{}_{\rho\lambda} - \Gamma^{\lambda}{}_{\rho\nu} \Gamma^{\rho}{}_{\mu\lambda}) \\
&= (\nabla_{\rho} \delta \Gamma^{\rho}{}_{\mu\nu} - \Gamma^{\rho}{}_{\rho\lambda} \delta \Gamma^{\lambda}{}_{\mu\nu} + \Gamma^{\lambda}{}_{\rho\mu} \delta \Gamma^{\rho}{}_{\lambda\nu} + \Gamma^{\lambda}{}_{\rho\nu} \delta \Gamma^{\rho}{}_{\lambda\mu}) - (\nabla_{\mu} \delta \Gamma^{\rho}{}_{\rho\nu} - \Gamma^{\rho}{}_{\mu\lambda} \delta \Gamma^{\lambda}{}_{\rho\nu} + \Gamma^{\lambda}{}_{\mu\rho} \delta \Gamma^{\rho}{}_{\lambda\nu} \\
&\quad + \Gamma^{\lambda}{}_{\mu\nu} \delta \Gamma^{\rho}{}_{\rho\lambda}) + \delta(\Gamma^{\lambda}{}_{\mu\nu} \Gamma^{\rho}{}_{\rho\lambda} - \Gamma^{\lambda}{}_{\rho\nu} \Gamma^{\rho}{}_{\mu\lambda}) \\
&= \nabla_{\rho} \, \delta \Gamma^{\rho}{}_{\mu\nu} - \nabla_{\mu} \, \delta \Gamma^{\rho}{}_{\rho\nu}.
\end{aligned}
\tag{8.29}
$$

In the second line of the above equation we have used the fact that $\delta \Gamma^{\rho}{}_{\mu\nu}$ is a tensor since it is the difference of two connections. Thus

$$\int d^n x \sqrt{-\mathrm{detg}} \; g^{\mu\nu} \delta R_{\mu\nu} = \int d^n x \sqrt{-\mathrm{detg}} \; g^{\mu\nu} (\nabla_\rho \delta \Gamma^\rho{}_{\mu\nu} - \nabla_\mu \, \delta \Gamma^\rho{}_{\rho\nu})$$
$$= \int d^n x \sqrt{-\mathrm{detg}} \; \nabla_\rho \, (g^{\mu\nu} \delta \Gamma^\rho{}_{\mu\nu} - g^{\rho\nu} \delta \Gamma^\mu{}_{\mu\nu}). \tag{8.30}$$

We compute also (with $\delta g_{\mu\nu} = -g_{\mu\alpha} g_{\nu\beta} \delta g^{\alpha\beta}$)

$$\delta \Gamma^\rho{}_{\mu\nu} = \frac{1}{2} g^{\rho\lambda} (\nabla_\mu \delta g_{\nu\lambda} + \nabla_\nu \, \delta g_{\mu\lambda} - \nabla_\lambda \, \delta g_{\mu\nu})$$
$$= -\frac{1}{2} (g_{\nu\lambda} \, \nabla_\mu \, \delta g^{\lambda\rho} + g_{\mu\lambda} \, \nabla_\nu \, \delta g^{\lambda\rho} - g_{\mu\alpha} g_{\nu\beta} \, \nabla^\rho \, \delta g^{\alpha\beta}). \tag{8.31}$$

Thus

$$\int d^n x \sqrt{-\mathrm{detg}} \; g^{\mu\nu} \delta R_{\mu\nu} = \int d^n x \sqrt{-\mathrm{detg}} \; \nabla_\rho \, (g_{\mu\nu} \, \nabla^\rho \, \delta g^{\mu\nu} - \nabla_\mu \, \delta g^{\mu\rho}). \tag{8.32}$$

By Stokes's theorem this integral is equal to the integral over the boundary of spacetime of the expression $g_{\mu\nu} \, \nabla^\rho \, \delta g^{\mu\nu} - \nabla_\mu \, \delta g^{\mu\rho}$ which is 0 if we assume that the metric and its first derivatives are held fixed on the boundary. The variation of the action reduces to

$$\delta S = \int d^n x \delta \sqrt{-\mathrm{detg}} \; g^{\mu\nu} R_{\mu\nu} + \int d^n x \sqrt{-\mathrm{detg}} \; \delta g^{\mu\nu} R_{\mu\nu}. \tag{8.33}$$

Next, we use the result

$$\delta \sqrt{-\mathrm{detg}} = -\frac{1}{2} \sqrt{-\mathrm{detg}} \; g_{\mu\nu} \delta g^{\mu\nu}. \tag{8.34}$$

Hence

$$\delta S = \int d^n x \sqrt{-\mathrm{detg}} \; \delta g^{\mu\nu} (R_{\mu\nu} - \frac{1}{2} g_{\mu\nu} R). \tag{8.35}$$

This will obviously lead to Einstein's equations in vacuum which is partially our goal. We want also to include the effect of matter which requires considering the more general actions of the form

$$S = \frac{1}{16\pi G} \int d^n x \; \sqrt{-\mathrm{detg}} \; R + S_M. \tag{8.36}$$

$$S_M = \int d^n x \; \sqrt{-\mathrm{detg}} \; \hat{\mathcal{L}}_M. \tag{8.37}$$

The variation of the action becomes

$$\delta S = \frac{1}{16\pi G} \int d^n x \sqrt{-\mathrm{detg}} \; \delta g^{\mu\nu} (R_{\mu\nu} - \frac{1}{2} g_{\mu\nu} R) + \delta S_M$$
$$= \int d^n x \sqrt{-\mathrm{detg}} \; \delta g^{\mu\nu} \left[\frac{1}{16\pi G} (R_{\mu\nu} - \frac{1}{2} g_{\mu\nu} R) + \frac{1}{\sqrt{-\mathrm{detg}}} \frac{\delta S_M}{\delta g^{\mu\nu}} \right]. \tag{8.38}$$

In other words,

$$\frac{1}{\sqrt{-\det g}} \frac{\delta S}{\delta g^{\mu\nu}} = \frac{1}{16\pi G}\left(R_{\mu\nu} - \frac{1}{2}g_{\mu\nu}R\right) + \frac{1}{\sqrt{-\det g}} \frac{\delta S_M}{\delta g^{\mu\nu}}. \tag{8.39}$$

Einstein's equations are therefore given by

$$R_{\mu\nu} - \frac{1}{2}g_{\mu\nu}R = 8\pi G T_{\mu\nu}. \tag{8.40}$$

The stress–energy–momentum tensor must therefore be defined by the equation

$$T_{\mu\nu} = -\frac{2}{\sqrt{-\det g}} \frac{\delta S_M}{\delta g^{\mu\nu}}. \tag{8.41}$$

As a first example, we consider the action of a scalar field in curved spacetime given by

$$S_\phi = \int d^n x \sqrt{-\det g}\left[-\frac{1}{2}g^{\mu\nu}\,\nabla_\mu\,\phi\,\nabla_\nu\,\phi - V(\phi)\right]. \tag{8.42}$$

The corresponding stress–energy–momentum tensor is calculated to be given by

$$T_{\mu\nu}^{(\phi)} = \nabla_\mu\,\phi\,\nabla_\nu\,\phi - \frac{1}{2}g_{\mu\nu}g^{\rho\sigma}\,\nabla_\rho\,\phi\,\nabla_\sigma\,\phi - g_{\mu\nu}V(\phi). \tag{8.43}$$

As a second example, we consider the action of the electromagnetic field in curved spacetime given by

$$S_A = \int d^n x \sqrt{-\det g}\left[-\frac{1}{4}g^{\mu\nu}g^{\alpha\beta}F_{\mu\nu}F_{\alpha\beta}\right]. \tag{8.44}$$

In this case the stress–energy–momentum tensor is calculated to be given by

$$T_{\mu\nu}^{(A)} = F^{\mu\lambda}F^\nu{}_\lambda - \frac{1}{4}g^{\mu\nu}F_{\alpha\beta}F^{\alpha\beta}. \tag{8.45}$$

The cosmological constant is one of the simplest matter actions that one can add to the Hilbert–Einstein action. It is given by

$$S_{cc} = -\frac{1}{8\pi G}\int d^4 x \sqrt{-\det g}\,\Lambda. \tag{8.46}$$

In this case the energy–momentum tensor and the Einstein equations read

$$T_{\mu\nu} = -\frac{\Lambda}{8\pi G}g_{\mu\nu}. \tag{8.47}$$

$$R_{\mu\nu} - \frac{1}{2}g_{\mu\nu}R + \Lambda g_{\mu\nu} = 0. \tag{8.48}$$

8.1.4 The vielbein formalism

First, we remark that the vielbein field is essentially the square root of the metric. Physically, it gives the local orientations ξ^m of freely falling frames in the gravitational field associated with the metric $g_{\mu\nu}$, i.e. ξ^m are orientations of the local inertial frames with respect to the coordinate axes x^μ of the curved spacetime manifold, viz

$$e_\mu^m = \frac{\partial \xi^m}{\partial x^\mu}. \tag{8.49}$$

The metric is the square toot of the vielbein which means that we have (η being the flat metric with signature $-1, +1, +1, +1$)

$$g_{\mu\nu} = e_\mu^m e_\nu^n \eta_{mn}. \tag{8.50}$$

The inverse of e_μ^m is denoted by e_m^μ, viz $e_m^\mu e_\mu^n = \eta_m^n$ and $e_\mu^m e_m^\nu = \eta_\mu^\nu$. Thus, we also have $g^{\mu\nu} = e_m^\mu e_n^\nu \eta^{mn}$.

The fundamental equation (8.116) can be derived in a straightforward way from the Clifford algebra of the Dirac matrices in the curved spacetime manifold which is of the usual form but only with the replacement $\eta \longrightarrow g$, i.e.

$$\{\gamma^\mu, \gamma^\nu\} = 2g^{\mu\nu}. \tag{8.51}$$

Indeed, equation (8.116) is obtained by replacing in the Clifford algebra (8.51) the ansatz (where γ^m are the flat spacetime Dirac matrices)

$$\gamma^\mu = \gamma^m e_m^\mu. \tag{8.52}$$

Two solutions e_μ^m and f_μ^m of (8.116) are related by a local Lorentz transformation $\Lambda \in SO(1, 3)$ (where $SO(1, 3)$ is the restricted Lorentz group with antisymmetric generators) as follows

$$f_\mu^m = \Lambda_k^m e_\mu^k, \quad \Lambda_k^m \eta_{mn}(\Lambda^T)_l^n = \eta_{kl}. \tag{8.53}$$

Obviously, the index μ refers to the curved coordinates x^μ of the manifold, whereas the index m refers to the flat coordinates ξ^m. A vector field v will then have components v^μ in the system x^μ and components v^m in the system ξ^m. They are related by means of the vielbein, i.e. $v^\mu = v^m e_m^\mu$ and $v^m = v^\mu e_\mu^m$.

The covariant derivative with respect to the curved index μ is defined by means of the parallel transport with respect of the affine connection Γ given by the Christoffel symbols, viz

$$\tilde{v}^\mu(x + \Delta x) = v^\mu(x) - \Delta x^\rho \Gamma_{\rho\sigma}^\mu v^\sigma \iff \tilde{v}^\mu(x) + \Delta x^\rho \nabla_\rho \tilde{v}^\mu(x) = v^\mu(x). \tag{8.54}$$

Similarly, the covariant derivative with respect to the flat index m is defined by means of the parallel transport with respect of the spin connection ω, viz

$$\tilde{v}^m(x + \Delta x) = v^m(x) - \Delta x^\rho \omega_{\rho n}^m v^n \iff \tilde{v}^m(x) + \Delta x^\rho \nabla_\rho \tilde{v}^m(x) = v^m(x). \tag{8.55}$$

Since we are dealing with the parallel transport of the same vector we must have the property that the parallel transport of the curved index from x to $x + \Delta x$ followed by its the projection to a flat index equal to the projection at x of the curved index to a flat index followed by the parallel transport of the flat index from x to $x + \Delta x$, viz

$$\tilde{v}^{\mu}(x + \Delta x)e_{\mu}^{m}(x + \Delta x) = \tilde{v}^{m}(x + \Delta x). \tag{8.56}$$

This gives immediately the so-called vielbein postulate which states that the vielbein e_{μ}^{m} is covariantly constant with respect to both connections affine Γ (which acts on the curved index μ) and spin ω (which acts on the flat index m), i.e.

$$\mathcal{D}_{\rho}e_{\mu}^{m} \equiv \partial_{\rho}e_{\mu}^{m} - \Gamma_{\rho\mu}^{\nu}e_{\nu}^{m} + \omega_{\rho n}^{m}e_{\mu}^{n} = 0. \tag{8.57}$$

This equation gives us the spin connection ω (which defines the spinor bundle) in terms of the affine connection Γ (which defines the tangent/co-tangent bundle) or vice versa. Clearly, the spin connection encodes the local Lorentz (special coordinate transformations) invariance of the theory, whereas the affine connection encodes diffeomorphism (general coordinate transformations) invariance.

Now, by requiring the length of the vector v to be invariant under the parallel transport we get by using the flat and the curved indices, respectively, the two equivalent results that the spin connection is antisymmetric (by using the components v^m) and the metric is covariantly constant (by using the components v^{μ}), viz

$$\omega_{\mu}^{mn} = -\omega_{\mu}^{nm}, \quad \omega_{\mu}^{mn} = \omega_{\mu k}^{m}\eta^{kn}. \tag{8.58}$$

$$\nabla_{\rho}\, g_{\mu\nu} \equiv \partial_{\rho}g_{\mu\nu} - \Gamma_{\rho\mu\nu} - \Gamma_{\rho\nu\mu}, \quad \Gamma_{\rho\mu\nu} = \Gamma_{\rho\mu}^{\lambda}g_{\lambda\nu}. \tag{8.59}$$

The spin connection will play the role of a local $SO(1, 3)$ gauge field representing local Lorentz invariance and hence this connection must be antisymmetric by construction and as a consequence the invariance of the length and the covariant constancy of the metric follow naturally from a local symmetry principle.

The Riemann curvature tensor can be computed from the commutator of two covariant derivatives. The action of this commutator on the vielbein field should vanish identically. We have then

$$[\mathcal{D}_{\rho}, \mathcal{D}_{\sigma}]e_{\mu}^{m} = 0. \tag{8.60}$$

From this equation we obtain the result that the Riemann curvature tensor in terms of Γ is equal to the Riemann curvature tensor in terms of ω, viz

$$R_{\rho\sigma\mu}^{\nu}(\Gamma)e_{\nu}^{m} = R_{\rho\sigma n}^{m}(\omega)e_{\mu}^{n}. \tag{8.61}$$

Here,

$$R_{\rho\sigma\mu}^{\nu}(\Gamma) = \partial_{\rho}\Gamma_{\sigma\mu}^{\nu} + \Gamma_{\rho\tau}^{\nu}\Gamma_{\sigma\mu}^{\tau} - (\rho \leftrightarrow \sigma). \tag{8.62}$$

$$R_{\rho\sigma n}^{m}(\omega) = \partial_{\rho}\omega_{\sigma n}^{m} + \omega_{\rho k}^{m}\omega_{\sigma n}^{k} - (\rho \leftrightarrow \sigma). \tag{8.63}$$

Hence we get

$$R_{\rho\sigma\mu\tau}(\Gamma) = R_{\rho\sigma\mu\tau}(\omega), \quad R_{\rho\sigma\mu\tau}(\Gamma) = -R_{\rho\sigma\tau\mu}(\Gamma). \tag{8.64}$$

The action of general relativity written in terms of the metric g (the Hilbert–Einstein action) is therefore the same as the action written in terms of the vielbein field e and the spin connection ω (the Palatini action).

Another important property satisfied by the affine connection Γ in general relativity is torsionless (the Christoffel symbol $\Gamma^\rho_{\mu\nu}$ is symmetric in its two lower indices μ and ν). In general, torsion is generated from the vielbein field ($T = de + \omega \wedge e$) in the same way that curvature is generated from the spin connection ($R = d\omega + \omega \wedge \omega$).

Also, similarly to the fact that the curvature tensor measures the gap, if one parallel transports a vector parallel to itself along a closed curved the torsion tensor measures the gap if one parallel transports one vector along another one minus the other way around.

As we have said, the affine Levi-Civita connection defines the tangent bundle which is the associated vector bundle corresponding to the $O(1, 3)$ bundle of orthonormal frames. In fact the affine connection is induced from the connection on the $O(1, 3)$ bundle of orthonormal frames and since the spacetime manifold is orientable this connection can be restricted to the $SO(1, 3)$ bundle of orthonormal frames and then lifted to a connection on the corresponding spinor bundle. Thus, the spin connection is indeed more fundamental than the affine connection in every respect.

Finally, we note that spinors provide a representation of the Lorentz group $SO(1, 3)$. Thus, they transform covariantly under Lorentz transformations with the covariant derivative given explicitly in terms of the spin connection ω by (with $\gamma_{\mu\nu} = i[\gamma^\mu, \gamma^\nu]/2$)

$$D_\mu\psi = \partial_\mu\psi + \frac{1}{4}\omega_\mu^{mn}\gamma_{mn}\psi. \tag{8.65}$$

The covariant Dirac action in a curved spacetime manifold (discovered by Wigner in 1929) is therefore given by (with $e = \sqrt{-\det(g_{\mu\nu})} = \det(e_\mu^m)$)

$$\mathcal{L}_D = -\frac{e}{2}\bar{\psi}\gamma^\mu D_\mu\psi. \tag{8.66}$$

8.1.5 ADM formulation and geometrodynamics

The ADM formulation is first put forward in the classic paper [1].

Spacetime is naturally assumed to be globally hyperbolic, which means that it is diffeomorphic to the direct product $\mathbb{R} \times \Sigma$ where Σ is a three-dimensional smooth manifold.

We consider then a foliation of the spacetime manifold given by the spatial Cauchy hypersurfaces Σ_t of constant time t. Let n^μ be the unit normal vector field to the hypersurfaces Σ_t which is given explicitly by

$$n^\mu = -N\frac{\partial t}{\partial x_\mu}, \quad n_\mu n^\mu = -1. \tag{8.67}$$

The normalization N is the lapse function which measures the rate of change of the proper time with respect to the coordinate time t as one moves normally to the hypersurfaces Σ_t. It is given explicitly by

$$N = -g_{\mu\nu}t^\mu n^\nu. \tag{8.68}$$

The time flow in this foliated spacetime will be given by a vector field t^μ which satisfies $t^\mu \nabla_\mu \, t = 1$, i.e.

$$t^\mu = \frac{\partial x^\mu}{\partial t}. \tag{8.69}$$

We decompose t^μ into its normal and tangential parts with respect to the hypersurface Σ_t as

$$t^\mu = Nn^\mu + N^\mu, \quad N^\mu = N^i e_i^\mu. \tag{8.70}$$

The e_i^μ are tangent vectors to the hypersurface Σ_t given by

$$e_i^\mu = \frac{\partial x^\mu}{\partial y^i}. \tag{8.71}$$

The y^i are coordinates on the hypersurface Σ_t, i.e. the coordinates x^μ are split as $x^\mu \longrightarrow y^\mu = (t, y^i)$, and N^i is the so-called shift vector which measures the shift of the local spatial coordinate system as one moves normally to the hypersurfaces Σ_t. It is given by

$$N^\mu = h^\mu{}_\nu t^\nu. \tag{8.72}$$

The three-dimensional metric h_{ij} is the induced metric on the hypersurface Σ_t given explicitly by

$$h_{ij} = g_{\mu\nu}e_i^\mu e_j^\nu = h_{\mu\nu}e_i^\mu e_j^\nu, \quad h_{\mu\nu} = g_{\mu\nu} + n_\nu n_\nu. \tag{8.73}$$

We define the inverse metric h^{ij} in the usual way, viz $h_{ij}h^{jk} = \delta_i^k$. More precisely, we compute (using also $h^{\mu\nu} = g^{\mu\nu} - n^{\mu\nu}$) the result $h_{\mu\nu}h^{\nu\alpha} = \delta_\mu^\alpha + n_\mu n^\alpha$.

We compute immediately that

$$\begin{aligned} dx^\mu &= \frac{\partial x^\mu}{\partial t}dt + \frac{\partial x^\mu}{\partial y^i}dy^i \\ &= t^\mu dt + e_i^\mu dy^i \\ &= (Ndt)n^\mu + (dy^i + N^i dt)e_i^\mu. \end{aligned} \tag{8.74}$$

Also,

$$\begin{aligned} ds^2 &= g_{\mu\nu}dx^\mu dx^\nu \\ &= g_{\mu\nu}\left[N^2 dt^2 n^\mu n^\nu + (dy^i + N^i dt)(dy^j + N^j dt)e_i^\mu e_j^\nu \right] \\ &= -N^2 dt^2 + h_{ij}(dy^i + N^i dt)(dy^j + N^j dt). \end{aligned} \tag{8.75}$$

The ADM metric is then given explicitly by (with $N_i = h_{ij}N^j$)

$$g_{\mu\nu} = \begin{pmatrix} -N^2 + N^i N_i & N_j \\ N_i & h_{ij} \end{pmatrix}. \tag{8.76}$$

The inverse ADM metric $g^{\mu\nu}$ is then given by

$$g^{\mu\nu} = \begin{pmatrix} -\dfrac{1}{N^2} & \dfrac{1}{N^2}N^j \\ \dfrac{1}{N^2}N^i & h^{ij} - \dfrac{1}{N^2}N^i N^j \end{pmatrix}. \tag{8.77}$$

We conclude that all information about the original four-dimensional metric $g_{\mu\nu}$ is contained in the lapse function N, the shift vector N^i and the three-dimensional metric h_{ij}. The lapse and the shift N and N^i are not dynamical variables but only Lagrange multipliers yielding under their respective variation the so-called Hamiltonian and diffeomorphism constraints which satisfy a Dirac algebra of first class constraints. A particular choice of N and N^i, i.e. a particular choice of foliation plays the role of a gauge fixing condition (called the time gauge) for the diffeomorphism group. In other words, invariance under general coordinate transformations which form the group of diffeomorphism is not lost but only fixed and in fact the diffeomorphism invariance of the theory is still precisely encoded in the Dirac algebra of the first class constraints of the theory.

Next we would like to rewrite the Hilbert–Einstein Lagrangian density in terms of the three-dimensional quantities N, N^i and h_{ij} and then compute the Hamiltonian density.

First we compute

$$\sqrt{-g}\, d^4x = N\sqrt{h}\, d^4y. \tag{8.78}$$

A central object in the discussion of how the hypersurfaces Σ_t are embedded in the four-dimensional spacetime manifold $\mathcal{M}$ is the extrinsic curvature $K_{\mu\nu}$. This is given essentially by: (1) comparing the normal vector n_μ at a point p and the parallel transport of the normal vector n_μ at a nearby point q along a geodesic connecting q to p on the hypersurface Σ_t, and then (2) projecting the result onto the hypersurface Σ_t. The first part is clearly given by the covariant derivative, whereas the projection is done through the three-dimensional metric tensor. Hence the extrinsic curvature must be defined by

$$\begin{aligned} K_{\mu\nu} &= -h_\mu^\alpha h_\nu^\beta \, \nabla_\alpha \, n_\beta \\ &= -h_\mu^\alpha \, \nabla_\alpha \, n_\nu. \end{aligned} \tag{8.79}$$

In the second line of the above equation we have used $n^\beta \, \nabla_\alpha \, n_\beta = 0$ and $\nabla_\alpha \, g_{\mu\nu} = 0$. We can check that $K_{\mu\nu}$ is symmetric and tangent, viz

$$K_{\mu\nu} = K_{\nu\mu}, \quad h_\mu^\alpha K_{\alpha\nu} = K_{\mu\nu}. \tag{8.80}$$

The next goal is to compute in terms of the three-dimensional quantities the scalar curvature R. We start from (where G is the Einstein tensor $G_{\mu\nu} = R_{\mu\nu} - Rg_{\mu\nu}/2$)

$$
\begin{aligned}
R &= - Rg_{\mu\nu}n^{\mu}n^{\nu} \\
&= - 2(R_{\mu\nu} - G_{\mu\nu})n^{\mu}n^{\nu} \\
&= - 2R_{\mu\nu}n^{\mu}n^{\nu} + R_{\mu\nu\alpha\beta}h^{\mu\alpha}h^{\nu\beta}.
\end{aligned}
\tag{8.81}
$$

We compute

$$
\begin{aligned}
R_{\mu\nu\alpha\beta}h^{\mu\alpha}h^{\nu\beta} &= h_{\beta\rho}R_{\mu\nu\alpha}{}^{\rho}h^{\mu\alpha}h^{\nu\beta} \\
&= g^{\beta\eta}g^{\kappa\sigma}\Big(h_{\kappa}^{\ \mu}h_{\eta}^{\ \nu}h_{\sigma}^{\ \alpha}R_{\mu\nu\alpha}{}^{\rho}h_{\rho}^{\ \theta}\Big)h_{\theta\beta} \\
&= g^{\beta\eta}g^{\kappa\sigma}\Big((3)R_{\kappa\eta\sigma}{}^{\theta} + K_{\kappa\sigma}K_{\eta}^{\theta} - K_{\eta\sigma}K_{\kappa}^{\theta}\Big)h_{\theta\beta} \\
&= g^{\kappa\sigma}\Big((3)R_{\kappa\eta\sigma}{}^{\theta} + K_{\kappa\sigma}K_{\eta}^{\theta} - K_{\eta\sigma}K_{\kappa}^{\theta}\Big)h_{\theta}^{\eta} \\
&= (3)R + K^2 - K_{\mu\nu}K^{\mu\nu}.
\end{aligned}
\tag{8.82}
$$

In the third line we have used the first Gauss–Codacci relation. Next, we compute

$$
R_{\mu\nu}n^{\mu}n^{\nu} = \nabla_{\mu}\left(Kn^{\mu} + n^{\nu}\nabla_{\nu}n^{\mu}\right) - K_{\mu\nu}K^{\mu\nu} + K^2.
\tag{8.83}
$$

The first term is a total divergence and hence it can be neglected. We get then the so-called ADM Lagrangian density

$$
\begin{aligned}
\mathcal{L}_{\text{ADM}} &= \sqrt{-g}\,R \\
&= \sqrt{h}\,N((3)R - K^2 + K_{\mu\nu}K^{\mu\nu}).
\end{aligned}
\tag{8.84}
$$

In the above equation $K^2 = (h_{\mu\nu}K^{\mu\nu})^2$. The extrinsic curvature $K_{\mu\nu}$ is the covariant analog of the time derivative of the metric. Indeed, by using the concept of the Lie derivative we can show after some more steps that (where D_{μ} is the three-dimensional covariant derivative)

$$
K_{\mu\nu} = - \frac{1}{2N}(\dot{h}_{\mu\nu} - D_{\mu}N_{\nu} - D_{\nu}N_{\mu}).
\tag{8.85}
$$

It is straightforward now to compute the conjugate momentum $\Pi^{\mu\nu}$ corresponding to the metric $h_{\mu\nu}$. We find

$$
\begin{aligned}
\Pi_{\mu\nu} &= \frac{\mathcal{L}_{\text{ADM}}}{\partial\dot{h}_{\mu\nu}} \\
&= - \sqrt{h}(K_{\mu\nu} - Kh_{\mu\nu}).
\end{aligned}
\tag{8.86}
$$

From this identity we can show that $h_{\mu\nu}\Pi^{\mu\nu} = 2\sqrt{h}\,h_{\mu\nu}K^{\mu\nu}$.

The ADM Hamiltonian density is then given by

$$
\begin{aligned}
\mathcal{H}_{\text{ADM}} &= \dot{h}^{\mu\nu}\Pi_{\mu\nu} - \mathcal{L}_{\text{ADM}} \\
&= - \sqrt{h}\,N^{(3)}R + 2D_{\mu}N_{\nu}.\,\Pi^{\mu\nu} + \frac{N}{\sqrt{h}}(\Pi_{\mu\nu}\Pi^{\mu\nu} - \frac{1}{2}\Pi^2) \\
&= \sqrt{h}\,N\left[-^{(3)}R + \frac{\Pi_{\mu\nu}\Pi^{\mu\nu}}{h} - \frac{\Pi^2}{2h}\right] + \sqrt{h}\,N_{\nu}\left[-2D_{\mu}\left(\frac{\Pi^{\mu\nu}}{\sqrt{h}}\right)\right] \\
&= \sqrt{h}\,NH_0 + \sqrt{h}\,N_iH^i.
\end{aligned}
\tag{8.87}
$$

In the last two lines we have dropped a total divergence since it only leads to a boundary term which is assumed to be negligible for large spatial surfaces encompassing spacetime. Indeed, the corresponding ADM Lagrangian density takes the form

$$\mathcal{L}_{\text{ADM}} = \dot{h}_{\mu\nu}\Pi^{\mu\nu} - \sqrt{h}\,NH_0 - \sqrt{h}\,N_iH^i. \tag{8.88}$$

The Hamiltonian is then obtained by integrating the Hamiltonian density over the hypersurface Σ_t. We get

$$\begin{aligned} H_{\text{ADM}} &= \int d^3y\mathcal{H}_{\text{ADM}} \\ &= \int d^3y\sqrt{h}\,NH_0 + \int d^3y\sqrt{h}\,N_iH^i \\ &= H(N) + D(\vec{N}). \end{aligned} \tag{8.89}$$

Finally, by varying the Lagrangian density with respect to the lapse function N and the shift vector N^μ we obtain the Hamiltonian and diffeomorphsim first class constraints given, respectively, by

$$H_0 \equiv -^{(3)}R + \frac{\Pi_{\mu\nu}\Pi^{\mu\nu}}{h} - \frac{\Pi^2}{2h} = 0, \tag{8.90}$$

and

$$H_i \equiv -2D_\mu\left(\frac{\Pi^{\mu\nu}}{\sqrt{h}}\right) = 0. \tag{8.91}$$

This vanishing should be properly understood not as identically vanishing but as weakly vanishing in the sense of Dirac, i.e. it vansihes only on physical states not any state. We have then the constraints

$$H(N) \simeq 0, \tag{8.92}$$

and

$$D(\vec{N}) \simeq 0. \tag{8.93}$$

The Hamiltonian constraint $H(N) \simeq 0$ constrains the Hamiltonian (and it generates the time flow of the theory which connects different hypersurfaces Σ_t), whereas the diffeomorphism constraint $D(\vec{N}) \simeq 0$ constrains the momentum of the theory (and generates diffeomorphism transformations on the hypersurfaces Σ_t themselves). These constraints are first class which means that they do close under the Poisson brackets, i.e. their Dirac algebra is given by [2]

$$\begin{aligned} \left\{D(\vec{N}), D(\vec{N}')\right\} &= 8\pi GD(\mathcal{L}_{\vec{N}}\vec{N}') \\ \left\{D(\vec{N}), H(N')\right\} &= 8\pi GH(\mathcal{L}_{\vec{N}}N') \\ \{H(N), H(N')\} &= 8\pi GD(q^{-1}(NdN' - N'dN)). \end{aligned} \tag{8.94}$$

In the above equation $\mathcal{L}$ is the Lie derivative and q is the pullback metric, i.e. $q \equiv h$. This algebra is universal in the sense that it encodes in a precise sense the diffeomorphism invariance of the theory (despite the explicit choice of the foliation, i.e. the explicit choice of the lapse function N and the shift vector N^i which should only be viewed as a gauge fixing choice).

In fact, any theory characterized by invariance under general coordinate transformations will contain Hamiltonian and diffeomorphism constraints satisfying precisley the above Dirac algebra (this always comes about from the arbitrary nature of the foliation and the associated arbirary choice of the lapse function N and shift vector N^i which necessarily appear as Lagrange multipliers in the action with singular Legender transformation).

We remark that the Hamiltonian vansihes also weakly which is also another universal property of theories with diffeomorphism invariance, which is the fact that there is no Hamiltonian in the dynamics of these theories (since there is no time really!!) but only Hamiltonian constraint.

In summary, the phase space of the Hamiltonian formulation of general relativity consists therefore of all pairs (h_{ij}, Π^{kl}) where the extrinsic curvature K^{kl} stands in place of the momentum Π^{kl} through the relation $\Pi_{\mu\nu} = -\sqrt{h}(K_{\mu\nu} - Kh_{\mu\nu})$. The fundamental Poisson bracket is given by

$$\left\{ h_{ij}(t, \vec{y_1}), \Pi^{kl}(t, \vec{y_2}) \right\} = \delta^3(\vec{y_1} - y_2)\delta_i^k \delta_j^l. \tag{8.95}$$

The starting point of the canonical quantization program of geometrodynamics is then the commutation relations

$$[\hat{h}_{ij}(t, \vec{y_1}), \hat{\Pi}^{kl}(t, \vec{y_2})] = i\hbar\delta^3(\vec{y_1} - \delta y_2)\delta_i^k \delta_j^l. \tag{8.96}$$

The operators $\hat{h}_{ij}(t, \vec{y_1})$ and $\hat{\Pi}^{kl}(t, \vec{y_2})$ are defined on physical states $\Psi(h_{ij})$ by

$$\hat{h}_{ij}\Psi(h_{ij}) = h_{ij}\Psi(h_{ij})$$
$$\hat{\Pi}^{kl}\Psi(h_{ij}) = -i\hbar\frac{\delta}{\delta h_{kl}}\Psi(h_{ij}). \tag{8.97}$$

The physical states $\Psi(h_{ij})$ are thoses states in the Hilbert space which are annihilated by the Hamiltonian and diffeomorphism constraints which are also implemented as operators, namely

$$\hat{H}(N)\Psi(h_{ij}) = 0, \tag{8.98}$$

and

$$\hat{D}(\vec{N})\Psi(h_{ij}) = 0. \tag{8.99}$$

8.1.6 The Palatini action and Ashtekar variables

The Hilbert–Einstein action of general relativity expressed in terms of the 4-dimensional metric g is equivalent to the Palatini action expressed in terms of the vielbein field e and the spin connection ω. In the first formulation the affine connection is used implicitly since it is determined by the metric tensor, whereas in the second formulation the spin connection is used explicitly since it is an independent dynamical variable. The Palatini action is of the form (G being Newton's constant)

$$S = \frac{1}{16\pi G} \int d^4 x \, \epsilon_{mnkl} \, \tilde{\eta}^{\mu\nu\alpha\beta} e_\mu^m e_\nu^n R_{\alpha\beta}^{kl}. \tag{8.100}$$

The indices m, n, ... are internal indices associated with the local $SO(1, 3)$ Lorentz group, whereas the indices μ, ν, ... are external indices associated with spacetime (and consequently with the local diffeomorphism group of general coordinate transformations). The tensor $\tilde{\eta}$ is the Levi-Civita tensor density corresponding to the curved indices μ, ν, ... , whereas ϵ is the flat Levi-Civita symbol. The curvature (from the previous results) is defined by the relation

$$R_{\rho\sigma\mu\nu} = R_{\rho\sigma n}^m e_\mu^n e_m^\alpha g_{\alpha\nu}. \tag{8.101}$$

Now we use the result (with $g = \det(g_{\mu\nu})$)

$$\tilde{\eta}^{\alpha\beta\mu\nu} \epsilon_{klmn} e_\mu^m e_\nu^n = \frac{\sqrt{-g}}{2} (e_k^\alpha e_l^\beta - e_k^\beta e_l^\alpha). \tag{8.102}$$

The Palatini action becomes then (recall that the Ricci curvature tensor and the Ricci scalar are defined by $R_{\mu\nu} = R_{\mu\rho\nu}^\rho$ and $R = g^{\mu\nu} R_{\mu\nu}$)

$$\begin{aligned} S &= \frac{1}{16\pi G} \int \sqrt{-g} \, d^4 x \, e_k^\alpha e_l^\beta R_{\alpha\beta}^{kl} \\ &= \frac{1}{16\pi G} \int \sqrt{-g} \, d^4 x R. \end{aligned} \tag{8.103}$$

This is the Hilbert–Einstein action.

Alternatively, quantization based on the Palatini action gives immediately geometrodynamics of Wheeler, DeWitt and others. Indeed, the conjugate momentum associated with the spin connection ω_μ^{mn} is found to be given by $\Pi_{mn}^\mu = \tilde{\eta}^{\mu\nu\alpha} \epsilon_{mnkl} e_\nu^k e_\alpha^l$. The theory has thus an additional (second class) constraint consisting in the fact that the momentum is decomposable as a product of two vielbein fields. By solving the second class constraint (using Dirac's formalism) we obtain new canonical variables in which the spin connection is lost as a dynamical variable and we end up again with geometrodynamics [3].

The revolutionary solution provided by Ashtekar (see [4] for a modern review and for the original references) consists in insisting that the Palatini action is the correct starting point but with the additional twist that the spin connection must be self-dual. In other words, we must replace in the Palatini action the real $SO(1, 3)$ spin connection ω_μ^{mn} by the complex self-dual connection A_μ^{mn} defined by

$$A_\mu^{mn} = \frac{1}{2G} (\omega_\mu^{mn} - \frac{i}{2} \epsilon^{mn}_{\ \ kl} \omega_\mu^{kl}). \tag{8.104}$$

The complex connection A_μ^{mn} is self-dual because it satisfies the self-dual condition

$$iA_\mu^{mn} = \frac{1}{2}\epsilon^{mn}{}_{kl}A_\mu^{kl}. \tag{8.105}$$

The Palatini action becomes

$$S = \frac{1}{16\pi G}\int d^4x\,\epsilon_{mnkl}\,\tilde{\eta}^{\mu\nu\alpha\beta}e_\mu^m e_\nu^n F_{\alpha\beta}^{kl}. \tag{8.106}$$

F is the curvature tensor of the self-dual connection A. Thus, it must be given by

$$F_{\alpha\beta m}^n = \partial_\alpha A_{\beta m}^n - \partial_\beta A_{\alpha m}^n + G^4 A_{\alpha m}^k A_{\beta k}^n - G^4 A_{\beta m}^k A_{\alpha k}^n. \tag{8.107}$$

We can check that the classical equations of motion derived from the self-dual Palatini action (8.106) are exactly equivalent to the classical equations of motion derived from the original Palatini action (8.100). In particular, the variation of the action (8.106) with respect to the connection A_μ^{mn} gives as equation of motion the result that A_μ^{mn} is the (self-dual part of the) spin connection ω_μ^{mn} which is compatible with the vielbein field e_μ^m, i.e. it is determined by the condition $\mathcal{D}_\rho e_\mu^m = 0$. Hence, the connection A_μ^{mn} is completely determined by the vielbein field e_μ^m. On the other hand, the variation of the action (8.106) with respect to the vielbein field e_μ^m gives as equation of motion the result that the spacetime metric $g_{\mu\nu} = e_\mu^m e_\nu^n \eta_{mn}$ solves Einstein's equations.

Thus the classical equations of motion derived from the self-dual Palatini action (8.106) are exactly equivalent to the classical equations of motion derived from the standard Palatini action (8.100). But this does not mean that the two actions are identical. Indeed, the difference between (8.106) and (8.100) is an imaginary term which is not a pure divergence but reproduces as a correction to the equation of motion the first Bianchi identity (the trace of the dual of the Riemann tensor vanishes) which thus holds automatically. This imaginary term leads, however, under the Legendre transform of the self-dual Palatini action to a different conjugate momentum which is linear instead of being quadratic in the vielbein field and that makes the self-dual Palatini action (8.106) distinctly different and quite superior to the standard Palatini action (8.100) which is nothing else but the Hilbert–Einstein action. See [4] and references therein.

8.1.7 General relativity as an $SU(2)$ gauge theory of self-dual spin connections

We start with the Hilbert–Einstein action with variable given only by the metric tensor $g_{\mu\nu}$ (the affine Levi-Civita connection $\Gamma_{\mu\nu}^\rho$ is not an independent variable here). Then by performing a Legendre transform and an ADM analysis the canonically conjugate variables are from the one hand the three-dimensional metric $h_{\mu\nu}$ or equivalently the vielbein fields (or triads) e_μ^m and from the other hand we have the corresponding conjugate momentum $\Pi_{\mu\nu}$ defined in terms of the extrinsic curvature $K_{\mu\nu}$ by the relation $\Pi_{\mu\nu} = -\sqrt{h}\,(K_{\mu\nu} - Kh_{\mu\nu})$. We are here only dealing with geometrodynamics where only first class constraints are involved.

If we start on the other hand with the standard Palatini action (8.100) with variables given by the real vielbein field e_μ^m and the real spin connection ω_μ^{mn} then the canonically conjugate variables are the spin connection ω_μ^{mn} and the conjugate momentum $\Pi_{mn}^\mu = \tilde{\eta}^{\mu\nu\alpha}\epsilon_{mnkl}e_\nu^k e_\alpha^l$. By solving the second class constraint (the momentum is decomposable as a product of two vielbein fields) we obtain new canonical variables in which the spin connection is lost as a dynamical variable and we end up again with geometrodynamics.

We also recall that the vielbein field e_μ^m is covariantly constant with respect to both the spin connection ω_μ^{mn} and the Levi-Civita connection $\Gamma_{\mu\nu}^\alpha$, viz

$$\mathcal{D}_\rho e_\mu^m = \partial_\rho e_\mu^m - \Gamma_{\rho\mu}^\nu e_\nu^m + \omega_{\rho n}^m e_\mu^n$$
$$= 0. \tag{8.108}$$

This compatibility condition gives the spin connection in terms of the Levi-Civita connection and the vielbein field.

However, the theory formulated in terms of the Ashtekar variables given by a self-dual spin connection A_μ^{mn} which is necessarily complex and a real vielbein field (or tetrads) e_μ^m with an action given by the self-dual Palatini action (8.106) is equivalent to a complex $SU(2)$ gauge theory of the self-dual spin connections. Indeed, after Legendre transform the canonically conjugate variables are found to be the self-dual spin connection A_μ^{mn} with a corresponding conjugate momentum Π_{mn}^μ which is also self-dual and furthermore is proportional to a single vielbein field not two and hence second class constraints are avoided.

Here it is technically simpler to start with complex general relativity since the connection is necessarily complex. So we start with a complex vielbein field e_μ^m and a complex self-dual $SO(1,3)$ connection A_μ^{mn} with an action given by the Palatini action (8.106). After Legendre transform we get as our canonically conjugate variables the connection A_μ^{mn} and the conjugate momentum Π_{mn}^μ which are both in the self-dual part of the complexified $so(1,3)$ Lie algebra.

The original spin connection ω_μ^{mn} is a real $SO(1,3)$ connection and recall that $SL(2,\mathbb{C})$ is the universal cover of $SO(1,3)$. The self-dual connection A_μ^{mn} belongs, however, to the complexified group $SO(1,3)_\mathbb{C}$. We have the Lie algebra isomorphisms

$$so(1,3)_\mathbb{C} = so(4)_\mathbb{C} = so(3)_\mathbb{C} \oplus so(3)_\mathbb{C}. \tag{8.109}$$

The first $so(3)_\mathbb{C}$ factor represents self-dual (chiral, right-handed) fields, whereas the second factor represents anti-self-dual (anti-chiral, left-handed) fields. The connection A_μ^{mn} is a complex connection (thus belonging to $so(1,3)_\mathbb{C}$ not to $so(1,3)$ like the connection ω_μ^{mn}) which is also self-dual (thus it belongs to the first factor $so(3)_\mathbb{C}$).

We are therefore dealing with an $so(3)_\mathbb{C}$-valued one-form and since the universal cover of $SO(3)$ is $SU(2)$ the connection A_μ^{mn} is in fact an $su(2)_\mathbb{C}$-valued one-form.

Using the above isomorphism between the self-dual subalgebra of the complexified Lie algebra $so(1,3)_\mathbb{C}$ and the complexfied Lie algebra $so(3)_\mathbb{C}$ we can map the canonically conjugate variables A_μ^{mn} and Π_{mn}^μ to the $so(3)_\mathbb{C}$-valued fields A_μ^n and Π_μ^n given, respectively, by

$$A_\mu^m = \frac{1}{2} A_{\mu kl}\,\epsilon^{klm}, \quad \Pi_\mu^m = \frac{1}{2}\Pi_{\mu kl}\,\epsilon^{klm}. \tag{8.110}$$

The self-dual connection A_μ^m is also called the chiral spin connection. As it turns out, the canonical momentum Π_μ^m is precisely the densitized vielbein field (or triad) given by

$$\Pi_\mu^m = \tilde{e}_\mu^m = \sqrt{h}\,e_\mu^m. \tag{8.111}$$

The Ashtekar variables are precisley the densitized triad $\tilde{e}_\mu^m$ and the self-dual connection A_μ^m. The self-dual Palatini action in terms of these variables takes the form

$$S = \int d^4x \left(-2i\tilde{e}_\mu^m \mathcal{L}_t A_m^\mu - 2i(t^\mu A_\mu^m)G_m + 2iN^\mu \mathcal{V}_\mu + \frac{N}{\sqrt{h}}\mathcal{S} \right). \tag{8.112}$$

The quantities G_m, $\mathcal{V}_\mu$ and $\mathcal{S}$ are explicitly given by

$$G_m = \mathcal{D}_\mu \tilde{e}_m^\mu, \quad \mathcal{V}_\mu = \tilde{e}_n^\nu F_{\mu\nu}^n, \quad \mathcal{S} = \epsilon_{ijk}\tilde{e}_i^\mu \tilde{e}_j^\nu F_{\mu\nu}^k. \tag{8.113}$$

The curvature $F_{\mu\nu}^k$ of the gauge field A_μ^k is explicitly given by

$$F_{\mu\nu}^l = \partial_\mu A_\nu^l - \partial_\nu A_\mu^l + G\epsilon_{lmn}A_\mu^m A_\nu^n. \tag{8.114}$$

This shows explicitly that we are indeed dealing with an $SU(2)$ gauge theory.

In the first term of the self-dual Palatini action (8.112) the operator $\mathcal{L}_t$ is the Lie derivative along the time direction and hence $\mathcal{L}_t A_m^\mu$ is the covariant time derivative of the field configuration A_m^μ along the vector field $t^\mu = Nn^\mu + N^\mu$ which defines the spacetime foliation with hypersurfaces Σ_t whose normal vector field is given by n^μ (N and N^μ are then the lapse function and the shift vector).

From the first term in the action (8.112) which is then of the form $p\dot{q}$ we can immediately conclude that the densitized triad $\tilde{e}_\mu^m$ is precisely the conjugate momentum p associated with the self-dual connection A_μ^m which acts exactly as the configuration variable q. Indeed, the fundamental Poisson brackets are of the form

$$\left\{ A_\mu^m(x), \tilde{e}_n^\nu(y) \right\} = \frac{i}{2}\delta_m^n \delta_\mu^\nu \delta^3(x - y). \tag{8.115}$$

In summary, we have gone from the ADM variables consisting of the three-dimensional metric $h_{\mu\nu}$ (or equivalently the densitized triads $\tilde{e}_\mu^m$) and the canonical momentum $\Pi_{\mu\nu}$ (or equivalently the extrinsic curvature K_μ^m defined by $K_\mu^m = K_{\mu\nu}e^{\nu m}$) to the complex Ashtekar variables consisting of $\tilde{e}_\mu^m$ and the connection A_μ^m. The relation between the self-dual connection A_μ^m and the original variables $\tilde{e}_\mu^m$ and K_μ^m is given explicitly by

$$GA_\mu^m = \Gamma_\mu^m - iK_\mu^m. \tag{8.116}$$

The spin connection Γ_μ^m which is compatible with the densitized triads $\tilde{e}_\mu^m$ is given obviously by the relation

$$\Gamma_\mu^m = \frac{1}{2}\omega_{\mu k l}\,\epsilon^{klm}. \tag{8.117}$$

See [5] and references therein.

At the end of all this we will naturally need to return to real (Lorentzian) general relativity and thus one must impose reality conditions. In terms of the geometrodynamic variables these reality conditions are simply the requirements that the three-dimensional metric $h_{\mu\nu} = e_\mu^m e_\nu^n \eta_{mn}$ and the extrinsic curvature $K_{\mu\nu}$ must be real. Let us emphasize here that the self-dual spin connection A_μ^{mn} given by equation (8.105) is necessarily complex (since the spacetime manifold is Lorentzian) and thus the reality conditions will not alter this fact. But, in Euclidean signature the self-dual connections are necessarily real and thus the reality conditions which are needed to be imposed on complex general relativity to recover the real phase space are the requirements that the triads must be real and the connections must also be real. In Lorentzian signature the connections will remain complex after imposing the reality conditions (which will cause other problems for the integration measure in the quantum theory).

The reality conditions can also be understood in a more illuminating way as follows.

We start with real general relativity, i.e. real vielbein field and real spin connection in the Palatini action. After Legendre transform we can take as our variables the densitized triads $\tilde{e}_\mu^m$ (instead of the three-dimensional metric $h_{\mu\nu}$) and the extrinsic curvature e_μ^m (instead of the momentum $\Pi_{\mu\nu}$).

On the real phase space $(q, p) \equiv (\tilde{e}_\mu^m, K_\mu^m)$ we perform then a complex canonical transformation which takes us to the complex Ashtekar variables (q, z) where z is the complex coordinate given by $z = f(q) - ip$. Explicitly, $f(q)$ is the spin connection Γ_μ^m which is determined by the densitized triads $\tilde{e}_\mu^m$ and z is precisely the self-dual connection $GA_\mu^m = \Gamma_\mu^m - iK_\mu^m$. As we have seen, $\tilde{e}_\mu^m$ and A_μ^m are canonically conjugate to each other, where the densitized triads are what play the role of the conjugate momentum in the Ashtekar variables (q, z) contrary to their role in the original real coordinates (q, p), i.e. $(q^A, p^A) \equiv (A_\mu^m, \tilde{e}_\mu^m)$.

The reality conditions are now given by the requirement that the three-dimensional metric $h_{\mu\nu} = e_\mu^m e_\nu^n \eta_{mn}$ is real and the requirement that $GA_\mu^m - \Gamma_\mu^m$ is pure imaginary.

In the Palatini action (8.112) which is written in terms of Ashtekar variables the second, third and fourth terms lead to the constraints. Indeed, the lapse function N, the shift vector N^μ and the component of the connection A_μ along the time direction, i.e. $t^\mu A_\mu^m$ are all Legendre multipliers and the variation of the action with respect to them will lead to the constraints

$$G_m = \mathcal{D}_\mu \tilde{e}_m^\mu = 0. \tag{8.118}$$

$$\mathcal{V}_\mu = \tilde{e}_n^\nu F_{\mu\nu}^n = 0. \tag{8.119}$$

$$\mathcal{S} = \epsilon_{ijk} \tilde{e}_i^\mu \tilde{e}_j^\nu F_{\mu\nu}^k = 0. \tag{8.120}$$

These seven first class constraints are simple polynomials in the basic variables (as opposed to what happens in geometrodynamics). And, they reduce the nine degrees of freedom of A_μ^i to the two degrees of freedom of the graviton.

The constraints (8.119) and (8.120) are the diffeomorphism and Hamiltonian constraints found in geometrodynamics which generate, respectively, spatial diffeomorphisms on each surface Σ_t and time evolution between different surfaces Σ_t and $\Sigma_{t+\delta t}$.

The first constraint (8.118) is the so-called Gauss constraint and it represents Gauss law in this gauge theory and generates local $SO(3)$ invariance of the triads. It arises from the fact that the time component of the connection A_μ is not a dynamical field. We are really dealing with an $SU(2)$ gauge theory on the the three-dimensional surfaces Σ_t with gauge field A_μ^m and since $\tilde{e}_m^\mu$ is the conjugate momentum it will act as the electric field E_m^μ with a quantized flux leading to quantized geometry and discretized spacetime (in the form of discrete spectra of areas and volumes) and also leads to the absence of gravitational singularities.

Thus, the Gauss constraint and the diffeomorphism constraints generate the local invariance group which is the semi-direct product of the local $SO(3)$ rotation group of the triads and the spatial diffeomorphism group on Σ_t. On the other hand, the scalar constraint is of the form $G^{\alpha\beta} p_\alpha p_\beta = 0$ where G is the supermetric and thus this constraint generates null geodesics motion in the configuration space of the connection.

Similarly to the constraints, the Hamiltonian and the equations of motion are all low order polynomials in the basic variables of Ashtekar.

8.1.8 The real $SU(2)$ gauge theory

The self-dual or chiral connection A_μ^m in Lorentzian signature is a complex $SU(2)$ gauge field which means in particular that the corresponding holonomies or Wilson loops (which define the obeservables of the quantum gauge theory) are non-compact, i.e. they belong to a non-compact subgroup $SL(2, \mathbb{C})_{sd}$ of $SL(2, \mathbb{C})$ generated by the self-dual part of the Lie algebra $sl(2, \mathbb{C})$. As a consequence, the path integrals defining the quantum theory are ill-defined and require a regularization in the form of a Wick rotation in the internal space which sends the non-compact group $SL(2, \mathbb{C})_{sd}$ to the compact group $SU(2)$, i.e. the chiral connection A_μ^m given by (8.116) is replaced with a real $SU(2)$ gauge field given by

$$G A_\mu^m = \Gamma_\mu^m + \beta K_\mu^m. \tag{8.121}$$

In other words, we replace the '-i' in (8.116) with a real parameter β in (8.121) called the Barbero–Immirzi parameter [6, 7]. The fundamental Poisson brackets become

$$\left\{ A_\mu^m(x), \tilde{e}_n^\nu(y) \right\} = -\frac{\beta}{2} \delta_m^n \delta_\mu^\nu \delta^3(x - y). \tag{8.122}$$

The choice of the parameter does not alter the Gauss and the spatial diffeomorphism constraints. But the Hamiltonian (which is a linear combination of the constraints) acquires an additional term, viz

$$H = \frac{\epsilon_{lmn}\tilde{e}_m^\mu \tilde{e}_n^\nu F_{\mu\nu}^l}{\sqrt{h}} + 2\frac{\beta^2 + 1}{\beta^2}\frac{\tilde{e}_m^\mu \tilde{e}_n^\nu - \tilde{e}_n^\mu \tilde{e}_m^\nu}{\sqrt{h}}(A_\mu^m - \Gamma_\mu^m)(A_\nu^n - \Gamma_\nu^n). \qquad (8.123)$$

The second term vanishes for $\beta = \pm 1$. This Hamiltonian was simplified by Thiemann. See, for example, his book [8].

8.1.9 Loop representation and spin networks

We have now a real $SU(2)$ gauge theory with a connection or gauge field A_μ^m living on a three-dimensional surface Σ_t. The classical configuration space (the space of all connections A_μ^m) is denoted by $\mathcal{A}$ while the quantum configuration space is denoted by $\bar{\mathcal{A}}$ will constitue of holonomies of connections along paths in Σ_t.

Each connection A_μ^m defines a holonomy $h_\alpha[A]$ along any oriented path on the surface Σ_t, i.e. α: $[s_0, s_1] = [0, 1] \longrightarrow \Sigma_t$ with an affine parameter s by the relation [9]

$$h_\alpha[A] = U(s_1, s_0) = \mathcal{P}\exp\left(-\int_0^1 ds\dot{\alpha}^\mu(s)A_\mu^m(\alpha(s))T_m\right). \qquad (8.124)$$

The $\mathcal{P}$ is the usual path ordering operation (operators with larger values of s are placed to the left of the operators with to smaller values of s) and T_m are the usual generators of $SU(2)$ which satisfy the Lie algebra

$$[T_m, T_n] = i\epsilon_{mnl}T_l. \qquad (8.125)$$

The set of all holonomies $h_\alpha[A]$ define the quantum configuration space $\bar{\mathcal{A}}$ in the same way that the set of all connections A_μ^m define the classical configuration space $\mathcal{A}$. A holonomy (called generalized connection in [10]) is a map on the space of all paths in Σ_t which assigns an element of the group $SU(2)$ to each path $\alpha(t)$ (in contrast to the connection which is a map on the hypersurface Σ_t which assigns an element of the Lie algebra $su(2)$ to each point on the surface).

The holonomy defines the parallel transport of a spinor in the background of the configuration A_μ^m along the curve $\alpha(t)$ between the start point $\alpha(0)$ and the end point $\alpha(1)$ and it measures the accumulated phase difference between the initial and final values of the spinor at the two points. More explicitly, under gauge transformations g we must have

$$h_\alpha \longrightarrow h_\alpha' = g^{-1}(\alpha(1))h_\alpha g(\alpha(0)). \qquad (8.126)$$

This defines the so-called generalized gauge transformations [10] which act on holonomies only at the endpoints of paths in contrast to ordinary gauge transformations which act on connections at every point of Σ_t. They are generated by the Gauss gauge constraint which can then be solved explicitly by using only Wilson loops which are holonomies traced out around closed paths (loops), viz

$$W_\alpha[A] = Trh_\alpha[A]. \tag{8.127}$$

These are gauge invariant by construction. Any Gauss gauge invariant function $\Psi[A]$ in the connection representation is then expanded in terms of Wilson loops as

$$\Psi[A] = \sum_\alpha \Psi[\alpha]W_\alpha[A]. \tag{8.128}$$

This is called the loop transform. The function $\Psi[\alpha]$ defines the loop representation and it is given by the inverse loop transform

$$\Psi[\alpha] = \int [dA]\Psi[A]W_\alpha[A]. \tag{8.129}$$

In the loop representation we can also solve the spatial diffeomorphism constraint by considering functions $\Psi[\alpha]$ which are invariant under diffeomorphisms of the loop α. These are knot invariants [11]. The Hamiltonian constraint is solved on the other hand by non-intersecting Wilson loops.

If we denote the set of all generalized gauge transformations by $\bar{\mathcal{G}}$ then the gauge invariant quantum configuration space must be given by the quotient $\bar{\mathcal{A}}/\bar{\mathcal{G}}$. This space can be viewed as a projective limit of a family of compact, smooth and finite dimensional configuration spaces $\{\bar{\mathcal{A}}_\alpha/\bar{\mathcal{G}}_\alpha\}$.

Each configuration space $\bar{\mathcal{A}}_\alpha/\bar{\mathcal{G}}_\alpha$ is labeled by a path or graph α characterized by N edges and V vertices (each edge e starts at a vertex v_{e_0} and ends at a vertex v_{e_1}). From the one hand, the space $\bar{\mathcal{A}}_\alpha$ is the space of generalized connections or holonomies over the graph α which consists of the mappings which assign to each edge of the graph an element of the group $SU(2)$, i.e. $\bar{\mathcal{A}}_\alpha$ is isomorphic to $SU(2)^N$. From the other hand, the space $\bar{\mathcal{G}}_\alpha$ is the space of generalized gauge transformations over the graph α which consists of all mappings which assign to each vertex an element of $SU(2)$. More precisely, the action of a given generalized gauge transformation g on an edge e of the graph α is given by $h_\alpha(e) \longrightarrow h'_\alpha(e) = g^{-1}(v_{e_1})h_\alpha(e)g(v_{e_0})$. Hence the space $\bar{\mathcal{G}}_\alpha$ is isomorphic to $SU(2)^V$ and as a consequence the configuration space $\bar{\mathcal{A}}_\alpha/\bar{\mathcal{G}}_\alpha$ is isomorphic to $SU(2)^{N-V}$.

The Hilbert space of state vectors on the configuration space $\bar{\mathcal{A}}_\alpha/\bar{\mathcal{G}}_\alpha$ (which is compact and finite dimensional) is precisely the space of square-integrable functions $\mathcal{H}_\alpha = L^2(\bar{\mathcal{A}}_\alpha/\bar{\mathcal{G}}_\alpha)$ where the measure is obviously induced by the usual Haar measure on $SU(2)$.

Elementary quanta of geometry are elements of $\mathcal{H}_\alpha$ and they are of the form [10]

$$\Psi_\alpha(h_\alpha) = \psi(h_\alpha(e_1), ..., h_\alpha(e_N)), \ \psi \in SU(2)^N. \tag{8.130}$$

In summary, the members of the family $\{\bar{\mathcal{A}}_\alpha/\bar{\mathcal{G}}_\alpha\}$ are all compact and finite dimensional spaces with corresponding Hilbert spaces $\mathcal{H}_\alpha = L^2(\bar{\mathcal{A}}_\alpha/\bar{\mathcal{G}}_\alpha)$ and as a consequence the projective limit $\bar{\mathcal{A}}/\bar{\mathcal{G}}$ which is also compact admits a regular Borel measure allowing the construction of a corresponding Hilbert space of square-integrable functions $\mathcal{H} = L^2(\bar{\mathcal{A}}/\bar{\mathcal{G}})$ where the measure is also induced by the usual Haar measure on $SU(2)$.

The Hilbert space $\mathcal{H} = L^2(\bar{\mathcal{A}}/\bar{\mathcal{G}})$ admits a more interesting decomposition as a direct sum of finite dimensional orthogonal Hilbert spaces $\mathcal{H}_{\alpha\vec{j}}$ characterized together with the graph α by a vector $\vec{j}$ of half-integers, i.e. $\vec{j} = (j_1, j_2, ..., j_N)$ where the integer j_i represents the irreducible representation of $SU(2)$ which labels the edge i of the graph α.

We have then [12]

$$\mathcal{H} = L^2(\bar{\mathcal{A}}/\bar{\mathcal{G}}) = \oplus_{\alpha\vec{j}} \mathcal{H}_{\alpha\vec{j}}. \tag{8.131}$$

$\mathcal{H}_{\alpha\vec{j}}$ are Hilbert spaces of spin network states on the configuration space $\bar{\mathcal{A}}_\alpha/\bar{\mathcal{G}}_\alpha$.

A spin network state $\Psi_{\alpha\vec{j},\vec{i}}(A)$ is then an element of $\mathcal{H}_{\alpha\vec{j}}$ which is characterized by: (1) the graph α, (2) the N irreducible representations j_i associated with the edges of the graph, and (3) the V intertwining operators i_i associated with the vertices of the graph.

More precisely, let ρ_e be the irreducible representation associated with the edge e, i.e. ρ_e is the homomorphism $\rho_e: SU(2) \longrightarrow \text{End}(V_e)$ where $\text{End}(V_e)$ is the group of endomorphisms of a vector space V_e. Then let $S(v)$ be the set of edges with the vertex v as a source (start point) and let $T(v)$ be the set of edges with the vertex v as a target (end point). An intertwining operator I_v is a linear endomorphism between the two vector spaces $\otimes_{e \in S(v)} V_e$ and $\otimes_{e \in T(v)} V_e$.

The intertwining operator I_v can also be understood as an invariant element of the representation $\otimes_{e \in S(v)} V_e \otimes \otimes_{e \in T(v)} V_e^\star$.

Thus, if $\vec{j}$ is a labeling of the edges of the graph α by irreducible representations of $SU(2)$, similarly, $\vec{i}$ is a labeling of the vertices of the graph α by intertwining operators from the tensor product of incoming representations to the tensor product of the outgoing representations [13].

For example, in Penrose spin networks [14] we consider trivalent graphs labeled by spins j satisfying the rule that if the spins of the edges at a given vertex are j_1, j_2 and j_3 then these three spins must satisfy the rules of conservation of angular momentum, i.e. the Clebsch–Gordon condition $|j_1 - j_2| \leqslant j_3 \leqslant j_1 + j_2$ must hold. This Clebsch–Gordon condition is a necessary and a sufficient condition for the existence of intertwining operators from $j_1 \otimes j_2$ to j_3.

The set of all spin network states with all possible graphs α, all assignments $\vec{j}$ of irreducible representations of $SU(2)$ to the edges, and all assignments $\vec{i}$ of intertwining operators to the vertices form the Hilbert space $\mathcal{H} = L^2(\bar{\mathcal{A}}/\bar{\mathcal{G}})$.

The spin network states $\Psi_{\alpha\vec{j},\vec{i}}(A)$ are eigenvectors of area and volume operators in the hypersurfaces Σ_t with discrete spectra [15].

For example, the area of a two-dimensional surface $\mathcal{S}$ in the three-dimensional hypersurface Σ_t is given by

$$A_\mathcal{S} = \int dx^1 dx^2 \sqrt{\det h^{(2)}}. \tag{8.132}$$

The metric $h_{ab}^{(2)}$ is the induced metric on the surface $\mathcal{S}$. This can be expressed in terms of the metric h_{ab}, then in terms of the densitized triads $\tilde{e}_m^\mu$ using the relation $\tilde{e}_m^\mu \tilde{e}^{\nu m} = h^{\mu\nu} \det h$.

In the quantum theory the densitized triads become operators $\widehat{\tilde{e}}_m^\mu$ which act (up to a numerical factor) as $\delta/\delta A_\mu^m$ and hence the area $A_{\mathcal{S}}$ becomes an area operator $\hat{A}_{\mathcal{S}}$ which admits the Wilson loops (8.127) as eigenvectors, viz [16]

$$A_{\mathcal{S}} W_\alpha[A] = 8\pi l_{\mathrm{P}}^2 \beta \sum_I \sqrt{j_I(j_I + 1)}\, W_\alpha[A]. \tag{8.133}$$

The sum is over all edges of the Wilson loop α that intersect the surface $\mathcal{S}$. The surface areas of $\mathcal{S}$ are then quantized given by $8\pi l_P^2 \beta \sum_I \sqrt{j_I(J_I + 1)}$ where j_I is the spin associated with the edge I, β is the Immirzi parameter and l_P is the Planck length.

8.2 BFSS matrix model and gauge/gravity duality

8.2.1 The action and symmetries

- The objective here is to explore the basics of the BFSS matrix model, known also as M-(atrix) theory, which is the oldest Yang–Mills matrix quantum mechanics proposed for quantum gravity or M-theory. More precisley, we aim at understanding the connection of this model to D0-branes, black holes, supermembranes and non-commutative geometry. For pedagogical expositions of this subject, see also [17–20].
- This model was proposed in 1996 by Tom Banks, Willy Fischler, Stephen Shenker, and Leonard Susskind [21]. It aimed at providing a non-perturbative definition of M-theory, an 11-dimensional theory that unifies all consistent versions of superstring theory.
- It is in fact related to other matrix models like the IKKT (Ishibashi–Kawai–Kitazawa–Tsuchiya) model, which provides a different non-perturbative formulation of string theory [22].
- The BFSS matrix model is also related to the BMN (Berenstein–Maldacena–Nastase) model [23]. The BMN model is a maximally supersymmetric deformation of the BFSS model. They are both significant frameworks in string theory and M-theory, each providing valuable insights into the non-perturbative dynamics of these theories.
- This model is derived from the dynamics of D0-branes in type IIA string theory. D0-branes are point-like objects whose interactions can be described by a matrix quantum mechanics. The D0-branes are described by $N \times N$ Hermitian matrices, where N is the light-cone quantized momentum (in unit of the inverse of the compactification radius R) which corresponds also to the number of D0-branes.
- It is formulated as a supersymmetric quantum mechanics of nine $N \times N$ Hermitian matrices given by a (0+1)-dimensional gauge theory with 16 supercharges.

- Lower-dimensional versions of this matrix quantum mechanics with a smaller number of matrices also exist with and without maximally supersymmetric mass deformations.
- The action S of the BFSS matrix model is given explicitly by

$$S = \int dt \ \mathrm{Tr}\left(\frac{1}{2R}(D_t X^i)^2 + \frac{R}{4}[X^i, X^j]^2 + \psi^T D_t \psi - iR\psi^T\gamma^i[X^i, \psi]\right). \quad (8.134)$$

Here:
  - R is the compactification radius.
  - X^i are nine $N \times N$ Hermitian bosonic matrices ($i = 1, ..., 9$).
  - $\psi(t)$ is a 16-component nine-dimensional spinor composed of $N \times N$ Hermitian fermionic matrices.
  - D_t is the covariant derivative $D_t = \partial_t - i[A, \cdot]$.
  - A is the gauge field.
  - γ^i are nine-dimensional Dirac matrices which satisfy the anticommutation relations:

$$\{\gamma^i, \gamma^j\} = 2\delta^{ij} \quad (8.135)$$

  - **Path integral:** The path integral is given by

$$Z = \int \prod_i \mathcal{D}X^i \prod_\alpha \mathcal{D}\psi_\alpha \exp(iS[X, \psi]). \quad (8.136)$$

  - **Gauge symmetry:** The local gauge group of the BFSS matrix model is $SU(N)$ which consists of $N \times N$ unitary matrices with determinant 1, i.e. physical quantities are invariant under the gauge transformations

$$X^i \longrightarrow UX^iU^{-1}, \quad \psi \longrightarrow U\psi U^{-1}. \quad (8.137)$$

  Here, U is a time-dependent unitary matrix.
  - **Supersymmetry:** This model has $\mathcal{N} = 16$ supersymmetry, providing a rich structure and protection against certain quantum corrections and ensures stability of the vacuum state. Supersymmetry plays a crucial role in the non-perturbative analysis, as it leads to cancellations of certain divergences and helps in the understanding of the spectrum and dynamics of the theory.

  Explicitly, we have the supersymmetric transformations

$$\delta X^i = \sqrt{R}\,\epsilon\gamma^i\psi \quad (8.138)$$

$$\delta\psi = i\sqrt{R}\left(\frac{1}{4}[X^i, X^j]\gamma^{ij} + \frac{1}{2}D_t X^i\gamma^i\right)\epsilon + \epsilon' \quad (8.139)$$

$$\delta A = R\epsilon\psi. \quad (8.140)$$

8.2.2 The non-perturbative dynamics

The non-perturbative dynamics of the BFSS matrix model revolves around the following axes:

- **M-theory and D0-branes:** The BFSS model is originally a proposal for M-theory quantization and D0-branes dynamics.
  - The BFSS model is conjectured to describe the discrete light-cone quantization (DLCQ) of M-theory in 11 dimensions.
  - It describes the dynamics of D0-branes which in type IIA string theory correspond to the light-cone quantized momentum modes in M-theory. The non-perturbative interactions between these D0-branes are captured by the commutator terms $[X^i, X^j]$ in the action. At low energies, the model reduces to a quantum mechanics of these D0-branes, with their bound states representing various string and M-theory objects.

- **Black holes and gauge/gravity duality:** The BFSS model is another instance of the holographic principle, i.e. of the gauge/gravity duality which is quite distinct yet intimately connected to the anti-de Sitter/conformal field theory (AdS/CFT) correspondence.
  - It connects to black hole physics through the holographic matrix degrees of freedom representing the microscopic states of black holes.
  - The large N limit of the BFSS matrix model provides a way to capture the gravitational dynamics of the 11-dimensional M-theory, similar to how the AdS/CFT correspondence relates a higher-dimensional gravity theory in AdS space to a lower-dimensional boundary CFT.
  - M-theory compactified on a circle results in a theory that can be described by the dynamics of D0-branes, whose interactions are captured by the BFSS model. This scenario is similar to how AdS/CFT handles the compactification of higher-dimensional theories to AdS spaces.

- **Supermembranes and non-commutative geometry:** The BFSS model has significant implications for supermembranes and non-commutative geometry.
  - In the BFSS model, the variables X_i are interpreted as the Fourier modes of the transverse coordinates of supermembranes in the light-cone gauge. This approach is consistent with the holographic principle, where bulk gravitational degrees of freedom (membranes) are encoded in lower-dimensional variables (matrices).
  - The non-commutativity of the matrices representing the positions of the D0-branes leads to intriguing phenomena in the BFSS matrix model, such as the emergence of spacetime from the dynamics of the matrices. Indeed, the matrices X^i can be interpreted as coordinates in a non-commutative space, leading to a better understanding of spacetime at the Planck scale.

In summary, we have in this model a rich and varied quantum physics centered around quantum gravity effects. In particular, we have:

- **Black holes:** The BFSS model provides a non-perturbative framework for studying black hole physics in M-theory. The bound states of D0-branes can form black hole solutions, and their dynamics can be analyzed within this model. This approach has been used to derive entropy and other thermodynamic properties of black holes, providing a microscopic understanding consistent with holographic principles.

- **Gauge/gravity duality:** The non-perturbative dynamics of the BFSS model contribute to the understanding of the gauge/gravity duality. The model's dynamics can be mapped onto gravitational systems in certain limits, providing a concrete realization of the holographic principle.
- **Supersymmetric ground state:** Determining the exact nature of the supersymmetric ground state and the spectrum of excitations is also a non-perturbative problem. Supersymmetry plays a critical role in constraining the possible states and ensuring stability.
- **Emergent geometry:** The matrices X^i can be interpreted as non-commutative coordinates in some non-perturbative phases of the theory, leading to a non-commutative emergent geometry, which becomes commutative in the large N limit. This non-commutative nature of the BFSS model and the corresponding emergent geometry provides a framework for understanding spacetime at the Planck scale, where classical geometry breaks down.
- **Continuum and large N limits:** The large N limit of the BFSS model is crucial for its connection to M-theory and other physical interpretations. Understanding the exact dynamics in this limit remains a significant challenge, requiring advanced computational techniques and theoretical insights. The continuum limit needs also to be better understood to fully capture the non-perturbative regime of the physics of M-theory and other interpretations.
- **Non-perturbative approach:** Computational approaches and numerical studies (e.g. lattice regularizations, Monte Carlo simulations, etc) are employed to access and study the non-perturbative regime of the BFSS matrix model.
  - **Lattice regularizations:** Non-perturbative effects in the BFSS model are often studied using numerical methods such as lattice simulations. These simulations start by going to the Euclidean formulation via a Wick rotation, then they discretize time by introducing a lattice a, and introduce a temperature by imposing periodic boundary conditions. This allows us to study the dynamics of the matrices on a lattice, allowing for the exploration of the model's behavior in the strong coupling regime.
  - **Monte Carlo methods:** Monte Carlo simulations are used to sample the path integral of the BFSS model, providing insights into the non-perturbative ground state and excitation spectrum. These methods help in understanding phase transitions, thermal properties, and the structure of the vacuum state in the model.

8.2.3 Dp-branes or how matrices arise

- A Dp-brane is a p-dimensional hypersurface on which open strings can end, leading to gauge fields and other degrees of freedom living on the brane. In ten-dimensional gravity, soliton-like solutions to classical equations of motion correspond to these p-branes, which possess finite tension (mass per unit spatial volume) and carry RR charges.

- **D-brane's position:** The D-brane is assumed to lie on a hyperplane defined by fixing the coordinates $x^{p+1} = 0$, ..., $x^9 = 0$. This means that the D-brane extends in the x^0, x^1, ..., x^p directions and is localized in the remaining spatial dimensions.

- **World-volume coordinates:** The coordinates x^0, ..., x^p parametrize the world-volume of the D-brane, which is the surface it occupies in spacetime.

- **D-branes as boundaries:** When open strings have their endpoints fixed on a D-brane, the D-brane acts as a boundary condition for the string. These boundary conditions are known as Dirichlet boundary conditions for the coordinates transverse to the D-brane (x^{p+1}, ..., x^9) and Neumann boundary conditions for the coordinates along the D-brane (x^0, ..., x^p).

- **Gauge field emergence:** The endpoints of open strings can move along the D-brane. The degrees of freedom associated with these endpoints give rise to a $U(1)$ gauge field A_a on the D-brane's world-volume which, in the low-energy effective theory, is the only relevant mode of the open string propagating on the world-volume of the D-brane.

- **Tangential components:** The components of the vector potential A_a, where $a = 0$, ..., p, are tangential to the D-brane. These components represent internal gauge fields on the D-brane. In other words, they describe how gauge fields behave along the dimensions in which the D-brane extends.

- **Transverse components:** The components A_i, where $i = p + 1$, ..., 9, appear as scalars from the viewpoint of the field theory on the D-brane. These components describe the position and fluctuations of the D-brane in the transverse directions.

- **Role of scalars:** The transverse components A_i are related to the coordinates X^i that describe the D-brane's fluctuations. The relationship is given by:

$$X^{i-p}(x) = 2\pi\alpha' A^i(x). \tag{8.141}$$

Here, $i = p + 1$, ..., 9. The coordinates $X^i(x)$ represent the position of the D-brane in the transverse directions and are dependent on the world-volume coordinates x.

- **Superposition of D-branes:** A state with an RR charge equal to N can be seen as a superposition of N D-branes. Each D-brane in this superposition is static and parallel, positioned at coordinates $x^{p+1} = X_I^1$, ..., $x^9 = X_I^{9-p}$, where $I = 1$, ..., N. Each D-brane has an associated $U(1)$ field, i.e. the gauge group of the resulting low-energy effective theory becomes $U(1)^N$.

- **String connections:** When there are N D-branes, the number of string degrees of freedom increases to N^2 rather than just N. This is because strings can start on one D-brane and end on another.

- **String tension:** However, since the strings have tension, the energy required to create a string between two branes I and J is proportional to the distance between them, i.e. the gauge field associated with this string is massive, with mass given by

$$M_{IJ} \approx T|X_I - X_J|. \tag{8.142}$$

Here, T is the string tension.

- **Coincident Dp-branes:** In the limit of N coincident D-branes, all N^2 fields become massless and we end up with a non-Abelian $U(N)$ gauge field A_a.
- **Dirac–Born–Infeld and Chern–Simons actions:** In the context of a Dp-brane, the action that describes the dynamics of the gauge field A_a includes the Dirac–Born–Infeld (DBI) action and the Chern–Simons (CS) term (which describes the coupling to the RR fields).
- **Yang–Mills supersymmetric theory:** The low-energy and weak-field effective action of the theory can be obtained by reducing the ten-dimensional supersymmetric $U(N)$ gauge theory to a $(p + 1)$-dimensional space. Explicitly, this is given by

$$S_{\text{eff}} = \frac{1}{g^2} \int d^{p+1}x \ \text{Tr}\left(-\frac{1}{4}F_{ab}F^{ab} - \frac{1}{2}(D_a X^i)(D^a X^i) + \frac{1}{4}[X^i, X^j]^2 + \frac{i}{2}\Psi\Gamma^a D_a \Psi - \frac{1}{2}\Psi\Gamma^i[X^i, \Psi]\right). \tag{8.143}$$

Here, we have:

- F_{ab} is the field strength tensor.
- D_a is the covariant derivative.
- X^i are the scalar fields representing the transverse coordinates of the D-branes.
- Ψ are the fermionic fields.
- **Non-commutative geometry:** The scalar fields X^i represent positions of the Dp-brane in the transverse directions. These coordinates are $N \times N$ Hermitian matrices and thus they are non-commutative coordinates.
- **Flat directions:** The potential energy of the matrices X^i is given by the Yang–Mills term and thus it vanishes if the matrices X^i commute, i.e. if they can be simultaneously diagonalized. These are the so-called flat directions (which are directions in the space of solutions where the potential energy does not change). This means that the classical background solutions of the theory are given by

$$X^i = \begin{pmatrix} X_1^i & 0 & \cdots & 0 \\ 0 & X_2^i & \cdots & 0 \\ \vdots & \vdots & \ddots & \vdots \\ 0 & 0 & \cdots & X_N^i \end{pmatrix}. \tag{8.144}$$

Here, X_I^i are the diagonal components representing the positions of the I-th D-brane.

- **Spontaneous gauge symmetry breaking:** When the scalar fields X^i acquire vacuum expectation values (VEVs) in this form, the gauge symmetry is spontaneously broken from $U(N)$ to $U(1)^N$. Each X_I^i corresponds to the position of a D-brane. The off-diagonal components of the scalar and gauge fields acquire masses proportional to the separation between the D-branes, as described previously.

8.2.4 D0-branes or how gravitons arise

- **Compactification:** Compactification is the process of reducing the number of dimensions in a theory by 'curling up' one or more dimensions into a compact space. In the context of M-theory, compactifying one spatial dimension on a circle of radius R leads to type IIA string theory.
- **Supersymmetry in 10D and 11D:** Both 10D type IIA string theory and 11D M-theory preserve a significant amount of supersymmetry. In 11D, the theory has 32 supercharges. Upon compactification, these supersymmetries reduce appropriately to match the supersymmetry structure of the 10D theory which 16 supercharges.
- **Planck length and string coupling:** The fundamental length scales involved are the 11D Planck length l_p and the 10D string length l_s. The relationship between these scales and the string coupling constant g_s is given by

$$R = g_s^{2/3} l_p, \quad l_s = g_s^{-1/3} l_p. \tag{8.145}$$

Here, R is the radius of the compactified dimension and $T = 1/2\pi\alpha' = 1/\pi l_s^2$ is the string tension.

M-theory is then seen as the strong coupling limit of type IIA string theory. Indeed, in the limit $g_s \longrightarrow \infty$ we have $R \longrightarrow \infty$, i.e. M-theory can be viewed as the strong coupling limit of type IIA string theory, where the compactification radius R becomes large.

- **Kaluza–Klein modes:** Massless states in M-theory correspond to an 11D graviton and its superpartners (a three-form field and a gravitino). Kaluza–Klein modes in 10D arise from compactification. The zero-modes corresponds to the massless fields present in type IIA supergravity, including the metric, dilaton, NS-NS two-form, Ramond–Ramond (RR) fields, and their fermionic partners.
- **D0-branes as massive Kaluza–Klein modes:** In 11D M-theory, the graviton has 11 components, corresponding to the metric tensor. Upon compactification, the momentum modes in the compact dimension correspond to D0-branes in type IIA string theory. These D0-branes can be interpreted as massive Kaluza–Klein modes, which are states with quantized momentum along the compactified dimension. Thus, D0-branes in 10D correspond to 11D gravitons.
- **Action:** The action for D0-branes in type IIA string theory can be derived from the reduction of the 11D supergravity action. The low-energy and weak-field effective Lagrangian of the theory can also be obtained by reducing the ten-dimensional supersymmetric $U(N)$ gauge theory to a one-dimensional space. Explicitly, we have

$$S_{\text{eff}} = \frac{1}{g^2} \int dx^0 \; \text{Tr}\left(\frac{1}{2}(D_0 X^i)^2 + \frac{1}{4}[X^i, X^j]^2 + \frac{i}{2}\bar{\Psi}\Gamma^0 D_0 \Psi - \frac{1}{2}\bar{\Psi}\Gamma^i[X^i, \Psi]\right). \tag{8.146}$$

- Here, the fermion sector is such that:
    - Ψ is a 32-component ten-dimensional spinor, i.e. Γ^i are ten-dimensional Dirac matrices.
    - We can choose the following representation

$$\Gamma^0 = i\sigma_2 \otimes \mathbf{1}_{16}, \ \Gamma^i = \sigma_1 \otimes \gamma^i, \ i = 1, \, ..., \, 9 \Rightarrow C_{10} = \sigma_1 \otimes C_9. \qquad (8.147)$$

   - The spinor Ψ satisfies the Majorana–Weyl condition. The 32-component spinor Ψ can then be rewritten in terms of a 16-component (effectively Majorana) spinor ψ as follows

$$\Psi = \sqrt{2i}\begin{pmatrix} \psi \\ 0 \end{pmatrix}, \ \psi = C_9^{-1}\psi^*, \ \psi^\dagger = \psi^T C_9. \qquad (8.148)$$

   - We can choose for simplicity the Majorana representation in which the gamma matrices γ^M are real and symmetric and $C_9 = \mathbf{1}_{16}$.
- The action becomes then given by

$$S_{\text{eff}} = \frac{1}{g^2} \int dx^0 \text{Tr}\left(\frac{1}{2}(D_0 X^i)^2 + \frac{1}{4}[X^i, \, X^j]^2 + \psi^T D_0 \psi - i\psi^T \gamma^i[X_i, \, \psi] \right). \qquad (8.149)$$

8.2.5 The DLCQ quantization

- **DLCQ:** In DLCQ, the light-cone coordinate x^- is compactified on a circle of radius R which leads to the quantization of the light-cone momentum p_-. The compactification introduces discrete momentum modes, which simplifies the analysis of the theory by breaking it down into simpler components.
- **IMF:** The infinite momentum frame (IMF) is a reference frame in which the momentum of the system along a chosen direction (typically the x^- direction) is taken to be infinitely large. This approach simplifies the dynamics by effectively reducing the degrees of freedom, focusing on the transverse components, which can be treated as perturbations compared to the longitudinal dynamics.
- **Comparison:** Thus, the IMF can be viewed as a particular case of DLCQ. DLCQ involves compactifying the light-cone coordinate and quantizing the momentum along this direction, while IMF takes the limit where this momentum becomes very large. Together, they provide a robust framework for understanding the non-perturbative aspects of M-theory and the dynamics of D0-branes in type IIA string theory. Indeed, both IMF and DLCQ result in similar forms of the Hamiltonian for the BFSS matrix model, focusing on the transverse dynamics of the matrices X^i. The primary difference lies in the treatment of the longitudinal momentum: continuous and very large in the IMF, and discrete in DLCQ.
- **Light-cone coordinates:** In M-theory, the eleven coordinates x^m (where $m = 0, 1, \, ..., \, 10$) are redefined into nine transverse coordinates x^i (where

$i = 1, \ldots, 9$) and two light-cone coordinates: x^+ (light-cone time) and x^- (longitudinal direction). Explicitly, we have

$$x^\pm = \frac{1}{\sqrt{2}}(x^0 \pm x^{10}).$$ (8.150)

Similarly, the momenta p^+ and p^- conjugate to x^- and x^+, respectively, are given by

$$p^\pm = \frac{1}{\sqrt{2}}(p^0 \pm p^{10}).$$ (8.151)

- **Metric and energy:** The metric is given through the quadratic form:

$$p^2 = 2p^+ p^- - p_\perp^2.$$ (8.152)

Here, $p_\perp$ represents the transverse momenta. Thus, the energy of massless modes in this frame is given by the non-relativistic formula:

$$E \equiv p^- = p_+ = \frac{p_\perp^2}{2p^+}.$$ (8.153)

- **Compactification:** The longitudinal coordinate x^- is compactified on a circle with radius R, implying that x^- is periodic with period $2\pi R$. Due to compactification, the longitudinal momentum p^+ is quantized:

$$p^+ = p_- = \frac{N}{R}.$$ (8.154)

Here, N is an integer representing the quantized momentum units.
- **Supersymmetry:** The DLCQ quantization preserves a significant portion of the supersymmetry of the original theory. In the light-cone gauge ($A_+ = 0$), the structure of the theory is simplified, maintaining 16 out of the 32 supercharges.
- **Simplification of dynamics:** The DLCQ quantization significantly simplifies the dynamics of M-theory by focusing on the transverse modes and considering discrete momenta.
- **Holographic principle:** The DLCQ quantization provides a natural setting for the holographic principle, where the higher-dimensional dynamics is encoded in a lower-dimensional matrix model, which is essential for the non-perturbative formulation of M-theory provided by the BFSS model.

8.2.6 DLCQ quantization of M-theory and BFSS model

- **M-theory:** M-theory is an 11-dimensional theory that unifies all five consistent superstring theories and supergravity in 11 dimensions. In the low-energy limit, M-theory reduces to 11-dimensional supergravity.

- **Type IIA String Theory**: Type IIA string theory is a 10-dimensional theory that includes both open and closed strings, with the strings being unoriented and non-chiral. It includes D-branes of various dimensions, including D0-branes, which are point-like objects which carry RR charge.
- **Super-graviton and D0-branes**: Kaluza–Klein modes in 10D arise from compactification. The zero-modes corresponds to the massless fields present in type IIA supergravity. The momentum modes (with $N \neq 0$) in the compact dimension correspond to D0-branes in type IIA string theory. Thus, D0-branes in 10D correspond to 11D gravitons with longitudinal momentum $p_- = 1/R$. The bound state of N D0-branes corresponds to a graviton with momentum $p_- = N/R$.
- **BFSS hypotheses**:
  - The first underlying hypothesis of the BFSS model is the statement that the fundamental degrees of freedom of M-theory, when formulated in the light-cone gauge, are D0-branes. The compactification radius R and the quantized momentum N directly relate to the parameters in the BFSS model. In particular, the number N of D0-branes represents the RR charge. This RR charge corresponds to the longitudinal momentum p_- in M-theory. As $R \longrightarrow \infty$, we must simultaneously take $N \longrightarrow \infty$ in order to keep $p_- = N/R$ fixed.
  - The second underlying hypothesis of the BFSS model is the statement that the low-energy and weak-field effective Yang–Mills action of the D0-branes describes in fact the complete quantum theory Lagrangian of M-theory in the light-cone gauge. The compactification in the DLCQ framework translates the dynamics of these D0-branes into a matrix quantum mechanics, with the matrices $X^i(t)$ representing the transverse coordinates of the D0-branes.
- **Action**: The BFSS model, which is also called M-(atrix) theory, is then given by the action derived previously for D0-branes. We obtain the BFSS action

$$S = \frac{1}{g^2} \int dx^0 \ \text{tr}\left(\frac{1}{2}(D_0 X^i)^2 + \frac{1}{4}[X^i, X^j]^2 + \frac{i}{2}\bar{\Psi}\Gamma^0 D_0 \Psi - \frac{1}{2}\bar{\Psi}\Gamma^i[X^i, \Psi] \right). \quad (8.155)$$

- **Lagrangian and equations of motion**: By reducing the Dirac algebra from 32 dimensions to 16 dimensions by using the Majorana–Weyl condition we obtain the Lagrangian (we set $g^2 = 1$ and re-insert the compactification radius R)

$$L = \frac{1}{g^2} \text{Tr}\left(\frac{1}{2R}(D_0 X^i)^2 + \frac{R}{4}[X^i, X^j]^2 + \psi^T D_0 \psi - iR\psi^T \gamma^i[X_i, \psi] \right). \quad (8.156)$$

The equations of motion satisfied by the matrices X^i read

$$D_0^2 X^i + [X^j, [X^j, X^i]] + \text{fermionic terms} = 0. \quad (8.157)$$

The equations of motion satisfied by the fermions ψ can simply be solved by $\psi = 0$ and hence the fermionic terms in the above equation drop out.

Both (quantized) supermembranes and (non-commutative) emergent geometry originate in the matrix theory from these equations of motion.

- **Hamiltonian:** The corresponding Hamiltonian reads

$$H = R\,\mathrm{Tr}\left(\frac{1}{2}P_i^2 - \frac{1}{4}[X^i,\,X^j]^2 - iR\psi^T\gamma^M[X_M,\,\psi]\right). \tag{8.158}$$

This is a positive-definite Hamiltonian because of its equality to the square of the supercharge. Its ground state is a subtle and complex topic due to the interplay between the discrete and continuous spectra, as well as the presence of flat directions.

- **Supermembranes:** The (quantized) supermembranes emerge from the matrix theory by taking the large N limit of the above Lagrangian, equations of motion and Hamiltonian.

- **Spectrum:** This consists of the ground state, a discrete spectrum, flat directions and a continuum spectrum.
    - **Ground state:** The BFSS model has a supersymmetric ground state, which is a stable zero-energy state. This state corresponds to the vacuum state in supergravity where no super-graviton excitations are present. This ground state is part of the discrete spectrum of the BFSS model. However, due to the presence of flat directions, there is a continuous spectrum of zero-energy states around this ground state. These flat directions correspond to configurations where the D0-branes are arbitrarily far apart, leading to a moduli space of vacua with zero potential energy.
    - **Discrete spectrum:** The discrete spectrum includes bound states of D0-branes which have quantized energy levels above the ground state. The discrete spectrum also consists of various excited states of the bound D0-branes above the ground state. This discrete spectrum is consistent with a super-graviton with longitudinal momentum $p_- = N/R$ which is the fundamental excitation in 11-dimensional supergravity.
    - **Flat directions:** These are directions in the configuration space where the potential energy remains constant (usually zero). These directions correspond to configurations where the matrices commute, leading to zero potential energy. Thus, these matrices can be simultaneously diagonalized where the eigenvalues represent the positions of the D0-branes. Clearly, the potential energy remains zero regardless of the separation of these eigenvalues, i.e. the D0-branes can be arbitrarily far apart (as the matrices can have large eigenvalues) without any energy cost. This leads to a continuous range of possible configurations, corresponding to a continuous spectrum of states.
    - **Continuum spectrum:** The flat directions lead, in quantum mechanics, to zero-energy states forming a continuous spectrum, which arise because the

wavefunctions corresponding to the D0-brane positions can spread out indefinitely along these flat directions. The continuum spectrum corresponds to the unbound states of free super-gravitons and their scattering states. The quantum fluctuations of flat directions can also lead to tunneling between different vacua or the condensation of D0-branes into bound states, moving the system from a continuous to a discrete state.

– **Supermembranes:** In the large N limit the Hamiltonian still has a continuous spectrum (zero-energy modes of the quantized super-membrane). This means that the continuum spectrum is not only due to the presence of flat directions but it is also due to supersymmetry.

This also means that the continuous spectrum in the large N limit corresponds to states where the degrees of freedom of the super-membrane represent unbound, extended configurations of the super-membrane. In other words, flat directions correspond to infinitely long membrane configurations.

This dual nature is crucial for a complete non-perturbative formulation of M-theory, accommodating both localized bound states (particle-like excitations corresponding to D0-branes) and extended configurations (which correspond to the continuum spectrum of the super-membrane interpreted as states of free super-gravitons).

8.2.7 Supermembranes and non-commutative geometry

- M-(atrix theory), i.e. the BFSS matrix model, in the large N limit can be related to the dynamics of supermembranes which are higher-dimensional generalization of a string.
- In fact, the BFSS matrix model can be thought of as describing a quantized super-membrane which is nothing else but a non-commutative space.
- This will be illustrated with the case of membranes with toroidal topology which corresponds to the non-commutative torus.
- We introduce the two unitary matrices g (lapse operator) and h (shift operator) in the general linear group $gl(N)$ by the relations:

$$hg = \exp\left(\frac{2\pi i}{N}\right)gh, \quad h^N = 1, \quad g^N = 1. \tag{8.159}$$

We work in the basis:

$$g|n\rangle = \exp\left(\frac{2\pi i n}{N}\right)|n\rangle, \quad h|n\rangle = |n - 1\rangle. \tag{8.160}$$

Here, we have $|0\rangle \equiv |N\rangle$.

- Any $N \times N$ matrix can be expressed as a function of g and h:

$$Z = \sum_{n,m=0}^{N-1} Z_{n,m} g^n h^m. \tag{8.161}$$

- As $N \to \infty$, the matrices g and h become operators in an infinite-dimensional Hilbert space, and the relations (8.159) can then be solved by coordinate and momentum operators q and p on the non-commutative torus $\mathbf{T}_\theta^2$ which are given by

$$g = \exp(iq), \quad h = \exp(ip), \quad [q, p] = \frac{2\pi i}{N}. \tag{8.162}$$

- Here, N plays the role of the non-commutativity parameter or the Planck constant $\frac{2\pi}{N}$. Thus, $N \to \infty$ is a commutative or classical limit where q and p become c-numbers. We have

$$\theta \equiv \frac{2\pi}{N}: \text{ noncommutativity parameter}$$
$$N \longrightarrow \infty: \text{ commutative limit}. \tag{8.163}$$

- This leads to the Weyl map between operators in the matrix model and functions on the non-commutative torus given by

$$Z \to Z(q, p) = \sum_{n,m=0}^{N-1} Z_{n,m} \exp(inq)\exp(imp). \tag{8.164}$$

- In this commutative/classical limit, commutators are replaced by Poisson brackets. More precisely, the matrix product between operators is replaced by the Moyal–Weyl star product between functions, viz

$$XY \to X*Y(q, p) = X(q, p)Y(q, p) + i\theta\partial_q X \partial_p Y + \ldots$$
$$[X, Y] \longrightarrow \{X, Y\}(q, p) = X*Y(q, p) - Y*X(q, p) = i\theta(\partial_q X \partial_p Y - \partial_p X \partial_q Y) + \ldots \tag{8.165}$$

- Similarly, the trace in the finite-dimensional Hilbert space of matrices becomes an integral over the phase space of the torus, i.e.

$$\operatorname{tr} Z = N \int_0^{2\pi} \frac{dpdq}{(2\pi)^2} Z(p, q). \tag{8.166}$$

- As a consequence, the matrix theory Lagrangian in the gauge $A = 0$ becomes, in the strict limit $N \longrightarrow \infty$, the super-membrane Lagrangian in the light-cone gauge given by

$$L = \frac{N}{g^2} \int \frac{dpdq}{(2\pi)^2} \left(\frac{1}{2}\dot{X}^2 + \frac{1}{4}\{X_i, X_j\}^2 + \frac{i}{2}\bar{\Psi}\Gamma^0\dot{\Psi} - \frac{1}{2}\bar{\Psi}\Gamma^i\{X^i, \Psi\} \right). \tag{8.167}$$

8.2.8 Conclusion

- The BFSS matrix model stands as a cornerstone in the quest for a non-perturbative formulation of M-theory.
- Offers profound insights into the nature of spacetime, quantum gravity, and string theory.
- Its rich mathematical structure and deep physical implications continue to inspire research in theoretical physics.

8.3 Causal dynamical triangulation and multitrace matrix models

Parts of this section have been reproduced with permission from [34]. Copyright (2017), World Scientific Publishing Co Pte Ltd.

Another powerful approach to the emergence of time and spacetime is Lorentzian causal dynamical triangulation or CDT [24–26]. In some sense, CDT can be be viewed as a more fundamental theory than LCQ. In the CDT approach spacetime is built out of four-simplices (generalization of two-simplices, i.e. triangles, to four dimensions) which are equipped with a flat Minkowski metric. The causality requirement singles out globally hyperbolic manifolds which admit a global proper-time foliation structure and as a consequence Wick rotation to Euclidean is meaningful. The Hilbert–Einstein action is given in this discrete setting by the Regge action [27]. The path integral is obtained as the sum over the set of all causal triangulations weighted with the Regge action. The parameters of the model are Newton's gravitational constant G and the cosmological constant Λ which appear as the parameters K_0 and K_4 in the Regge action. The model also depends on two more parameters given by the lengths of time-like and spatial-like links a_t and a_s, respectively. We have $a_t^2 = \alpha a_s^2$ where the asymmetry factor $\alpha < 0$ appears as a parameter Δ in the Regge action.

CDT is intimately related to Horava–Lifhsitz (HL) gravity [28–30] which, like CDT, also assumes global time foliation and introduces anisotropy between space and time but in such a way as to achieve power-counting renormalizability of quantum gravity. This theory is effectively a generalization to gravity of the d-dimensional Lifhsitz scalar field theory given by the Lifhsitz–Landau free energy density [31]

$$S = a_2\phi^2 + a_4\phi^4 + \cdots + c_2(\partial_\alpha\phi)^2 + d_2(\partial_\beta\phi)^2 + e_2(\partial_\beta^2\phi)^2 + \cdots \qquad (8.168)$$

The anisotropy is introduced by the distinction between the indices $\beta = 1, ..., m$ and $\alpha = m + 1, ..., d$. The three phases present in the Lifhsitz scalar field theory theory are: helicoidal $(|\partial_t\phi(x)| < 0)$, paramagnetic $(\phi(x) = 0)$ and ferromagnetic $(|\phi(x)| > 0)$. The phase diagram is depicted in figure 8.1(a) together with the phase diagrams of causal dynamical triangulation and non-commutative scalar phi-four theory.

In particular, the phase structure of Lorentzian causal dynamical triangulation is given in figure 8.1(b). The cosmological constant K_4, which controls the total volume, is fixed at its critical value and the phase diagram is then drawn in the plane

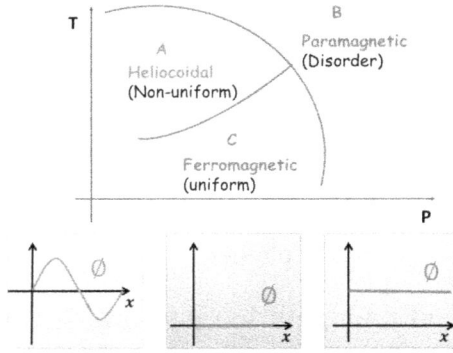

(a) Landau-Lifshitz scalar field theory.

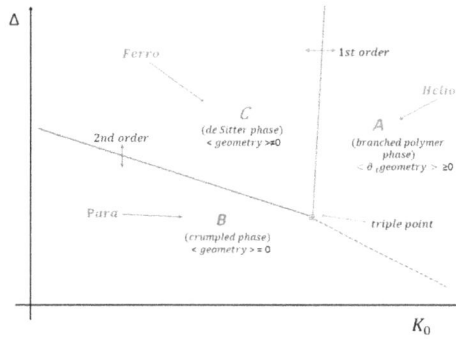

(b) Causal dynamical triangulation.

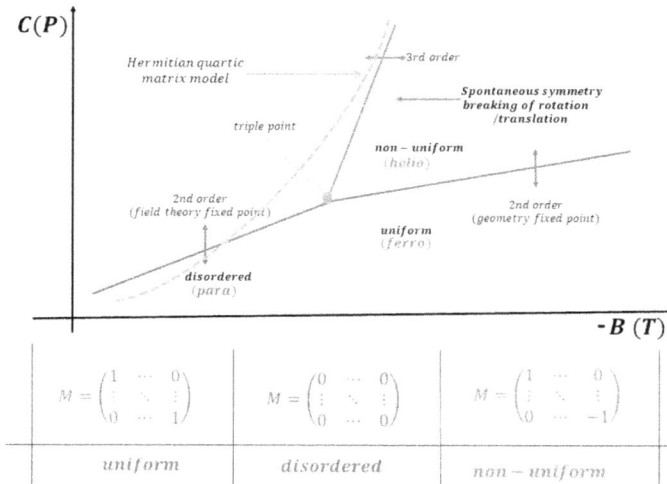

(c) Multitrace matrix model.

Figure 8.1. The phase diagrams of causal dynamical triangulation, Lifshitz scalar field theory and multitrace matrix model.

$K_0 - \Delta$ where K_0 is proportional to the inverse bare gravitational coupling constant G while Δ is effectively equal to the asymmetry factor α.

There are three distinct phases in the CDT approach given explicitly by [32, 33]:

- The de Sitter spacetime phase C. This is the analog of the ferromagnetic phase ($d_2 > 0$, $a_2 < 0$) of the Lifshitz scalar field theory or the ordered phase in non-commutative scalar field theory (see below).
- The crumpled phase B where neither space nor time have any extent and therefore there is no geometry. This is the analog of the paramagnetic phase ($d_2 > 0$, $a_2 > 0$) of the Lifshitz scalar field theory or the disordered phase in non-commutative scalar field theory (see below).
- The branched polymer phase A where the geometry oscillates in time. This is the analog of the helicoidal phase ($d_2 < 0$) in Lifshitz scalar field theory or the non-uniform ordered phase in non-commutative scalar field theory (see below).

The transition from A to C is first order, whereas the transition from B to C could be either first order or second order and as a consequence there is a possibility of a continuum limit. Similarly, the transition between the ferromagnetic (C) and paramagnetic (B) phases in the Lifshitz scalar field theory although usually second order it could be first order. There is a strikingly similar situation occurring in non-commutative scalar field theory (see below).

Also in both theories CDT and HL, the spectral dimension at short distances is 2 and only becomes 4 at large distances and the anisotropy between space and time disappear in CDT in the de Sitter spacetime phase while in HL it disappears at low energies.

In this final paragraph we wish to propose a new theory of emergent geometry based on the multitrace matrix model discussed originally in [34, 35]. This approach can be viewed as the fundamental theory of the BFSS matrix model. This theory is in fact a generalization of the Hermitian quartic matrix model [36, 37] to multitrace Hermitian matrix models which can thus be understood as approximations of non-commutative scalar field theory [38]. The corresponding phase structure is very similar to the case found in causal dynamical triangulation and the reason is easily understood in the effective Landau–Lifhsitz scalar field theory (8.168) which, as it turns out, also describes the phase structure of non-commutative scalar field theory and multitrace matrix models [39, 40].

References

[1] Deser S 2008 Arnowitt–Deser–Misner formalism *Scholarpedia* **3** 7533
[2] Thiemann T 2007 Loop quantum gravity: an inside view *Approaches to Fundamental Physics* (Lecture Notes in Physics vol 721) (Springer) pp 185–263
[3] Ashtekar A, Balachandran A P and Jo S 1989 The CP problem in quantum gravity *Int. J. Mod. Phys.* A **4** 1493
[4] Ashtekar A 2015 Ashtekar variables *Scholarpedia* **10** 32900
[5] Pullin J 1994 Knot theory and quantum gravity in loop space: a primer *AIP Conf. Proc.* **317** 141–90 (For a concise summary see also the Wikipedia page)

[6] Barbero J G 1995 Real Ashtekar variables for Lorentzian signature space times *Phys. Rev.* D **51** 5507–10

[7] Immirzi G 1997 Real and complex connections for canonical gravity *Class. Quant. Grav.* **14** L177–181

[8] Thiemann T 2007 *Modern Canonical Quantum General Relativity* (Cambridge University Press)

[9] Loll R 1994 Gauge theory and gravity in the loop formulation *Lect. Notes Phys.* **434** 254–88

[10] Ashtekar A 1996 Geometric issues in quantum gravity *The Geometric Universe: Science, Geometry, and the Work of Roger Penrose* ed Roger Penrose, S Huggett, L Mason, K Tod, S Tsou and N Woodhouse *Proc. Symp. Geometric Issues in the Foundations of Science (Oxford, 25–29 June)*

[11] Rovelli C and Smolin L 1988 Knot theory and quantum gravity *Phys. Rev. Lett.* **61** 1155

[12] Ashtekar A, Reuter M and Rovelli C 2014 From general relativity to quantum gravity arXiv:1408.4336 https://doi.org/10.48550/arXiv.1408.4336

[13] Baez J C 1996 Spin network states in gauge theory *Adv. Math.* **117** 253–72

[14] Penrose R 1971 Angular momentum; an approach to combinatorial spacetime *Quantum Theory and Beyond* ed T Bastin (Cambridge: Cambridge University Press)

[15] Rovelli C and Smolin L 1995 Spin networks and quantum gravity *Phys. Rev.* D **52** 5743–59

[16] Gambini R and Pullin J 2011 *A first course in loop quantum gravity* (Oxford University Press)

[17] Zarembo K L and Makeenko Y M 1998 An introduction to matrix superstring models *Phys. Usp.* **41** 1–23

[18] Hanada M 2016 What lattice theorists can do for superstring/M-theory *Int. J. Mod. Phys.* A **31** 1643006

[19] BFSS matrix model in nlab https://ncatlab.org/nlab/show/BFSS.matrix.model

[20] Ydri B 2018 Review of m(atrix)-theory, type IIB matrix model and matrix string theory arXiv:1708.00734 https://doi.org/10.48550/arXiv.1708.00734

[21] Banks T, Fischler W, Shenker S H and Susskind L 1997 M theory as a matrix model: a conjecture *Phys. Rev.* D **55** 5112–28

[22] Ishibashi N, Kawai H, Kitazawa Y and Tsuchiya A 1997 A large N reduced model as superstring *Nucl. Phys.* B **498** 467–91

[23] Berenstein D E, Maldacena J M and Nastase H S 2002 Strings in flat space and pp waves from $N = 4$ super Yang-Mills *J. High Energy Phys.* **4** 013

[24] Ambjorn J, Jurkiewicz J and Loll R 2005 Reconstructing the Universe *Phys. Rev.* D **72** 064014

[25] Ambjørn J, Janik R, Westra W and Zohren S 2006 The emergence of background geometry from quantum fluctuations *Phys. Lett.* B **641** 94

[26] Ambjorn J, Gorlich A, Jurkiewicz J and Loll R 2008 Planckian birth of the quantum de Sitter universe *Phys. Rev. Lett.* **100** 091304

[27] Regge T 1961 General relativity without coordinates *Nuovo Cim.* **19** 558

[28] Horava P 2009 Spectral dimension of the Universe in quantum gravity at a Lifshitz point *Phys. Rev. Lett.* **102** 161301

[29] Horava P 2009 Membranes at quantum criticality *J. High Energy Phys.* **0903** 020

[30] Horava P 2009 Quantum gravity at a Lifshitz point *Phys. Rev.* D **79** 084008

[31] Goldenfeld N 1992 *Lectures on Phase Transitions and the Renormalization Group* (Frontiers of Physics vol 85) (Basic Books)

[32] Ambjorn J, Gorlich A, Jordan S, Jurkiewicz J and Loll R 2010 CDT meets Horava-Lifshitz gravity *Phys. Lett.* B **690** 413

[33] Gorlich A 2011 Causal dynamical triangulations in four dimensions arXiv:1111.6938 https://doi.org/10.48550/arXiv.1111.6938

[34] Ydri B, Soudani C and Rouag A 2017 Quantum gravity as a multitrace matrix model *Int. J. Mod. Phys.* A **32** 1750180

[35] Ydri B, Khaled R and Soudani C 2022 Quantized non-commutative geometry from multitrace matrix models *Int. J. Mod. Phys.* A **37** 2250052

[36] Brezin E, Itzykson C, Parisi G and Zuber J B 1978 Planar diagrams *Commun. Math. Phys.* **59** 35

[37] Shimamune Y 1982 On the phase structure of large N matrix models and gauge models *Phys. Lett.* **108B** 407

[38] O'Connor D and Saemann C 2007 Fuzzy scalar field theory as a multitrace matrix model *JHEP* **0708** 066

[39] Saemann C 2010 The multitrace matrix model of scalar field theory on fuzzy CP^n *SIGMA* **6** 050

Sämann C 2015 Bootstrapping Fuzzy Scalar Field Theory *JHEP* **1504** 44

[40] Tekel J 2013 Random matrix approach to scalar fields on fuzzy spaces *Phys. Rev.* D **87** 085015

Tekel J 2014 Uniform order phase and phase diagram of scalar field theory on fuzzy $\mathbb{C}P^n$ *J. High Energy Phys.* **1410** 144

Tekel J 2015 Phase structure of fuzzy field theories and multitrace matrix models *Acta Phys. Slov.* **65** 369

IOP Publishing

Lectures on General Relativity, Cosmology and Quantum Black Holes (Second Edition)

Badis Ydri

Appendix A

Differential geometry primer

A.1 Manifolds

A.1.1 Maps, open set and charts

Definition 1: A map ϕ between two sets M and N, viz ϕ: $M \longrightarrow N$ is a rule which takes every element of M to exactly one element of N, i.e it takes M into N.

This is a generalization of the notion of a function. The set M is the domain of ϕ while the subset of N that M gets mapped into the image of ϕ. We have the following properties:

- An injective (one-to-one) map is a map in which every element of N has at most one element of M mapped into it. Example: $f = e^x$ is injective.
- A surjective (onto) map is a map in which every element of N has at least one element of M mapped into it. Example: $f = x^3 - x$ is surjective.
- A bijective (and therefore invertible) map is a map which is both injective and surjective.
- A map from R^m to R^n is a collection of n functions ϕ^i of m variables x^i given by

$$\phi^i(x^1,\ldots,x^m) = y^i, \quad i = 1,\ldots, n. \tag{A.1}$$

- The map ϕ: $R^m \longrightarrow R^n$ is a C^p map if every component ϕ^i is at least a C^p function, i.e. if the pth derivative exists and is continuous. A C^∞ map is called a smooth map.
- A diffeomorphism is a bijective map ϕ: $M \longrightarrow N$ which is smooth and with an inverse ϕ^{-1}: $N \longrightarrow M$ which is also smooth. The two sets M and N are said to be diffeomorphic which means essentially that they are identical.

doi:10.1088/978-0-7503-5824-8ch9

Definition 2: An open ball centered around a point $y \in R^n$ is the set of all points $x \in R^n$ such that $|x - y| < r$ for some $r \in R$ where $|x - y|^2 = \sum_{i=1}^{n}(x_i - y_i)^2$. This is clearly the inside of a sphere S^{n-1} in R^n of radius r centered around the point y.

Definition 3: An open set $V \subset R^n$ is a set in which every point $y \in V$ is the center of an open ball which is inside V. Clearly an open set is a union of open balls. Also, it is obvious that an open set is the inside of an $(n - 1)$-dimensional surface in R^n.

Definition 4: A chart (coordinate system) is a subset U of a set M together with a one-to-one map $\phi: U \longrightarrow R^n$ such that the image $V = \phi(U)$ is an open set in R^n. We say that U is an open set in M. The map $\phi: U \longrightarrow \phi(U)$ is clearly invertible.

Definition 5: A C^∞ atlas is a collection of charts $\{(U_\alpha, \phi_\alpha)\}$ which must satisfy the two conditions:

- The union is M, viz $\bigcup_\alpha U_\alpha = M$.
- If two charts U_α and U_β intersect then we can consider the maps $\phi_\alpha \circ \phi_\beta^{-1}$ and $\phi_\beta \circ \phi_\alpha^{-1}$ defined as

$$\phi_\alpha \circ \phi_\beta^{-1}: \phi_\beta(U_\alpha \cap U_\beta) \longrightarrow \phi_\alpha(U_\alpha \cap U_\beta),$$
$$\phi_\beta \circ \phi_\alpha^{-1}: \phi_\alpha(U_\alpha \cap U_\beta) \longrightarrow \phi_\beta(U_\alpha \cap U_\beta). \tag{A.2}$$

Clearly $\phi_\alpha(U_\alpha \cap U_\beta) \subset R^n$ and $\phi_\beta(U_\alpha \cap U_\beta) \subset R^n$. These two maps are required to be C^∞, i.e. smooth.

It is clear that definition 4 provides a precise formulation of the notion that a manifold 'will locally look like R^n', whereas definition 5 provides a precise formulation of the statement that a manifold 'will be constructed from pieces of R^n (in fact the open sets U_α) which are sewn together smoothly'.

A.1.2 Manifold: definition and examples

Definition 6: A C^∞ n-dimensional manifold M is a set M together with a maximal atlas, i.e. an atlas which contains every chart that is compatible with the conditions of definition 5. This requirement means in particular that two identical manifolds defined by two different atlases will not be counted as different manifolds.

Example 1: The Euclidean spaces R^n, the spheres S^n and the tori T^n are manifolds.

Example 2: Riemann surfaces are two-dimensional manifolds. A Riemann surface of genus g is a kind of two-dimensional torus with a g holes. The two-dimensional torus has genus $g = 1$, whereas the sphere is a two-dimensional torus with genus $g = 0$.

Example 3: Every compact orientable two-dimensional surface without boundary is a Riemann surface and thus is a manifold.

Example 4: The group of rotations in R^n (which is denoted by $SO(n)$) is a manifold. Any Lie group is a manifold.

Example 5: The product of two manifolds M and M' of dimensions n and n' respectively is a manifold $M \times M'$ of dimension $n + n'$.

Example 6: Let us consider the circle S^1. Let us try to cover the circle with a single chart (S^1, θ) where $\theta : S^1 \longrightarrow R$. The image $\theta(S^1)$ is not open in R if we include both $\theta = 0$ and $\theta = 2\pi$ since clearly $\theta(0) = \theta(2\pi)$ (the map is not bijective). If we do not include both points then the chart does not cover the whole space. The solution is to use (at least) two charts.

Example 7: We consider a sphere S^2 in R^3 defined by the equation $x^2 + y^2 + z^2 = 1$. First let us recall the stereographic projection from the north pole onto the plane $z = -1$. For any point P on the sphere (excluding the north pole) there is a unique line through the north pole $N = (0, 0, 1)$ and $P = (x, y, z)$ which intersects the $z = -1$ plane at the point $p' = (X, Y)$. We have immediately

$$X = \frac{2x}{1 - z}, \quad Y = \frac{2y}{1 - z}. \tag{A.3}$$

The first chart will be therefore given by the subset $U_1 = S^2 - \{N\}$ and the map

$$\phi_1(x, y, z) = (X, Y) = \left(\frac{2x}{1 - z}, \frac{2y}{1 - z} \right). \tag{A.4}$$

The stereographic projection from the south pole onto the plane $z = 1$. Again for any point P on the sphere (excluding the south pole) there is a unique line through the south pole $N' = (0, 0, -1)$ and $P = (x, y, z)$ which intersects the $z = 1$ plane at the point $p' = (X', Y')$. Now we have

$$X' = \frac{2x}{1 + z}, \quad Y' = \frac{2y}{1 + z}. \tag{A.5}$$

The second chart will be therefore given by the subset $U_2 = S^2 - \{N'\}$ and the map

$$\phi_2(x, y, z) = (X', Y') = \left(\frac{2x}{1 + z}, \frac{2y}{1 + z} \right). \tag{A.6}$$

The two charts (U_1, ϕ_1) and (U_2, ϕ_2) cover the whole sphere. They overlap in the region $-1 < z < +1$. In this overlap region we have the map

$$(X', Y') = \phi_2 \circ \phi_1^{-1}(X, Y). \tag{A.7}$$

We compute first the inverse map ϕ_1^{-1} as

$$x = \frac{4}{4 + X^2 + Y^2} X, \quad y = \frac{4}{4 + X^2 + Y^2} Y, \quad z = -\frac{4 - X^2 - Y^2}{4 + X^2 + Y^2}. \tag{A.8}$$

Next by substituting in the formulas of X' and Y' we obtain

$$X' = \frac{4X}{X^2 + Y^2}, \quad Y' = \frac{4Y}{X^2 + Y^2}. \tag{A.9}$$

This is simply a change of coordinates.

A.1.3 Vectors and directional derivative

In special relativity Minkowski spacetime is also a vector space. In general relativity spacetime is a curved manifold and is not necessarily a vector space. For example the sphere is not a vector space because we do not know how to add two points on the sphere to get another point on the sphere. The sphere which is naturally embedded in R^3 admits at each point P a tangent plane. The notion of a 'tangent vector space' can be constructed for any manifold which is embedded in R^n. As it turns out manifolds are generally defined in intrinsic terms and not as surfaces embedded in R^n (although they can: Whitney's embedding theorem), and as such the notion of a 'tangent vector space' should also be defined in intrinsic terms, i.e. with reference only to the manifold in question.

Directional derivative: There is a one-to-one correspondence between vectors and directional derivatives in R^n. Indeed the vector $v = (v^1, ..., v^n)$ in R^n defines the directional derivative $\sum_\mu v^\mu \partial_\mu$ which acts on functions on R^n. These derivatives are clearly linear and satisfy the Leibniz rule. We will therefore define tangent vectors on a general manifold as directional derivatives which satisfy linearity and the Leibniz rule. Remark that the directional derivative $\sum_\mu v^\mu \partial_\mu$ is a map from the set of all smooth functions into R.

Definition 7: Let now $\mathcal{F}$ be the set of all smooth functions f on a manifold M, viz $f: M \longrightarrow R$. We define a tangent vector v at the point $p \in M$ as the map $v: \mathcal{F} \longrightarrow R$ which is required to satisfy linearity and the Leibniz rule, viz

$$v(af + bg) = av(f) + bv(g),$$
$$v(fg) = f(p)v(g) + g(p)v(f), \quad a, b \in R, \quad f, g \in \mathcal{F}. \tag{A.10}$$

We have the following results:
- For a constant function $(h(p) = c)$ we have from linearity $v(c^2) = cv(c)$, whereas the Leibniz rule gives $v(c^2) = 2cv(c)$ and thus $v(c) = 0$.
- The set V_p of all tangents vectors v at p form a vector space since $(v_1 + v_2)(f) = v_1(f) + v_2(f)$ and $(av)(f) = av(f)$ where $a \in R$.
- The dimension of V_p is precisely the dimension n of the manifold M. The proof goes as follows. Let $\phi: O \subset M \longrightarrow U \subset R^n$ be a chart which includes the point p. Clearly for any $f \in \mathcal{F}$ the map $f \circ \phi^{-1}: U \longrightarrow R$ is smooth since both f and ϕ are smooth maps. We define the maps $X_\mu: \mathcal{F} \longrightarrow R, \mu = 1,..., n$ by

$$X_\mu(f) = \frac{\partial}{\partial x^\mu}(f \circ \phi^{-1})|_{\phi(p)}. \tag{A.11}$$

Given a smooth function $F: R^n \longrightarrow R$ and a point $a = (a^1,..., a^n) \in R^n$ then there exists smooth functions H_μ such that for any $x = (x^1,..., x^n) \in R^n$ we have the result

$$F(x) = F(a) + \sum_{\mu=1}^{n}(x^\mu - a^\mu)H_\mu(x), \quad H_\mu(a) = \frac{\partial F}{\partial x^\mu}|_{x=a}. \tag{A.12}$$

We choose $F = f \circ \phi^{-1}$, $x \in U$ and $a = \phi(p) \in U$ we have

$$f \circ \phi^{-1}(x) = f \circ \phi^{-1}(a) + \sum_{\mu=1}^{n}(x^\mu - a^\mu)H_\mu(x). \tag{A.13}$$

Clearly $\phi^{-1}(x) = q \in O$ and thus

$$f(q) = f(p) + \sum_{\mu=1}^{n}(x^\mu - a^\mu)H_\mu(\phi(q)). \tag{A.14}$$

We think of each coordinate x^μ as a smooth function from U into R, viz $x^\mu: U \longrightarrow R$. Thus the map $x^\mu \circ \phi: O \longrightarrow R$ is such that $x^\mu(\phi(q)) = x^\mu$ and $x^\mu(\phi(p)) = a^\mu$. In other words

$$f(q) = f(p) + \sum_{\mu=1}^{n}(x^\mu \circ \phi(q) - x^\mu \circ \phi(p))H_\mu(\phi(q)). \tag{A.15}$$

Let now v be an arbitrary tangent vector in V_p. We have immediately

$$v(f) = v(f(p)) + \sum_{\mu=1}^{n} v(x^\mu \circ \phi - x^\mu \circ \phi(p)) H_\mu \circ \phi(q)|_{q=p}$$

$$+ \sum_{\mu=1}^{n} (x^\mu \circ \phi(q) - x^\mu \circ \phi(p))|_{q=p} v(H_\mu \circ \phi) \tag{A.16}$$

$$= \sum_{\mu=1}^{n} v(x^\mu \circ \phi) H_\mu \circ \phi(p).$$

But

$$H_\mu \circ \phi(p) = H_\mu(a) = \frac{\partial}{\partial x^\mu}(f \circ \phi^{-1})|_{x=a} = X_\mu(f). \tag{A.17}$$

Thus

$$v(f) = \sum_{\mu=1}^{n} v(x^\mu \circ \phi) X_\mu(f) \Rightarrow v = \sum_{\mu=1}^{n} v^\mu X_\mu, \quad v^\mu = v(x^\mu \circ \phi). \tag{A.18}$$

This shows explicitly that the X_μ satisfy linearity and the Leibniz rule and thus they are indeed tangent vectors to the manifold M at p. The fact that an arbitrary tangent vector v can be expressed as a linear combination of the n vectors X_μ shows that the vectors X_μ are linearly independent, span the vector space V_p and that the dimension of V_p is exactly n.

Coordinate basis: The basis $\{X_\mu\}$ is called a coordinate basis. We may pretend that

$$X_\mu \equiv \frac{\partial}{\partial x^\mu}. \tag{A.19}$$

Indeed if we work in a different chart ϕ' we will have

$$X'_\mu(f) = \frac{\partial}{\partial x'^\mu}(f \circ \phi'^{-1})|_{x'=\phi'(p)}. \tag{A.20}$$

We compute

$$X_\mu(f) = \frac{\partial}{\partial x^\mu}(f \circ \phi^{-1})|_{x=\phi(p)}$$

$$= \frac{\partial}{\partial x^\mu} f \circ \phi'^{-1}(\phi' \circ \phi^{-1})|_{x=\phi(p)}$$

$$= \sum_{\nu=1}^{n} \frac{\partial x'^\nu}{\partial x^\mu} \frac{\partial}{\partial x'^\nu}(f \circ \phi'^{-1}(x'))|_{x'=\phi'(p)} \tag{A.21}$$

$$= \sum_{\nu=1}^{n} \frac{\partial x'^\nu}{\partial x^\mu} X'_\nu(f).$$

The tangent vector v can be rewritten as

$$v = \sum_{\mu=1}^{n} v^{\mu} X_{\mu} = \sum_{\mu=1}^{n} v'^{\mu} X'_{\mu}. \tag{A.22}$$

We conclude immediately that

$$v'^{\nu} = \sum_{\nu=1}^{n} \frac{\partial x'^{\nu}}{\partial x^{\mu}} v^{\mu}. \tag{A.23}$$

This is the vector transformation law under the coordinate transformation $x^{\mu} \longrightarrow x'^{\mu}$.

Vectors as directional derivatives: A smooth curve on a manifold M is a smooth map from R into M, viz $\gamma: R \longrightarrow M$. A tangent vector at a point p can be thought of as a directional derivative operator along a curve which goes through p. Indeed a tangent vector T at $p = \gamma(t) \in M$ can be defined by

$$T(f) = \frac{d}{dt}(f \circ \gamma(t))|_p. \tag{A.24}$$

The function f is $\in \mathcal{F}$ and thus $f \circ \gamma: R \longrightarrow R$. Given a chart ϕ the point p will be given by $p = \phi^{-1}(x)$ where $x = (x^1, \ldots, x^n) \in R^n$. Hence

$$\gamma(t) = \phi^{-1}(x). \tag{A.25}$$

In other words the map γ is mapped into a curve $x(t)$ in R^n. We have immediately

$$T(f) = \frac{d}{dt}(f \circ \phi^{-1}(x))|_p = \sum_{\mu=1}^{n} \frac{\partial}{\partial x^{\mu}}(f \circ \phi^{-1}(x)) \frac{dx^{\mu}}{dt}\Big|_p = \sum_{\mu=1}^{n} X_{\mu}(f) \frac{dx^{\mu}}{dt}\Big|_p. \tag{A.26}$$

The components T^{μ} of the vector T are therefore given by

$$T^{\mu} = \frac{dx^{\mu}}{dt}\Big|_p. \tag{A.27}$$

A.1.4 Dual vectors and tensors

Definition 8: Let V_p be the tangent vector space at a point p of a manifold M. Let V_p^* be the space of all linear maps ω^* from V_p into R, viz $\omega^*: V_p \longrightarrow R$. The space V_p^* is the so-called dual vector space to V_p where addition and multiplication by scalars are defined in an obvious way. The elements of V_p^* are called dual vectors.

The dual vector space V_p^* is also called the cotangent dual vector space at p (also the vector space of one-forms at p). The elements of V_p^* are then called cotangent dual vectors. Another nomenclature is to refer to the elements of V_p^* as covariant vectors, whereas the elements of V_p are referred to as contravariant vectors.

Dual basis: Let X_μ, $\mu = 1,\dots, n$ be a basis of V_p. The basis elements of V_p^* are given by vectors $X^{\mu *}$, $\mu = 1,\dots, n$ which are defined by

$$X^{\mu *}(X_\nu) = \delta_\nu^\mu. \tag{A.28}$$

The Kronecker delta is defined in the usual way. The proof that $\{X^{\mu *}\}$ is a basis is straightforward. The basis $\{X^{\mu *}\}$ of V_p^* is called the dual basis to the basis $\{X_\mu\}$ of V_p. The basis elements X_μ may be thought of as the partial derivative operators $\partial/\partial x^\mu$ since they transform under a change of coordinate systems (corresponding to a change of charts $\phi \longrightarrow \phi'$) as

$$X_\mu = \sum_{\nu=1}^n \frac{\partial x'^\nu}{\partial x^\mu} X_\nu'. \tag{A.29}$$

We immediately deduce that we must have the transformation law

$$X^{\mu *} = \sum_{\nu=1}^n \frac{\partial x^\mu}{\partial x'^\nu} X^{\nu *\prime}. \tag{A.30}$$

Indeed we have in the transformed basis

$$X^{\mu *\prime}(X_\nu') = \delta_\nu^\mu. \tag{A.31}$$

From this result we can think of the basis elements $X^{\mu *}$ as the gradients dx^μ, viz

$$X^{\mu *\prime} \equiv dx^\mu. \tag{A.32}$$

Let $v = \sum_\mu v^\mu X_\mu$ be an arbitrary tangent vector in V_p, then the action of the dual basis elements $X^{\mu *}$ on v is given by

$$X^{\mu *}(v) = v^\mu. \tag{A.33}$$

The action of a general element $\omega^* = \sum_\mu \omega_\mu X^{\mu *}$ of V_p^* on v is given by

$$\omega^*(v) = \sum_\mu \omega_\mu v^\mu. \tag{A.34}$$

Recall the transformation law

$$v'^\nu = \sum_{\nu=1}^n \frac{\partial x'^\nu}{\partial x^\mu} v^\mu. \tag{A.35}$$

Again we conclude the transformation law

$$\omega_\nu' = \sum_{\nu=1}^n \frac{\partial x^\mu}{\partial x'^\nu} \omega_\mu. \tag{A.36}$$

Indeed we confirm that

$$\omega^*(v) = \sum_\mu \omega_\mu' v'^\mu. \tag{A.37}$$

Double dual vector space: Let now V_p^{**} be the space of all linear maps v^{**} from V_p^* into R, viz $v^{**}:V_p^* \longrightarrow R$. The vector space V_p^{**} is naturally isomorphic (an isomorphism is one-to-one and onto map) to the vector space V_p since to each vector $v \in V_p$ we can associate the vector $v^{**} \in V_p^{**}$ by the rule

$$v^{**}(\omega^*) = \omega^*(v), \quad \omega^* \in V_p^*. \tag{A.38}$$

If we choose $\omega^* = X^{\mu*}$ and $v = X_\nu$ we get $v^{**}(X^{\mu*}) = \delta_\nu^\mu$. We should think of v^{**} in this case as $v = X_\nu$.

Definition 9: A tensor T of type (k, l) over the tangent vector space V_p is a multilinear map form $(V_p^* \times V_p^* \times \cdots \times V_p^*) \times (V_p \times V_p \times \cdots \times V_p)$ (with k cotangent dual vector space V_p^* and l tangent vector space V_p) into R, viz

$$T: V_p^* \times V_p^* \times \cdots \times V_p^* \times V_p \times V_p \times \cdots \times V_p \longrightarrow R. \tag{A.39}$$

The vectors $v \in V_p$ are therefore tensors of type $(1, 0)$, whereas the cotangent dual vectors $v \in V_p^*$ are tensors of type $(0, 1)$. The space $\mathcal{T}(k, l)$ of all tensors of type (k, l) is a vector space (obviously) of dimension $n^k \cdot n^l$ since $\dim V_p = \dim V_p^* = n$.

Contraction: The contraction of a tensor T with respect to its ith cotangent dual vector and jth tangent vector positions is a map $C: \mathcal{T}(k, l) \longrightarrow \mathcal{T}(k-1, l-1)$ defined by

$$CT = \sum_{\mu=1}^{n} T(\ldots, X^{\mu*}, \ldots; \ldots, X_\mu, \ldots). \tag{A.40}$$

The basis vector $X^{\mu*}$ of the cotangent dual vector space V_p^* is inserted into the ith position, whereas the basis vector X_μ of the tangent vector space V_p is inserted into the jth position.

A tensor of type $(1, 1)$ can be viewed as a linear map from V_p into V_p since for a fixed $v \in V_p$ the map $T(., v)$ is an element of V_p^{**} which is the same as V_p, i.e. $T(., v)$ is a map from V_p into V_p. From this result it is obvious that the contraction of a tensor of the type $(1, 1)$ is essentially the trace and as such it must be independent of the basis $\{X_\mu\}$ and its dual $\{X^{\mu*}\}$. Contraction is therefore a well-defined operation on tensors.

Outer product: Let T be a tensor of type (k, l) and 'components' $T(X^{1*}, \ldots, X^{k*}; Y_1, \ldots, Y_l)$ and T' be a tensor of type (k', l') and components $T'(X^{k+1*}, \ldots, X^{k+k'*}; Y_{l+1}, \ldots, Y_{l+l'})$. The outer product of these two tensors which we denote $T \otimes T'$ is a tensor of type $(k + k', l + l')$ defined by the 'components' $T(X^{1*}, \ldots, X^{k*}; Y_1, \ldots, Y_l)T'(X^{k+1*}, \ldots, X^{k+k'*}; Y_{l+1}, \ldots, Y_{l+l'})$.

Simple tensors: Simple tensors are tensors obtained by taking the outer product of cotangent dual vectors and tangent vectors. The $n^k \cdot n^l$ simple tensors $X_{\mu_1} \otimes \cdots \otimes X_{\mu_k} \otimes X^{\nu_1*} \otimes \cdots \otimes X^{\nu_l*}$ form a basis of the vector space $\mathcal{T}(k, l)$. In other words, any tensor T of type (k, l) can be expanded as

$$T = \sum_{\mu_i} \sum_{\nu_i} T^{\mu_1 \cdots \mu_k}{}_{\nu_1 \ldots \nu_l} X_{\mu_1} \otimes \cdots \otimes X_{\mu_k} \otimes X^{\nu_1 *} \otimes \cdots \otimes X^{\nu_l *}. \tag{A.41}$$

By using $X^{\mu *}(X_\nu) = \delta^{\mu\nu}$ and $X_\mu(X^{\nu *}) = \delta^{\mu\nu}$ we calculate

$$T^{\mu_1 \cdots \mu_k}{}_{\nu_1 \ldots \nu_l} = T(X^{\mu_1 *} \otimes \cdots \otimes X^{\mu_k *} \otimes X_{\nu_1} \otimes \cdots \otimes X_{\nu_l}). \tag{A.42}$$

These are the components of the tensor T in the basis $\{X_\mu\}$. The contraction of the tensor T is now explicitly given by

$$(CT)^{\mu_1 \cdots \mu_{k-1}}{}_{\nu_1 \ldots \nu_{l-1}} = \sum_{\mu=1}^{n} T^{\mu_1 \cdots \mu \ldots \mu_{k-1}}{}_{\nu_1 \ldots \mu \ldots \nu_{l-1}} \tag{A.43}$$

The outer product of two tensors can also be given now explicitly in the basis $\{X_\mu\}$ in a quite obvious way.

We conclude by writing down the transformation law of a tensor under a change of coordinate systems. The transformation law of $X_{\mu_1} \otimes \cdots \otimes X_{\mu_k} \otimes X^{\nu_1 *} \otimes \cdots \otimes X^{\nu_l *}$ is obviously given by

$$X_{\mu_1} \otimes \cdots \otimes X_{\mu_k} \otimes X^{\nu_1 *} \otimes \cdots \otimes X^{\nu_l *} = \sum_{\mu_i} \sum_{\nu_i} \frac{\partial x'^{\mu_1'}}{\partial x^{\mu_1}} \cdots \frac{\partial x'^{\mu_k'}}{\partial x^{\mu_k}} \frac{\partial x^{\nu_1}}{\partial x'^{\nu_1'}} \cdots \frac{\partial x^{\nu_l}}{\partial x'^{\nu_l'}} X_{\mu_1'}$$

$$\otimes \cdots \otimes X_{\mu_k'} \otimes X^{\nu_1' *}$$

$$\otimes \cdots \otimes X^{\nu_l' *}. \tag{A.44}$$

Thus we must have

$$T = \sum_{\mu_i} \sum_{\nu_i} T'^{\mu_1 \cdots \mu_k}{}_{\nu_1 \ldots \nu_l} X_{\mu_1'} \otimes \cdots \otimes X_{\mu_k'} \otimes X^{\nu_1' *} \otimes \cdots \otimes X^{\nu_l' *}. \tag{A.45}$$

The transformed components $T'^{\mu_1 \cdots \mu_k}{}_{\nu_1 \ldots \nu_l}$ are defined by

$$T'^{\mu_1' \cdots \mu_k'}{}_{\nu_1' \ldots \nu_l'} = \sum_{\mu_i} \sum_{\nu_i} \frac{\partial x'^{\mu_1'}}{\partial x^{\mu_1}} \cdots \frac{\partial x'^{\mu_k'}}{\partial x^{\mu_k}} \frac{\partial x^{\nu_1}}{\partial x'^{\nu_1'}} \cdots \frac{\partial x^{\nu_l}}{\partial x'^{\nu_l'}} T^{\mu_1 \cdots \mu_k}{}_{\nu_1 \ldots \nu_l}. \tag{A.46}$$

A.1.5 Metric tensor

A metric g is a tensor of type $(0, 2)$, i.e. a linear map from $V_p \times V_p$ into R with the following properties:

- The map $g: V_p \times V_p \longrightarrow R$ is symmetric in the sense that $g(v_1, v_2) = g(v_2, v_1)$ for any $v_1, v_2 \in V_p$.
- The map g is nondegenerate in the sense that if $g(v, v_1) = 0$ for all $v \in V_p$ then one must have $v_1 = 0$.
- In a coordinate basis where the components of the metric are denoted by $g_{\mu\nu}$ we can expand the metric as

$$g = \sum_{\mu,\nu} g_{\mu\nu} dx^\mu \otimes dx^\nu. \tag{A.47}$$

This can also be rewritten symbolically as

$$ds^2 = \sum_{\mu,\nu} g_{\mu\nu} dx^\mu dx^\nu. \tag{A.48}$$

- The map g provides an inner product on the tangent space V_p which is not necessarily positive definite. Indeed given two vectors v and w of V_p, their inner product is given by

$$g(v, w) = \sum_{\mu,\nu} g_{\mu\nu} v^\mu w^\nu. \tag{A.49}$$

By choosing $v = w = \delta x = x_f - x_i$ we see that $g(\delta x, \delta x)$ is an infinitesimal squared distance between the points f and i. Hence the use of the name 'metric' for the tensor g. In fact $g(\delta x, \delta x)$ is the generalization of the interval (also called line element) of special relativity $ds^2 = \eta_{\mu\nu} dx^\mu dx^\nu$ and the components $g_{\mu\nu}$ are the generalization of $\eta_{\mu\nu}$.

- There exists a (non-unique) orthonormal basis $\{X_\mu\}$ of V_p in which

$$g(X_\mu, X_\nu) = 0, \quad \text{if } \mu \neq \nu \text{ and } g(X_\mu, X_\nu) = \pm 1, \quad \text{if } \mu = \nu. \tag{A.50}$$

The number of plus and minus signs is called the signature of the metric and is independent of choice of basis. In fact the number of plus signs and the number of minus signs are separately independent of choice of basis.

A manifold with a metric which is positive definite is called Euclidean or Riemannian, whereas a manifold with a metric which is indefinite is called Lorentzian or Pseudo-Riemannian. Spacetime in special and general relativity is a Lorentzian manifold.

- The map $g(., v)$ can be thought of as an element of V_p^*. Thus the metric can be thought of as a map from V_p into V_p^* given by $v \longrightarrow g(., v)$. Because of the nondegeneracy of g, the map $v \longrightarrow g(., v)$ is one-to-one and onto and as a consequence it is invertible. The metric provides thus an isomorphism between V_p and V_p^*.

- The nondegeneracy of g can also be expressed by the statement that the determinant $g = \det(g_{\mu\nu}) \neq 0$. The components of the inverse metric will be denoted by $g^{\mu\nu} = g^{\nu\mu}$ and thus

$$g^{\mu\rho} g_{\rho\nu} = \delta^\mu_\nu, \quad g_{\mu\rho} g^{\rho\nu} = \delta^\nu_\mu. \tag{A.51}$$

The metric $g_{\mu\nu}$ and its inverse $g^{\mu\nu}$ can be used to raise and lower indices on tensors as in special relativity.

A.2 Curvature

A.2.1 Covariant derivative

Definition 10: A covariant derivative operator ∇ on a manifold M is a map which takes a differentiable tensor of type (k, l) to a differentiable tensor of type $(k, l + 1)$ which satisfies the following properties:

- Linearity:

$$\nabla(\alpha T + \beta S) = \alpha \, \nabla \, T + \beta \, \nabla \, S, \quad \alpha, \beta \in R, \quad T, S \in T(k, l). \tag{A.52}$$

- Leibniz rule:

$$\nabla(T \otimes S) = \nabla \, T \otimes S + T \otimes \nabla \, S, \quad T \in T(k, l), \quad S \in T(k', l'). \tag{A.53}$$

- Commutativity with contraction: In the so-called index notation a tensor $T \in T(k, l)$ will be denoted by $T^{a_1 \cdots a_k}{}_{b_1 \ldots b_l}$ while the tensor $\nabla T \in T(k, l + 1)$ will be denoted by $\nabla_c \, T^{a_1 \cdots a_k}{}_{b_1 \ldots b_l}$. The almost obvious requirement of commutativity with contraction means that for all $T \in T(k, l)$ we must have

$$\nabla_d (T^{a_1 \cdots c \ldots a_k}{}_{b_1 \ldots c \ldots b_l}) = \nabla_d \, T^{a_1 \cdots c \ldots a_k}{}_{b_1 \ldots c \ldots b_l}. \tag{A.54}$$

- The covariant derivative acting on scalars must be consistent with tangent vectors being directional derivatives. Indeed for all $f \in \mathcal{F}$ and $t^a \in V_p$ we must have

$$t^a \, \nabla_a \, f = t(f). \tag{A.55}$$

- Torsion free: For all $f \in \mathcal{F}$ we have

$$\nabla_a \, \nabla_b \, f = \nabla_b \, \nabla_a \, f. \tag{A.56}$$

Ordinary derivative: Let $\{\partial/\partial x^\mu\}$ and $\{dx^\mu\}$ be the coordinate bases of the tangent vector space and the cotangent vector space, respectively, in some coordinate system ψ. An ordinary derivative operator ∂ can be defined in the region covered by the coordinate system ψ as follows. If $T^{\mu_1 \cdots \mu_k}{}_{\nu_1 \ldots \nu_l}$ are the components of the tensor $T^{a_1 \cdots a_k}{}_{b_1 \ldots b_l}$ in the coordinate system ψ, then $\partial_\sigma T^{\mu_1 \cdots \mu_k}{}_{\nu_1 \ldots \nu_l}$ are the components of the tensor $\partial_c T^{a_1 \cdots a_k}{}_{b_1 \ldots b_l}$ in the coordinate system ψ. The ordinary derivative operator ∂ satisfies all the above five requirements as a consequence of the properties of partial derivatives. However, it is quite clear that the ordinary derivative operator ∂ is coordinate dependent.

Action of covariant derivative on tensors: Let ∇ and $\tilde{\nabla}$ be two covariant derivative operators. By condition 4 of definition 10 their action on scalar functions must coincide, viz

$$t^a \nabla_a f = t^a \tilde{\nabla}_a f = t(f). \tag{A.57}$$

We compute now the difference $\tilde{\nabla}_a(f\omega_b) - \nabla_a (f\omega_b)$ where ω is some cotangent dual vector. We have

$$\tilde{\nabla}_a(f\omega_b) - \nabla_a (f\omega_b) = \tilde{\nabla}_a f. \, \omega_b + f\, \tilde{\nabla}_a \, \omega_b - \nabla_a \, f. \, \omega_b - f \nabla_a \, \omega_b$$
$$= f(\tilde{\nabla}_a\omega_b - \nabla_a \, \omega_b). \tag{A.58}$$

The difference $\tilde{\nabla}_a\omega_b - \nabla_a \, \omega_b$ depends only on the value of ω_b at the point p although both $\tilde{\nabla}_a\omega_b$ and $\nabla_a \, \omega_b$ depend on how ω_b changes as we go away from the point p since they are derivatives. The proof goes as follows. Let ω'_b be the value of the cotangent dual vector ω_b at a nearby point p', i.e. $\omega'_b - \omega_b$ is zero at p. Thus by equation (A.12) there must exist smooth functions $f_{(\alpha)}$ which vanish at the point p and cotangent dual vectors $\mu_b^{(\alpha)}$ such that

$$\omega'_b - \omega_b = \sum_\alpha f_{(\alpha)} \mu_b^{(\alpha)}. \tag{A.59}$$

We compute immediately

$$\tilde{\nabla}(\omega'_b - \omega_b) - \nabla (\omega'_b - \omega_b) = \sum_\alpha f_{(\alpha)}(\tilde{\nabla}_a\mu_b^{(\alpha)} - \nabla_a \, \mu_b^{(\alpha)}). \tag{A.60}$$

This is 0 since by assumption $f_{(\alpha)}$ vanishes at p. Hence we get the desired result

$$\tilde{\nabla}_a\omega'_b - \nabla_a \, \omega'_b = \tilde{\nabla}_a \, \omega_b - \nabla_a \, \omega_b. \tag{A.61}$$

In other words, $\tilde{\nabla}_a\omega_b - \nabla_a \, \omega_b$ depends only on the value of ω_b at the point p. Putting this differently we say that the operator $\tilde{\nabla}_a - \nabla_a$ is a map which takes cotangent dual vectors at a point p into tensors of type (0, 2) at p (not tensor fields defined in a neighborhood of p) which is clearly a linear map by condition 1 of definition 10. We write

$$\nabla_a \, \omega_b = \tilde{\nabla}_a \, \omega_b - C^c{}_{ab}\omega_c. \tag{A.62}$$

The tensor $C^c{}_{ab}$ stands for the map $\tilde{\nabla}_a - \nabla_a$ and it is clearly a tensor of type (1, 2). By setting $\omega_a = \nabla_a \, f = \tilde{\nabla}_a f$ we get

$$\nabla_a \, \nabla_b \, f = \tilde{\nabla}_a \, \tilde{\nabla}_b - C^c{}_{ab}\nabla_c \, f. \tag{A.63}$$

By employing now condition 5 of definition 10 we get immediately

$$C^c{}_{ab} = C^c{}_{ba}. \tag{A.64}$$

Let us consider now the difference $\tilde{\nabla}_a(\omega_b t^b) - \nabla_a (\omega_b t^b)$ where t^b is a tangent vector. Since $\omega_b t^b$ is a function we have

$$\tilde{\nabla}_a(\omega_b t^b) - \nabla_a (\omega_b t^b) = 0. \tag{A.65}$$

From the other hand we compute

$$\tilde{\nabla}_a(\omega_b t^b) - \nabla_a (\omega_b t^b) = \omega_b(\tilde{\nabla}_a t^b - \nabla_a t^b + C^b{}_{ac}t^c). \tag{A.66}$$

Hence we must have

$$\nabla_a t^b = \tilde{\nabla}_a t^b + C^b{}_{ac}t^c. \tag{A.67}$$

For a general tensor $T^{b_1 \cdots b_k}{}_{c_1 \ldots c_l}$ of type (k, l) the action of the covariant derivative operator will be given by the expression

$$\nabla_a T^{b_1 \cdots b_k}{}_{c_1 \ldots c_l} = \tilde{\nabla}_a T^{b_1 \cdots b_k}{}_{c_1 \ldots c_l} + \sum_i C^{b_i}{}_{ad} T^{b_1 \cdots d \ldots b_k}{}_{c_1 \ldots c_l}$$
$$- \sum_j C^d{}_{ac_j} T^{b_1 \cdots b_k}{}_{c_1 \ldots d \ldots c_l}. \tag{A.68}$$

The most important case corresponds to the choice $\tilde{\nabla}_a = \partial_a$. In this case $C^c{}_{ab}$ is denoted $\Gamma^c{}_{ab}$ and is called Christoffel symbol. This is a tensor associated with the covariant derivative operator ∇_a and the coordinate system ψ in which the ordinary partial derivative ∂_a is defined. By passing to a different coordinate system ψ' the ordinary partial derivative changes from ∂_a to ∂'_a and hence the Christoffel symbol changes from $\Gamma^c{}_{ab}$ to $\Gamma'^c{}_{ab}$. The components of $\Gamma^c{}_{ab}$ in the coordinate system ψ will not be related to the components of $\Gamma'^c{}_{ab}$ in the coordinate system ψ' by the tensor transformation law since both the coordinate system and the tensor have changed.

A.2.2 Parallel transport

Definition 11: Let C be a curve with a tangent vector t^a. Let v^a be some tangent vector defined at each point on the curve. The vector v^a is parallelly transported along the curve C if and only if

$$t^a \nabla_a v^b |_{\text{curve}} = 0. \tag{A.69}$$

We have the following consequences and remarks:
- We know that

$$\nabla_a v^b = \partial_a v^b + \Gamma^b{}_{ac}v^c. \tag{A.70}$$

Thus

$$t^a(\partial_a v^b + \Gamma^b{}_{ac}v^c) = 0. \tag{A.71}$$

Let t be the parameter along the curve C. The components of the vector t^a in a coordinate basis are given by

$$t^\mu = \frac{dx^\mu}{dt}. \tag{A.72}$$

In other words,

$$\frac{dv^\nu}{dt} + \Gamma^\nu{}_{\mu\lambda} t^\mu v^\lambda = 0. \tag{A.73}$$

From the properties of ordinary differential equations we know that this last equation has a unique solution. In other words, we can map tangent vector spaces V_p and V_q at points p and q of the manifold if we are given a curve C connecting p and q and a derivative operator. The corresponding mathematical structure is called connection. In some usage the derivative operator itself is called a connection.

- By demanding that the inner product of two vectors v^a and w^a is invariant under parallel transport we obtain the condition

$$t^a \nabla_a (g_{bc} v^b w^c) = 0 \Rightarrow t^a \nabla_a g_{bc}. \, v^b w^c$$
$$+ g_{bc} w^c. \, t^a \nabla_a v^b + g_{bc} v^b. \, t^a \nabla_a w^c = 0. \tag{A.74}$$

By using the fact that v^a and w^a are parallelly transported along the curve C we obtain the condition

$$t^a \nabla_a g_{bc}. \, v^b w^c = 0. \tag{A.75}$$

This condition holds for all curves and all vectors and thus we get

$$\nabla_a g_{bc} = 0. \tag{A.76}$$

Thus given a metric g_{ab} on a manifold M the most natural covariant derivative operator is the one under which the metric is covariantly constant.

- It is a theorem that given a metric g_{ab} on a manifold M, there exists a unique covariant derivative operator ∇_a which satisfies $\nabla_a g_{bc} = 0$. The proof goes as follows. We know that $\nabla_a g_{bc}$ is given by

$$\nabla_a g_{bc} = \tilde{\nabla}_a g_{bc} - C^d{}_{ab} g_{dc} - C^d{}_{ac} g_{bd}. \tag{A.77}$$

By imposing $\nabla_a g_{bc} = 0$ we get

$$\tilde{\nabla}_a g_{bc} = C^d{}_{ab} g_{dc} + C^d{}_{ac} g_{bd}. \tag{A.78}$$

Equivalently,

$$\tilde{\nabla}_b g_{ac} = C^d{}_{ab} g_{dc} + C^d{}_{bc} g_{ad}. \tag{A.79}$$

$$\tilde{\nabla}_c g_{ab} = C^d{}_{ac} g_{db} + C^d{}_{bc} g_{ad}. \tag{A.80}$$

Immediately we conclude that

$$\tilde{\nabla}_a g_{bc} + \tilde{\nabla}_b g_{ac} - \tilde{\nabla}_c g_{ab} = 2C^d{}_{ab} g_{dc}. \tag{A.81}$$

In other words,

$$C^d{}_{ab} = \frac{1}{2} g^{dc} (\tilde{\nabla}_a g_{bc} + \tilde{\nabla}_b g_{ac} - \tilde{\nabla}_c g_{ab}). \tag{A.82}$$

This choice of $C^d{}_{ab}$ which solves $\nabla_a\, g_{bc} = 0$ is unique. In other words the corresponding covariant derivative operator is unique.

- Generally a tensor $T^{b_1\,\dots\,b_k}{}_{c_1\,\dots\,c_l}$ is parallelly transported along the curve C if and only if

$$t^a\, \nabla_a\, T^{b_1\,\dots\,b_k}{}_{c_1\,\dots\,c_l}|_{\text{curve}} = 0. \tag{A.83}$$

A.2.3 The Riemann curvature

Riemann curvature tensor: The so-called Riemann curvature tensor can be defined in terms of the failure of successive operations of differentiation to commute. Let us start with an arbitrary tangent dual vector ω_a and an arbitrary function f. We want to calculate $(\nabla_a\, \nabla_b - \nabla_b\, \nabla_a)\omega_c$. First we have

$$\nabla_a\, \nabla_b\, (f\omega_c) = \nabla_a\, \nabla_b\, f\cdot \omega_c + \nabla_b\, f \nabla_a\, \omega_c + \nabla_a\, f \nabla_b\, \omega_c + f \nabla_a\, \nabla_b\, \omega_c. \tag{A.84}$$

Similarly,

$$\nabla_b\, \nabla_a\, (f\omega_c) = \nabla_b\, \nabla_a\, f\cdot \omega_c + \nabla_a\, f \nabla_b\, \omega_c + \nabla_b\, f \nabla_a\, \omega_c + f \nabla_b\, \nabla_a\, \omega_c. \tag{A.85}$$

Thus,

$$(\nabla_a\, \nabla_b - \nabla_b\, \nabla_a)(f\omega_c) = f(\nabla_a\, \nabla_b - \nabla_b\, \nabla_a)\omega_c. \tag{A.86}$$

We can follow the same set of arguments which led from (A.58) to (A.62) to conclude that the tensor $(\nabla_a\, \nabla_b - \nabla_b\, \nabla_a)\omega_c$ depends only on the value of ω_c at the point p. In other words, $\nabla_a\, \nabla_b - \nabla_b\, \nabla_a$ is a linear map which takes tangent dual vectors into tensors of type $(0, 3)$. Equivalently, we can say that the action of $\nabla_a\, \nabla_b - \nabla_b\, \nabla_a$ on tangent dual vectors is equivalent to the action of a tensor of type $(1, 3)$. Thus we can write

$$(\nabla_a\, \nabla_b - \nabla_b\, \nabla_a)\omega_c = R_{abc}{}^d\omega_d. \tag{A.87}$$

The tensor $R_{abc}{}^d$ is precisely the Riemann curvature tensor.

Action on tangent vectors: Let now t^a be an arbitrary tangent vector. The scalar product $t^a\omega_a$ is a function on the manifold and thus

$$(\nabla_a\, \nabla_b - \nabla_b\, \nabla_a)(t^c\omega_c) = 0. \tag{A.88}$$

But

$$(\nabla_a\, \nabla_b - \nabla_b\, \nabla_a)(t^c\omega_c) = (\nabla_a\, \nabla_b - \nabla_b\, \nabla_a)t^c\cdot \omega_c + t^c\cdot (\nabla_a\, \nabla_b - \nabla_b\, \nabla_a)\omega_c. \tag{A.89}$$

In other words,

$$(\nabla_a\, \nabla_b - \nabla_b\, \nabla_a)t^d = -R_{abc}{}^d t^c \tag{A.90}$$

Generalization of this result and the previous one to higher tensors is given by

$$(\nabla_a\, \nabla_b - \nabla_b\, \nabla_a)T^{d_1\,\dots\,d_k}{}_{c_1\,\dots\,c_l} = -\sum_{i=1}^{k} R_{abe}{}^{d_i} T^{d_1\,\dots\,e\dots d_k}{}_{c_1\,\dots\,c_l}$$
$$+ \sum_{i=1}^{l} R_{abc_i}{}^{e} T^{d_1\,\dots\,d_k}{}_{c_1\,\dots\,e\dots c_l}. \tag{A.91}$$

Properties of the curvature tensor: We state without proof the following properties of the curvature tensor:

- Anti-symmetry in the first two indices:

$$R_{abc}{}^d = -R_{bac}{}^d.$$
(A.92)

- Anti-symmetrization of the first three indices yields 0:

$$R_{[abc]}{}^d = 0, \quad R_{[abc]}{}^d = \frac{1}{3}(R_{abc}{}^d + R_{cab}{}^d + R_{bca}{}^d).$$
(A.93)

- Anti-symmetry in the last two indices:

$$R_{abcd} = -R_{abdc}, \quad R_{abcd} = R_{abc}{}^e g_{ed}.$$
(A.94)

- Symmetry if the pair consisting of the first two indices is exchanged with the pair consisting of the last two indices:

$$R_{abcd} = R_{cdab}.$$
(A.95)

- Bianchi identity:

$$\nabla_{[a}R_{bc]d}{}^e = 0, \quad \nabla_{[a}R_{bc]d}{}^e = \frac{1}{3}(\nabla_a R_{bcd}{}^e + \nabla_c R_{abd}{}^e + \nabla_b R_{cad}{}^e).$$
(A.96)

Ricci and Einstein tensors: The Ricci tensor is defined by

$$R_{ac} = R_{abc}{}^b.$$
(A.97)

It is not difficult to show that $R_{ac} = R_{ca}$. This is the trace part of the Riemann curvature tensor. The so-called scalar curvature is defined by

$$R = R_a{}^a.$$
(A.98)

By contracting the Bianchi identity and using $\nabla_a g_{bc} = 0$ we get

$$g_e{}^c(\nabla_a R_{bcd}{}^e + \nabla_c R_{abd}{}^e + \nabla_b R_{cad}{}^e) = 0 \Rightarrow \nabla_a R_{bd}$$
$$+ \nabla_e R_{abd}{}^e - \nabla_b R_{ad} = 0.$$
(A.99)

By contracting now the two indices b and d we get

$$g^{bd}(\nabla_a R_{bd} + \nabla_e R_{abd}{}^e - \nabla_b R_{ad}) = 0 \Rightarrow \nabla_a R - 2 \nabla_b R_a{}^b = 0.$$
(A.100)

This can be put in the form

$$\nabla^a G_{ab} = 0.$$
(A.101)

The tensor G_{ab} is called Einstein tensor and is given by

$$G_{ab} = R_{ab} - \frac{1}{2}g_{ab}R. \tag{A.102}$$

Geometrical meaning of the curvature: The parallel transport of a vector from point p to point q is actually path-dependent. This path-dependence is directly measured by the curvature tensor as we will now show.

We consider a tangent vector v^a and a tangent dual vector ω_a at a point p of a manifold M. We also consider a curve C consisting of a small closed loop on a two-dimensional surface S parameterized by two real numbers s and t with the point p at the origin, viz $(t, s)|_p = (0, 0)$. The first leg of this closed loop extends from p to the point $(\Delta t, 0)$, the second leg extends from $(\Delta t, 0)$ to $(\Delta t, \Delta s)$, the third leg extends from $(\Delta t, \Delta s)$ to $(0, \Delta s)$ and the last leg from $(0, \Delta s)$ to the point p. We parallel transport the vector v^a but not the tangent dual vector ω_a around this loop.

We form the scalar product $\omega_a v^a$ and compute how it changes under the above parallel transport. Along the first stretch between $p = (0, 0)$ and $(\Delta t, 0)$ we have the change

$$\delta_1 = \Delta t \frac{\partial}{\partial t}(v^a \omega_a)|_{(\Delta t/2, 0)}. \tag{A.103}$$

This is obviously accurate up to correction of the order Δt^3. Let T^a be the tangent vector to the line segment connecting $p = (0, 0)$ and $(\Delta t, 0)$. It is clear that T^a is also the tangent vector to all the curves of constant s. The above change can then be rewritten as

$$\delta_1 = \Delta t T^b \nabla_b (v^a \omega_a)|_{(\Delta t/2, 0)}. \tag{A.104}$$

Since v^a is parallelly transported we have $T^b \nabla_b v^a = 0$. We have then

$$\delta_1 = \Delta t v^a T^b \nabla_b \omega_a |_{(\Delta t/2, 0)}. \tag{A.105}$$

The variation δ_3 corresponding to the third line segment between $(\Delta t, \Delta s)$ and $(0, \Delta s)$ must be given by

$$\delta_3 = -\Delta t v^a T^b \nabla_b \omega_a |_{(\Delta t/2, \Delta s)}. \tag{A.106}$$

We have then

$$\delta_1 + \delta_3 = \Delta t [v^a T^b \nabla_b \omega_a |_{(\Delta t/2, 0)} - v^a T^b \nabla_b \omega_a |_{(\Delta t/2, \Delta s)}]. \tag{A.107}$$

This is clearly 0 when $\Delta s \longrightarrow 0$ and as a consequence parallel transport is path-independent at first order. The vector v^a at $(\Delta t/2, \Delta s)$ can be thought of as the parallel transport of the vector v^a at $(\Delta t/2, 0)$ along the curve connecting these two points, i.e. the line segment connecting $(\Delta t/2, 0)$ and $(\Delta t/2, \Delta s)$. By the previous remark parallel transport is path-independent at first order, which means that v^a at $(\Delta t/2, \Delta s)$ is equal to v^a at $(\Delta t/2, 0)$ up to corrections of the order of Δs^2, Δt^2 and $\Delta s \Delta t$. Thus

$$\delta_1 + \delta_3 = \Delta t v^a [T^b \nabla_b \omega_a |_{(\Delta t/2, 0)} - T^b \nabla_b \omega_a |_{(\Delta t/2, \Delta s)}]. \tag{A.108}$$

Similarly $T^b \nabla_b \ \omega_a$ at $(\Delta t/2, \Delta s)$ is the parallel transport of $T^b \nabla_b \ \omega_a$ at $(\Delta t/2, 0)$ and hence up to first order we must have

$$T^b \nabla_b \ \omega_a \mid_{(\Delta t/2,\, 0)} - T^b \nabla_b \ \omega_a \mid_{(\Delta t/2,\, \Delta s)} = -\Delta s S^c \nabla_c (T^b \nabla_b \ \omega_a). \qquad \text{(A.109)}$$

The vector S^a is the tangent vector to the line segment connecting $(\Delta t/2, 0)$ and $(\Delta t/2, \Delta s)$ which is the same as the tangent vector to all the curves of constant t. Hence

$$\delta_1 + \delta_3 = -\Delta t \Delta s v^a S^c \nabla_c (T^b \nabla_b \ \omega_a). \qquad \text{(A.110)}$$

The final result is therefore

$$\begin{aligned}
\delta(v^a \omega_a) &= \delta_1 + \delta_3 + \delta_2 + \delta_4 \\
&= \Delta t \Delta s v^a [T^c \nabla_c (S^b \nabla_b \ \omega_a) - S^c \nabla_c (T^b \nabla_b \ \omega_a)] \\
&= \Delta t \Delta s v^a [(T^c \nabla_c S^b - S^c \nabla_c T^b) \nabla_b \ \omega_a + T^c S^b (\nabla_c \nabla_b - \nabla_b \nabla_c) \omega_a] \\
&= \Delta t \Delta s v^a T^c S^b R_{cba}{}^d \omega_d.
\end{aligned} \qquad \text{(A.111)}$$

In the third line we have used the fact that S^a and T^a commute. Indeed the commutator of the vectors T^a and S^a is given by the vector $[T, S]^a$ where $[T, S]^a = T^c \nabla_c \ S^a - S^c \nabla_c \ T^a$. This must vanish since T^a and S^a are tangent vectors to linearly independent curves. Since ω_a is not parallelly transported we have $\delta(v^a \omega_a) = \delta v^a . \ \omega_a$ and thus one can finally conclude that

$$\delta v^d = \Delta t \Delta s v^a T^c S^b R_{cba}{}^d. \qquad \text{(A.112)}$$

The Riemann curvature tensor measures therefore the path-dependence of parallelly transported vectors.

Components of the curvature tensor: We know that

$$(\nabla_a \nabla_b - \nabla_b \nabla_a)\omega_c = R_{abc}{}^d \omega_d. \qquad \text{(A.113)}$$

We know also

$$\nabla_a \ \omega_b = \partial_a \omega_b - \Gamma^c{}_{ab}\omega_c. \qquad \text{(A.114)}$$

We compute then

$$\begin{aligned}
\nabla_a \ \nabla_b \ \omega_c &= \nabla_a(\partial_b \omega_c - \Gamma^d{}_{bc}\omega_d) \\
&= \partial_a(\partial_b \omega_c - \Gamma^d{}_{bc}\omega_d) - \Gamma^e{}_{ab}(\partial_e \omega_c - \Gamma^d{}_{ec}\omega_d) - \Gamma^e{}_{ac}(\partial_b \omega_e - \Gamma^d{}_{be}\omega_d) \\
&= \partial_a \partial_b \omega_c - \partial_a \Gamma^d{}_{bc}. \ \omega_d - \Gamma^d{}_{bc}\partial_a \omega_d - \Gamma^e{}_{ab}\partial_e \omega_c \\
&\quad + \Gamma^e{}_{ab}\Gamma^d{}_{ec}\omega_d - \Gamma^e{}_{ac}\partial_b \omega_e + \Gamma^e{}_{ac}\Gamma^d{}_{be}\omega_d.
\end{aligned} \qquad \text{(A.115)}$$

And

$$(\nabla_a \nabla_b - \nabla_b \nabla_a)\omega_c = (\partial_b \Gamma^d{}_{ac} - \partial_a \Gamma^d{}_{bc} + \Gamma^e{}_{ac}\Gamma^d{}_{be} - \Gamma^a{}_{bc}\Gamma^d{}_{ae})\omega_d. \qquad \text{(A.116)}$$

We get then the components

$$R_{abc}{}^d = \partial_b \Gamma^d{}_{ac} - \partial_a \Gamma^d{}_{bc} + \Gamma^e{}_{ac}\Gamma^d{}_{be} - \Gamma^e{}_{bc}\Gamma^d{}_{ae}. \qquad \text{(A.117)}$$

A.2.4 Geodesics

Parallel transport of a curve along itself: Geodesics are the straightest possible lines on a curved manifold. Let us recall that a tangent vector v^a is parallelly transported along a curve C with a tangent vector T^a if and only if $T^a \nabla_a v^b = 0$. A geodesics is a curve whose tangent vector T^a is parallelly transported along itself, viz

$$T^a \nabla_a T^b = 0. \tag{A.118}$$

This reads in a coordinate basis as

$$\frac{dT^\nu}{dt} + \Gamma^\nu{}_{\mu\lambda} T^\mu T^\lambda = 0. \tag{A.119}$$

In a given chart ϕ the curve C is mapped into a curve $x(t)$ in R^n. The components T^μ are given in terms of $x^\mu(t)$ by

$$T^\mu = \frac{dx^\mu}{dt}. \tag{A.120}$$

Hence

$$\frac{d^2 x^\nu}{dt^2} + \Gamma^\nu{}_{\mu\lambda} \frac{dx^\mu}{dt} \frac{dx^\lambda}{dt} = 0. \tag{A.121}$$

This is a set of n coupled second order ordinary differential equations with n unknown $x^\mu(t)$. Given appropriate initial conditions $x^\mu(t_0)$ and $dx^\mu/dt |_{t=t_0}$ we know that there must exist a unique solution. Conversely, given a tangent vector T_p at a point p of a manifold M there exists a unique geodesics which goes through p and is tangent to T_p.

Length of a curve: The length l of a smooth curve C with tangent T^a on a manifold M with Riemannian metric g_{ab} is obviously given by

$$l = \int dt \sqrt{g_{ab} T^a T^b}. \tag{A.122}$$

The length is parametrization independent. Indeed we can show that

$$l = \int dt \sqrt{g_{ab} T^a T^b} = \int ds \sqrt{g_{ab} S^a S^b}, \quad S^a = T^a \frac{dt}{ds}. \tag{A.123}$$

In a Lorentzian manifold, the length of a space-like curve is also given by this expression. For a time-like curve for which $g_{ab} T^a T^b < 0$ the length is replaced with the proper time τ, which is given by $c\tau = \int dt \sqrt{-g_{ab} T^a T^b}$. For a light-like (or null) curve for which $g_{ab} T^a T^b = 0$ the length is always 0. Geodesics in a Lorentzian manifold cannot change from time-like to space-like or null and vice versa since the norm is conserved in a parallel transport. The length of a curve which changes from space-like to time-like or vice versa is not defined.

Geodesics extremize the length as we will now show. We consider the length of a curve C connecting two points $p = C(t_0)$ and $q = C(t_1)$. In a coordinate basis the length is given explicitly by

$$l = \int_{t_0}^{t_1} dt \sqrt{g_{\mu\nu} \frac{dx^\mu}{dt} \frac{dx^\nu}{dt}}. \tag{A.124}$$

The variation in l under an arbitrary smooth deformation of the curve C which keeps the two points p and q fixed is given by

$$
\begin{aligned}
\delta l &= \frac{1}{2} \int_{t_0}^{t_1} dt \left(g_{\mu\nu} \frac{dx^\mu}{dt} \frac{dx^\nu}{dt} \right)^{-\frac{1}{2}} \left(\frac{1}{2} \delta g_{\mu\nu} \frac{dx^\mu}{dt} \frac{dx^\nu}{dt} + g_{\mu\nu} \frac{dx^\mu}{dt} \frac{d\delta x^\nu}{dt} \right) \\
&= \frac{1}{2} \int_{t_0}^{t_1} dt \left(g_{\mu\nu} \frac{dx^\mu}{dt} \frac{dx^\nu}{dt} \right)^{-\frac{1}{2}} \left(\frac{1}{2} \frac{\partial g_{\mu\nu}}{\partial x^\sigma} \delta x^\sigma \frac{dx^\mu}{dt} \frac{dx^\nu}{dt} + g_{\mu\nu} \frac{dx^\mu}{dt} \frac{d\delta x^\nu}{dt} \right) \\
&= \frac{1}{2} \int_{t_0}^{t_1} dt \left(g_{\mu\nu} \frac{dx^\mu}{dt} \frac{dx^\nu}{dt} \right)^{-\frac{1}{2}} \left(\frac{1}{2} \frac{\partial g_{\mu\nu}}{\partial x^\sigma} \delta x^\sigma \frac{dx^\mu}{dt} \frac{dx^\nu}{dt} - \frac{d}{dt}(g_{\mu\nu} \frac{dx^\mu}{dt}) \delta x^\nu + \frac{d}{dt}(g_{\mu\nu} \frac{dx^\mu}{dt} \delta x^\nu) \right).
\end{aligned}
\tag{A.125}
$$

We can assume without any loss of generality that the parametrization of the curve C satisfies $g_{\mu\nu}(dx^\mu/dt)(dx^\nu/dt) = 1$. In other words, choose dt^2 to be precisely the line element (interval) and thus $T^\mu = dx^\mu/dt$ is the 4-velocity. The last term in the above equation becomes obviously a total derivative which vanishes by the fact that the considered deformation keeps the two end points p and q fixed. We get then

$$
\begin{aligned}
\delta l &= \frac{1}{2} \int_{t_0}^{t_1} dt \delta x^\sigma \left(\frac{1}{2} \frac{\partial g_{\mu\nu}}{\partial x^\sigma} \frac{dx^\mu}{dt} \frac{dx^\nu}{dt} - \frac{d}{dt}(g_{\mu\sigma} \frac{dx^\mu}{dt}) \right) \\
&= \frac{1}{2} \int_{t_0}^{t_1} dt \delta x^\sigma \left(\frac{1}{2} \frac{\partial g_{\mu\nu}}{\partial x^\sigma} \frac{dx^\mu}{dt} \frac{dx^\nu}{dt} - \frac{\partial g_{\mu\sigma}}{\partial x^\nu} \frac{dx^\nu}{dt} \frac{dx^\mu}{dt} - g_{\mu\sigma} \frac{d^2 x^\mu}{dt^2} \right) \\
&= \frac{1}{2} \int_{t_0}^{t_1} dt \delta x^\sigma \left(\frac{1}{2} \left(\frac{\partial g_{\mu\nu}}{\partial x^\sigma} - \frac{\partial g_{\mu\sigma}}{\partial x^\nu} - \frac{\partial g_{\nu\sigma}}{\partial x^\mu} \right) \frac{dx^\mu}{dt} \frac{dx^\nu}{dt} - g_{\mu\sigma} \frac{d^2 x^\mu}{dt^2} \right) \\
&= \frac{1}{2} \int_{t_0}^{t_1} dt \delta x_\rho \left(\frac{1}{2} g^{\rho\sigma} \left(\frac{\partial g_{\mu\nu}}{\partial x^\sigma} - \frac{\partial g_{\mu\sigma}}{\partial x^\nu} - \frac{\partial g_{\nu\sigma}}{\partial x^\mu} \right) \frac{dx^\mu}{dt} \frac{dx^\nu}{dt} - \frac{d^2 x^\rho}{dt^2} \right) \\
&= \frac{1}{2} \int_{t_0}^{t_1} dt \delta x_\rho \left(-\Gamma^\rho{}_{\mu\nu} \frac{dx^\mu}{dt} \frac{dx^\nu}{dt} - \frac{d^2 x^\rho}{dt^2} \right).
\end{aligned}
\tag{A.126}
$$

The curve C extremizes the length between the two points p and a if and only if $\delta l = 0$. This leads immediately to the equation

$$\Gamma^\rho{}_{\mu\nu} \frac{dx^\mu}{dt} \frac{dx^\nu}{dt} + \frac{d^2 x^\rho}{dt^2} = 0. \tag{A.127}$$

In other words the curve C must be a geodesic. Since the length between any two points on a Riemannian manifold (and between any two points which can be connected by a space-like curve on a Lorentzian manifold) can be arbitrarily long, we conclude that the shortest curve connecting the two points must be a geodesic as it is an extremum of length. Hence the shortest curve is the straightest possible curve. The converse is not true. A geodesic connecting two points is not necessarily the shortest path.

The proper time between any two points which can be connected by a time-like curve on a Lorentzian manifold can be arbitrarily small and thus the curve with greatest proper time (if it exists) must be a time-like geodesic as it is an extremum of proper time. However, a time-like geodesic connecting two points is not necessarily the path with maximum proper time.

Lagrangian: It is not difficult to convince ourselves that the geodesic equation can also be derived as the Euler–Lagrange equation of motion corresponding to the Lagrangian

$$L = \frac{1}{2} g_{\mu\nu} \frac{dx^\mu}{dt} \frac{dx^\nu}{dt}. \tag{A.128}$$

In fact given the metric tensor $g_{\mu\nu}$ we can write explicitly the above Lagrangian and from the corresponding Euler–Lagrange equation of motion we can read off directly the Christoffel symbols $\Gamma^\rho{}_{\mu\nu}$.

www.ingramcontent.com/pod-product-compliance
Lightning Source LLC
Chambersburg PA
CBHW082122210326
41599CB00031B/5844